Springer Geography

The Springer Geography series seeks to publish a broad portfolio of scientific books, aiming at researchers, students, and everyone interested in geographical research.

The series includes peer-reviewed monographs, edited volumes, textbooks, and conference proceedings. It covers the major topics in geography and geographical sciences including, but not limited to; Economic Geography, Landscape and Urban Planning, Urban Geography, Physical Geography and Environmental Geography.

Springer Geography — now indexed in Scopus

Anu Rai • Supratim Karmakar
Suman Chatterjee • Jitendra Kumar Pandey
Editors

Contemporary Social Physics

Decoding Social Behaviour with Advanced Geospatial Tools

Springer

Editors
Anu Rai
Department of Geography
School of Basic and Applied Sciences
Adamas University
Kolkata, West Bengal, India

Suman Chatterjee
Department of Geography
School of Basic and Applied Sciences
Adamas University
Kolkata, West Bengal, India

Supratim Karmakar
Department of Geography
Adamas University
Kolkata, West Bengal, India

Jitendra Kumar Pandey
Himalaya Institute for Learning
and Leadership
UPES- Dehradun, Uttarakhand, India

ISSN 2194-315X ISSN 2194-3168 (electronic)
Springer Geography
ISBN 978-3-031-77595-6 ISBN 978-3-031-77596-3 (eBook)
https://doi.org/10.1007/978-3-031-77596-3

This Springer imprint is published by the registered company Springer Nature Switzerland AG
The registered company address is: Gewerbestrasse 11, 6330 Cham, Switzerland

Foreword

The rapid urbanization in the twenty-first century demands innovative and interdisciplinary approaches to deciphering the complex dynamics of human behaviour within spatial contexts. This anthology, *Contemporary Social Physics: Decoding Social Behaviour with Advanced Geospatial Tools*, co-edited by Anu Rai, Supratim Karmakar, Suman Chatterjee, and Jitendra Kumar Pandey, offers a timely and insightful exploration of how the fusion of geospatial technologies and social physics can illuminate societal patterns and contribute to sustainable urban development.

This book is remarkable for its dual emphasis on foundational theory and real-world applications. It provides a comprehensive overview of the emerging discipline of geospatial social physics while presenting an array of case studies that bridge the gap between academic research and practical implementation. The editors delve into critical global issues, such as migration, healthcare accessibility, crime mapping, and environmental management, offering actionable strategies to address these pressing challenges.

The methodological synergy between social physics and geospatial tools represents a pioneering leap forward. Social physics applies data-driven models to analyse collective human behaviour, while geospatial technologies add a spatial dimension, enabling researchers to unravel the complexities of urban systems. This combination provides a robust framework for understanding and tackling urban challenges, from resource optimization and social equity to climate resilience and governance innovation.

This book aligns closely with the Sustainable Development Goals, especially Goal 11 on Sustainable Cities and Communities, adopted by the United Nations Members States in 2015. By demonstrating how advanced geospatial tools can contribute to creating equitable and resilient societies, this co-edited volume provides critical insights for policymakers, urban planners, practitioners, and researchers.

Based on my rich professional experience on leading the provision of policy advisory and technical support services to cities and urban regions in developing countries whilst working with international organizations, such as the World Bank, UN-Habitat, the Asian Development Bank, and the United Nations Development Programme, I can attest to the growing importance of designing interdisciplinary solutions for sustainable urban development. This book exemplifies the innovative thinking required to navigate the socio-economic, environmental, and governance complexities of our rapidly urbanizing world.

I commend the editors for their dedication to advancing this field and for curating a diverse range of perspectives. *Contemporary Social Physics* not only enriches academic discourse but also offers practical solutions that are both innovative and scalable. This book may inspire further exploration and collaboration at the intersection of social physics, geospatial technologies, and urban sustainability.

Director, Research Center for Sustainable Development and Innovation
School of Global Studies, Thammasat University
Bangkok, Thailand

Bharat Dahiya

Extraordinary Professor, School of Public Leadership
Stellenbosch University
Western Cape, South Africa

Preface

The rapidly evolving dynamics of modern societies demand innovative approaches to understanding the complex interplay between human behaviour and spatial environments. Geospatial social physics, an emerging interdisciplinary field, offers a unique framework for exploring these phenomena. By combining the analytical rigor of social physics with the precision of geospatial analysis, this approach provides new insights into the intricate patterns of human behaviour within diverse spatial contexts. This book, *Contemporary Social Physics: Decoding Social Behaviour with Advanced Geospatial Tools*, advances the field through a rich collection of theoretical explorations, practical methodologies, and case studies.

The volume begins with *Part I: Advancement of Social Physics with Geospatial Theory Integration*, which lays the foundation for this interdisciplinary field. Chapter "Unveiling the Nature of Social Physics" introduces the core principles of social physics, while Chapter "Mapping the Convergence of Social Physics and Geospatial Analysis: A Bibliometric Study" highlights its convergence with geospatial analysis through a bibliometric study. Together, these chapters provide a comprehensive overview of the emerging trends and research directions that define this field.

Part II: Population Dynamics and Social Behaviour examines the spatial dimensions of historical and contemporary population movements. Chapters "Historical Influx of East Pakistan Refugees and Their Settlement in West Bengal (India)", "Reconstructing Historical Landscape Based on Paleochannel Investigations and Archaeological Discoveries in Debalgarh-Anulia of West Bengal, India", "Identification of Spaces of Political Violence in Purulia District, West Bengal: A Geospatial Discourse", "Framing a Historical Landscape: Role of Tea and Railways in the Creation of 'Colonial Assam' in the Nineteenth and First Half of the Twentieth Century", "India-Bangladesh Border Dispute: An Indian Perspective", and "Effects of COVID-19 on Migration in Murshidabad District of West Bengal" cover a broad range of topics, including the settlement of East Pakistan refugees in West Bengal, paleochannel investigations in Debalgarh-Anulia, and the role of tea and railways in shaping Assam's socio-political landscape. The section also delves into spaces of political violence in Purulia District, the complexities of the India-Bangladesh border dispute, and the impact of the COVID-19 pandemic on migration patterns in Murshidabad, West Bengal. These chapters demonstrate how geographical analysis can reconstruct historical narratives and provide critical insights into socio-political and demographic dynamics.

Part III: Social Learning and Environmental Management focuses on the role of social learning and geospatial tools in promoting environmental sustainability and resilience. Chapters "Access to Maternal and Child Healthcare in India: Mapping of Regional and Social Variation", "Strategic Reference for Adopting Sustainable Agricultural Practices Through Social Learning in Indian Agriculture for Environmental Management", "Role of Anthropogenic Activities on the Riverbank Instability: A Geospatial Analysis on Part of the Bhagirathi-Hugli River, India", "Paddy Yield Modelling Using Selected Physical and Socio-economic Parameters: A Synergy Between Geospatial and Machine Learning Approaches", and "The Advent of Tuberose Floriculture: Agricultural Growth and Its Impact on Farmers' Socio-economic Well-Being" explore key topics such as access to maternal and child healthcare, the adoption of sustainable agricultural practices, and the impact of human activities on riverbank stability along the Bhagirathi-Hugli River. The section also examines paddy yield modelling using geospatial and machine learning techniques and highlights the socio-economic benefits of tuberose floriculture for farmers. These studies underscore the transformative potential of integrating geospatial tools into environmental and agricultural management.

Part IV: Dynamics of Social Behaviour in Spatial Contexts investigates how social behaviour interacts with spatial heterogeneity. Chapters "Spatial Heterogeneity and Pandemic Dynamics: Geospatial Insights from Kolkata During Post-Lockdown COVID-19 Spread and Management", "The Impact of Evolving Contemporary Social Behaviour on Geoheritage Site from a Geolocation Perspective: A Tale of Queen City of Himalayas—Darjeeling", "Unraveling Crime in Kolkata: A Socio-economic Perspective", and "Analysis of Cyber Risk in Finance: The State-of-the-Art and Future Research Trends" explore diverse themes, including the management of post-lockdown COVID-19 dynamics in Kolkata, the impact of geo-heritage on tourism in Darjeeling, socio-economic perspectives on crime in Kolkata, and the growing risk of cyber threats in the financial sector. These analyses provide a deeper understanding of how spatial and social factors influence human behaviour in varied contexts.

The final section, *Part V: Fostering Sustainability through Social Physics*, addresses sustainability challenges in urban and peri-urban settings. Chapters "Aligning Peri-Urban Areas Toward Sustainable Development Goals: A Case of Pratapsasan—A Peri-Urban Area of Bhubaneswar City", "Design of GIS-Based Optimization Approach for Transportation Route Selection in Municipal Solid Waste Management", "Deconstructing the Impact of Government Interventions Among the Marginalized Populations Using GIS: A Study of Burgula Cluster, Telangana", and "Selection of Potential Site for Solid Waste Disposal Using Geospatial Techniques and Analytic Hierarchy Process (AHP)" present innovative approaches such as aligning peri-urban areas with the United Nations' Sustainable Development Goals, optimizing transportation routes for municipal solid waste management, assessing government interventions in marginalized communities, and identifying suitable sites for solid waste disposal using geospatial techniques and analytic hierarchy processes. This section emphasizes the critical role of geospatial social physics in driving sustainable development and urban planning. This book is a collective effort to deepen our understanding of human behaviour in spatial contexts and its implications for addressing socio-environmental challenges. By integrating cutting-edge geospatial tools with the principles of social physics, the contributors provide fresh perspectives and actionable insights.

We have designed this volume for academics, researchers, urban planners, and policymakers who are eager to explore interdisciplinary approaches to complex societal issues. We hope the methodologies and case studies presented here will inspire further research and practical applications in this exciting and rapidly developing field.

We express our sincere gratitude to the contributors for their dedication and expertise and to the readers for their interest in this work. Together, we can foster a deeper understanding of human behaviour and its spatial dimensions to create more sustainable and equitable solutions for our shared future.

Kolkata, West Bengal, India
Anu Rai
Supratim Karmakar
Suman Chatterjee
Dehradun, Uttarakhand, India
Jitendra Kumar Pandey

Contents

About the Editors

Anu Rai is Assistant Professor of Geography in School of Basic and Applied Sciences (SoBAS) of Adamas University. Being specialized in Regional Planning and Development, Dr. Rai is much interested in exploring the anatomy of spatial events from the perspective of neighborhood analyst. Mapping infrastructure accessibility and availability for vulnerable communities; modeling spatial interactions and resultant space reconstruction; and developing more viable, transparent, and real option strategies are her core domain of investigation.

With keen interests in the issues of healthcare accessibility, availability, and patient-migration in search of better opportunity for healthcare services, Dr. Rai has authored book entitled "Medical Tourism in Kolkata, Eastern India" published by Springer Nature Switzerland AG, Nov 2019. This book examines the global influence and scope of medical tourism with an emphasis on the city of Kolkata in Eastern India as an emerging destination at the regional scale. Through a geographical research perspective, the book discusses the importance of the phenomenon of medical tourism including recent trends, policies, and scale studies to develop sustainable strategies for medical tourism at particular micro destinations.

Supratim Karmakar is Assistant Professor of Geography in School of Basic and Applied Science of Adamas University, Kolkata. Having a specialization in Urban and Transport Planning, Dr. Karmakar has a keen interest in Human Geography. He has completed his doctoral research in the area of Geography of Crime. As an extension of his research work, he continued to delve in the sphere of gender studies from the perspectives of gender-related crimes. Dr. Karmakar has publications related to crime mapping techniques with special focus on identification of crime hotspots. Evolution of geography of crime throughout the world and the change in perspectives are his major areas of interest.

Suman Chatterjee is an assistant professor in the Department of Geography at the Adamas University. Prior to that, he completed his PhD as junior and senior research fellow at Department of Geography, Jadavpur University. His research interests are sustainable environmental and public health management in India and neighboring countries and applications of geospatial techniques in environmental and public health management.

Jitendra Kumar Pandey completed his PhD in Chemistry (Polymer Science), 2007 from National Chemical Laboratory-Pune. His research areas are in bio/nano-composites, degradation and stabilization of polymer composites, nano-material for water treatment, and synthesis, characterization, and analysis of reinforced composites. He has 70+ research articles, 6 books, 17 patents to his credit. He was associated in Research and Development in University of Petroleum and Energy Studies, Dehradun; Brain Pool Fellow Seoul National University, South Korea; the University of Tokushima, Japan; and Max Planck Institute of Colloid and Interfaces, Golm, Germany, in various roles prior joining Adamas University.

List of Figures

List of Tables

Part I
Advancement of Social Physics with Geospatial Theory Integration

Unveiling the Nature of Social Physics

Supratim Karmakar and Dhruba Jyoti Mudoi

1 Introduction

Human society is an intricate machinery which runs on complicated social processes. The various junctures of societal change show how human society is evolving as a dynamic body. With time, it grows and changes its nature. Society becomes vibrant with the dynamic behaviour and actions of human beings which in turn gives new formations of the social spheres of human interactions. As mentioned as a machinery, human society collectively follows some sort of commonality of behaviour (Wilder et al., 2018; Brunsting et al., 2024). Social physics comes into play in this context to deal with the mechanism of functionality of human society. Physics as the domain of knowledge works on the basis of preconceived and scientifically proven laws. These laws govern the universal functioning of nature. Social physics is the extension of these laws in the realm of social interactions of human beings. It is almost impossible to restrict human behaviour and decision-making process within the limits of certain laws of physics, but from a wider viewpoint, if society is considered a single unit of functionality, then some common mechanisms can be found (Bruhn et al., 2020; Kentrop et al., 2020; Cassimiro et al., 2024). Migration, criminal behaviour, climate change adaptations, epidemics, urban development and transport networking are some major thrust areas where the laws of physics can be applied and the foundation of social physics can be vividly identified. Social physics outlines the mechanisms of financial decision-making in modern society. Social physics is not only dealing with the mechanism of interactions between human beings and different segments of society, but it is trying to delve

S. Karmakar (✉)
Department of Geography, Adamas University, Kolkata, West Bengal, India

D. J. Mudoi
Department of Geography, Gauhati University, Guwahati, Assam, India

A. Rai et al. (eds.), *Contemporary Social Physics*, Springer Geography,
https://doi.org/10.1007/978-3-031-77596-3_1

into the deeper area of social forces that make these processes run in a consistent way (Hilken et al., 2020; Zhang, 2020). The sudden breaks and exceptions are also a part of the explanation of social physics. Mass population behaviours can be very well explained with the help of social physics.

Social physics is an umbrella term which incorporates ideas from various disciplines. Human psychology, statistics, sociological theories, urban infrastructural developments, network analysis and model building are some key concerns under social physics. The works on social physics mainly move through the alleys of these mentioned approaches to the study of society. The amalgamation of physical laws with humanistic experiential activities is one pivotal part of social physics. Human experiences have a myriad impact on social dynamics. Human experiences shape the human mind which lays a cornerstone for any decision-making affair. Multiple layers of human experience make every human being unique in their nature (Konovalov & Ruff, 2022). That poses a challenge to social physics to formulate the human behavioural pattern. The pattern analysis for human behaviour as well as decision-making is the very basic approach to social physics. Until the pattern of human behaviour cannot be identified, further progress in explaining human interactions cannot be made. The basic structure of the human decision-making process needs to be decoded for drawing any conclusion towards the collective social behaviour of human beings. Psycho-mathematical models are highly applicable in the understanding of human behaviour. Time-distance-number are the numerical aspects of human behavioral modelling. Social physics combines subjective experiences with numerical derivations which bring some analogy between human behaviours and natural process mechanisms (Galam, 2022; Perelló et al., 2024).

2 Evolution of Social Physics

The concept of social physics has evolved with the passage of time culminating the ideas from sociology, physics, statistics and data science. The term social physics was coined by August Comte in the nineteenth century. The idea of social physics fully bloomed with its modern interpretation through the works of researchers in the twentieth century and twenty-first century. August Comte initially developed the idea because of a hidden desire to bridge the methodological gap between natural science and social science (Bellomo et al., 2022). Comte opined that there must be some science of society also just like natural science. Through his positivist philosophy, he wanted to formulate some universal law which would govern human society. His idea of social physics was not a quantitative one but more of a philosophical approach to studying social events. August Comte is a key figure in the development of positivism. The positivist approach was being implemented for the discussion of social physics (Schoeller et al., 2018). Both social physics and positivism share an emphasis on empirical observation and the application of scientific methods of social phenomena.

Durkheim, building on Comte's ideas, conducted empirical studies to identify social facts and laws. Emile Durkheim, a renowned critical social theorist, also advocated in favour of the approach of social physics. He emphasized that social phenomena should be studied with the application of scientific methods which are being used in the study of natural science. He emphasized on the study of social events with the help of empirical evidences which can be coupled with a systematic approach. Durkheim conducted both quantitative and qualitative research to understand social phenomena (Schmaus, 1985). He explored the concept of social solidarity, investigating how different types of solidarity (mechanical and organic) contribute to social orders. His research highlighted the role of social integration in preventing anomie, a state of normlessness and breakdown of social cohesion (Chiriac, 2014). His functionalist perspective posited that each element of society serves a function and contributes to the overall stability and functioning of the social system. Durkheim's emphasis on using scientific methods in sociology had a lasting impact on the discipline (Bulgaru, 2013). His approach laid the groundwork for future sociologists to conduct systematic and empirical research.

The modern concept of social physics as we understand it today is mostly on the basis of the works of Alex Pentland, a computer scientist and professor at MIT. Pentland started using the idea of social physics to explain his research work on human behaviour through quantitative and computational analysis and application of big data. Social physics has become closely aligned with the field of computational social science, which uses computational methods and quantitative techniques to analyze large-scale social phenomena. Machine learning and network analysis have become an integral part of social physics (Hälterlein, 2021). The availability of vast amounts of data generated by digital technologies and social media platforms has fueled the growth of social physics. Researchers use this data to identify human behavioural patterns and ultimately reach any conclusion for understanding social phenomena. In the twenty-first century, social physics found its application in the fields of urban planning, financial marketing, economics, public policy, healthcare and business organization. Financial predictions, system analysis of the business domains and optimization of decision-making in human behavioural sciences use the principles of social physics. With the increasing resilience of personal data for social physics research, ethical and privacy concerns are cropping up day by day. Social researchers now have to ensure that the benefits of using large sensitive data are balanced with respect for individual privacy and ethical standards. The journey of social physics can be summarized from a philosophical approach to a quantitative multidisciplinary field highly depending on the computational modelling of big data (Fig. 1).

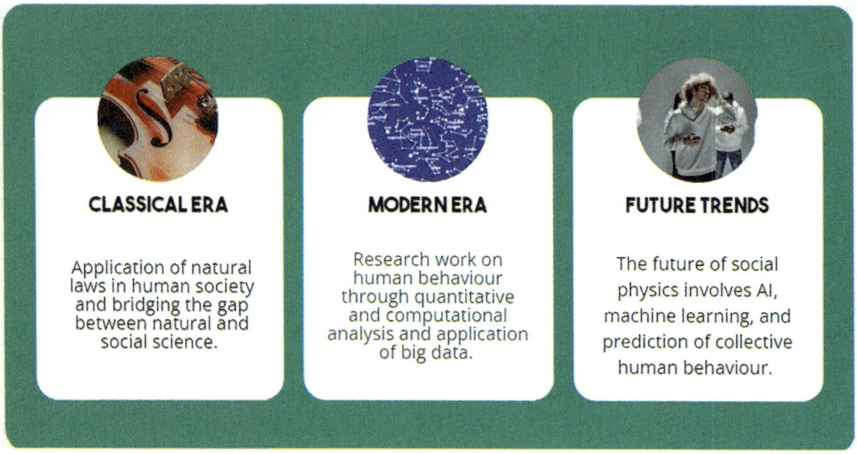

Fig. 1 Evolution of social physics

3 Major Approaches of Social Physics

The philosophical dimension of social physics encompasses a range of considerations related to the nature of human behaviour, the structural social systems and the underlying principles that govern social interactions. As a field that draws inspiration from physics to understand social phenomena, social physics raises philosophical questions about determinism, reductionism, ethics and *the existence of individual persons in the collective form of society*. This raises the philosophical question about the extent to which human behaviour is deterministic and whether societies can be accurately predicted based on observable patterns (Treanor et al., 2024). The tension between reductionism and holism is present in social physics. Reductionism *aims to underline complicated social behaviours by dividing them into fragmented parts*, akin to the reductionist approach in physics. From another perspective, holism *is focused on the significance of dealing* with social systems as a whole, considering emergent properties that cannot be solely explained by analyzing individual components. The application of social physics to understand and predict social behaviour brings up questions about the role of free will. To what extent are the individuals free to make choices, and how much are their actions influenced or even determined by social, economic and cultural factors?

Social physics emerges with questions related to the ontology of social entities and structures. Philosophical discussions may revolve around the nature of social reality, the existence of social entities beyond individual actions and the ontological status of social phenomena (Al-Ababneh, 2020; Wątróbski, 2020). The interdisciplinary nature of social physics prompts philosophical reflections on epistemology—the nature and limits of knowledge. How can insights from physics, data science and social science be integrated *to cater to some* more *detailed knowledge* of social phenomena? The epistemic boundaries of application physics in social structures are determined by social physics. Social physics faces challenges in balancing subjective and objective elements in its analysis. Philosophical discussions may explore the role of subjective experiences, interpretations and meanings in social life alongside the objective measurement and quantification of social phenomena. Social physics often deals with *the discourse of active social organizations*. Philosophical considerations arise concerning *the scenario of change in social realms*, the predictability of social dynamics and the implications for societal development. The philosophical dimension of social physics emerges with the idea that reality is socially constructed. This perspective emphasizes that social phenomena are not only discovered but actively created through social interactions and shared meanings. Philosophical reflections on social physics include considerations of the ultimate goals of societal understanding.

Mathematics plays a crucial role in the field of social physics by providing the tools and methods necessary to analyze, model and understand complex social systems. Graphs and networks are used to represent social structures, such as social networks, collaboration networks, or communication networks. Graph theory provides mathematical frameworks for analyzing the structures and properties of these networks, including measures of centrality, connectivity and clustering (D'Oca et al., 2017). In agent-based modelling, individuals or entities are modelled as agents with specific characteristics and behaviours. Differential equations can be employed to describe how these agents' attributes change over time, allowing for the simulation of dynamic social systems. Social systems are often complex and exhibit non-linear dynamics. Non-linear differential equations and chaos theory can be used to model and understand the behaviour of such systems over time. Mathematics is used in spatial analysis, employing the GIS tool to study the spatial distribution of phenomena. This is crucial in areas like urban planning, where understanding spatial patterns is essential. Advanced mathematical algorithms, including machine learning techniques, are applied to analyze large datasets and make predictions about future social trends.

From the evolution of social physics, it can be understood that the journey of social physics is full of methodological leaps. It has some myriad key approaches to focus on. Firstly, social physics is heavily dependent on data collection and analysis. With the advancement of technology, particularly big data and artificial intelligence, researchers can deal with large amounts of data on human behaviour and social events. That will pave the way to social network modelling and communication patterns. Social physics seeks to model human behaviour mathematically, treating individuals as particles in a physical system. By analyzing large datasets, researchers aim to identify regularities and trends in human interactions, similar to how natural systems are studied in traditional physics. One of the objectives of social physics is to make predictions about collective human behaviour based on patterns observed in data. This predictive capability can have implications in various fields, including economics, policy-making and urban and regional planning. Social physics often incorporates concepts from network theory to understand how individuals are connected within social structures (Cuomo et al., 2021). The analysis of social networks helps to reveal the flow of information, the influence of key individuals and the emergence of collective phenomena. Social physics has practical applications in diverse areas such as business, healthcare and transportation. For instance, it can be used to optimize team performance in organizations and improve public health issues through both organizational behaviour and policy-making. Transport efficiency can be altered for betterment through computational modelling.

4 Theoretical Underpinning

Few theories have been proposed within the realm of social physics, offering frameworks for understanding the complexities of human behaviour. Proposed by Professor Alex Pentland, the theory of collective intelligence posits that groups of individuals exhibit an intelligence that emerges from their collaborative interactions and communication patterns to understand how collective intelligence influences decision-making and problem-solving in groups. Social network theory is a fundamental concept in social physics that explores the structure and dynamics of social relationships (Wahid-Ul-Ashraf et al., 2019). It views individuals as nodes and social connections as edges in a network. The theory examines how information, influence and behaviours flow within these networks, highlighting the impact of social ties on individual and collective behaviour. Temporal patterns of human activities and interactions can be explained through the notion of social physics. It explores the rhythms and regularities in behaviour, such as daily and weekly activity cycles, and aims to uncover universal laws governing human dynamics. This implication helps to understand how individuals allocate their time and energy across various activities.

Theories of migration in population studies apply principles from social physics. Statistical mechanics are being used to study human mobility patterns. Researchers analyze large datasets of individuals' movements to identify regularities and predictability in human mobility, understanding how people move within an urban environment to study spatial interactions. Theories of cultural diffusion also focus on the principles of social physics. Cultural dynamics focus on how opinions and beliefs spread within populations. Modelled after physical systems, these theories explore the mechanism by which individuals influence each other's opinions and how consensus and polarization emerge. Opinion dynamics has applications in understanding information dissemination, public opinion formation and the dynamics of social influence. Originally proposed by sociologist Everett Rogers, the dispersal of innovation theory deals with the fact of how new notions, thoughts and performances blow out within a society. It identifies factors that influence the adoption of innovation itself, *avenues of communication and social arenas* in which the innovation is introduced (Huynh et al., 2023).

Proposed by sociologist Ronald Burt, the structural hole theory emphasizes the importance of structure in social systems. It suggests that *entities who connect structural gaps in a chain of linkages*, connecting otherwise disconnected groups, gain access to unique information and opportunities. This theory has implications for social capital and innovation within networks.

5 Social Physics and Geography

Social physics has a number of applications in geography, offering insights into human behaviour and social interactions within geographic spaces. Social physics found its major applications in urban planning and design. Social physics helps to analyze how people use and move within urban spaces. This information is crucial for urban planners to optimize the layout of cities, ensuring efficient use of space and improved accessibility (Aubin, 2014). By understanding commuting patterns and transportation networks, social physics aids in designing effective and sustainable transportation systems. This includes optimizing public transport routes and traffic flow based on observed human behaviour.

Social physics contributes to understanding patterns of land use within a geographic area. This information is valuable for policymakers and city planners to make informed decisions about zoning regulations and resource allocation (Leypunskiy et al., 2018). Analyzing human activity patterns can help in identifying geographic hotspots of social interactions. This knowledge is beneficial for businesses, identification of crime events, law enforcement and city management (Lozano et al., 2017).

Social physics can provide insights into cultural dynamics by examining how people interact and share information in different geographic areas. This helps geographers understand the spread of cultural traits, languages and ideas. Studying social interactions contributes to the concept of cultural landscapes, where the physical environment reflects the cultural practices and social behaviours of a community (Kebede, 2023).

Social physics is connected to the mapping of social networks within a geographic context. This information is valuable for understanding community structures and relationships. Social physics aids in creating dynamic maps that represent real-time changes in human behaviour and activity, such as crowd movements during events or changes in traffic patterns with the change of urban growth and alteration in transport networks (Wasim et al., 2023).

Social physics can help in analyzing the impact of human activities on the environment within a specific geographical area. This includes the study of resource consumption, waste generation and environmental sustainability as a whole. Social physics heavily contributes to understating the patterns of mass migration and population distribution over space (Amazeen, 2018). Using computational models and data analytics, social physics contributes to the development of predictive models for population changes. This includes forecasting population growth, understanding demographic shifts and predicting the impact of various factors on population dynamics. Social physics helps to identify and analyze community structures within populations. By examining interactions and connections, researchers can delineate community boundaries and understand how these structures influence demographic behaviour. This information is essential for the demographers and policymakers in planning for housing, healthcare and educational facilities. Social physics aids in developing effective evacuation plans during natural disasters by understanding how people move and make decisions in high-stress situations. This contributes to more effective disaster management. Social physics enriches the field of geography by providing a data-driven understanding of human behaviour within geographic areas. This knowledge is invaluable for making informed decisions in urban planning, cultural geography, environmental geography and various other geographical applications.

6 Some Drawbacks

While social physics holds promise in understanding and predicting human behaviour in social systems, it also faces some drawbacks. Social physics models may oversimplify the complexity of human behaviour by reducing it to mathematical equations and statistical patterns. Human actions are influenced by various factors, including cultural, historical and psychological elements which may not be adequately captured by quantitative models (Spreng, 2017). The reliance on large-scale data, often sourced from social media and other digital platforms, raises significant ethical concerns regarding privacy and consent. Social physics research that involves the analysis of personal data must be conducted with careful consideration of privacy implications and ethical standards. Data used in social physics research may be subject to biases, reflecting existing societal inequalities and prejudices. If not properly addressed, these biases can be perpetuated in the models, leading to inaccurate or unfair predictions and reinforcing social disparities (Perc, 2019).

Social physics models may struggle to account for the rich contextual nuances that shape human behaviour. Cultural, historical and situational factors may play a crucial role in determining social dynamics and ignoring these aspects can lead to a shallow understanding of the phenomena under study. Social systems and dynamics can change rapidly in response to various stimuli. Social physics models may struggle to keep pace with the evolving nature of these systems, especially when faced with sudden events or shifts in societal norms (Barnes & Wilson, 2014).

Establishing causation in social physics research is challenging. Correlations between variables may be identified, but discerning the causal relationships underlying these correlations can be complex. This information can hinder the ability to make accurate predictions and draw meaningful conclusions.

7 Future Avenues

The future of social physics holds exciting prospects as advancements in technology, data analytics and interdisciplinary collaboration continue to shape the field. As computational power and machine learning algorithms continue to advance, social physicists will have more sophisticated tools for analyzing and interpreting large data sets. This includes the development of predictive models that can forecast social phenomena and behaviour with greater accuracy. The integration of social physics principles with artificial intelligence technologies could lead to the development of intelligent systems that better understand and respond to human behaviour (Peng et al., 2018). This may have applications in areas such as personalized education, healthcare and customer service. The future of social physics will likely involve a heightened focus on ethical considerations and responsible data use. Researchers and practitioners will need to navigate the balance between extracting valuable insights from data and ensuring privacy and ethical standards. Social physics can play a crucial role in understanding and managing public health crises (Manda et al., 2020; Gheisari et al., 2021). By analyzing patterns of human interaction, mobility and information flow, social physics may contribute better strategies for disease prevention, early detection and response. Social physics will continue to inform urban planning and the development of smart cities. By analyzing data on human movement, transport facilities and social interactions, cities can optimize infrastructure, *make the betterment of civic infrastructures and uplift the holistic quality of life for citizens*. Social physics may increasingly contribute to the development of evidence-based social policies. By understanding the dynamics of social systems, policymakers can design interventions that address societal challenges more effectively, whether in education, poverty alleviation or crime prevention. The future of social physics will likely involve increased collaboration across diverse disciplines, including sociology, computer science, economics and psychology. Interdisciplinary research terms can bring a holistic perspective to understand complex social phenomena. Social physics may contribute to the development of human and artificial intelligence collaboration models, where machines and humans work together to solve complex problems (La Torre et al., 2021). This collaboration could lead to innovations in creativity, decision-making and problem-solving. Social physics practitioners may explore how cultural differences impact social dynamics and behaviour. Understanding global patterns of interaction and communication can contribute to improved cross-cultural understanding and cooperation. The integration of social physics ideas into educational curricula could prepare future generations for data-driven decision-making and a deeper understanding of social systems. This education may extend to various fields, including social sciences (Provost & Fawcett, 2013; Shahat Osman & Elragal, 2021). In essence, the future of social physics is likely to be shaped by ongoing technological advancements, ethical considerations and the application of insights gained from the analysis of large-scale social data to address complex societal challenges. As the field continues to evolve, it holds great potential for contributing to a more informed and interconnected world.

8 Conclusion

Social physics, as a subject, holds immense promise and potential for reshaping our understanding of human behaviour and societal dynamics. Rooted in the idea that social phenomena can be studied and understood through quantitative and scientific methods, social physics has the capacity to revolutionize the social sciences and bring about transformative changes in various fields. The analytical depth allows for the identification of underlying principles that govern human behaviour, enabling us to move beyond anecdotal observations and delve into the intricacies of societal dynamics.

Social physics has the potential to enhance our predictive capabilities regarding social trends and incidences. The advent of machine learning and artificial intelligence allows researchers to develop models that can forecast changes in social systems, predict the diffusion of ideas and anticipate shifts in collective behaviour. The predictive power can be invaluable for the policymakers, businesses and communities seeking to proactively address emerging challenges and opportunities. The interdisciplinary nature of social physics is another compelling aspect of its prospect. Drawing on insights from physics, sociology, computer science and other disciplines, social physics fosters collaboration and cross-fertilization of ideas. The interdisciplinary approach enables researchers to tackle complex problems from multiple angles, leading to a more comprehensive and holistic understanding of social phenomena.

In conclusion, the prospects of social physics as a subject are undeniably exciting. Its potential to deepen our understanding of human behaviour, predict social trends, foster interdisciplinary collaboration and drive innovation positions it as a transformative force in the social sciences. As researchers, policymakers and society at large continue to explore and harness the power of social physics, it is essential to approach its development with a commitment to ethical considerations, ensuring that its benefits are realized in ways that promote the well-being and agency of individuals and communities.

References

Al-Ababneh, M. (2020). Linking ontology, epistemology and research methodology. *Science & Philosophy, 8*(1), 75–91.

Amazeen, P. G. (2018). From physics to social interactions: Scientific unification via dynamics. *Cognitive Systems Research, 52*, 640–657.

Aubin, D. (2014). "Principles of mechanics that are susceptible of application to society": An unpublished notebook of Adolphe Quetelet at the root of his social physics. *Historia Mathematica, 41*(2), 204–223.

Barnes, T. J., & Wilson, M. W. (2014). Big data, social physics, and spatial analysis: The early years. *Big Data & Society, 1*(1), 26–27.

Bellomo, N., Esfahanian, M., Secchini, V., & Terna, P. (2022). What is life? Active particles tools towards behavioral dynamics in social-biology and economics. *Physics of Life Reviews, 43*, 189.

Bruhn, A. L., Wehby, J. H., & Hasselbring, T. S. (2020). Data-based decision making for social behavior: Setting a research agenda. *Journal of Positive Behavior Interventions, 22*(2), 116–126.

Brunsting, N. C., Stark, K., Bettini, E., Lane, K. L., Royer, D. J., Common, E. A., & Rock, L. M. (2024). Self-efficacy, burnout, and intent to leave for teachers of students with emotional and behavioral disorders. *Behavioral Disorders, 49*(2), 75–90.

Bulgaru, I. (2013). Emile Durkheim's sociological thinking evolution from a pedagogical perspective. *Procedia-Social and Behavioral Sciences, 76*, 262–266.

Cassimiro, L., Cecchini, M. A., Cipolli, G. C., & Yassuda, M. S. (2024). Age, but not education, affects social decision-making in the ultimatum game paradigm. *Arquivos de Neuro-Psiquiatria, 82*(07), 001–009.

Chiriac, H. C. (2014). Descriptive imaginary and the epistemological profile of modern social sciences. *Procedia-Social and Behavioral Sciences, 149*, 169–173.

Cuomo, M. T., Tortora, D., Foroudi, P., Giordano, A., Festa, G., & Metallo, G. (2021). Digital transformation and tourist experience co-design: Big social data for planning cultural tourism. *Technological Forecasting and Social Change, 162*, 120345.

D'Oca, S., Chen, C. F., Hong, T., & Belafi, Z. (2017). Synthesizing building physics with social psychology: An interdisciplinary framework for context and occupant behavior in office buildings. *Energy Research & Social Science, 34*, 240–251.

Galam, S. (2022). Physicists, non-physical topics, and interdisciplinarity. *Frontiers in Physics, 10*, 986782.

Gheisari, M., Najafabadi, H. E., Alzubi, J. A., Gao, J., Wang, G., Abbasi, A. A., & Castiglione, A. (2021). OBPP: An ontology-based framework for privacy-preserving in IoT-based smart city. *Future Generation Computer Systems, 123*, 1–13.

Hälterlein, J. (2021). Epistemologies of predictive policing: Mathematical social science, social physics and machine learning. *Big Data & Society, 8*(1), 20539517211003118.

Hilken, T., Keeling, D. I., de Ruyter, K., Mahr, D., & Chylinski, M. (2020). Seeing eye to eye: Social augmented reality and shared decision making in the marketplace. *Journal of the Academy of Marketing Science, 48*, 143–164.

Huynh, M. T., Nippa, M., & Aichner, T. (2023). Big data analytics capabilities: Patchwork or progress? A systematic review of the status quo and implications for future research. *Technological Forecasting and Social Change, 197*, 122884.

Kebede, B. D. (2023). Geographical imagination as portrayed in contemporary Ethiopian creative works that feature the Battle of Adwa. *Social Sciences & Humanities Open, 8*(1), 100446.

Kentrop, J., Kalamari, A., Danesi, C. H., Kentrop, J. J., van Ijzendoorn, M. H., Bakermans-Kranenburg, M. J., & van der Veen, R. (2020). Pro-social preference in an automated operant two-choice reward task under different housing conditions: Exploratory studies on pro-social decision making. *Developmental Cognitive Neuroscience, 45*, 100827.

Konovalov, A., & Ruff, C. C. (2022). Enhancing models of social and strategic decision making with process tracing and neural data. *Wiley Interdisciplinary Reviews: Cognitive Science, 13*(1), e1559.

La Torre, D., Colapinto, C., Durosini, I., & Triberti, S. (2021). Team formation for human-artificial intelligence collaboration in the workplace: A goal programming model to foster organizational change. *IEEE Transactions on Engineering Management, 70*(5), 1966–1976.

Leypunskiy, E., Kıcıman, E., Shah, M., Walch, O. J., Rzhetsky, A., Dinner, A. R., & Rust, M. J. (2018). Geographically resolved rhythms in twitter use reveal social pressures on daily activity patterns. *Current Biology, 28*(23), 3763–3775.

Lozano, M. G., Schreiber, J., & Brynielsson, J. (2017). Tracking geographical locations using a geo-aware topic model for analyzing social media data. *Decision Support Systems, 99*, 18–29.

Manda, P., SayedAhmed, S., & Mohanty, S. D. (2020, June). Automated ontology-based annotation of scientific literature using deep learning. In *Proceedings of the international workshop on semantic Big Data* (pp. 1–6).

Peng, S., Yu, S., & Mueller, P. (2018). Social networking big data: Opportunities, solutions, and challenges. *Future Generation Computer Systems, 86*, 1456–1458.

Perc, M. (2019). The social physics collective. *Scientific reports, 9*(1), 16549.
Perelló, J., Larroya, F., Bonhoure, I., & Peter, F. (2024). Citizen science for social physics: Digital tools and participation. *The European Physical Journal Plus, 139*(7), 572.
Provost, F., & Fawcett, T. (2013). Data science and its relationship to big data and data-driven decision making. *Big Data, 1*(1), 51–59.
Schmaus, W. (1985). Hypotheses and historical analysis in Durkheim's sociological methodology: A Comtean tradition. *Studies in History and Philosophy of Science Part A, 16*(1), 1–30.
Schoeller, F., Perlovsky, L., & Arseniev, D. (2018). Physics of mind: Experimental confirmations of theoretical predictions. *Physics of Life Reviews, 25*, 45–68.
Shahat Osman, A. M., & Elragal, A. (2021). Smart cities and big data analytics: A data-driven decision-making use case. *Smart Cities, 4*(1), 286–313.
Spreng, D. (2017). On physics and the social in energy policy. *Energy Research & Social Science, 26*, 112–114.
Treanor, M., Samuel, B., & Nelson, M. J. (2024). Prototyping slice of life: Social physics with symbolically grounded LLM-based generative dialogue. In *Proceedings of the 19th international conference on the foundations of digital games* (Vol. 1-4).
Wahid-Ul-Ashraf, A., Budka, M., & Musial, K. (2019). How to predict social relationships—Physics-inspired approach to link prediction. *Physica A: Statistical Mechanics and its Applications, 523*, 1110–1129.
Wasim, M., Al-Obeidat, F., Moreira, F., Gul, H., & Amin, A. (2023). Forecasting networks links with Laplace characteristic and geographical information in complex networks. *Procedia Computer Science, 224*, 357–364.
Wątróbski, J. (2020). Ontology learning methods from text-an extensive knowledge-based approach. *Procedia Computer Science, 176*, 3356–3368.
Wilder, D. A., Lipschultz, J. L., King, A., Driscoll, S., & Sigurdsson, S. (2018). An analysis of the commonality and type of preintervention assessment procedures in the Journal of Organizational Behavior Management (2000–2015). *Journal of Organizational Behavior Management, 38*(1), 5–17.
Zhang, L. (2020). Foreign ink: Student mobility, overseas training and Chinese geography, 1912–1952. *Journal of Historical Geography, 68*, 44–54.

Supratim Karmakar, currently working as an Assistant Professor, at the Department of Geography, Adamas University completed his Ph.D. from the Department of Geography, Visva Bharati in 2017. Prior to that, Dr. Karmakar served as Post Graduation Coordinator at Bag Bazar Women's College and Guest Lecturer at Vidyasagar College. Dr. Karmakar has published 17 research articles in peer-reviewed journals and 11 book chapters. He coedited a book on Health Geography in 2023. His areas of interest are geography of crime, gender studies, urban ecology and folk culture studies.

Dhruba Jyoti Mudoi obtained his B.A. from Dibrugarh University in the year 2020 and his M.A. in geography with a specialization in social geography from Gauhati University, Guwahati in the year 2022. He also qualified for the UGC-NET in January 2024. He has published a couple of research papers and book chapters and presented several papers at national and international conferences. His primary research interests include social geography, urban geography, tribal studies and environmental geography.

Mapping the Convergence of Social Physics and Geospatial Analysis: A Bibliometric Study

Anu Rai

1 Introduction

Social physics, originally conceptualized by Auguste Comte, seeks to mathematically model human behavior and social interactions (Pentland, 2014). By leveraging the power of data, this approach aims to better understand patterns in how people communicate, move, and interact within societies. On the other hand, geospatial analysis, particularly through the use of geographic information systems (GISs), focuses on the spatial dimensions of data, analyzing how human activities and behaviors are influenced by geography and location (Goodchild, 2010). When these two fields are brought together, they may offer a robust framework for examining complex social phenomena within the context of physical space (Pentland, 2014).

Recent technological advancements—such as mobile data, big data analytics, and AI—have greatly accelerated the convergence of social physics and geospatial analysis. With the availability of spatial data from sources like smartphones, GPS, and social media, researchers can now track human movements and behaviors in real time, enabling deeper insights into how society operates on a spatial level (Blondel et al., 2015). This has led to new research opportunities, especially in areas such as population studies, urban planning, public health, social learning and environmental adaptation, and sustainable development (Buckee et al., 2020; Sokadjo & Atchade, 2020).

As the challenges we face become increasingly complex—ranging from pandemics and urbanization to climate change—understanding the interplay between social systems and spatial environments has become more crucial now than ever in the history of academia. The integration of mathematical models of social physics

A. Rai (✉)
Department of Geography, School of Basic and Applied Sciences, Adamas University, Kolkata, West Bengal, India
e-mail: anu.rai@adamasuniversity.ac.in

A. Rai et al. (eds.), *Contemporary Social Physics*, Springer Geography,
https://doi.org/10.1007/978-3-031-77596-3_2

within the spatial arena of geography may empower scholars and researchers to better understand human behavior in real-world, spatially dynamic environments (Shah et al., 2022). This chapter conducts a bibliometric analysis using the VOSviewer over dimension database to explore the evolving literature at this intersection and highlight the most influential authors, themes, and emerging trends shaping this field.

2 Theoretical Foundations

Social physics, at its core, seeks to understand social behavior through data-driven models that resemble physical systems. Human interactions, like the particles in physics, follow certain patterns that can be quantified and predicted (Pentland, 2014). This approach has found applications across various domains, from analyzing urban mobility to studying how ideas spread in social networks. For example, Alex Pentland's work has shown how the flow of ideas within a social system can be mathematically modeled in much the same way as physical systems (Pentland, 2021). Moreover, recent studies leveraging large datasets have revealed the predictability of human behavior, such as urban movement patterns (Alessandretti et al., 2020).

On the other hand, geospatial analysis is the use of spatial data and GIS to explore geographic patterns and relationships. It has become essential in fields like urban planning, environmental science, and public health, helping researchers visualize and analyze spatially distributed data (Goodchild, 2020). By incorporating GIS with social physics, researchers can gain a richer understanding of how space influences social behavior. This integration has enabled more sophisticated studies on urban mobility, migration patterns, and health risks, providing valuable insights into how social and spatial factors interact (Tatem, 2017). Bringing together social physics and geospatial analysis creates powerful tools for studying human behavior in space and time. The combination of mathematical models and spatial data allows for more detailed predictions about human activity in urban environments. For instance, researchers have used social network models and GIS data to forecast how people move through cities, predict traffic congestion, and even model the spread of infectious diseases (Bengio et al., 2021).

The COVID-19 pandemic highlighted the importance of this integration, as researchers used mobility data from smartphones and social interaction models to track the virus's spread (Buckee et al., 2020). By mapping human movements and interactions in real time, they were able to predict where outbreaks were likely to happen and implement public health interventions accordingly. Further, the use of agent-based models, which simulate interactions between individuals within a spatial environment, has further enhanced this interdisciplinary approach. These models allow researchers to examine how small-scale decisions and behaviors influence large-scale social patterns, providing valuable insights into issues like urban planning and public health (Helbing & Balietti, 2012). The integration of social physics and geospatial analysis is unlocking new possibilities for understanding human behavior within spatial environments. By combining mathematical models of social behavior with the spatial insights provided by geospatial tools and methods, researchers can develop more sophisticated models of complex social systems. This growing field has already proven invaluable in addressing challenges such as pandemics, urbanization, and climate change. Through a bibliometric analysis using VOSviewer, this chapter explores the evolving intellectual landscape at the intersection of social physics and geospatial analysis, identifying key trends and future directions for research.

3 Aims and Objective of Study

The objective of this study is to investigate the intellectual convergence between social physics and geospatial analysis using bibliometric methods, with the goal of mapping key research trends and identifying gaps in the literature. Through an analysis of academic publications, this study aims to reveal how these two fields intersect, focusing on influential works, leading authors, and emerging themes. Furthermore, it seeks to address underexplored areas where the integration of spatial and social dynamics can offer deeper insights into critical issues such as population dynamics, migration, the reconstruction of historical landscapes, social learning, disease and pandemic management, healthcare systems, agriculture, tourism, crime patterns, and sustainable development.

4 Methodology

Data for this study was sourced from the Dimensions database, a robust platform for academic research, using targeted filters to gather literature on the convergence of social physics and geospatial analysis. At first stage of the search, key search terms included "Social Physics," "Geospatial Analysis," "GIS," "Spatial Networks," "Spatial Modeling," "Spatial Behavior," and "Spatial Analysis in Behavioral Science." The search covered publications from 1970 to 2024 to ensure both historical and contemporary perspectives.

Filters based on the *ANZSRC 2020* classification system were applied, focusing on

- *FoR 4206* (public health)
- *FoR 4202* (epidemiology)
- *FoR 4905* (statistics)
- *FoR 44* (human society)
- *FoR 37* (earth sciences)
- *FoR 4602* (artificial intelligence)
- *FoR 42* (health sciences)
- *FoR 46* (information and computing sciences)

The study also targeted emerging fields aligned with the Sustainable Development Goals (SDGs). Using these filters, 2500 relevant articles were downloaded. The Dimensions database has proven valuable for its comprehensive data, facilitating bibliometric analysis across multiple disciplines (Herzog et al., 2020; Hook et al., 2021).

In the next round of search, an attempt was made to specifically analyze the interaction between social physics research and the geospatial domain, and a more refined filtering approach was employed. This search utilized precise filters based on publication type and research categories to ensure relevance. The key filters were applied as follows:

- *Keywords*: "Social Physics," "Geospatial Analysis," "GIS," "Spatial Networks," "Spatial Modeling," "Spatial Behavior," and "Spatial Analysis in Behavioral Science"
- *Fields of research (ANZSRC 2020)*: 4406 human geography, 37 earth sciences, 44 human society, and 4608 human-centered computing
- *Publication type*: edited books, articles, monographs, proceedings, and chapters

The analysis revealed the following number of publications over the decades:

- *1800 publications* from *2014 to the present*, indicating a significant rise in interest and research output in recent years
- *396 publications* from *2003 to 2013*, reflecting a modest engagement during this period
- *36 publications* from *1994 to 2004*, suggesting a nascent stage of research interaction between social physics and the geospatial domain.

In total, 2230 publications were identified, illustrating a trend of increasing scholarly activity and a re-growing recognition of the importance of integrating social physics with geospatial analysis in addressing complex social phenomena.

The database downloaded from Dimensions was further processed using VOSviewer, a software tool specifically designed for visualizing and analyzing bibliometric networks. VOSviewer allows researchers to create visual representations of relationships among authors, citations, countries, and keywords, making it easier to identify patterns, trends, and clusters within the data. Co-authorship analysis provides insights into collaborative networks among researchers, revealing influential authors and their contributions to the field. Co-citation analysis, on the other hand, helps in understanding the intellectual structure of the research area by highlighting how frequently two documents are cited together, indicating a thematic relationship. Additionally, keyword analysis reveals the evolution of research topics and the emergence of new themes over time. Several researchers have applied VOSviewer, which effectively supports the mapping of academic literature, facilitating the exploration of citation patterns and collaborative networks (Van Eck & Waltman, 2010; Wang et al., 2016). Its ability to generate various types of visualizations, such as network maps and density visualizations, makes it a valuable tool for researchers aiming to gain insights into the dynamics of academic disciplines.

5 Results and Findings

5.1 Co-authorship Networks

Co-authorship analysis, performed with a threshold of at least two documents per author and a minimum of ten citations, identified 166 published works meeting these criteria. Out of these, 47 co-authorship network links were uncovered, reflecting significant collaboration patterns across various academic contributions. As visualized in Fig. 1, the co-authorship network revealed the existence of eight distinct clusters, highlighting prominent collaborations and influential researchers within the fields of social physics and geospatial analysis. Eight major clusters were identified, with key contributors such as Batty, Michael, Dodge and Somayeh, Gao and Song, and Crooks emerging as central figures in the network. Notably, Batty has long been recognized for his contributions to urban modeling and spatial analysis (Batty, 2013). His work on cities as complex systems and the use of computational models for understanding urban dynamics has greatly influenced both social physics and geospatial studies. Similarly, Michael and Goodchild, often regarded as pioneers in GIScience, have contributed extensively to spatial data infrastructures and geospatial analysis (Goodchild, 2007). Gao and Song have been identified as the most impactful researchers in the contemporary era. Their work on integrating GIS and social network analysis to examine spatial behaviors and modeling has attracted considerable attention. Gao's research on human mobility patterns and urban dynamics using geospatial techniques has been particularly influential (Gao & Liu, 2021). The recognition of Gao and Song as leading figures reflects the growing importance of analyzing human spatial behaviors through geospatial data and network analysis, offering new avenues for understanding social phenomena in both physical and virtual spaces. Crooks (2017) has contributed significantly to the development of agent-based modeling and spatial simulations, which are integral to

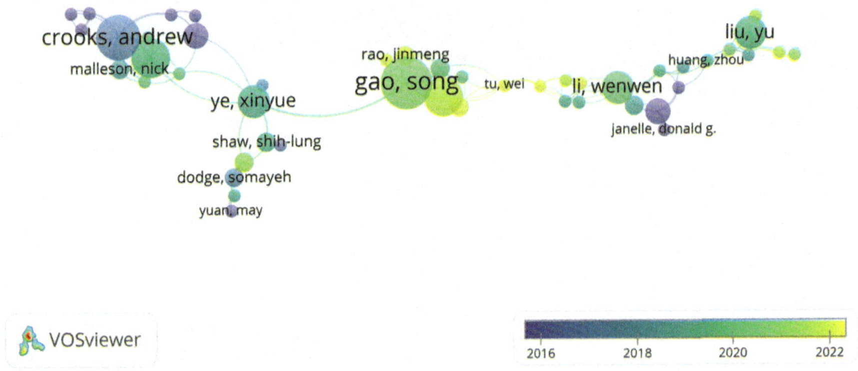

Fig. 1 Co-authorship network mapping

social physics. These tools allow for the exploration of complex social systems, interactions, and emergent behaviors in both urban and rural contexts.

5.2 *Citation Networks*

The overlay visualization of the citation network, based on average citations per year, identifies several key researchers at the intersection of social physics and geospatial analysis (Fig. 2). Nigel Gilbert leads the field with extensive citations for his work in agent-based modeling and social simulation, particularly in understanding complex social behaviors through computational methods (Gilbert, 2008, 2019). His contributions have been foundational in bridging social theory with simulation models, making him a significant figure in social physics. Yu Zheng ranks high in citations due to his pioneering research on urban computing and human mobility, where he integrates spatial data and big data analytics to enhance understanding of urban dynamics and location-based services (Zheng et al., 2014). His work has reshaped approaches to spatial behavior modeling in urban environments. Further impactful researchers include Happenstall and Alison, recognized for their use of agent-based models in socioeconomic and demographic processes (Happenstall et al., 2011). Their simulations have provided new insights into the geospatial dynamics of social behaviors. Similarly, Tang and Xinyue have gained substantial recognition for their work on spatial network analysis, contributing to a deeper understanding of urban spatial patterns and their applications in the social sciences (Tang et al., 2016; Xinyue et al., 2020).

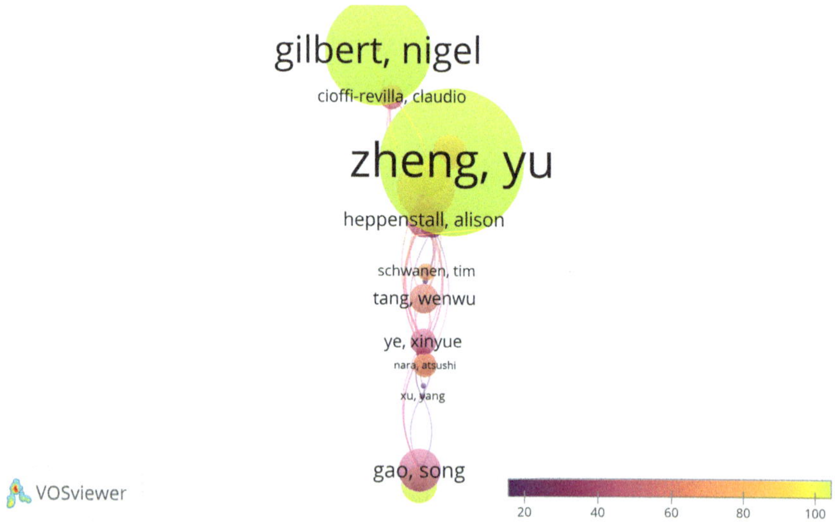

Fig. 2 Overlay network mapping based on average citations per year

6 Co-citation Network Mapping

Country-wise mapping of research as shown in Fig. 3 highlights the substantial impact of studies originating from the USA, UK, China, Germany, India, and Italy, forming key clusters in the global research network at the intersection of social physics and geospatial analysis. Referring to Figs. 3 and 4, it is observed that the USA stands out with institutions such as Harvard University and Columbia University, contributing significantly to fields like spatial modeling and complex systems theory (Batty, 2013; Goodchild, 2007). In the UK, University College London (UCL), under the leadership of scholars like Michael Batty, has become a center for research in urban systems and agent-based modeling (Batty, 2018).

China plays a crucial role in urban computing and spatial behavior modeling, particularly through Peking University and leading researchers like Yu Zheng, who are leveraging advances in big data and artificial intelligence to tackle urban and spatial dynamics (Zheng et al., 2014). In Germany, prominent institutions like the Max Planck Institute and Humboldt University focus on spatial network analysis and complex social systems (Helbing et al., 2015), while India has made significant strides in public health GIS and epidemiology, particularly through the efforts of the Indian Institutes of Technology (IITs) (Ghosh, 2020). Italy has also emerged as a contributor, especially in migration studies and spatial behavior analysis, with key work being done at Sapienza University (Barbieri et al., 2019). Emerging clusters around institutions such as Arizona State University, George Mason University, and Peking University highlight growing research in spatial networks and behavioral science (Crooks et al., 2017).

Fig. 3 Network mapping based on country

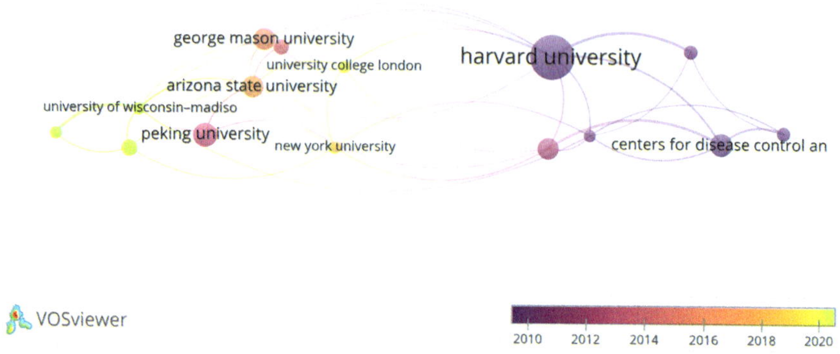

Fig. 4 Overlay network mapping based on the highest number of documents published with maximum citations

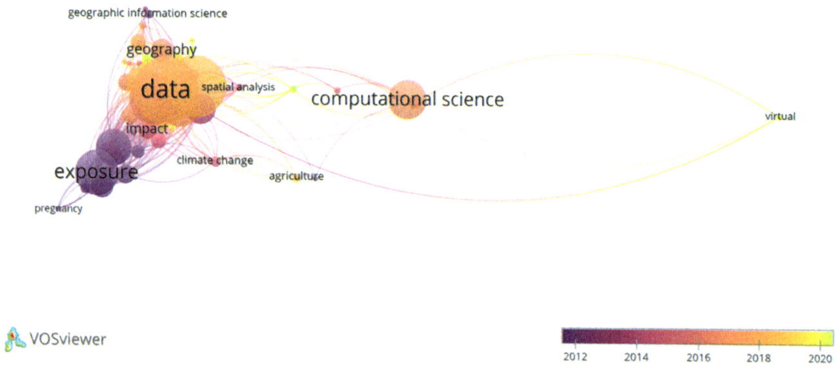

Fig. 5 Overlay analysis of major keyword occurrences

6.1 Keyword Co-occurrence

The keyword analysis in this study utilizes the full counting method. To refine the data, a threshold of a minimum of ten keyword occurrences was established. Only the most relevant keywords were retained, and less significant terms were excluded from further analysis. Out of these, the top 100 keywords were selected, enabling a robust mapping of key thematic trends within the intersection of social physics and geospatial analysis.

The analysis identified prominent keywords such as "city," "data," "GIScience," "Geographic Information System," "climate change," "spatial analysis," and "computational science." As presented in Fig. 5, these terms frequently co-occurred, forming dense clusters that illustrate the integration of geospatial methods with computational tools to analyze social and environmental phenomena. The presence of terms like "city" and "data" highlights the growing emphasis on urban studies and big data, particularly in understanding the dynamics of urban spaces through geospatial techniques (Batty, 2013). Similarly, spatial analysis and computational science clusters indicate increasing research focus on applying computational models to investigate spatial patterns in both social behaviors and environmental changes (Goodchild, 2007). The strong association with climate change underlines the relevance of geospatial analysis in addressing environmental issues, particularly in modeling the impacts of climate variations on urban and rural landscapes (Zheng et al., 2014). Additionally, the convergence of GIScience with computational networks and augmented reality (AR) suggests the development of advanced methodologies for real-time spatial visualization and analysis. This is particularly significant in fields such as urban planning, environmental monitoring, and public health management, where AR and other digital technologies provide innovative ways to interact with spatial data (Crooks et al., 2017). These trends not only indicate a convergence of GIS and social physics but also point to underexplored areas where deeper integration of these domains could yield valuable insights, particularly in areas such as healthcare, disaster management, and sustainable development.

7 Social Physics at Intersection of Geospatial Data Science

In the subsequent phase of our research, a more nuanced approach was adopted to examine the interplay between social physics and the geospatial domain. To ensure the relevance and specificity of the findings, we implemented refined filtering techniques based on publication type and research categories. The key filters applied in this analysis included.

- *Keywords*: "Social Physics," "Geospatial Analysis," "GIS," "Spatial Networks," "Spatial Modeling," "Spatial Behavior," and "Spatial Analysis in Behavioral Science"
- *Fields of research (ANZSRC 2020)*: 4406 human geography, 37 earth sciences, 44 human society, and 4608 human-centered computing

Through this refined filtering process, a total of six distinct clusters emerged, characterized by their co-authorship and citation patterns (Fig. 6). Notably, key figures such as Goodchild and Michael, Rossmo and Kim, Swiaczny and Franck Arbelbide, and Sylvia were found to represent separate clusters, reflecting their significant contributions to the integration of social physics and geospatial analysis.

This clustering highlights the diversity of research themes and collaborative networks within the field. For instance, Goodchild's work has been instrumental in advancing geographic information science (GIScience) and its applications to social phenomena (Goodchild, 2007). Similarly, Rossmo and Kim have focused on spatial crime analysis, bringing insights from social physics into criminology (Rossmo, 2000; Kim & Rossmo, 2017). Swiaczny and Franck Arbelbide have contributed to understanding the spatial dynamics of human behavior, thereby reinforcing the intersection of these domains (Swiaczny et al., 2014).

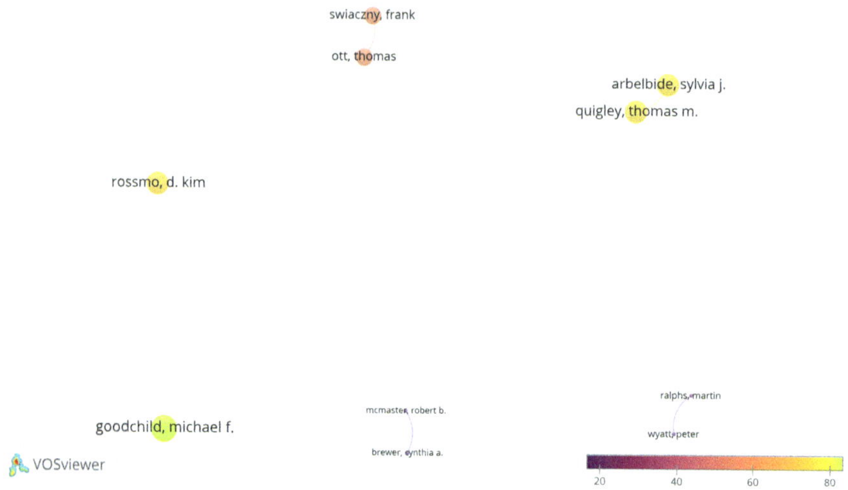

Fig. 6 Network mapping of authors in select fields of research (human geography, earth sciences, human society, and human-centered computing)

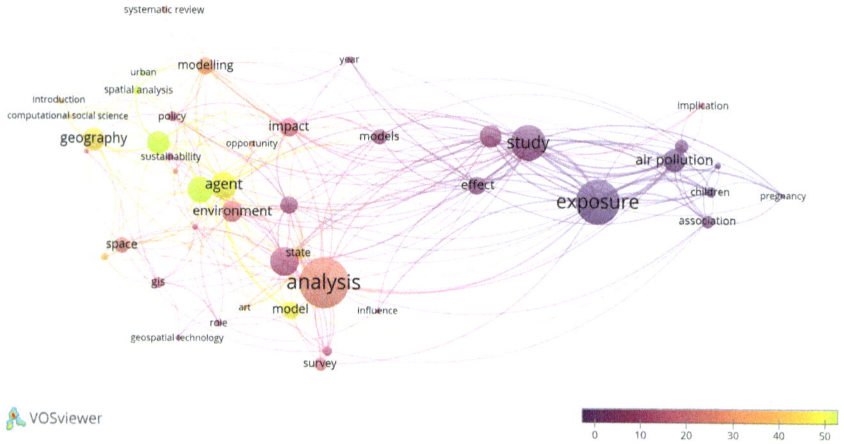

Fig. 7 Mapping keyword co-occurrences of authors in select fields of research (human geography, earth sciences, human society, and human-centered computing)

Referring to Fig. 7, the co-occurrence of terms in the literature highlights significant themes at the intersection of social physics and geospatial analysis, particularly focusing on agent-based modeling, urbanization, geospatial technology, modeling, and exposure studies. Urbanization remains a key area of research, influenced by various socioeconomic and environmental factors, with geospatial technologies providing essential data for visualizing spatial changes (Batty, 2013). Agent-based modeling (ABM) serves as a crucial method for simulating complex urban systems, allowing researchers to analyze individual agent behaviors and emergent patterns

(Railsback & Grimm, 2012). Modeling techniques are fundamental for examining theoretical concepts and predicting outcomes in social systems (Fischer et al., 2016). Exposure studies assess environmental impacts on health and behavior, benefiting from the integration of spatial analysis and agent-based modeling to understand health disparities in urban areas (Gao et al., 2018). Together, these terms reflect the intricate interplay between social phenomena and spatial dynamics, fostering interdisciplinary research to address complex urban challenges.

8 Conclusion and Future Research

Despite the promising potential of integrating social physics and geospatial analysis, there are still significant gaps in the literature regarding how fully these two fields can be leveraged to address complex social and spatial phenomena. Most existing studies focus on specific applications, such as urban mobility or disease spread, but fail to explore the broader implications of combining these fields for understanding diverse socio-spatial dynamics (Alessandretti et al., 2020; Buckee et al., 2020). While there is growing interest in using large datasets for modeling human behavior, many models remain limited in their scope, often overlooking the full range of spatial and temporal factors that influence social systems (Pentland, 2014). A lack of integration between high-resolution spatial data and the dynamic nature of social systems is also observed. For example, while GIS can effectively model geographic patterns, it often lacks the temporal depth required to understand how social behaviors evolve over time in different contexts (Goodchild, 2020). Similarly, while social physics models can predict how ideas and behaviors spread within social networks, they often do not account for the influence of geographical factors such as physical barriers, urban infrastructure, or environmental conditions (Helbing & Balietti, 2012). This creates a need for more comprehensive frameworks that can merge spatial and social dynamics seamlessly. Many studies use aggregated mobility or social interaction data, which can obscure important individual-level variations in behavior (Blondel et al., 2015). This has been particularly evident during the COVID-19 pandemic, where aggregated mobility data provided useful insights but failed to capture the nuances of individual decision-making, social inequalities, and regional differences (Bengio et al., 2021; Sokadjo & Atchade, 2020). Bridging this gap requires integrating fine-grained, real-time data with predictive models that can account for both spatial and social heterogeneity.

In fact, there is still a limited understanding of how spatial data and social physics can inform policy-making and long-term sustainability efforts. Most studies focus on short-term applications, such as crisis management or traffic prediction, but there is a gap in research that explores how these interdisciplinary insights can support broader goals like climate adaptation, equitable urban development, or resilience planning (Shah et al., 2022). Therefore, while significant progress has been made in the convergence of social physics and GIS, more research is needed to fully realize the potential of this integration. This requires addressing the aforementioned gaps, particularly the integration of spatial-temporal data, fine-grained individual behavior analysis, and long-term policy implications (Pentland, 2021). By doing so, future studies could provide a more holistic understanding of how human societies function in both social and geographical contexts.

AI Declaration This chapter has undergone grammatical proofreading and language refinement using AI-based tools. However, the core research work, including data collection, data analysis, and result representation, has been conducted solely by the authors. The intellectual content, conclusions drawn, and interpretations presented within this chapter are entirely the product of the authors' own efforts and research. The AI was not involved in the generation, validation, or interpretation of the research data or findings.

References

Alessandretti, L., Aslak, U., & Lehmann, S. (2020). The scales of human mobility. *Nature, 587*(7834), 402–407.

Barbieri, P., Bonifazi, C., Heins, F., & Strozza, S. (2019). Migration and spatial distribution: Italy's evolving landscape. *Journal of Urban Affairs, 41*(3), 472–487. https://doi.org/10.1080/07352166.2019.1555648

Batty, M. (2013). *The new science of cities*. MIT Press. https://doi.org/10.7551/mitpress/9583.001.0001

Batty, M. (2018). *Inventing future cities*. MIT Press. https://doi.org/10.7551/mitpress/11311.001.0001

Bengio, Y., Janda, R., Yu, Y. W., Ippolito, D., Jarvie, M., Pilat, D., & Obryk, R. (2021). The need for privacy with public digital contact tracing during the COVID-19 pandemic. *The Lancet Digital Health, 3*(5), e264–e269.

Blondel, V. D., Decuyper, A., & Krings, G. (2015). A survey of results on mobile phone datasets analysis. *EPJ Data Science, 4*(1), 10.

Buckee, C. O., Balsari, S., Chan, J., Crosas, M., Dominici, F., Gasser, U., et al. (2020). Aggregated mobility data could help fight COVID-19. *Science, 368*(6487), 145–146.

Crooks, A. T. (2017). Agent-based modeling and geospatial simulation for understanding complex systems. *International Journal of Geographical Information Science, 31*(1), 1–4.

Crooks, A. T., Castle, C. J. E., & Batty, M. (2017). Key challenges in agent-based modeling for geo-spatial simulation. *Journal of Artificial Societies and Social Simulation, 20*(4). https://doi.org/10.18564/jasss.3520

Fischer, F., Kauffmann, M., & Figueiredo, M. (2016). Modeling social dynamics with agent-based models: A review of current research. *Computers, Environment and Urban Systems, 58*, 1–15.

Gao, S., & Liu, Y. (2021). Understanding urban dynamics and human mobility through the lens of geospatial big data. *Computers, Environment and Urban Systems, 84*, 101–106.

Gao, Y., Ranjit, N., & Huang, J. (2018). Integrating spatial analysis with health exposure assessment: A case study on air quality and health outcomes. *Environmental Research, 166*, 314–321.

Ghosh, A. (2020). Geospatial health and epidemiology in India: Present status and future prospects. *Geospatial Health, 15*(1), 23–35. https://doi.org/10.4081/gh.2020.818

Gilbert, N. (2008). *Agent-based models*. Sage Publications.

Gilbert, N. (2019). Computational social science: Agent-based modeling in social physics. *Annual Review of Sociology, 45*, 21–42.

Goodchild, M. F. (2007). Citizens as sensors: The world of volunteered geography. *GeoJournal, 69*(4), 211–221. https://doi.org/10.1007/s10708-007-9111-y

Goodchild, M. F. (2010). Twenty years of progress: GIScience in 2010. *Journal of Spatial Information Science, 1*, 3–20.

Goodchild, M. F. (2020). How well do we really know the world? Uncertainty in GIScience. *Journal of Spatial Information Science*, 97–102. http://dx.doi.org/10.5311/JOSIS.2020.20.664

Happenstall, A. J., Birkin, M. H., & Evans, A. J. (2011). Agent-based models of socioeconomic and demographic systems. *Computational and Mathematical Organization Theory, 17*(2), 202–223.

Helbing, D., & Balietti, S. (2012). How to do agent-based simulations in the future: From modeling social mechanisms to emergent phenomena and interactive systems design. In *Agent-based models of geographical systems* (pp. 25–70). Springer.

Helbing, D., Yu, W., & Rauhut, H. (2015). Data-driven modeling of social dynamics. *Proceedings of the National Academy of Sciences, 112*(4), 1080–1086. https://doi.org/10.1073/pnas.1415085112

Herzog, C., Hook, D. W., & Konkiel, S. (2020). Dimensions: Bringing down barriers between scientometricians and data. *Quantitative Science Studies, 1*(1), 387–395.

Hook, D. W., Porter, S. J., & Herzog, C. (2021). Dimensions: Building context for search and evaluation. *Frontiers in Research Metrics and Analytics, 3*, 23.

Kim, Y., & Rossmo, D. (2017). Spatial analysis in crime prevention and control. *Crime Science, 6*(1), 1–13.

Pentland, A. (2014). *Social physics: How good ideas spread-the lessons from a new science.* Penguin Press.

Pentland, A. (2021). Social physics: Understanding human society through data. *Nature Human Behaviour, 5*(8), 989–996.

Railsback, S. F., & Grimm, V. (2012). Agent-based modeling for natural resource management: Introduction to the special feature. *Ecological Modelling, 220*(14), 1790–1794.

Rossmo, D. K. (2000). *Geographic profiling*. CRC Press.

Shah, S., Merchant, H. A., & Merchant, A. A. (2022). Addressing climate change through interdisciplinary collaboration. *Sustainability, 14*(10), 6240.

Sokadjo, Y. M., & Atchade, M. P. (2020). The influence of passenger air traffic on the spread of COVID-19 in the world. *Transportation Research Interdisciplinary Perspectives, 8*, 100213.

Swiaczny, F., Arbelbide, F., & Hatz, B. (2014). Spatial behavior in human geography: A new perspective. *Human Geography, 7*(1), 27–39.

Tang, Y., et al. (2016). Spatial network analysis and applications in urban systems. *Journal of Geographical Systems, 18*(4), 319–336.

Tatem, A. J. (2017). WorldPop, open data for spatial demography. *Scientific Data, 4*, 170004.

Van Eck, N. J., & Waltman, L. (2010). Software survey: VOSviewer, a computer program for bibliometric mapping. *Scientometrics, 84*(2), 523–538.

Wang, Y., Zhang, L., & Wang, Y. (2016). Visualizing the knowledge structure of social media research: A bibliometric analysis. *Computers in Human Behavior, 62*, 172–179.

Xinyue, Y., et al. (2020). Spatial interaction models and urban form: A critical review. *Progress in Human Geography, 44*(2), 215–235.

Zheng, Y., Capra, L., Wolfson, O., & Yang, H. (2014). Urban computing: Concepts, methodologies, and applications. *ACM Transactions on Intelligent Systems and Technology (TIST), 5*(3), 1–55. https://doi.org/10.1145/2629592

Dr. Anu Rai is an Assistant Professor of Geography in the School of Basic and Applied Sciences (SoBAS) of Adamas University. Specializing in Regional Planning and Development, Dr. Rai is very interested in exploring the anatomy of spatial events from the perspective of a neighborhood analyst. Mapping infrastructure accessibility and availability for vulnerable communities, modeling spatial interactions and resultant space reconstruction, and developing more viable, transparent, and real option strategies are her core domains of investigation.

With keen interests in the issues of healthcare accessibility, availability, and patient migration in search of a better opportunity for healthcare services, Dr. Rai has authored a book entitled "*Medical Tourism in Kolkata, Eastern India*" published by Springer Nature Switzerland AG, Nov 2019. This book examines the global influence and scope of medical tourism with an emphasis on the city of Kolkata in Eastern India as an emerging destination at the regional scale. Through a geographical research perspective, the book discusses the importance of the phenomenon of medical tourism including recent trends, policies, and scale studies to develop sustainable strategies for medical tourism at particular micro-destinations.

Part II
Population Dynamics and Social Behaviour

Historical Influx of East Pakistan Refugees and Their Settlement in West Bengal (India)

Sayantika Baral and Tuhin Ghosh

1 Introduction

Migration is a dynamic phenomenon which depends on several factors. Socio-economic imbalances, political turmoil, demographic patterns, etc. exaggerated the migration flow primarily. Apart from those, sudden environmental disasters and natural calamities also created a rapid flow of displacements from one place to another. In the quest for a higher standard of living, financial security, and educational opportunities, people started to move towards the densely populated regions. This process led to a new challenge for the destination countries as it created rapid pressure on their limited lands and resources (Sarma, 2015). Conflicts arise between the existing population and new immigrants, and to overcome these consequences, the government must adopt several strategic techniques.

India was the leading Asian country in the world hosting a majority of the immigrants from Bangladesh, Pakistan, and Sri Lanka in 2001. As per the migration corridor, the India-Bangladesh migration corridor secured the fourth position globally, constituting around 6.1 million migrants (World Bank, 2015; Sarma, 2015). This also placed India in securing its position among the top ten migration destination countries of the world, and the majority of the migrants originated from its neighbouring countries (World Migration Report, 2010; Sarma, 2015). According to the report from United Nations Department of Economic and Social Affairs (UN-DESA) in 2020, around 4.8 million immigrants were residing in India, and it secured the fourteenth position among the top fifteen migration destination countries of the world (UN DESA, 2020).

S. Baral (✉) · T. Ghosh
School of Oceanographic Studies, Jadavpur University, Kolkata, India
e-mail: tuhinghosh@jadavpuruniversity.in

A. Rai et al. (eds.), *Contemporary Social Physics*, Springer Geography,
https://doi.org/10.1007/978-3-031-77596-3_3

2 Background of the Study

Migration from Bangladesh is a normal phenomenon in India, and it started in the later part of the nineteenth century. This displacement was considered an intra-country movement between the two regions as India was undivided and under the control of British territory during this time. The population movement between India and Bangladesh peaked during the 1940s when India gained independence and was divided into two major countries (Sarma, 2015).

The partition of India was a major tragic event in history. It fragmented the geographical territory of India and also divided the community based on their religion. The partition involved the division of three major states: Assam, Bengal, and Punjab based on their dominance of the Hindu and Muslim populations. The prime reason for this division was the formation of Muslim-dominated Pakistan and Hindu-dominated India. Based on religion, the British provinces of Bengal were divided into two regions, i.e., the eastern and western sections. These sections were known as East Bengal and West Bengal, respectively. The former region was administrated by West Pakistan, separated by 1000 miles from each other and later incorporated with adjacent India (Alexander et al., 2016). British provinces of India became independent on 15 August 1947 with the transfer of power from the British to India. Before that in March 1946, British Prime Minister Clement Atlee announced that there would be a plan to hand over the power between the two leading organisations of Indian politics. The Hindu-dominated Congress party and Muslim-dominated Muslim League took this opportunity and wanted to form a separate government for the country. But this decision took a major turn when the British Cabinet Mission proposed a plan on 16 May 1946 to form an interim government with the representatives of Congress, the Muslim League, and other parties. According to the British plan, the Indian Congress got one extra seat than the Muslim League in the government which was opposed later, and they also created a resolution on 29 July 1946. This resolution also invited Direct Action Day on 16 August 1946 all over India. Despite having proper assurances from the Muslim League, this incident became violent in Kolkata, an event known as the 'Great Calcutta Killing'. This event took the lives of 5 to 10 thousand people, and around 15 thousand were severely injured. Both political leaders blamed each other for this brutal incident ('SciencesPo', 2007). The reaction to this incident made the people of East Pakistan furious. False propaganda about the Great Calcutta Killing was spread among the innocent Muslims of East Pakistan, claiming that Hindus intentionally killed the Muslims during the Kolkata riots. This news and its consequences created major violence among Muslims. They started to take the matter seriously. Organized murder, looting, rapes, abduction, conversion to Islam, and arson of Hindu property by the Muslim community were common in nature (Sinha & Dasgupta, 2011). During this time, it was required to look after the matter immediately. To achieve peace and harmony among the people of Bengal, several organizations known as Hindu Mahasabha, Bharat Sebasram Sangha, the Indian National Army, Arya Samaj, Communist Party of India, Prabartak Sangha, The Indian National Congress, Gita

Press, etc. worked together (Das, 2004), rescued the victims, and provided them proper reliefs. Thirty relief organisations and six medical missions were involved in their services (Nayyar, 1968). Initially, Gandhiji took the matter and started his journey towards Noakhali on 6 November 1946 to trickle down the adverse conditions. Despite his presence, several incidents took place in some parts of Noakhali ('Noakhali Riots', n.d.). It was also declared in a rally in Comilla that Gandhi's presence in Noakhali had created bitterness among the Muslims, and this resentment increased day by day. As a result, Muslims started to boycott Gandhi's peace meetings and became non-cooperative towards him (Majumder, 1975). Finally, he decided to put an end to his mission on 2 March 1947 and moved to Bihar. After one month, he received a telegram from Noakhali Congress workers in which it was written that the lives of the Hindus were not secured and that the situation was worsening over time. Gandhiji stated that due to unfavourable and unavoidable conditions, Hindus either should leave East Pakistan for India or leave their lives to fate (The New York Times, 8 April 1947). Such kind of words were also delivered by Pandit Jawaharlal Nehru, in which he declared that if any Bengali Hindus felt that their lives were not secured in East Pakistan, they might be leaving their country, and the Indian government would help them by providing suitable shelter and security (Bandyopadhyay, 1970). This tragic Noakhali and Tippera riots created tension among the Hindus of East Pakistan, and they decided to leave their country and move towards the adjacent country for a secure future. It was the Noakhali and Tippera riots which initiated the refugee movement from East Pakistan to India, and it continued up to the Bangladesh Liberation War of 1971. Their journey from East Pakistan to West Bengal occurred through several stages which were also influenced by several factors. This study mainly emphasizes the refugee influx from East Pakistan between 1946 and 1971 in West Bengal as after the Bangladesh Liberation War in 1971, the Government of India did not consider the words 'refugee migration' or 'partition migration' (Bandyopadhyay, 1970). Their arrival in India was simply recognized as 'immigrants from outside India' from the 1981 Census, and no documentation was found in favour of their partition refugee status after 1971.

3 Study Area

Geographically, West Bengal is located in the eastern part of India, extended from 85° 45′ E to 89° 45′ E and 23° 30′ N to 27° 25′ N. Presently, the state is surrounded by Nepal, Bhutan, Sikkim, Jharkhand, Bihar, and Bangladesh (which was part of the Indian subcontinent and known as East Bengal, later renamed as East Pakistan and finally Bangladesh) (Fig. 1).

From 1946–1971, the entire state and its 16 districts experienced the impact of refugee flow. After the completion of the Bangladesh Liberation War in 1971, displacements between East Pakistan and West Bengal continued, though these continuations were not the result of any fragmentation or war. These displaced people or refugees from 1946 to 1971 were not fully reliant on the government for their

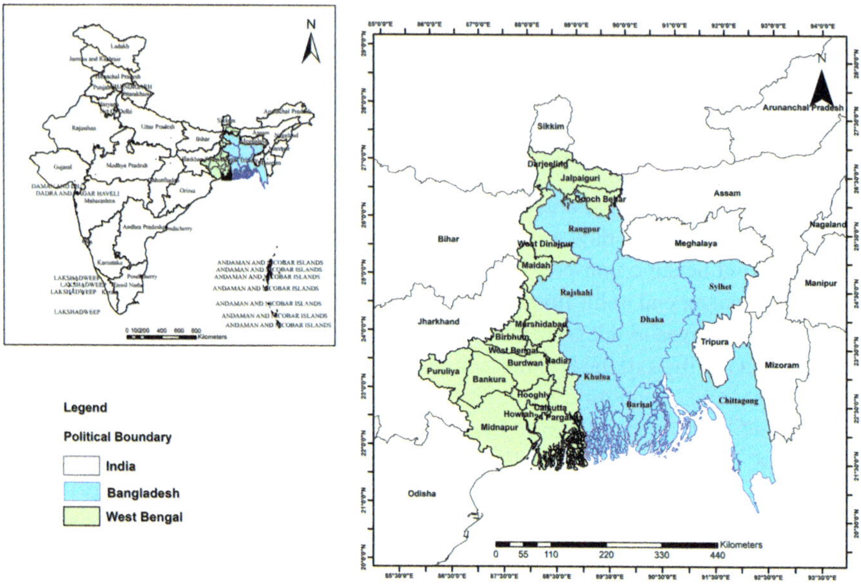

Fig. 1 Location map of the study area

survival; rather, some of them voluntarily took some initiatives for their rehabilitation. The division of Bengal and the movement of refugees from East Pakistan to West Bengal were distinct incidents that also affected West Bengal's economy. The majority of the Hindu refugees crossed the border and entered West Bengal through the eastern part. They generally chose railways, roadways, and waterways for their displacement. Some well-established migrants also chose airways for their movement. The detailed descriptions of their arrivals and their patterns of displacement in and around West Bengal are mentioned later.

4 Significance of the Study

Large-scale immigration is not a current scenario in West Bengal; rather, it started before the official formation of West Bengal on the political map. Indian Independence and the Bengal partition initiated a massive flow of migration between East Bengal (later renamed Bangladesh) and West Bengal. This journey was officially continued until the Bangladesh Liberation War in 1971. The study mainly highlights the temporal variation of the refugee movement in West Bengal between 1946 and 1971 as well as some government initiatives to control the refugee flows.

This time frame was used for partition migration as during 1946–1947, India got independence resulting in the fragmentation of undivided Bengal into West Bengal and East Bengal. These migrants were known as refugees as they left their country of origin due to some unavoidable circumstances, and the Indian government decided to look after them as per the definition of UNHCR (United Nations High Commissioner for Refugees) (Dasgupta, 2016). This partition movement continued until 1971 as a result of the independence of East Pakistan from its original part of Pakistan (in 1947, a separate country of Pakistan was formed after fragmenting from the Indian subcontinent) and appeared on the world political map as Bangladesh. After 1971, the movement between West Bengal and Bangladesh did not stop; rather, it increased over time. However, their movement was not considered a consequence of any particular event. As a result, they were not given any status of refugees or partition migrants and so on (from census 1981, they were treated as immigrants born outside India and immigrants residing outside India). They are simply recognised as immigrants.

5 Methodology

This study is entirely based on secondary sources. Many published articles, books, government reports, newspaper articles, dissertation reports, and, finally, census data are used to gather information related to the Bengal partition and the resultant refugee movement in West Bengal. As the study is confined to the immigration of East Pakistan refugees in West Bengal between 1946 and 1971, the main motive was to highlight their displacement in West Bengal over time and also try to explain the role of the government in tackling refugee situations which might be changed over the period. This study also used some Internet-based narratives to understand the background of the events. Geographically, this study of partition migration concentrated on the area between West Bengal and Bangladesh. After prolonged studies of different secondary sources, this spatial displacement was represented in maps by using ArcGIS software. Spatial data had been collected from published documents. Google Earth is used to detect various locations in both Bangladesh and West Bengal, and these locations are transferred to ArcGIS software. After the preparation of maps (Figs. 2 and 3), the final output was done in ArcGIS software.

6 Results and Discussion

6.1 Partition of Bengal and Initial Government Initiatives to Control the Refugee Issues

The exodus of Hindus from East Pakistan to West Bengal was initiated after the Noakhali and Tippera riots in 1946, and it became immense after independence. They crossed the international border and poured into Bengal in an ever-increasing number which also trickled down over time. Refugees chose the nearest border

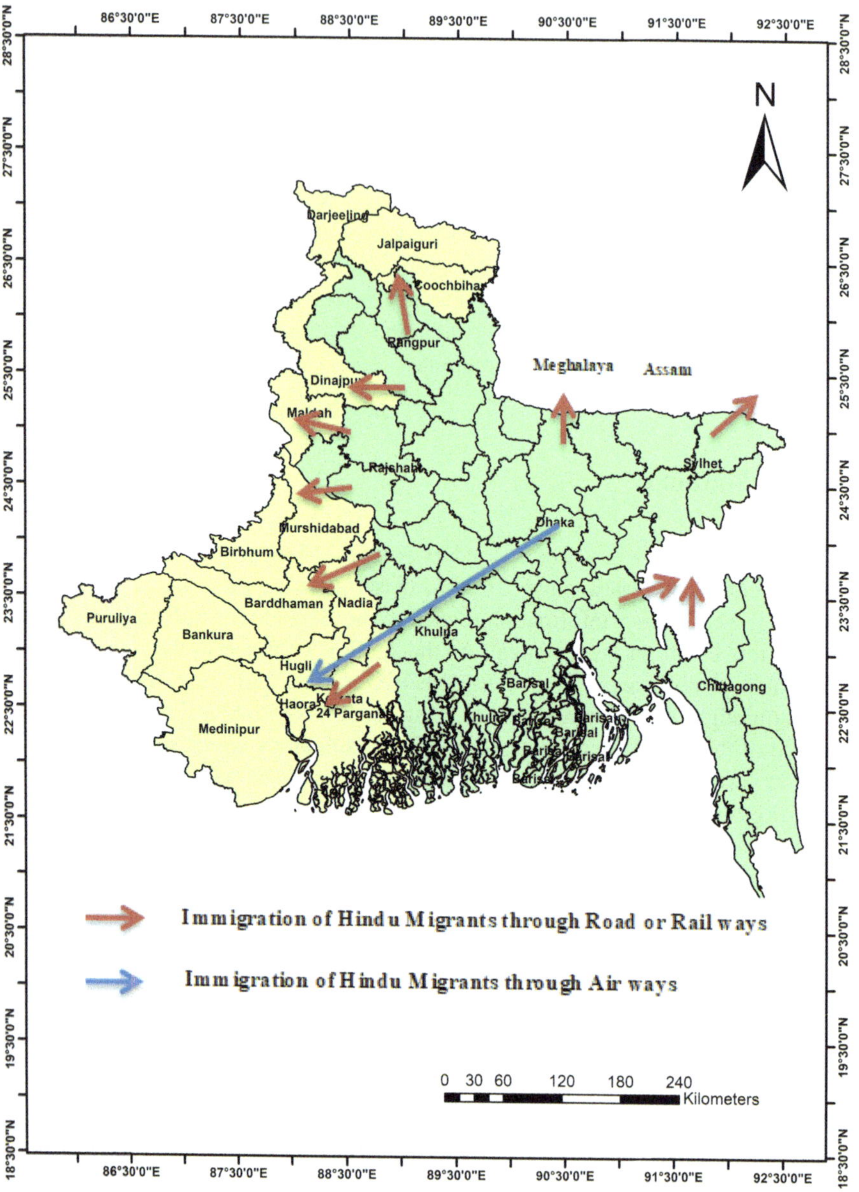

Fig. 2 Arrival of Hindu migrants from East Pakistan to West Bengal and the other states

regions of India for their displacement. The Indian state of Tripura was occupied by the people of Chittagong and the Tripura district of East Pakistan. Simultaneously, people from Mymensingh and Rangpur chose the North Assam, Cooch Behar, and Jalpaiguri districts of West Bengal, respectively. People originating from Rajshahi

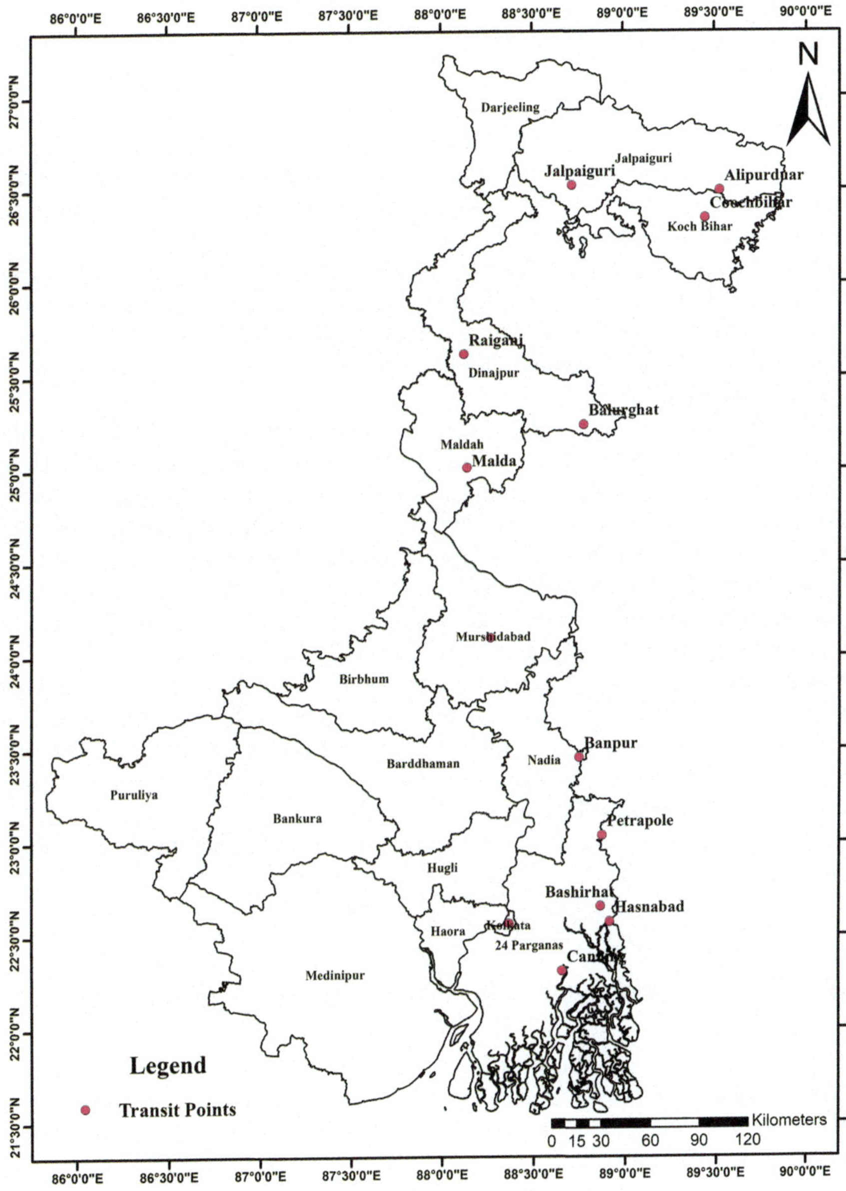

Fig. 3 Entry and transit points of refugees in West Bengal. (Source: Chatterjee, 2007. Map drawn by the researcher)

and East Dinajpur of East Pakistan entered West Bengal and settled in Malda and West Dinajpur, respectively (Fig. 2). The Indian portion of Banpur and Bongaon was occupied by the refugees from the central and southwestern part of East Pakistan, respectively (Dasgupta, 2016).

A group of higher-class refugees used airways and directly moved from Dhaka to Kolkata (Fig. 2). The government opened a temporary health centre at Kolkata airport.

At that time, rail connectivity was not available in the Madarihat subdivision of Faridpur district in East Pakistan. This issue was solved by Dr. Bidhan Chandra Roy, then Chief Minister of West Bengal. He organized 15 steamers to rescue the refugees. These steamers picked up the refugees and dropped them at the Kolkata port, from where they were shifted to the transit camps* by train.

Every day, around 15,000 refugees arrived at the shipyard; the government documented their family histories, gave them border slips to identify refugees, and transferred them to the camps. Babu Ghat and Shalimer Ghat, located on the eastern and western sides of the Bhagirathi River, were also used to drop the refugees (Bandyopadhyay, 1970).

Some transit points were used to receive new migrants and provided them with temporary accommodation. It was located in Banpur, Petrapole, Hasnabad, Sealdah, Basirhat, Balurghat, Malda, Cooch Behar, Raiganj, Alipurduar, Jalpaiguri, Canning, Berhampore, etc. (Chatterji, 2007) (Fig. 3). After receiving the border slips, the refugees were treated as bonafide refugees, and they started to receive all kinds of governmental supports. One of the important transit points was opened in Sealdah where refugees arrived by train. They gathered there and received the government doles until they were shifted to new camps. Station premises became their meeting points for them. They formed makeshift camps there, and hundreds of them were living together. Apart from food and shelter, sometimes, refugees were given vaccinations for cholera and tuberculosis at the transit points. They were sent to the transit camps which were located in different parts of the state and finally transferred to the relief camps (Dasgupta, 2016).

*A temporary meeting point for the refugees from where they shifted into different camps and colonies.

6.2 Noakahli and Tippera Riots and First Phase of Migration

The migration never seized except for a brief period. It continued through several stages. Chakrabarty (1999) elaborately explained the phases of displacement in which the first phase of exodus started in 1946. The Noakhali and Tippera disturbances of East Pakistan which were the result of the Great Calcutta Killing in 1946 were the prime reason for this phase of migration. The exodus of Hindus and the transfer of their assets began during this time which slowed down after the announcement of Indian Independence by the British government. A group of Hindu people left East Pakistan due to the direct consequences of the communal riots. Mental torture and emotional harassment subjected to Hindus by the Muslim community were some of the prime reasons for this phase of migration. This was the time for the upper- and middle-class Hindus to start their journey towards a new country where their co-religionists were a majority in numbers. There were three types of refugees: one group of refugees belonged to the upper class. They simply rented or purchased houses at their own expense and did not want to be dependent on anyone for their survival. The next group of people were from middle- or lower-middle-class families. They had insufficient funds for their survival but had enough courage which forced them to occupy the empty and deserted houses. They also wanted to be independent for their survival (Chakrabarty, 1999). Sometimes, they grabbed vacant lands and started to live there. Land speculators also took this opportunity and purchased vacant space and houses at low prices from the owners and sold them to wealthy migrants at high prices (Amrit Bazar Patrika, 24 August 1947). In some cases, the refugees could not buy the lands at their expense but rather occupied them forcefully. Around eighty families forcefully occupied the fifty flats in the Lalbagan area of East Upper Circular Road in Kolkata as a result of a large amount of salami (Amrit Bazar Patrika, 14 October 1947). The left-wing leaders formed a committee known as Santi Sena or Peace Brigade. They tried to save the refugees from eviction by the landowners (Amrit Bazar Patrika, 31 October 1947) and also helped them to occupy the vacant lands for their accommodation (Satpathi, 2013). In this way, a forcibly occupied colony or squatter colony was formed. The third group of people came from the economically lower class, and they were completely dependent on the government (Chakrabarty, 1999).

After partition, the migration flow accelerated for a time being and the number of migrants reached 2.46 lakhs at the end of 1947 (Sengupta, 2017). Some documents estimated the value of migration at 3.44 lakhs in 1947 (Partition of Bengal, 1947). Police action in Hyderabad the name of Operation Polo (Times of India (TOI), 15 September 2019) again accelerated the refugee influx from 18 September 1948.

Before partition, Hyderabad was a Hindu princely state, ruled by the Nizam. During partition, all the princely states were given equal chances to determine their future and incorporated either India or Pakistan according to their choice. Most of the princely states decided their fate and united with their favourite nation except the Nizam. He wanted Hyderabad as an independent state within India and wanted to do it with the help of Razakars (they were a private military troop owned by Qasim Razvi, a powerful politician in the princely state of Hyderabad). They also had a secret mission to be incorporated with Pakistan (Moraes, 2007). Finally, Nizam signed on Standstill Agreement on 29 November 1947 which was denoted to maintain the existing administrative arrangement of the British, and Nizam would continue with the government of India (Menon, 1956). The agreement was breached due to several reasons which infuriated the government of India, and they gave an ultimatum on 31 August 1948. Finally, Hyderabad was incorporated into the Indian Territory on 17 September 1948 (Raghavan, 2010). This pogrom created tension among the upper- and middle-class Bengali Hindus of East Pakistan, and they started to leave their country in search of permanent settlement in West Bengal (Chakrabarty, 1999).

According to the Commissioner of Relief and Rehabilitation in West Bengal, the Central Rehabilitation Minister estimated that the number of migrants would reach a million by March 1948 which again swelled to 1.1 million by June 1948. Among

Table 1 The rate of admission in the government camps in 1948

Month	Number of refugees accommodated in government relief camps
July	8408
August	19,109
September	18,638
October	9182

Source: Bandyopadhyay (1970), Chakrabarty (1999)

them, 3.5 lakhs were urban middle class, 5.5 lakhs were rural middle class, little more than 1 lakhs belonged to the agriculturist, and little less than 1 lakhs were artisans (Amrita Bazar Patrika, 12 March 1948). Every day, 10,000 to 12,000 refugees left East Pakistan and reached West Bengal until mid-1948 (Chaudhuri, 1983). The following Table 1 reflects the same.

Around 13 to 15 lakh people were staying in West Bengal by the end of 1948 (Chaudhuri, 1983). This phase of the influx was confined to the Hindus, and their migration was more frequent than Muslims. In 1948, a little less than 6 lakh Hindus entered West Bengal. By April 1949, their numbers increased to 19.5 lakh. Among these, 9.70 lakhs were living in various camps and colonies in Kolkata and its surroundings, and 6.5 lakhs had occupied many places in West Bengal (Amrita Bazar Patrika, 27 August 1949; Sengupta, 2017). By the end of 1949, the number of camp dwellers had reached seventy thousand, and after that, it began to settle down with the completion of the first phase of migration (Chakrabarty, 1999).

6.3 *Oppression on Namasudra Community, Second Phase of Refugee Migration, and Impact of Delhi Pact*

Economically well-off people from East Pakistan purchased the vacant lands at higher prices and started to colonize them. As the flow was mainly one-way, it was difficult for the relief department to search for new vacant lands. The number of people increased day by day, while the central government also pressured the state government to look after the refugee matter and sanctioned 60 lakh rupees for settling the issue. Dr. Bidhan Chandra Roy, then Chief Minister of West Bengal, introduced a new department for rehabilitation which was earlier conducted by the relief department. The government was also concerned about the medical facilities and reserved some beds in Jadavpur Tuberculosis Hospital for the refugees (Bandyopadhyay, 1970).

During this time, the government also decided to permanently close down all the relief camps, while a group of refugees still was unable to receive any kind of rehabilitation benefits. Those male-headed and widow-supervised families were incapable of looking after their families. Some had old and firm people, and some had paralyzed members who were unable to relocate by their own choices. The government then classified the camp refugees into two sectors: one group of people was able to receive every kind of rehabilitation benefit and another group of refugees was unable to receive such benefits and rather became liable to the government. The latter was known as permanent liabilities (P.L.s). The government calculated that among seventy thousand camp refugees, there were seven thousand five hundred permanent liabilities, and the remaining sixty-two thousand five hundred were waiting for their permanent rehabilitation. After the first phase of migration, only permanent liability (P.L.) camps remain open for those people (Bandyopadhyay, 1970).

The second phase of the exodus started in January 1950 when extensive violence occurred in the Namasudra community by the Muslims and the incident started in the Khulna district of East Pakistan. It further extended over a larger area of Rajshahi, Dacca, and Faridpur (Singh, 2003). The news circulated that Hindu families were again oppressed by the Muslim community in the Bagherhat district of East Pakistan which again exaggerated the displacement of Hindus from East Pakistan. Government assumption about this phase of migration was temporal which would be settled down soon. The central government started to look after the matter seriously and organized a meeting with the government of East Pakistan to settle the issue. After attending the meeting, the Chief Secretary of West Bengal concluded that due to the riots, the number of upcoming refugees would be around 10 lakhs. West Bengal government then decided to open some temporary relief camps at two border stations of Bongaon and Darshana and also introduced signature slips as proof of refugee. Later, this slip was known as a border slip (Bandyopadhyay, 1970). This phase of violence was mainly confined to the lower-class peasants and artisans (Singh, 2003). After one month of this massacre, around 30 thousand Hindus left Khulna and reached India (Nehru, 1992). During this time, not only Hindus of East Pakistan left their country but Muslims of West Bengal also started to change their position from Hindu-dominated India to Muslim-dominated East Pakistan (Banerjee, 1950; Chaudhuri, 1983). The following Table 2 denotes the rate of admission of East Pakistan refugees into the government camps.

During this phase of migration, refugees poured into Bengal-like waves, and their numbers reached around 56 thousand within the first fifty-three days of 1950. This number further increased to 1.5 lacks within the first 3 months, and at that time, the daily number of refugees was around 15 thousand (Chakrabarty, 1999). The above table only indicates the number of refugees who were given shelter in the government-sponsored relief camps.

Table 2 The rate of admission in the government camps in 1950

Month	Number of refugees accommodated in government relief camps
January	1150
February	1002
March	75,596
April	14,960
May	27,440

Source: Bandyopadhyay (1970), Chakrabarty (1999)

After arriving in West Bengal, the refugees were taken to Sealdah station which was again crowded since the second week of March 1950 (Amrita Bazar Patrika, 21 April 1950). On 7 April 1950, the number of refugees squatting at the Sealdah station was 8 thousand (TOI, 8 April 1950). The number further increased to 9 thousand by 27 June (TOI, 28 June 1950), and it reached 12 thousand on 22 July 1950 (TOI, 23 July 1950). In the first week of August, these numbers were slightly reduced to 10 thousand (TOI, 8 August 1950). The government also decided to reopen the closed camps to check the refugee flow (Chakrabarty, 1999).

It was estimated that on the third week of July 1950, the number of refugees was little less than 25 thousand which slightly reduced to 23 thousand in the following week of July 1950 (TOI, 26 July 1950; Sengupta, 2017). The rate of migration flow trickled down after the Nehru-Liaquat treaty. This treaty was popularly known as Delhi Pact which was signed on 8 April 1950 between Indian Prime Minister Jawaharlal Nehru and Pakistan Prime Minister Liaquat Ali Khan. The objective of this treaty was to encourage all the refugees from both countries to return to their native land. This treaty assured the migrants that the present government would return the complete proprietary rights of their immovable properties that they had left behind during migration. This treaty also dealt with the return of abducted and molested women, confirmation of minority rights, etc. (Chandra, 2008; Chakrabarty, 1999). Despite this effort, the migration never ceased; rather, it continued over time. From April 1950 to September 1950, every month, around 25 refugees were given shelter in various relief camps in West Bengal (Chakrabarty, 1999). So this agreement had a minimal effect on the migrants. The rate of admission of East Pakistan refugees in government camps also reflected the scenario (Table 3).

After the declaration of the Delhi Pact, both Hindus and Muslims started to return to their country. A large number of Muslims who left West Bengal and Assam also re-entered these states. According to the Committee of Review of Rehabilitation (COR) report of West Bengal, the number of returned persons was larger than their dispersal from East Pakistan to West Bengal and vice versa (Chakrabarty, 1999). The same view was observed by the Government of India, Parliamentary Debates (1951), p. 439), where it was stated that after the Delhi Pact, the flow of returned migrants increased in both directions. Dasgupta (2016) mentioned the reversed picture between April 1950 and November 1950 in which the numbers of displaced

Table 3 The rate of admission in the government camps

Year	Month	Number of refugees admitted to government relief camps
1950	October	8754
	November	9543
	December	6549
1951	January	4399
	February	4215
	March	5117

Source: Bandyopadhyay (1970), Chakrabarty (1999)

Table 4 Migration of displaced persons in both directions

From West Bengal to East Pakistan (From 9 April 1950 to 4 November 1950)	
Hindus	Muslims
1,158,715 (Returnees)	656,982 (Displaced)
From East Pakistan to West Bengal (From 9 April 1950 to 4 November 1950)	
Hindus	Muslims
1,826,276 (Displaced)	519,523 (Returnees)

Source: Dasgupta (2016)

Table 5 Movement of refugees in East India between May and July 1951

For the week ending on:	From East Bengal to West Bengal		From West Bengal to East Bengal	
	Hindu refugees	Muslim returnees	Hindu returnees	Muslim refugees
6 May 1951	20,479	12,795	34,280	12,689
13 May 1951	26,552	12,374	37,409	12,247
20 May 1951	25,070	12,462	39,490	13,747
27 May 1951	27,188	12,140	40,191	12,886
3 June 1951	30,123	12,565	37,123	13,134
10 June 1951	32,110	12,789	34,258	12,291
17 June 1951	33,490	11,941	29,163	11,563
24 June 1951	44,825	11,929	29,146	10,949
1 July 1951	32,943	9671	21,408	8571
8 July 1951	37,911	11,173	26,190	11,613
15 July 1951	36,582	12,891	27,938	12,448
22 July 1951	26,096	11,941	23,197	12,619
29 July 1951	29,643	10,044	20,086	12,161
Total	403,012	154,715	399,879	156,918

Source: Government of India, Constituent Assembly Debates (1951), p. 1470), Dasgupta (2016)

Hindus from East Pakistan were greater than their returnees from West Bengal. The following Table 4 reflects the same.

So this treaty created hope among the refugees, and they started to go back to their land. Around 12 lakh Hindu refugees left West Bengal for East Pakistan (Chakrabarty, 1999), but how many of them re-migrated to West Bengal during the following phases was not properly documented.

The available information stated that initially, the refugee influx was greater than their returnees in both directions, but the reversed picture was observed from May 1951, and it continued up to 10 June 1951. After that, refugee influx again increased than their returnees. Table 5 reflects the same.

The above information stated that initially, the Nehru Pact had a great impact on the refugees, and they started to remigrate to their land. From May 1951, it was noticed that a huge number of Hindu refugees returned to their native land which also exceeded their arrivals. But soon they realized that their new lives in their motherland might not be secure, and they changed their decision and tried to remain fixed on their earlier decision to stay in West Bengal. Despite this movement in both directions, the State Rehabilitation Department declared that there were 20 lakh immigrants in West Bengal by the end of 1951 (Ministry of Rehabilitation, 1952; Dasgupta, 2016). The government assumed that migration might stop after the declaration of the Delhi Pact, but again, a wave of refugees poured into West Bengal with the initiation of the third phase of migration.

6.4 *Introduction of the Passport System and Third Phase of Refugee Migration*

After 1951, the introduction of the unilateral passport system created anxiety among the Hindu refugees of East Pakistan. It was introduced on 15 October 1952, and it created tensions among the Hindus of East Pakistan. They assumed that this decision might seize their movement towards West Bengal; therefore, they had to stay in East Pakistan permanently. The Indian government also followed similar steps and introduced the passport system for all Pakistani citizens entering India, but the Pakistan government first took this initiative to check the smuggling activities. Indian press speculated that the main motive of the Pakistani government was financially weakening the Hindu people of East Pakistan who shifted to India. This strategy made it impossible for them to access their lands in East Pakistan which might have generated more evacuees' property for the Pakistan government (Cochrane, 2009). From January 1952 to June 1952, the average rate of refugees admitted into the government camps was around 2 thousand (Chakrabarty, 1999). The rate of migration never ceased as there were 25 lakh refugees in West Bengal until May 1952 (Chaudhuri, 1983). Table 6 reflects the refugee influx just before and after the introduction of the passport system.

The next phase of migration started in 1960–1961, but the refugee flow did not stop between 1953 and 1960. In 1955 and 1956, the refugee flow increased due to communal riots that occurred in East Pakistan. The Ministry of Rehabilitation initially published meticulous data in the official reports monthly from 1953 to 1957.

Table 6 Rate of admission in the government camps in 1952

Month	The rate of accommodation in government camps
July	11,600
August	7800
September	10,654
October	31,753
November	1715
December	755

Source: Bandyopadhyay (1970), Chakrabarty (1999)

Table 7 Influx of displaced persons from East Pakistan (1953 to March 1957) to West Bengal

Month	1953	1954	1955	1956	1957
January	6164	4769	17,058	19,206	2005
February	7582	6792	24,983	50,593	2470
March	9756	7035	30,187	55,381	1266
April	8870	6849	19,376	22,752	–
May	7882	7685	20,738	30,493	–
June	6123	7300	23,816	29,754	–
July	6268	6989	25,271	38,033	–
August	5330	8612	15,412	47,065	–
September	3774	11,289	10,966	16,411	–
October	5336	11,824	16,148	5344	–
November	4041	13,232	13,342	2265	–
December	4997	25,555	21,734	2329	–
Total	76,123	117,931	239,031	319,626	5741

Source: Ministry of Home Affairs, Government of India, Annual Report (1958), p. 39), Dasgupta (2016)

The following Table 7 represents the refugee influx from East Pakistan during this time.

The above data highlights that the introduction of the passport system could not even slow down the immigration flow in West Bengal. It gradually increased between 1953 and 1955 and peaked in March 1956. After that, it suddenly fell from October 1956 due to the introduction of migration certificates by the Indian government to check the migratory flow and the occurrence of communal disturbances in various parts of East Pakistan (Chakrabarty, 1999; Dasgupta, 2016). The same data was also published by the Relief and Rehabilitation Department of West Bengal in which the influx of refugees from January 1953 to July 1956 was also documented.

Both the tables represent the changes in refugee influx from January 1953 to March 1957 (Table 7) and January 1953 to July 1956 (Table 8) in which discrepancy in data was also observed in each month. It might have occurred during data enumeration.

6.5 *Communal Disturbances and Fourth Phase of Displacement*

Between 1958 and 1963, around 55 thousand Hindu refugees came to West Bengal. At that time, 38 lakh refugees were staying in West Bengal until 1963. Due to communal disturbances in Khulna, Dacca, Jessore, Chittagong, and other areas, around 18 thousand persons crossed the border and settled in West Bengal in 1964 (Maity, 1964; Chaudhuri, 1983). The inflow of refugees continued in 1965 and 1966 when the number of refugees reached around 40 lakhs in West Bengal alone. According to

Table 8 Month-wise break up of refugee influx to West Bengal

Month	1953	1954	1955	1956
January	5248	4077	15,674	17,011
February	5961	5710	22,848	42,360
March	7507	5821	26,503	15,167
April	6900	6002	15,070	18,039
May	6032	6656	18,190	34,657
June	4798	6354	21,146	24,734
July	5026	6208	22,957	27,442
August	4147	8127	13,813	–
September	3223	10,644	1971	–
October	4379	10,352	13,757	–
November	3212	11,073	11,535	–
December	4214	22,776	18,709	–
Total	60,627	103,800	209,573	179,410

Source: Relief and Rehabilitation of Displaced Persons in West Bengal (1956), Roy Chaudhury (2009)

government records, there were 42.59 lakh people in West Bengal up to 1970 (Chaudhuri, 1983). Officially, the final phase of migration started in 1971 during the Bangladesh Liberation War.

6.6 *Bangladesh Liberation War and Final Phase of Refugee Migration*

After the partition of India, the eastern province of Bengal was incorporated into Pakistan's territory in which the former region was distinct in its culture, economy, and political ideology. Their own identity created major turmoil among the inhabitants of East Pakistan. In 1970, the situation deteriorated when Awami League, the largest political party in East Pakistan, won the national election. This party was led by Sheikh Mujibur Rahaman. Out of 169 seats, the Awami League won 167 seats and was ready to form a government. Zulfikar Ali Bhutto, the Leader of the Pakistan People's Party, refused Rahaman as Prime Minister and proposed two Prime Ministers for both wings of Pakistan. This proposal created tension in East Pakistan to violate the 'one unit scheme', and Mujibar called a national strike. Finally, a meeting was conducted, and it was decided to form a coalition government in which Mujibar Rahaman became Premier and Zulfikar Ali Bhutto became President (Hasan, 2000). The military was not aware of this decision. On 7 March 1971, Rahaman delivered a speech among the people of his region and concluded it by saying that their struggle was for their freedom and their independence. These words motivated the people of East Pakistan, and they started a nationalist movement for their independence. West Pakistan military 'Junta' introduced a planned

Table 9 Number of refugees in West Bengal, 1971 Census

District	No. of refugees	Percentage
24 Parganas (North and South)	1,650,000	22.3
Nadia	1,500,750	20.3
Calcutta	900,000	12.9
Cooch Behar	442,000	5.9
Dinajpur (North and South)	292,500	3.3
Jalpaiguri	249,000	3.3
Burdwan	240,000	3.2
Bankura	1,415,750	19.1
Hugli	159,000	2.1
Haora	144,000	1.8
Murshidabad	135,000	1.8
Malda	127,500	1.7
Medinipur	63,000	0.8
Darjeeling	48,000	0.6
Birbhum	15,000	0.2
Purulia	975	0.2
Total	7,382,475	100.0

Source: Census Report (1971), Dasgupta (2016)

action named 'operation searchlight' to restrain the independence movement on 25 March 1971 (Bose, 2005). According to Pakistan's news, they arrested Rahaman on 25–26 March 1971. Finally, on 26 March 1971, M.A. Hannan, leader of the Awami League in Chittagong announced independence over the radio. This day was officially considered the Independence Day for East Pakistan, and the newly formed country was renamed Bangladesh. Indira Gandhi, the Prime Minister of India, also openly addressed the newly formed region as Bangladesh in July 1971. The conflict finally ended on 16 December 1971 (Bangladesh Liberation War, 2023). During this phase, 10 million people were crossing the border and reaching India (Nicholas, 1972; Vyas, 2001), while around 30 million people were migrating internally (Totten et al., 2008). Out of 10 million refugees, there were eight million belonging to the Hindu community (Kennedy, 1971; Vyas, 2001).

The 1971 Census (Table 9) also published the data which represents the district-wise refugee concentration. It was observed that refugee concentration was highest in the 24 Parganas and Nadia districts. But, during this time, the concentration of refugees in Kolkata slightly decreased from 1958 as most of the vacant lands and job opportunities became saturated. At the same time, the government also shifted many refugees outside of Bengal due to the lack of proper rehabilitation facilities in West Bengal. The central and state governments of India started to count the displaced migrants from East Pakistan at the beginning of the 1950s. The prime motive of this calculation was to provide adequate relief and shelter among the Hindus. But the numbers of actual migrants changed from one source to another.

From the above table, it may be stated that the refugee influx was confined to all districts; among them, 24 Parganas, Nadia, and Bankura received the maximum number of refugees. Ministry of Rehabilitation (MOR) also published a report in 1971 in which they enumerated the refugee influx in West Bengal (Table 10).

Dasgupta (2016) reviewed several governmental reports and stated that the West Bengal government formed the Refugee Rehabilitation Committee (RRC) in December 1978 to prepare a report on refugee-related problems. The committee reviewed the census report of 1971 and stated that West Bengal had a 73.82 lakh refugee population. Then, RRC started their survey and estimated that 74.4 lakh refugees were living in West Bengal. Around 2.1 lakh refugees were living outside the government-sponsored camps. If the total number of refugees was counted, it was estimated that around 76.5 lakh or nearly 80 lakh refugees were living in West Bengal in 1981. This was the year of the completion of the RRC report.

There was a difference between the report of RRC and MOR. The RRC found that there were some mistakes in the MOR estimation. According to MOR, 6 lakh refugees were crossing the border and settled in West Bengal between 1964 and 1971. Dasgupta (2016) stated that the values were not properly matched with the actual values as more numbers of refugees had come and taken shelter in West Bengal. It was also stated that during the estimation, a large number of squatter colonies were overlooked by the surveyors. Many displaced refugees who did not take shelter in the government-sponsored camps or who chose Muslim evacuees' properties for their shelter, forcefully occupied the vacant lands, or squatted on the open lands or government barracks were not counted by the MOR surveyors. RRC stated that MOR only counted a little more than three-fourths of the total refugees in West Bengal (Government of West Bengal, 1981). The same observation was found in the census report. Their survey also excluded people living in unauthorized camps, staying with their relatives, or accommodated with their initiatives (Dasgupta, 2016). Dasgupta (2016) assumed the reason behind the inconsistency of data enumerated by MOR. According to him, MOR was quoting a smaller number of refugees than the actual values due to some political reasons. He also opined that MOR might have tried to convince the central government that the refugees' issues were not so critical in the eastern section of the country; rather, more emphasis would be given to the western section. Limited funds and resources would be enough to control the situation of the eastern section. The ruling party also assumed that the vote banks of Western India were more crucial than Eastern India (Dasgupta, 2016).

Dasgupta (2016) also referred to the words of Atulya Ghosh, a prominent Congress leader who stated that by the end of 1963, around 48.64 lakh refugees were residing in West Bengal, and among them, 1.54 lakh left the state and started to live in the other state. In tune with the central government, Ghosh also added that the refugees who were starting to live in other states created a problem among the native residents. Their presence created congestion in the rural and urban areas and disrupted their daily lives. He also stated the role of the state government in which they agreed to the decision of the central government regarding the refugee issue. It was decided by the central government that migrants would be given rehabilitation benefits if they shifted to the other states, as West Bengal was already saturated with the availability of lands and other resources (The Statesman, 12 June, 1955).

Table 10 Refugee influx in West Bengal and other states, 1946–1971

Year	Refugee influx in West Bengal (in lakh)
1946–1952	25.18
1953	0.16
1954	1.04
1955	2.12
1956	2.47
1957	0.04
1958 (until 31 March)	0.01
1958 (1 April to 31 December)	0.04
1959	0.05
1960	0.09
1961	0.1
1962	0.13
1963	0.14
1964	4.19
1965	0.18
1966	0.04
1967	0.05
1968	0.04
1969	0.04
1970	2.23
1971 (up to 25 March 1971)	0.07
Total	38.41

Source: Ministry of Rehabilitation, Government of India (1981), Nakatani (1994), Dasgupta (2016)

7 Limitations of the Study

The study is based on secondary sources, and many archival reports and documents are used to gather the data. The outcomes of these reports are not well consistent. There are several contrasting views between central government and state government reports regarding the numbers of incoming and outgoing refugees in West Bengal. Newspaper reports and census data also reflect the same.

8 Conclusion

The independence of India and resultant fragmentation were one of the most trajectory incidents in the world. The division of Bengal and its consequences not only affected the inhabitants of India, but people from the adjacent country also experienced the changes. India gained independence, and it resulted in the division of the eastern section of India into two different regions. Despite having the same cultural identity in both Bengal, there was a single regional factor that instigated the partition. After partition, the migration started from East Pakistan in which the majority of them were Hindu. The migration occurred in several phases which initiated from the Noakhali and Tippera riots in 1946 and continued up to 1971 when the Bangladesh Liberation War occurred. This tragic displacement created major population pressure in West Bengal. Central and state governments initially accepted the migration flows of Bengali refugees and arranged temporary accommodations for them. Apart from that, refugees were also provided doles and financial support. Gradually, the flow of migration had increased which led the government to change its perception towards refugees. The government decided to send additional numbers of refugees outside the state due to the unavailability of vacant lands in West Bengal. Desertion of refugees from the outside camps and their return journey towards West Bengal made the situation worse. The government had to acquire additional lands for their settlements. Apart from the government-sponsored colonies, some forcefully occupied colonies or squatter colonies were developed around the state. Refugees' agitation, processions, rallies, and the government's action to control these movements were regular incidents in West Bengal. Apart from these negative consequences, there were some positive sides which influenced the economy of West Bengal. The inhabitants of East Pakistan were very much energetic in their culture. During the time of displacement, they carried their culture and distributed it among the people of West Bengal. The refugees of the Ranikuthi camp welcomed the Governor of West Bengal through their traditional rituals. Peasants coming from East Pakistan also were very skilled in jute cultivation. Good quality jute was grown in the eastern part of Bengal, and it was further processed in West Bengal due to abundant numbers of jute mills located on both banks of the Hugli River. This partition adversely affected the jute industry in West Bengal. Most of the jute mills of West Bengal were facing a major shortage of quality raw materials. These uprooted peasants revived the jute industry in West Bengal as they started to cultivate this cash crop with the same techniques and procedures which was used earlier in East Pakistan.

References

Abha Maity. (1964). Thirty-eighth session. *Assembly Proceedings, XXXVIII*(1), 392. (Incomplete Citation).

Alexander, C., Chatterji, J., & Jalais, A. (2016). *The Bengal diaspora: Rethinking Muslim migration*. Routledge.

Bandyopadhyay, H. (1970). *Udbastu*. Sahitya Samsad.

Bose, S. (2005). Anatomy of violence: Analysis of civil war in East Pakistan in 1971. *Economic and Political Weekly*, 4463–4471.

Census Report 1951.

Census Report 1971.

Chakrabarty, P. (1999). *The marginal men*. Naya Udyog.

Chandra, B. (2008). *India since independence*. Penguin UK.

Chatterji, J. (2007). 'Dispersal' and the failure of rehabilitation: Refugee camp-dwellers and squatters in West Bengal. *Modern Asian Studies, 41*(5), 995–1032.

Chaudhuri, P. (1983). Refugees in West Bengal: A study of the growth and distribution of refugee settlements within the CMD.
Cochrane, I. (2009). The causes of the Bangladesh War. Lulu.com.
Das, S. (2004). *Noakhali:1946* (p. 13). Sahitya Prakash. ISBN 984-465-373-8.
Dasgupta, A. (2016). *Displacement and exile: The state-refugee relations in India*. Oxford University Press.
Government of India, Constituent Assembly Debates. (1951). p. 1470 (Incomplete Citation).
Government of West Bengal (1981):2 (Incomplete Citation).
Governor. (1972). Fifty-second session. *Assembly Proceedings, 52*(1), 13. (Incomplete Citation).
Hassan, Doctor of Philosophy (Ph.D.), Dr. Professor Mubashir (2000). 'Zulfikar Ali Bhutto: All Power to People Democracy and Socialism to People!'. The Mirage of Power. Oxford University. United Kingdom: Dr. Professor Mubashir Hassan, Professor of Civil Engineering at the University of Engineering and Technology and the Oxford University Press, pp. 50–90. ISBN 978-0-19-579300-0.
Kennedy, E. M. (1971). *Crisis in South Asia*. US Government Printing Office.
Majumder, R. C. (1975). বাংলা দেশের ইতিহাস [History of Bangla Desh] (in Bengali). 4 (p. 428). General Printers and Publishers.
Menon, V. P. (1956). *The story of the integration of the Indian states*. Orient Longmans Ltd.
Ministry of Home Affairs, Government of India, Annual Report. (1958). p. 39 (Incomplete Citation).
Ministry of Rehabilitation, Government of India. (1981:39) (Incomplete Citation).
Ministry of Rehabilitation, Government of India. (1952: 35) (Incomplete Citation). Compiled by Univ. of Chicago group of scholars.
Moraes, F. (2007). *Jawaharlal Nehru*. Jaico Publishing House.
Nakatani, T. (1994). Away from home: The movement and settlement of refugees from East Pakistan in West Bengal, India. *Journal of the Japanese Association for South Asian Studies, 12*, 2000.
Nayyar, P. (1968). *Mahatma Gandhi: The last phase* (Vol. I).
Nehru, J. (1992). *Selected works of Jawaharlal Nehru: Pt. 1. 15 November 1949-8 April 1950* (Vol. 14). Jawaharlal Nehru Memorial Fund.
Nicholas, M. (1972). *Bangladesh: The birth of a nation*.
Raghavan, S (2010). *War and peace in modern India*. Palgrave Macmillan. ISBN 978-1-137-00737-7.
Relief and Rehabilitation of Displaced Persons in West Bengal (Calcutta: Home [Pub.] Department, Government of West Bengal, (1956) (Incomplete Citation).
Roy Chaudhury, A. (2009). Citizens, Non-Citizens, and in the Camps Lives.
Sarma, M. (2015). *A study of migration from Bangladesh to Assam, India and its impact*. Doctor of Philosophy (PhD) Thesis, The University of Adelaide.
Satpathi, S. (2013). Mobilizing women the experience of the left in West Bengal from 1947 to 1964.
Sengupta, A. (2017). *The railway refugees: Sealdah, the 1950s–1960s* (Occasional Paper). Institute of Development Studies Kolkata.
Sibnath Banerjee (1950, February 16). First session. *Assembly Proceedings, II*(1), 28 (Incomplete citation).
Singh, N. K. (2003). *Encyclopaedia of Bangladesh* (Set of 30 Vols.) (Vol. 1). Anmol Publications PVT. LTD.
Sinha, D. C., & Dasgupta, A. (2011). *1946: The Great Calcutta Killings and Noakhali Genocide, A Historical Study*. Tuhina Prakashani.
Totten, S., Bartrop, P. R., & Jacobs, S. L. (2008). *Dictionary of Genocide* (Vol. 1). ABC-CLIO.
The United Nations Department of Economic and Social Affairs (UN DESA), 2020.
Vyas, S. (2001). *Hindu genocide in East Pakistan*. Hindu Library of History.
World Bank, 2015.
World Migration Report, 2010.
The New York Times. Associated Press. 8. 04. 1947. P. 23.
Amrita Bazar Patrika, 12.03. 1948.

Amrita Bazar Patrika, 14.10.1947.
Amrita Bazar Patrika, 21.04.1950.
Amrita Bazar Patrika, 24.08.1947.
Amrita Bazar Patrika, 27.08. 1949.
Amrita Bazar Patrika, 31.10.1947.
The Statesman, 12. 06. 1955.
Times of India (TOI), 23.07.1950.
Times of India (TOI), 26.07.1950.
Times of India (TOI), 28. 06. 1950.
Times of India (TOI), 8. 04.1950.
Times of India (TOI), 8. 08. 1950.
Bangladesh Liberation War. Retrieved on June 7, 2023 from https://en.wikipedia.org/wiki/Bangladesh_Liberation_War-
Noakhali Riots. (n.d.). Retrieved on April 10, 2019 from https://en.wikipedia.org/wiki/Noakhali_riots-
Partition of Bengal. (1947). Retrieved on April 11, 2019 from https://en.wikipedia.org/wiki/Partition_of_Bengal_(1947)-
Sciences Po. (2007). Retrieved on April 9, 2019 from https://www.sciencespo.fr/mass-violence-war-massacre resistance/en/document/Calcutta-riots1946-

Sayantika Baral is a Research Scholar at the School of Oceanographic Studies, Jadavpur University, India. She has completed her MSc from Presidency University and MPhil from the School of Oceanographic Studies, Jadavpur University. Her focused areas of study are population and welfare geography, displacement, refugee migration, and partition movement. As a UGC-JRF Scholar, she has attended national and international seminars and published articles.

Tuhin Ghosh, PhD, is a Professor and Present Director of the School of Oceanographic Studies, Jadavpur University, India. He has expertise in coastal geomorphology, disaster management, climate change impacts, adaptation strategies, and human migration. He has published over a hundred articles, a few books, and edited volumes.

He acted as a Lead Author of the 'IPCC Special Report on Ocean and Cryosphere Chapter 4: Sea Level Rise and Implications for Low Lying Islands, Coasts and Communities'.

Reconstructing Historical Landscape Based on Paleochannel Investigations and Archaeological Discoveries in Debalgarh-Anulia of West Bengal, India

Biswajit Roy, Sanjoy Kumar Sadhukhan, and Premangshu Chakrabarty

1 Introduction

Historical geography is nothing but the geography of the past period, the application of which lies in the reconstitution of past landscapes. In writing historical geography, the outlook should be geographical (Pandey, 1963). However, for ancient Bengal, since the contribution of geographers to date is not noteworthy, there remain several research gaps on the site and situation of ancient settlements, particularly those that are depleted. From the account of Curtius, it is evident that around the third quarter of the fourth century B.C., Gangaridai, with its port city Ganga, and Pracyarastra, with its port city Tamralipti, were the two independent states (Hood, 1994), the united power of which influenced the decision of Alexander, the king of Macedonia, to return. A research gap exists on the location of the port city Ganga and Tamralipti port despite the efforts of the historians (Chakrabarty, 2011), but the presence of such a settlement on the cultural landscape of the past is beyond any doubt, which is referred to as Gangaridai by Greeks and Gangaridai by Romans. This paper is an attempt to resolve the issue in the paradigm of social physics with the following objectives:

(i) To evaluate the locational significance of the Debalgarh-Anulia archaeological belt concerning the alignment of paleo-channels of the existing drainage network.

B. Roy
Department of Geography, Pijush Kanti Mukherjee Mahavidyalaya,
Alipurduar, West Bengal, India

S. K. Sadhukhan · P. Chakrabarty (✉)
Department of Geography, Visva-Bharati University, Santiniketan, West Bengal, India
e-mail: premangshu.chakrabarty@visva-bharati.ac.in

A. Rai et al. (eds.), *Contemporary Social Physics*, Springer Geography,
https://doi.org/10.1007/978-3-031-77596-3_4

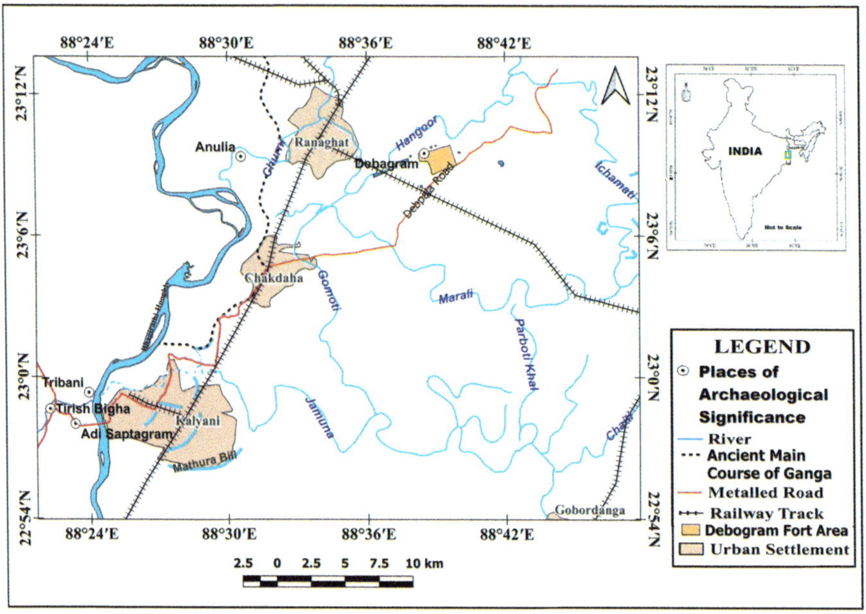

Fig. 1 Location map of the study area. (Authors)

(ii) To make a critical review of the hypothesis regarding footprints of the Gangaridai civilization near the confluence of the five channels as described by Ptolemy in the Bengal delta.

The present paper deals with the river dynamics of Ganga between Anulia and Tribeni. Tribeni (literally means tri-junction of rivers) was the offtake point of Jamuna and Saraswati from Ganga, but the origin of Jamuna could not be traced today due to river shifting through time even by paleo-channel investigation. Thousands of artifacts have been unearthed around a 10–15 km radius of Anulia, particularly at Debagram nearby (Fig. 1). The ruins of the fort of Debalgarh are a very interesting structure in which four ancient watch towers stand in four directions, covering almost one square kilometer area surrounded by an earthen rampart and deep entrenchment. The latitude of both Anulia and Debalgarh is the same (23° 9′ 30″ N), representing an essence of the continuity of the same settlement in the past, supported by the discovery of a large number of artifacts, particularly from the Buddhist legacy in the region. This raises a research question on the presence of a Buddhist establishment in the area in the remote past.

2 Materials and Methods

Geoinformatics and the study of the evolution of past settlements apparently may seem like different fields, altogether dissimilar, unless social physics contributed a paradigm of new social science. Social physics is devoted to the search for isomorphism between social phenomena and physical phenomena. Language of isomorphism (Stewart, 1950) has been adopted to focus on item-to-item correspondence between two or more fields, as concepts from physics correspond to items and relationships within the world (Barnes & Wilson, 2014). With the application of spatial analysis and GIS, social physics becomes a driving force while shifting the approach of historical geography from microgeography (description seeking) toward macrogeography (law seeking). This paper is an attempt to apply laws of location in mapping past cultural landscapes. Improvement in image processing techniques for paleo-channel analysis provides the opportunity for the evolution of historical geography as a law-giving science. New social physics (Urry, 2004) further witnessed the rise of mesogeography, seeking knowledge about new processes that evolve within the spatial context (Miller, 2017). Apart from careful textual analysis, paleographic and cartographic analyses are among the principal tools used (Goheen, 2015). Social physics thus extends the philosophical foundation in the adoption of a hypothesis-testing framework in comprehensive studies of depleted settlements. It is social physics that governed the evolution of terrestrial human settlements as a stand-alone discipline following the prediction of intellectual formation that it would be applied to everything (Zipf, 1942).

To reconstruct the past landscape in the realm of mesogeography, it is necessary to study the evolution of the ancient river courses. The geomorphic history of the study area reveals that the process of delta formation has been completed (Chakrabarti & Nag, 2015). Many tributaries have become extinct, leaving meander cut-off behind. Regular sedimentation has increased the height of the base level, the river has abandoned its past courses, and the local drainage pattern has been modified chronologically. As the delta formation in this part of Gangetic Bengal is the subject of dynamism, an attempt has been made to evaluate its impact with a focus on the discovery of geoarchaeological evidence of past civilizations. It is noteworthy to mention that historical geography in India has been developed by historians and archaeologists following their own methodologies which are different from the methods adopted by geographers while dealing with the cultural landscape of the past. The data used are as follows:

(i) *Historiographic data: Tabaqat-I-Nasiri* written by Minhaj-ud-din in the year 1256, translated by H.G. Raverty (1873); 'The List of Ancient Monuments in Bengal' (1896) by PWD, Govt. of India; *Pavandutam*, a masterpiece of Sanskrit literature contributed by Dhoyi, the court poet of Laksmanasenain the twelfth century; and *Ramacharitam* by Sandhyakar Nandi edited by Sri Ayodhyanath Bidyabinod (1936) are the major data sources among the historical accounts of contemporary period for analysis.
(ii) *Epigraphic data:* The copper plate of Laksmanasena was discovered from Anulia (Inscription of Bengal by Nani Gopal Majumder, 1929 and Indian Epigraphy by D.C. Sircar, 1965).
(iii) *Archaeological data:* Thousands of artifacts have been recovered from the Debalgarh archaeological site including black and red ware (Chattyopadhyay, 2018), black slipped and grey ware potteries, sculptures, and ruins of

architectural structures (1896, PWD). These archaeological objects have been helpful in analyzing the past cultural landscape over centuries.

(iv) *Toponymic data:* The place names are vital sources in reconstructing the image of the past places. For the present study, the following names are noteworthy (Table 1).

To understand the nature of deltaic landform evolution and regional variations, aforementioned literature references from different historical periods have been considered. In addition, the myths, legends, and tales about the locational aspects have been taken into serious consideration. The present study was inspired by a popular story on Bakhtiyar Khilji, a Turko-Afghan military general who invaded the capital of the Hindu king of the Sena dynasty in Bengal in the year 1201. A thorough field investigation has been made to find the probable route of his arrival in Bengal and which was his actual place of invasion because there is already a dispute on the location of the Sena capital Vijaypura established by the first Sena king Bijay Sena from Karnataka. According to historiographic data derived from the writings of Dhoyi, a twelfth-century Sanskrit language poet (Chakroborty, 1926), the location of Vijaypura was in the region where Jamuna offtakes from Bhagirathi (Roy, 1959). This contradicts the popular hypothesis that Nabadwip was the capital of the Sena dynasty, attacked by Bhaktiyar Khilji, and no archeological evidence of the presence of such a capital city has been discovered yet in Nabadwip. The confusion might be caused by the ruins of Ballal Dhipi, located near Nabadwip, which has been declared by the archeological survey of India as an ancient Buddhist establishment. The name Ballal Dhipi was for the effort made by Ballal Sena, a powerful king of the Sena dynasty, to convert the Buddhist establishment into a Hindu place of worship (Ghosh, 1988). An alternative hypothesis has been developed that the fortified structure of Debalgarh was the defense site of Vijaypura (Roy, 2021b). Myths about the Gangaradi are also taken into consideration concerning

Table 1 Toponymic traces of Buddhist past

Place name	Origin	Significance
Anulia (23° 09′ N, 88° 31′ E)	Analgram	Analgram is a name of Buddhist influence; the place was later under the rule of the Sena dynasty as evident from the presence of a copper plate inscription of Laksmanasena there (Roy, 2021b. p 42).
Debgram (23° 09′ N, 88° 39′ 30″ E)	Debgram/ Debal rajargarh	Debagram has been identified as a provincial capital of Bikramraj who helped Pala king Rampala (1072–1126 A.D.) against the Kaivartya revolt (Basu, 1914).
Nandighat (23° 08′ 57″ N, 88° 32′ 02″ E)	NathGhat	Near Anulia, Nandighat is still a settlement beside the riverbank of Curni and previously a noteworthy center of Tantra and Nath Yogis (Roy, 2021b, p 68).
Buddhakol (23° 08′ 45″ N, 88° 31′ 40″ E)	Budhhakol	Near Anulia, Budhhakol is still an ancient settlement beside the river Ganga's paleo-channel, where several artifacts related to Buddhist rituals from the Pala-Sena period have been discovered (Roy, 2021b, p 68).

Source: Literature Survey and Field Visit, November (2022)

archaeological and historical data. Epigraphic data plays a vital role in analyzing the antiquities and dimensions of the past civilization in the study area. The locations of various archaeological discovery sites have been analyzed with a focus on river dynamics of this region using LISS IV and Google Earth data that depict the alignment of past courses of the rivers, particularly Ganga and its tributaries and distributaries in the area that constituted a network of water transport for the ancient civilization during its lifetime functioning as arteries. The study has been supplemented by the toponomic data, which highlights glimpses of the cultural characteristics of the past.

The methodology of the study is quasi-qualitative, incorporating literature triangulation, application of remote sensing, and digital cartography in the GIS (geographical information system) domain with extensive use of GPS, Abney level, and clinometers for field survey to reconstruct the geography of the past considering the present landscape characteristics as the key of the past, following the uniformitarian principle in landscape analysis (Thornberry, 1957). As the pre-field task, the meander cut-off or paleo-channel of the Ganges has been pointed out on the map, distinguishing the several stretches. The length and stretch of the past course, along with minor channels, have been enumerated. Attempts have also been made to locate the ancient drainage system with particular reference to the alluvium deposition, slope change, and present flow dynamics. The study focuses on changing the course of the Ganga to discover the location of the civilization said to have existed during Ptolemy (150 A.D.). Claudius Ptolemy was one of the pioneer traveler-geographers who described the past riverine network of the Gangetic Delta. Ptolemy obtained the data from secondary sources that the Ganges split into five principal channels and merged into the Bay of Bengal. He has named the five channels and identified their position with latitudinal and longitudinal coordinates. From his description, it seems that the ancient Gangaridai civilization was formed at the second confluence, namely, "Mega" (Rudra, 2019).

3 Result and Discussions

Debalgarh archaeological site (23° 9′ 30″ N latitude and 88° 39′ 11″ E longitude) in the Nadia district of West Bengal is a potential archaeological tourism site of West Bengal that needs the attention of historians, archeologists, and tourism promoters. Thousands of artifacts have been found through decades during agricultural practices and construction works at this site, some of which are kept in the museum of "Debagram Debal Raja Puratatwao Lokosanskriti Sangha," an organization of local people keen on the preservation of heritage. Representatives from the Asiatic Society and veteran archaeologists from different universities/institutions have already inspected the site and expressed great concern regarding its importance (Chattapadhay, 2023). The Asiatic Society, in its prestigious journal, has already referred to this site (Fig. 2) as a "new archaeo-tourist destination of Bengal" (Banerjee, 2018). Early historical and typical Gupta-aged potteries, coins of King

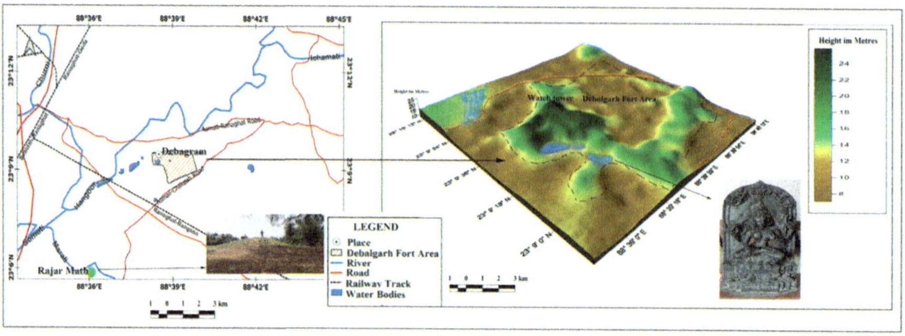

Fig. 2 Debalgarh and its surroundings as archaeo-tourist destination. (Authors)

Samudragupta (335–375 A.D.), and bricks of different sizes and shapes with floral and geometric designs are evidence of multiple layers of civilizations in the region up to the Sultanate period.

Anulia is presently a village (322,050 population according to census 2011) about 10 km west of Debalgarh on the bank of Churni in another interesting site for the mound locally called Dhanagar from which an icon of Vishnu was discovered (Chattopadhyay, 2018). Paleo-channel investigation reveals that the place of discovery of this outstanding sculpture representing post-Gupta and early Pala iconography was once the meeting point of Churni and the past course of Ganga. Though the icon is worshipped as Hindu Basudeva (a form of Lord Vishnu) from which the place name Basudevtala was derived, the sculpture (at the center of Fig. 3) has the possibility of being Avalokiteswara. Contact with the Hellenic countries on the sea route might be responsible for Hellenistic and Roman influence on the sculptures of ancient India (Banerjee, 1981).

It is noteworthy to mention that several Buddhist monasteries existed in South Bengal during the rule of the Pala dynasty (seventh to twelfth century A.D.), a few of which has been discovered by archeologists. Characteristically, such monasteries were situated on the bank of significant rivers, and while reconstructing the historical landscape based on paleo-channel investigation and archeological discoveries, it may be put forward as a hypothesis that it was the location of one of such depleted monasteries where this sculpture was worshiped as Avalokiteswara, a popular Buddhist deity who refused their own salvation in order to extend services to ordinary people. The deity had a trans-cultural appeal from which the Hindus discovered the shadow of Lord Vishnu within the deity assimilating as Basudava in the fold of Hinduism. Toponymically, the village name Buddhakol (Fig. 4), less than 2 km from Basudevtala, further strengthens the hypothesis of the location of an ancient Buddhist structure near the confluence of the past course of Ganga and Churni in the study region from where a vast number of artifacts are also discovered. In this connection, it is noteworthy to mention that the popular idea that prevails as Churni is an artificial canal is notoriously misleading. Churni was the waterway of ancient Bengal that served the Buddhist civilization, as evidenced by archeological discoveries in its banks (Roy, 2021a).

Fig. 3 Buddhist artifacts from Anulia-Debalgarh archaeological site. (Source: captured by authors during field survey 2021–2022)

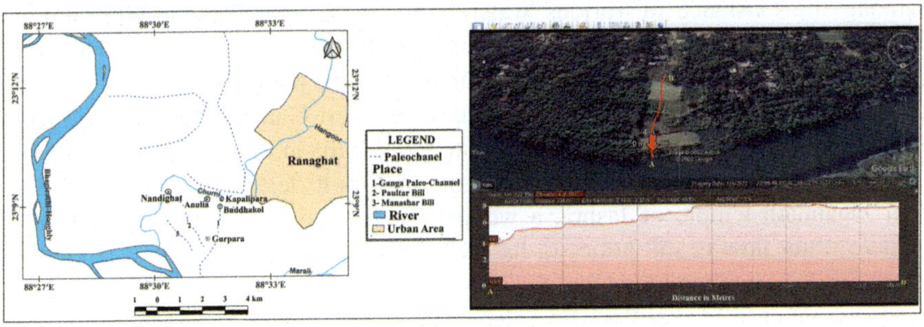

Fig. 4 Probable site of a Buddhist monastery on the bank of river Churni. (Authors)

An intensive field study along the paleo-channel of Ganga has been made to relate the archeological discoveries with the ancient river course focusing on its following stretches:

(a) Anulia (Basudevtala Site) to Shimurali (Abandoned Port Trust Office of the Britishers)

In between Ranaghat and Chakdaha, the past course of the Ganges stretches almost twenty km in length with 200–300 m in width. The past course of the Ganga stretches from northeast to southwest direction at Anulia (Fig. 1). It has been observed that the present channel of the Ganga flows almost five to seven km away from the past one, following nearly the same slope direction. At Gurpura village under Ranaghat block I, a steady downward slope toward the past course of the Ganges can easily be marked. Within a few meters, almost three to four meters decreases (23,009′ 05″ N to 23,009′ 09″ N latitude and 88,032′ 19″ E to 88,032′ 20″ E longitude), marking the location of the residual of the erosional bank of the past course of the Ganges. It is further observed that the past channel flows become prominent in the rainy season.

The discovery of a large number of Buddhist artifacts along the bank of Churni, especially its confluence with the paleo-channel of Ganga, begs to differ from the popular hypothesis on Churni as an artificial course. Part of it might be evidence of early efforts of river linking, but the paleo-channel (5–7 km eastward from the present course) compels to draw the attractions with the evidence of artifacts discovered along it. As the river existed in the past, traces of human habitation are found below the soil. The study of the toponyms of village names such as "Buddhakol," Nandighat (previously known as Nathghat), and "Kapalipara" with paleo-channel investigation (Fig. 4) reveals that during the gradual shifting of river Ganga toward the west, the river has followed at least three different channels within a couple of centuries interval. The place names had been derived from the century-long tradition of Buddhist culture and Tantra practices. This area provides maximum evidence of early civilization, a glimpse of which (Table 2) manifests the continuity of civilization from pre-Gupta to the Islamic period (third to thirteenth century A.D.).

In the Ghughia region of Chakdaha, previously known as Chakradahas (daha means water body), a network (called chakra in Bengali) of water bodies is found among the existence of various water bodies, e.g., the "Paultar bill" and the "Manashar bill" are cut-offs of the channels of the past. Local villagers have confirmed that in monsoon season, surface runoff finds the path of the ancient course of Ganga at Chakdaha region, which is known as Buri Ganga (Buri means old). A temple protected by the Archeological Survey of India at Palpara, the sacred grove at Ajgubitala and Manasapota, is among the sites on its left bank, while the historical Jagannath temple of Jashra, a wooden bridge (near Chakdaha crematorium, where water flow is still noticed), has been mapped (Fig. 5). The past riverbed is now fertile agricultural land, but Sukhsagar in Chakdaha is the waterbody that persists as evidence of past flow. The place names, i.e., Jahajghata (*jahaj* means ship and *ghata* is the place where it rests) on one side of it and Manasapota (*pota* locally means port) on the opposite side, are toponymic data supporting trade and commerce-based riverine economy once flourished in the area.

The abandoned office of the Calcutta Port Trust at Shimurali bears the legacy of a river port (Fig. 6) on the past course of the Ganges, even during the British period. As the Ganga gradually shifted toward the west, the port died. It is noteworthy to mention that a catastrophic flood in the year 1789 drastically changed the course of the Ganges, devastating the entire Sukhsagar region (Chattopadhyay, 1988).

Table 2 Archaeological discoveries corresponding to the time scale

Picture of evidences	Historical age	Archeological evidence and features
	Pre Gupta	Pot shards, pottery
	Gupta period	Gold coin of Sumudra Gupta, pottery, terracotta beads, terracotta beads with Bramhi inscriptions
	Post Gupta (eighth to ninth century)	Silver coin (Sahi-jital) of Sri Sthalapatidev, black slipped pottery, architectural materials, and brick with floral motif
	Pala Sena period	Black slipped pottery, terracotta beads, terracotta seals, architectural materials, stone sculptures, and metal objects
	Islamic period	Silver coin, laminated bricks, and slipped pottery

Source: prepared by authors during the field survey, 2021–2022

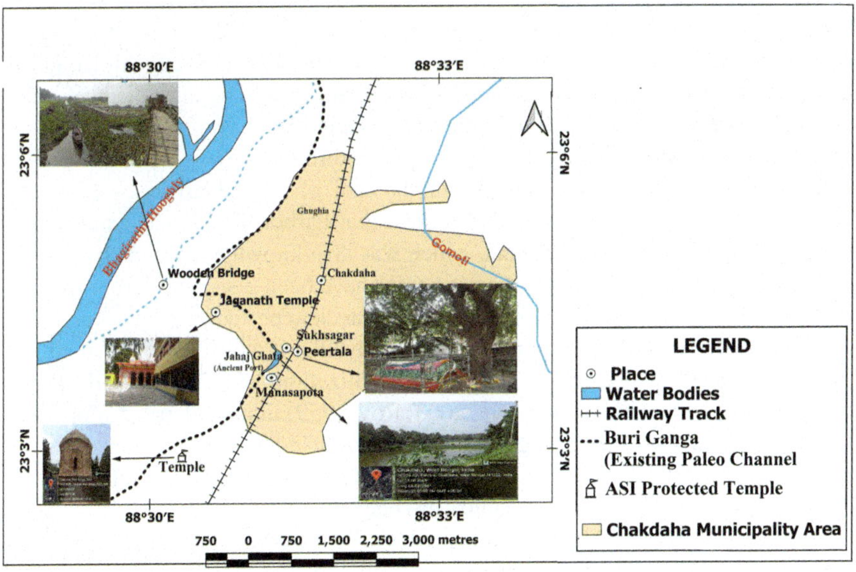

Fig. 5 Significant sites along the past course. (Authors)

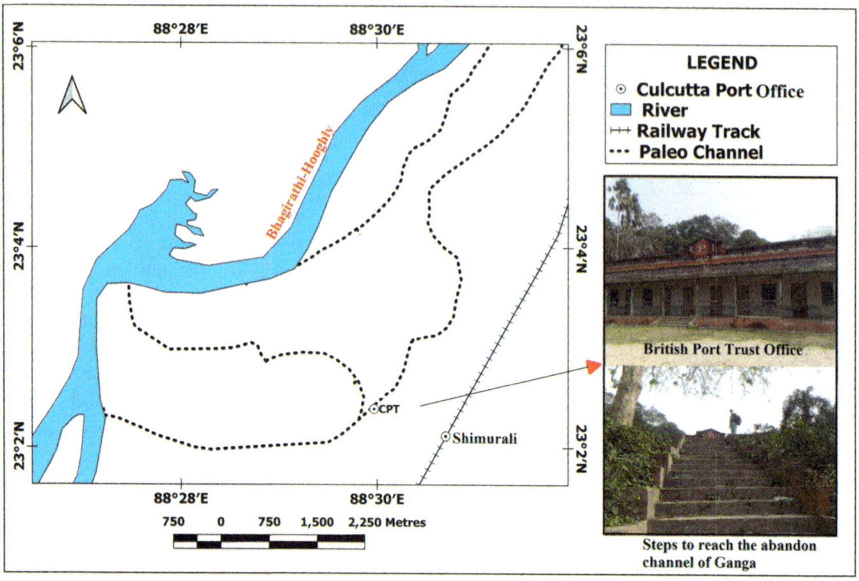

Fig. 6 Past flow path on which port was situated. (Authors)

(b) Shimurali to Tribeni Confluence

Jamuna, a left bank distributary channel of Bhagirathi, is a moribund river with high sinuosity that ultimately joins Icchamati. At its offtake point, an annual fair called "Dubopara Mayer Mela" is held yearly. The word Dubopara indicates water logging, but it is presently devoid of any upland flow from Bhagirathi (Chakrabarti & Nag, 2015). The waterbodies found in the Chandmari area at Kalyani, Kulia Bill, and Mathura Bill (Fig. 7) are cut-offs of the main flow. The alignment of paleo-channels, as found from satellite images and ground truth verification in the field, reveals the probability of the past flow of Jamuna through the Vidyadhari course that facilitated the rise of Chandraketugarh on the bank of Vidyadhari.

It is noteworthy to mention in this context that Parboti Khal once connected the parallel flow of Marali and Jamuna, which was also known as Jamuna by itself. A number of artifacts with Buddhist connections have been discovered from its banks that indicate it was a significant flow path in the ancient period and represents a river network involving Ganga, Marali, Jamuna, Vidyadhari, and Ichamati-like streams that once served as arteries of Gangaridai civilization. Many renowned scholars (Chakravarti, 2002; Chattopadhyaya, 2003; Roy, 1959) indicated the relationship of Chandraketugarh with the Gangaridai civilization. Under such circumstances, this river was one of the five mouths of the Ganga identified by Greek scholars.

The westward shifting of the Ganges has been identified as a major reason for the death of the Vidyadhari River (Rudra, 2019). To explore such identification, the offtake point of Jamuna from Buriganga is further identified (23° 04′ 35″ N and 88°

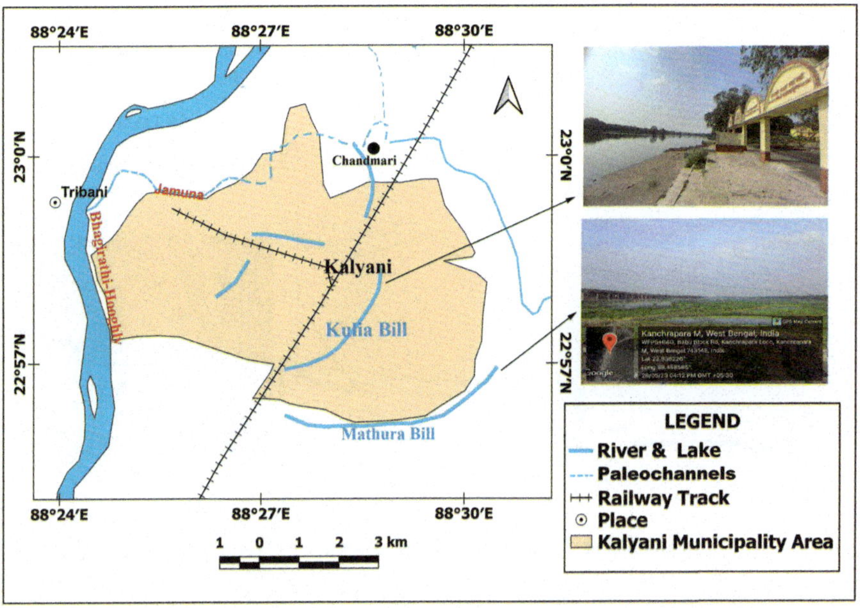

Fig. 7 River dynamics as evident from cut-offs

32′ 10″ E), which has been shifted westward since Ganga has changed its course leaving the Buriganga path and moved further west since Jamuna is the distributary of Ganga and thereby compelled to shift with the main system. As a consequence of such shiftings, however, it is not possible to identify the offtake point of the river from Ganga today, whereas the offtake of Saraswati, another distributary, could be identified (Fig. 8).

The survey along the paleo-channel of Ganga up to the past confluence of Ganga-Jamuna and Saraswati (due to the tri-junction, where the name Tribeni is derived) traces the imprint of early civilizations, the root of which might be the famous Gangaridai civilization (Ghosh, 1988; Halder, 1988). The latitude and longitude enumerated by Ptolemy in the second century A.D. from secondary information might be controversial (Chattopadhyay, 1983), but the most important fact is that Ptolemy marked a mouth incorporating five channels while writing on Gangaridai, and there is scope of investigation along those channels in the study region that bear traces of various past civilizations. The discovery of the Debalgarh archaeological site and its archaeological findings raise serious issues regarding the areal extent of the Gangaridai civilization, the location of its capital, and the ancient port of "Gange." The very close proximity of Debalgarh and Chandraketugarh (Fig. 9), the evidence of the riverine economy, and the establishment of a significant political and economic center in both cases have added a new dimension to the arena of geo-archaeological research.

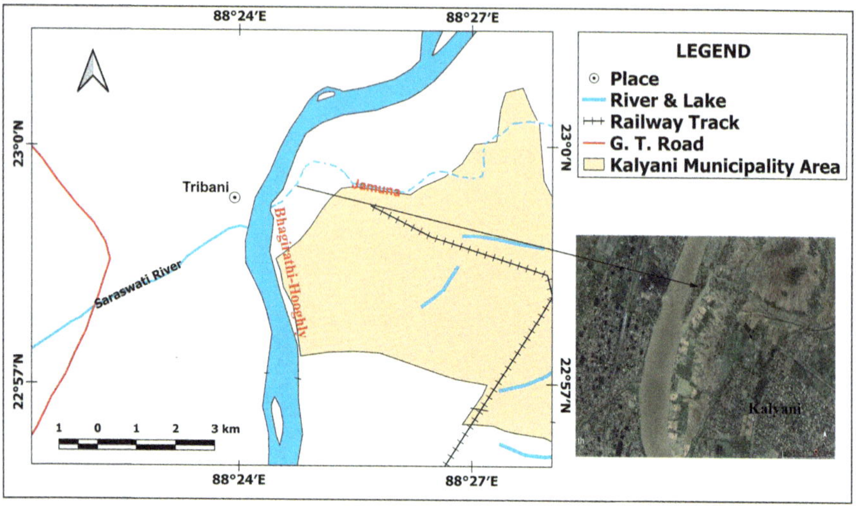

Fig. 8 Evidence of channel shifting in Ganga Jamuna confluence. (Authors)

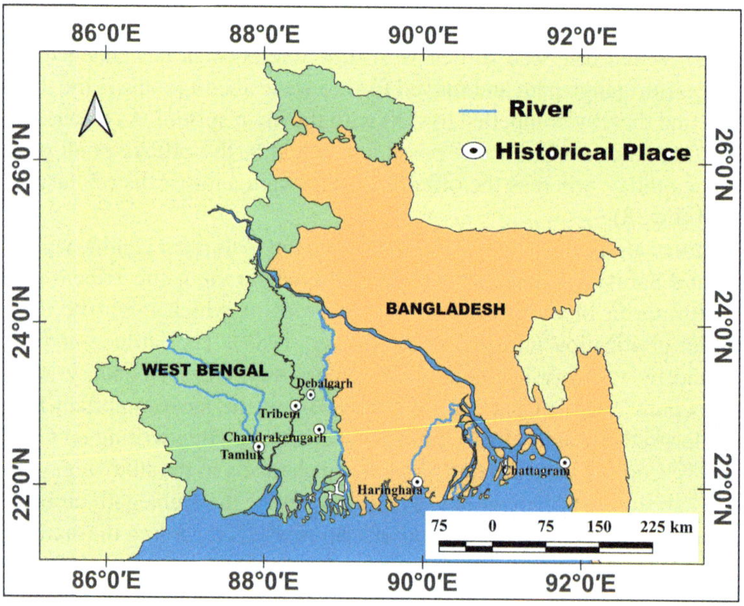

Fig. 9 Proximity between early settlements in South Bengal. (Authors)

4 Conclusion

The river dynamics and tectonic activities that took place in the region have been taken into special consideration while studying the region's geoarchaeology (Dasgupta, 1993) from a geographical perspective. The earliest reference to the Debalgarh site can be found in the famous literature *Ramacharita* by Sandhyakar Nandi. This literature reference exhibits that King Vikram Raj ruled the entire region during the region of Pala King Rampala (1082–1124 A.D.). Therefore, it is evident that this region became a significant political and economic nerve center during the Pala period, the root of which might be laid in the Gangaridai period as prominent from circumferential evidence. It appears that Gangaridai had multiple nuclei, and Chandraketugarh and Debalgarh are among the nuclei, as evident from the river geography of the past. The rivers have always been used as a means of communication and trade. It is obvious from the study that the past course of Ganga near Tribeni through Anulia was the cradle of ancient civilization that sustained Debalgarh-Anulia, the recent discoveries from which arises the necessities of writing a new archeological geography following the concepts and practices of social physics.

References

Banerjee, G. N. (1981). *Hellenism in ancient India.* Munshiram Manoharlal Publishers Pvt. Ltd.

Banerjee, K. A. (2018). Debagram: A new destination in the archaeotourist map of West Bengal. *The Asiatic Society of Kolkata, XLVII, 7*, 09.

Barnes, T. J., & Wilson, W. (2014). *Big data, social physics, and spatial analysis: The early years* (pp. 1–14). Original Research Article.

Basu, N. (1914). *Banger Jatiya Itihash, (Rajanyakanda)* (pp. 198–199). Biswakosh Press.

Bidyabinod, A. (1936). *Ramacharitam by Sandhyakar Nandi* (p. 45). Dibyasmriti Samiti.

Chakrabarti, P. S., & Nag, S. (2015). *Rivers of West Bengal: Changing scenario. Geoinformatics and RS cell* (p. 110). WBSCST, DST, Govt. of West Bengal.

Chakrabarty, P. (2011). Historical geography of Ganga Delta: Some aspects. *ILEE, 34*(1), 507–512.

Chakravarti, R. (2002). *Trade and traders in early Indian society* (pp. 129–130). Manohar.

Chakroborty, C. (1926). *Pavandutam, Sahitya Parishat Series: Sanskrit Sahitya Parishad* (pp. 25–26).

Chattapadhay, K. (2023, June 11). Debol Rajar Gor. *Anandabazar E-Paper.* https://www.anandabazar.com/rabibashoriyo/story-of-the-royal-family-ofdebalgarh/cid/1436682

Chattopadhay, D. (1988). *Culture of Bengal through ages: Some aspects* (p. 7). The University of Burdwan.

Chattopadhyay, B. D. (1983). Presidential address: Political processes and structure of polity in early medieval India: Problems of perspective. In *Proceedings of the Indian History Congress* (Vol. 44, pp. 25–63). Indian History Congress.

Chattopadhyay, R. K. (2018). *The archaeology of coastal Bengal* (pp. 35–36). Oxford University Press.
Chattopadhyaya, B. (2003). *Studying early India: Archeology* (p. 12). Texts and Historical Issues, Permanent Black.
Chattyopadhyay, R. K. (2018). *The archaeology of coastal Bengal* (p. 35). Oxford.
Dasgupta, S. (1993). Tectono-geologic framework of the eastern Gangetic fore deep. *Geological Survey of India Special Publication, 31*, 61–69.
Ghosh, P. K. (1988). *Gangaridi O Bangavumi*. J.L. Prakashani.
Goheen, P. G. (2015). Methodology in historical geography: The 1970s in review. *Historical Methods: A Journal of Quantitative and Interdisciplinary History*, 37–41.
Govt. of India. (1896). *List of ancient monuments in Bengal* (pp. 116–118). Public Works Dept., Bengal Secretariat Press.
Haldar, N. (1988). *Gangaridi: Alochona O Parjalochona*. Dey Book Store.
Hood, J. W. (1994). *History of the Bengal people: Ancient period* (pp. 297–298). Orient Longman.
Majumder, N. (1929). *Inscription of Bengal* (Vol. III, pp. 81–91). The Varendra Research Society.
Miller, J. H. (2017). Geographic information science II: Mesogeography: Social physics, GIS science and the quest for geographic knowledge. *Progress in Human Geography*, 1–10.
Pandey, M. S. (1963). *The historical geography and topography of Bihar*. Motilal Banarsidass.
Raverty, H. G (1873). *The Tabakat-i-Nasiri:ed and trans*. Gilbert and Rivington.
Roy, N. R. (1959). *Bangalir Itihash: Adiparba*. Dey publishing.
Roy, B. (2021a). Nadi Tumi kotha Hoyte Asiachho: Vu-protnottwer Aloke Churni. *Purabrittwa, 2*, 277–287.
Roy, B. (2021b). *Debalgarh-Anulia Pratnakshetra: Haranoek Rajdhanir Sandhane*. Digital Press, Alipurdwar Court.
Rudra, K. (2019). *Banglar Nadikatha* (pp. 50–51). Sahitya Sangsad.
Sircar, D. C. (1965). *Indian Epigraphy* (pp. 417–418). Motilal Banarsidass Pvt. Ltd..
Stewart, J. Q. (1950). The development of social physics. *American Journal of Physics, 18*, 239–253.
Thornbury, W. D. (1957). *Principles of geomorphology*. John & Wiley.
Urry, J. (2004). Small worlds and the new social physics. *Global Networks, 4*(2), 109–130.
Zipf, G. K. (1942). The unity of nature, list-action and natural social science. *Sociometry, 5*, 48–62.

Dr. Biswajit Roy [DoB: August 04, 1982] is an Assistant Professor and the Head of the Department of Geography at Pijush Kanti Mukherjee Mahavidyalaya, located in Sonapur, Alipurduar, West Bengal. He completed a M.A. in Geography (2005) and has been awarded a Ph.D. in Geography (2014) from the University of Calcutta on micro-level planning and development. Dr. Roy has authored and edited several books on geoarchaeology and founded a regional center in 2017 for geoarchaeological research, along with a rural museum in Gangnapur, Nadia District, West Bengal.

Dr. Sanjoy Kumar Sadhukhan [DoB: May 09, 1986], M.Sc., Ph.D. He completed his post-graduation from the University of Kalyani in 2010 with a specialization in the Geography of Tourism. He has been recently awarded a Ph.D. degree in Geography from Visva-Bharati on the aspects of community participation in tourism.

Prof. Premangshu Chakrabarty [DoB: August 25, 1972], M.A., Ph.D. (Visva-Bharati). Specializing in cultural geography, pilgrimage, and tourism geography, he has contributed more than eighty-five research articles in reputed journals and presented his research outcomes at various international and national conferences. He has more than two decades of teaching and research experience. Fourteen scholars have already been awarded Ph.D. under his supervision, and three are presently working under him for their Ph.D.

Identification of Spaces of Political Violence in Purulia District, West Bengal: A Geospatial Discourse

Ashit Modak, Supratim Karmakar, and Kasturi Mukherjee

1 Introduction

The relationship between politics and society is fundamental and intricate, as politics shapes the structures, processes, and dynamics of societies, while societal forces influence political institutions and behaviours. This interplay is central to understanding how political systems function and how societies evolve. In democratic societies, politics is a means through which individuals participate in governance. Elections, political parties, and civic engagements are mechanisms through which citizens express their preferences and influence decision-making. Political institutions, such as legislatures, executives, and judiciaries, are designed to represent and serve the interests of society. The functioning of these institutions directly impacts the well-being and rights of citizens. Political decisions and policy-making significantly impact societal well-being. Policies related to education, healthcare, social welfare, and economic development directly affect the quality of life for individuals and communities. Political systems play a crucial role in determining how resources are distributed within society. Taxation policies, social programs, and economic reforms influence the distribution of wealth and opportunities. Societal diversity, including factors such as ethnicity, religion, and cultural identity, plays a crucial role in political dynamics. Political systems must navigate the complexities of multiculturalism and pluralism to ensure equitable representation and policies.

The relationship between politics and spaces is intricate and multifaceted, with the physical and symbolic aspects of spaces playing a crucial role in shaping political dynamics. From public spaces to government buildings to virtual spaces online, the places where political activities unfold have a profound impact on how individuals engage with, perceive, and participate in the political process (Dawson, 2020).

A. Modak · S. Karmakar (✉) · K. Mukherjee
Department of Geography, Adamas University, Kolkata, West Bengal, India
e-mail: kasturi.mukherjee@adamasuniversity.ac.in

A. Rai et al. (eds.), *Contemporary Social Physics*, Springer Geography,
https://doi.org/10.1007/978-3-031-77596-3_5

Public squares and gathering places often serve as symbolic locations for political expression and civic engagement. Public spaces serve as a great influence on local politics and the impact of political ideology on general people's minds. The display of political hoardings with the proper agenda keeps a great influence on common people (Szołucha, 2021; Mandel, 2024). Electoral politics is highly dependent on the public space display of political motives and agendas of different political parties. In the present era, virtual spaces are becoming more powerful and more effective in terms of spreading political ideologies among a larger section of people without actually coming in touch with physical reality. Social media platforms and online forums serve as spaces where political ideas are disseminated, public opinion is shaped, and movements are mobilized. As political activities move online, issues of cyber security and digital governance become critical (Thakur, 2024). The protection of digital spaces from cyber threats and ensuring the integrity of online political processes are growing concerns.

Political violence, a menace that has plagued societies throughout history, manifests itself in various forms, ranging from riots and protests to terrorism and assassinations, and in the realm of criminal activities, political violence stands out as a unique and potent threat, not only because of the physical harm it inflicts but also due to its profound impact on the democratic fabric of the nation. A sense of social injustice can be a powerful catalyst for political violence. When marginalized, their groups feel disenfranchised or oppressed, and they may resort to violence as a means of expressing their grievances and demanding change. Deep-seated ethnic or religious divisions within a society can fuel political violence (Oostveen, 2022; Hassan, 2023).

1.1 Political Violence and Social Physics

Political violence, a manifestation of social unrest and conflict, often exhibits distinct spatial patterns that can be analyzed through the lenses of social physics. Social physics draws inspiration from the physical sciences to study collective human behaviour. Social physics posits that social phenomena can be analyzed and predicted using quantitative methods akin to those employed in natural sciences. The approach involves understanding social forces, interactions, and patterns through the ideas of spatial modelling and statistics (Bruhn et al., 2020). By combining principles of physics with social science methodologies, researchers can unravel the intricate dynamics of where and why political violence occurs. This paper delves into the spatiality of political violence in the Purulia district of West Bengal using a social physics approach, exploring the interconnectedness of social elements and physical spaces. Social physics allows for the identification of hot spots which is a quantitative technique for identification of the concentration of violent activities within a particular location (Kebede, 2023; Piazza, 2023). By analyzing the historical data and employing statistical models, researchers can pinpoint regions where social, economic, and political factors converge to create conditions conducive to violence. Agent-based modelling, a technique employed in social physics, involves simulating the behaviour of individual agents within a given space. In the context of political violence, this group of individuals or groups influence the spatial distribution of conflict (Brathwaite & Park, 2023). Social physics allows for the examination of spatio-temporal dynamics, considering how political violence evolves over both space and time. Analyzing patterns over different temporal scales helps in identifying trends and understanding the underlying causes of conflict. While social physics offers valuable insights into the spatiality of political violence, challenges exist. Contextual nuances, cultural factors, and the inherent complexity of human behaviour may not be fully captured by quantitative models. Additionally, ethical considerations must be taken into account when applying social physics to sensitive issues like political violence (Gunasekara, 2023; Wika & Abdullahi, 2023).

1.2 Study Area

Purulia district is one of the districts of West Bengal, India. Purulia is designated as one of the 250 most backward districts of India. It is located on the western side of the state and is famous for its diverse geographical characteristics in terms of topography, rich cultural heritage, and historical significance. Purulia district is part of Jungle Mahal (forest land). It is historically characterized by dense forests, tribal populations, and socio-economic challenges. This area is inhabited by tribes like Santhals, Munda, and Lodhas. Jungle Mahal gained national attention due to the Maoist insurgency in the area. The dense forests provided shelter for the Maoist rebels, leading to security concerns (Khalko, 2015; Mahato, 2021; Mahapatra, 2021). Purulia shares borders with Jharkhand in the west, specifically with West Singhbhum district in Jharkhand, and Bankura and Paschim Medinipur district in the east. To the south of Purulia lies Birbhum district. Purulia's northwestern border is shared with Jharkhand and specifically with Bokaro district. Bardhhaman district lies in the north of the district (Fig. 1). The neighbouring districts contribute to the overall socio-economic and cultural dynamics of the region. The borders also influence factors such as trade, transportation, and cultural exchange between the districts (Mahato, 2023; Mahadani, 2022). The district is characterized by undulating terrain, hills, and plateaus. It is a part of the larger Chota Nagpur Plateau, which extends into neighbouring states. Purulia district is divided into four sub-divisions: Purulia Sadar, Raghunathpur, Jhalda, and Manbazar. Purulia Sadar consists of six CD blocks, namely, Purulia I, Purulia II, Hura, Puncha, Arsha, and Balarampur. Jhalda subdivision is divided into four CD blocks, such as Jhalda I, Jhalda II, Baghmundi, and Jaipur. Raghunathpur subdivision is made up of six CD blocks: Para, Raghunathpur I, Raghunathpur II, Neturia, Santuri, and Kashipur. Manbazar subdivision consists of five CD blocks such as Manbazar I, Manbazar II, Barabazar, Bandwan, and Puncha. Purilia town is the district headquarters. There are 21 police stations in Purulia district as per the information of 2022. Purulia district has a population of 2,930,115. The population density of the district is 468 inhabitants per

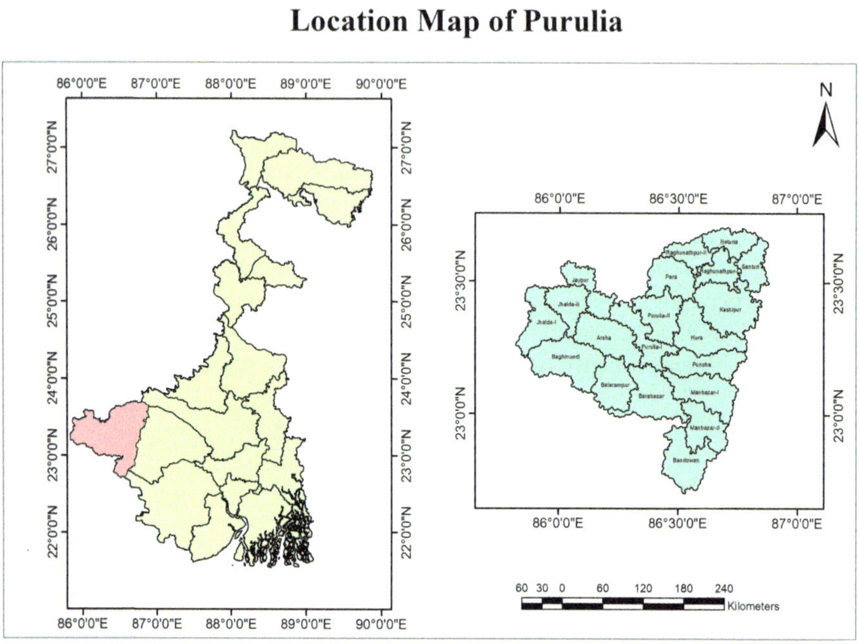

Fig. 1 Location map of Purulia

square kilometre with a population growth rate of 15.43% (Census of India, 2011) between 2001 and 2011.

Purulia has a sex ratio of 955 females per 1000 males and a literacy rate of 65.38% where 12.74% of the population lives in urban areas. The percentage of Scheduled Castes and Scheduled Tribe populations of the district are 19.37% and 18.45%, respectively (Census of India, 2011).

2 Literature Review

A systematic review of existing literatures has been done to understand the underlying mechanism of political violence. The Elsevier database has been used for review purposes (Table 1). The initial search was started with keywords like politics, violence, political behaviour, organized crime, and riots. Firstly, 347 papers were fetched on the basis of keywords selected and sorted on the basis of citation. After applying the filter, the papers were narrowed down to the papers that dealt with particularly the Indian scenario of political violence. It resulted in the elimination of 123 papers. Then, the second level of filtering was done after going through the methodology which is related to quantitative techniques as per the requirement of the social physics approach. It resulted in the selection of 54 papers which are

Table 1 Review of literature on political rioting

Steps	Procedure	Results
Data gathering	Data search on Elsevier based on keywords and cited papers	Bibliographical information of potentially relevant papers (347)
Data screening	If thus the abstract contains a focus on the Indian scenario of political violence	Total of 214 potential papers identified
Data cleaning	If the methodology is not related to the social physics approach	Total number of 160 papers are eliminated
Paper classification	54 papers are classified using three review categories Impact of political violence on mass population behaviour Spatial pattern analysis of political violence Influencing factors behind organized political violence	Total number of 54 papers are reviewed

actually relevant for the discussion of the Indian scenario of political violence with an approach close to social physics.

Political violence is deep in the human mind. The political field in India is often seen as an area which is very risky for these sorts of violent activities. In the Indian scenario, the elections are not only a democratic process, but it is highly massacred by the political violence at the time of elections (Adhikari et al., 2023; Ahen et al., 2023; Bochenkova et al., 2023). These sorts of political acts of violence have a huge impact on the general lives of the commoners. Women and children are not very much involved in active politics in India as well as in West Bengal. But their lives are highly impacted by the political violence. Each and every year, a large number of men were brutally affected by the political violence. Physical harm leading to permanent disability, imprisonment, and even loss of life are some consequences which have to be faced by the victims and offenders of political violence (Chowdhury et al., 2022; Eger, 2021; Legg, 2020; Makate & Nyamuranga, 2023). On the other hand, the social and economic impacts of political violence are far-reaching. Destruction of public properties is one of the commonest ways of showing political rage in India. It leads to a huge amount of financial loss also. In India, political violence leads to even the destruction of roads, railway lines, telephone, and telegraph lines which hampers overall public life and brings it to a standstill (Kobayashi et al., 2022; Joseph et al., 2021; Franklin & Herek, 2022; Ghosh, 2022).

The spatiality of political violence refers to the geographical dimension and distribution of violent incidents within a political context. It involves the analysis of where political violence occurs, its spatial patterns, and the relationship between the physical space and the acts of violence. Understanding the spatiality of political violence is crucial for policymakers, researchers, and security forces to devise effective strategies for conflict prevention, management, and resolution (Shafique et al., 2017; Prathap, 2022; Kounadi, 2015; Piętak et al., 2016). Spatial analysis often involves the use of GIS to map and analyze the patterns of political violence. GIS allows researchers to visualize the distribution of violent incidents, identify hot spots, and explore spatial relationships. Hot spot analysis helps in identifying specific areas where political violence is concentrated. These hot spots may be regions, cities, neighbourhoods, or even specific streets. Examining clusters of violent incidents helps to identify patterns and understand whether violence tends to cluster in specific spatial configurations (Murray & Grubesic, 2013; Boivin & de Melo, 2019; Hipp, 2016). Political violence often exhibits different spatial patterns in urban and rural areas. Urban violence may be characterized by protests, clashes, or targeted attacks in specific city centres or neighbourhoods. In rural areas, political violence may be linked to land disputes, agrarian issues, or conflicts over resources. Understanding the spatial dynamics in rural settings is critical for effective conflict resolution (Ratcliffe, 2005; Fuentes & Hernández, 2013; Linning, 2015). In case of political violence involving multiple regions or countries, understanding cross-border dynamics is essential. Conflict may spill over across borders, necessitating regional cooperation for resolution. Different regions within a country may experience political violence differently. Geospatial technology assists in monitoring and analyzing political violence near borders. Analyzing regional variation helps in interventions to address specific local grievances and conditions (Calas, 2007; Calas, 2008; Ahmad et al., 2018). Analyzing the spatial distribution of conflict zones helps in designating areas requiring special attention and resources. In post-conflict scenarios, understanding the spatial challenges and opportunities for reconstruction is vital for rebuilding infrastructure and fostering social cohesion (Slater, 2011; Pires & Crooks, 2017; Tiratelli, 2018). Geospatial technology enables the analysis of spatial networks involved in political violence, including the movement of armed groups, supply routes, and communication patterns. Satellite imagery and remote sensing technologies can provide valuable spatial data for analyzing changes in land use, population movement, and infrastructure development in conflict zones (Dhattiwala, 2017; Chainey & Tompson, 2012; Kindynis, 2014; Manning, 2008). Understanding the spatiality of political violence involves a multidimensional analysis that integrates geography, sociology, and political science. By examining where and how political violence manifests spatially, guardianship strategies based on surveillance can be devised (Jeppesen, 2022; Manning, 2017; and Bohstedt, 2016).

The physical geography and terrain of an area can influence the dynamics of political violence. Mountains, forests, or water bodies may serve as natural barriers or facilitators for the movement of groups involved in violence. Conflicts over resources, water minerals, may have spatial dimensions. Understanding resource distribution helps in predicting and managing conflicts related to resource access. Political violence may be spatially linked to ethnic and cultural divisions. Understanding the social fabric and cultural dynamics of an area is crucial for comprehending the spatiality of conflicts (Massey, 2019; Nath & Chowdhury, 2019; Treadwell et al., 2013).

2.1 *Objectives*

The study on the spatiality of political riots in Purulia district mainly aims to identify the locational hot spots of political riots. This identification of hot spots will reveal the nature of the political behaviour of people who are involved in these activities. Political riots symbolize the collective behavioural pattern of the masses. Social physics tends to examine the forces behind this kind of behaviour. Another objective of the paper is to identify the mean centre of rioting. The mean centre of any criminal activity defines how centripetal force acts on the masses and controls their behaviour spatially. People tend to commit riots in some particular areas making them more vulnerable than the other areas. The overall objective of the paper deals with the issue of collective behaviour in terms of political riots and their spatial concentration and trend of movement.

3 Methodology

The study of political violence has been done from the perspective of social physics. Social physics emphasizes on the study of social phenomena from the point of view of quantification. The spatial pattern of political violence has been studied on the basis of recorded cases of political violence. The data have been collected from the Office of the Superintendent of Police, Purulia district. The data have been displayed as location-specific point data. Particular years of 2005, 2010, 2013, and 2022 have been chosen for the spatio-temporal analysis. The data for political riots in the entire West Bengal have been provided by the State Crime Records Bureau.

2010 and 2013 are two important years for the entire West Bengal, because of the Legislative Assembly election in 2011. The election of 2011 marked a huge change in the political scenario of West Bengal as the ruling party has been changed which brought lots of political turmoil as a result. The *spatial distribution analysis of political riots* and murders due to political riots on the district level of West Bengal has been done through a simple bar graph (Fig. 2).

For the analysis of spatial severity of political violence in West Bengal and Purulia, hot spot analysis has been done district-wise and location-wise, respectively. *Crime hot spots* are those areas which show dense clusters of criminal incidences. Hot spot analysis involves the use of GIS and statistical tools to analyze crime data. It helps to visualize crime patterns and identify areas with statistically significant clusters of criminal incidents. This analytical approach helps legal agencies and police personnel to take proper surveillance measures effectively. Crime prevention strategies and crime interventions can be properly implemented through the identification of the hot spots.

Identification and Application: The Getis-Ord Gi* statistic helps identify and map areas with high crime concentrations. This tool is based on the idea that geographically close events are likely related. It is used in fields like geography,

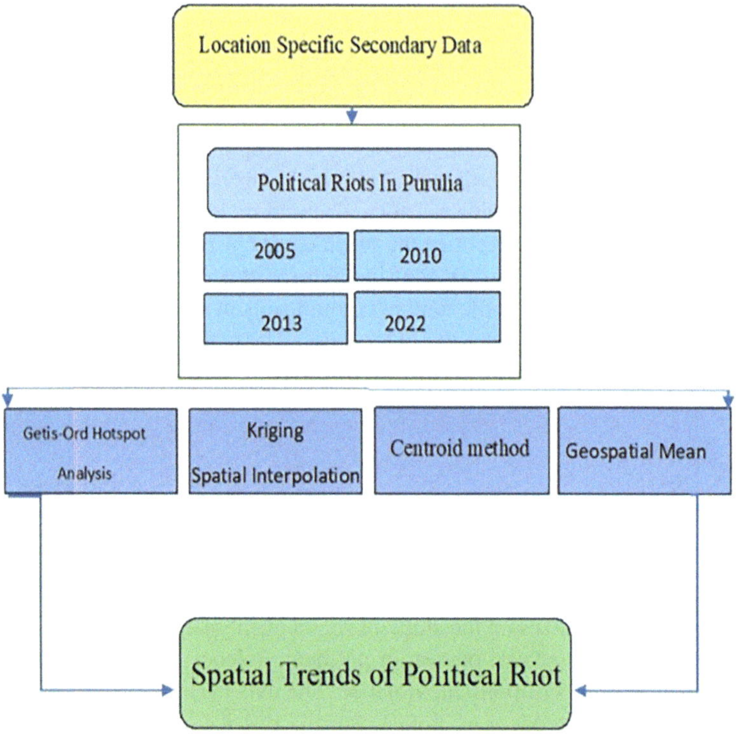

Fig. 2 Methodology

criminology, public health, and environmental science to analyze spatial data (Jana & Sar, 2016; Kim & Choi, 2017).

Method and Calculation: As a hot spot analysis method, Getis-Ord Gi* uses a z-score (GiZ score) to measure spatial clustering. It calculates the local spatial autocorrelation of each data point using the values of neighbouring points and compares this to the global spatial autocorrelation of the dataset.

Interpretation of Scores: The GiZ score for each data point indicates the type of spatial concentration: high positive GiZ scores mark hot spots, high negative GiZ scores denote cold spots, and scores near zero indicate random distribution. Equation 1 is used for this calculation:

$$G_i^* = \frac{\sum_{j=1}^{n} w_{ij} x_j - \bar{X} \sum_{j=1}^{n} w_{ij}}{S \sqrt{\frac{\left\{ \sum_{j=1}^{n} w_{ij}^2 - \left(\sum_{j=1}^{n} w_{ij} \right)^2 \right\}}{n-1}}} \tag{1}$$

where xjxj represents the attribute value for event jj, wi,jwi,j denotes the spatial weight between events ii and jj, nn stands for the total number of events, XX is the mean calculated using Eq. 2, and SS is the standard deviation calculated using Eq. 3:

$$\bar{X} = \frac{\sum_{j=1}^{n} x_j}{n} \quad (2)$$

$$S = \sqrt{\frac{\sum_{j=1}^{n} x_j^2}{n} - \left(\bar{X}\right)^2} \quad (3)$$

In this research, hot spot analysis was carried out using ArcGIS 10.5 software to examine the variation in hot spot and cold spot locations of political violence based on data (number of crimes) reported in the police stations of Purulia in the years 2005, 2010, 2013, and 2022.

Ordinary Kriging (OK): Ordinary Kriging is a geostatistical technique used for spatial interpolation of crime data based on specific locations. It uses a spatial and temporal covariance model to ensure statistically unbiased estimates. Kriging leverages knowledge of the underlying spatial relationships within the dataset to predict values for unobserved points. This knowledge is captured in the variogram, which measures the dissimilarity between observations. By utilizing the semi-variogram, which considers both distance and direction, Kriging can account for direction-dependent variability (Chitiga et al., 2023; Maiti & Mitra, 2023).

The variogram model parameters, including the sill, range, and nugget, are used to assign optimal weights for spatial prediction through cross-validation. In the general Kriging model, it is assumed that the data has a constant mean (μ), while the errors ($\varepsilon(S)$) exhibit spatial dependence (Eq. 4). By incorporating this information, Kriging provides an interpolation estimate that minimizes prediction errors and effectively incorporates the underlying spatial structure. This leads to robust and precise estimation of surface values in geostatistical analysis (Sahbeni & Székely, 2022).

$$Z(s) = \mu(s) + \varepsilon(s) \quad (4)$$

where $Z(s)$$Z(s)$ represents the variable of interest,

$\mu(s)$$\mu(s)$ denotes the deterministic trend, and
$\epsilon(S)$ refers to the random, auto-correlated errors.

OK assumes a constant, unknown mean and estimates this mean value within the studied neighbourhood. For this research, none of the Kriging hyperparameters, including the optimal range for the variogram, were adjusted, adhering to the default settings recommended by ArcGIS.

The *mean centre* is the mean of the geographical latitude and longitude coordinates of all features in the scope of the study. The resulting point is the mean centre, which can be used to represent the central tendency of the data (Nugroho et al., 2021). The mean centre is calculated with the following Eqs. 5 and 6.

$$\text{Mean } X = \frac{\Sigma x}{n} \tag{5}$$

$$\text{Mean } Y = \frac{\Sigma y}{n} \tag{6}$$

where Σx is the sum of all x coordinates, Σy is the sum of all y coordinates, and n is the total number of data points.

The mean centre, in conjunction with a standard deviational ellipse (SDE), is utilized to comprehend the patterns and trends of various crime types. The SDE serves as a statistical instrument depicting the dispersion of bivariate data on a two-dimensional graph. It involves outlining an ellipse around a scatter plot of the data, encompassing a certain proportion of data points (Rikala, 2023; Zhao et al., 2023). The dimensions and configuration of the ellipse indicate the extent of dispersion in the data. Commonly employed in geospatial research, the standard deviational ellipse aids in visualizing data distribution and patterns. It finds utility across diverse fields for analyzing and interpreting two-dimensional data dispersion. In this investigation, researchers have identified political violence hot spots for 2005, 2010, 2013, and 2022 to chart the spatial deployment patterns of various criminal activities across these years. This analysis aims to scrutinize the shifts in the mean centre and standard deviation ellipse concerning criminal activities and pinpoint migration trajectories. SDE is estimated using the following Eq. 7:

$$\begin{pmatrix} \text{var}(x) & \text{cov}(x,y) \\ \text{cov}(y,x) & \text{var}(y) \end{pmatrix} = \frac{1}{n} \begin{pmatrix} \sum_{i=1}^{n} \tilde{x}^2 & \sum_{i=1}^{n} \tilde{x}_i \tilde{y}_i \\ \sum_{i=1}^{n} \tilde{x}_i \tilde{y}_i & \sum_{i=1}^{n} \tilde{y}_i^2 \end{pmatrix} \tag{7}$$

where:

$$\text{var}(x) = \frac{1}{n}\sum_{i=1}^{n}(x_i - \bar{x})^2 = \frac{1}{n}\sum_{i=1}^{n}\tilde{x}_i^2$$

$$\text{cov}(x,y) = \frac{1}{n}\sum_{i=1}^{n}(x_i - \bar{x})(y_i - \bar{y}) = \frac{1}{n}\sum_{i=1}^{n}\tilde{x}_i\tilde{y}_i$$

$$\text{var}(y) = \frac{1}{n}\sum_{i=1}^{n}(y_i - \bar{y})^2 = \frac{1}{n}\sum_{i=1}^{n}\tilde{y}_i^2$$

The *geographical centroid method* is being used to calculate the centre point of the study area, which is most accessible for the offenders to commit or organize politically violent activities. The centroid method is essentially the average location of all the points in a given area (78 sample points) and is commonly used for mapping and spatial representation (Ru & Zhang, 2017; Sinha & Ashwini, 2021). The geographical centroid coordinates (C_x, C_y) of a set of points (x_i, y_i) are calculated as follows (Eq. 8):

$$C_x = \frac{\sum_i^n x_i}{n}$$

$$C_y = \frac{\sum_i^n y}{n} \tag{8}$$

where,

C_x and C_y are the coordinates of the centroid.
n is the total number of points of political violence.
x_i and y_i are the coordinates of each point.

The methods that have been used (hot spot analysis, mean centre and SDE identification, and Kriging) for the examination of the situation of political rioting in Purulia are related to the concept of social physics. Social physics tends to identify the spatial pattern of certain collective activities of masses (Lee & Son, 2020). These methods are based on the laws of concentration of events in certain sections of an area. These types of tendencies of concentration deal with some social laws which govern collective behaviour. Social consensus and collective behaviour are two major components of discussion in social physics (Zhang & Gläscher, 2020; Tesar, 2020).

3.1 *Political Violence in West Bengal*

The state of West Bengal has a long-time legacy of political violence in the form of political riots. West Bengal has experienced significant political violence over the last 50 years. The landscape of political violence in West Bengal is marked by a complex interplay of ideological conflicts, party rivalries, and historical socio-political factors. In the era of 1960s, the Naxalite movement emerged as an armed revolutionary movement inspired by leftist ideology. It sought to address the issues of land reform and social inequality (Mukhopadhyay & Mukhopadhyay, 1994; Donner, 2004; Roy, 2013). Uprise of leftist or Marxist ideology in Bengal persisted for more than three decades. Then, in 2011, the change in the ruling government brought many political clashes between opponent parties before, during and after the elections (Kumar, 2013; Chatterjee & Basu, 2017; Das, 2020). Ideological rivalries,

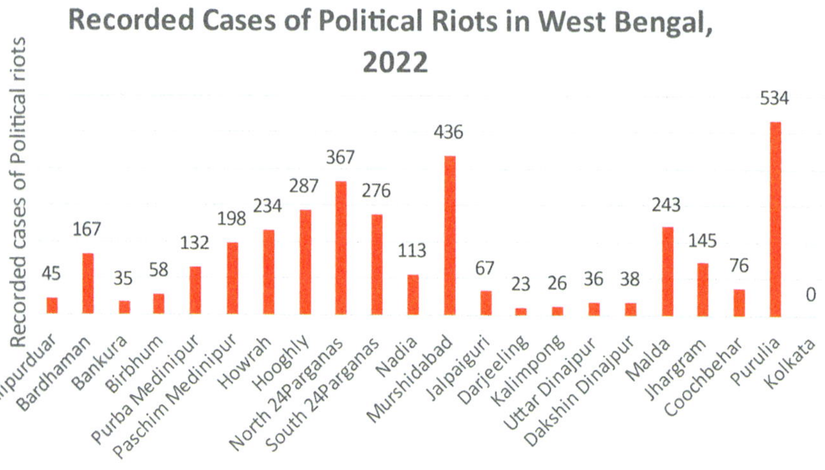

Source: Office Of the Superintendent of Police, Chowk Bazar, Purulia

Fig. 3 Political riots in West Bengal

electoral competitions, land disputes, labour unrest, and communal tensions are some of the major factors contributing to the political violence in West Bengal (Joshi & McGrath, 2015; Puthillam et al., 2021).

The district-level variation in terms of the number of recorded cases can be seen throughout the state. The varied nature of the physical set-up, location, demographic characteristics, and socio-political scenario make the districts variable in terms of political violence. Hot spot analysis (for the year 2022) based on the recorded cases of political riots has been done for the entire state of West Bengal. According to the data provided by the State Crime Records Bureau, West Bengal, the highest recorded cases of political riots in the year 2022 have been found in Purulia (534 cases), followed by Murshidabad (436 cases) and North 24 Parganas (367 cases) districts. Kolkata being the capital city is showing zero political riots in the year 2022.

From the hot spot analysis, the same picture can be depicted regarding the political riots. Purulia and Murshidabad are the hot spots for political rioting. The district of Purulia shows a 95% confidence level, while the district of Murshidabad shows a 90% level of confidence. There is no cold spot for political rioting in West Bengal for the year 2022 (Figs. 3 and 4).

3.2 *Hot Spots of Political Violence*

Hot spots are those points which are showing a higher occurrence of political violence in Purulia in comparison to the other surrounding points. A total of 78 points or locations have been used for the identification of the hot spots which are almost

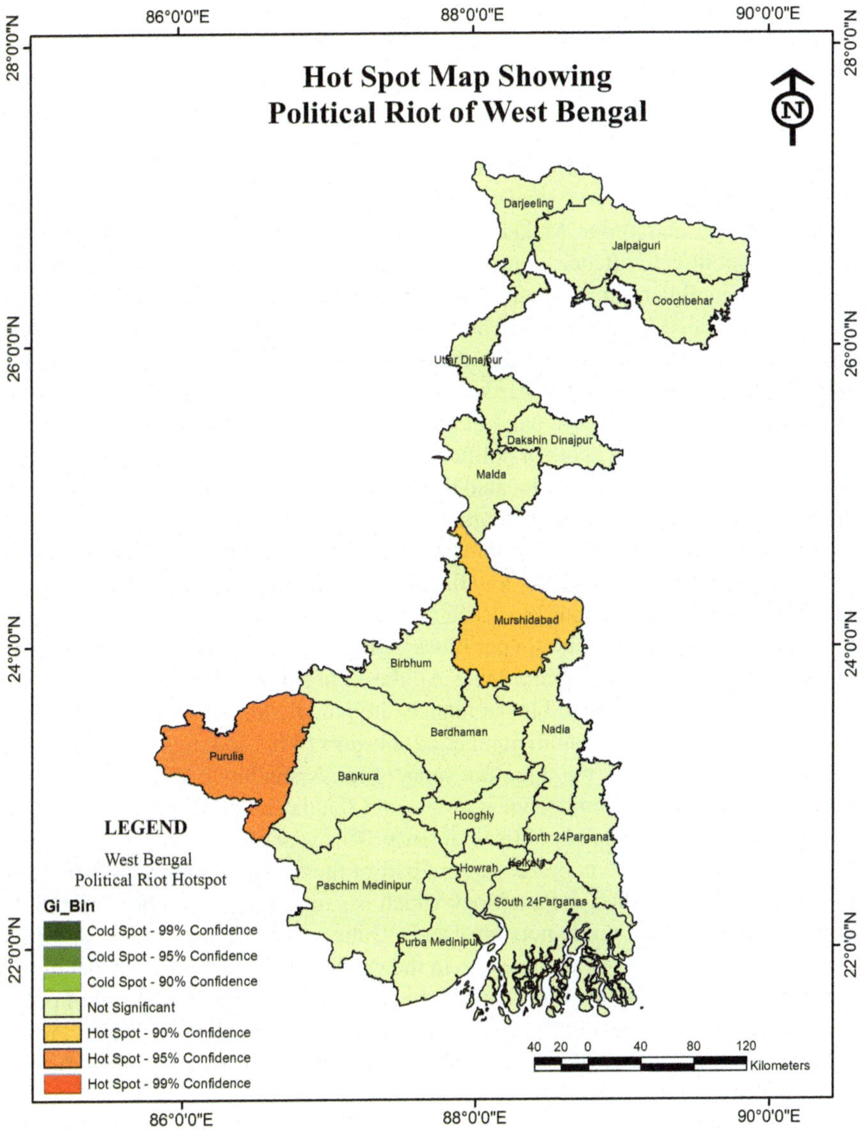

Fig. 4 Hot spot map of political riots, West Bengal

evenly distributed among the entire district of Purulia. The hot spot analysis has been done for 4 years, such as 2005, 2010, 2013, and 2022. These years have been chosen based on the availability of the data. 2010 and 2013 are those 2 years within which a big political shift of power has taken place in West Bengal. The ruling party has been changed during this time through election. Political violence has a connection with elections as already discussed. Through methods like hot spot analysis, the spatial pattern and its change over time can be identified through quantification which denotes the implication of social physics in explaining the socio-political phenomena like election and violence.

The first analysis for the year 2005 showed that maximum points (74 points) are not significant as a hot spot of political violence. Only four points have been identified as a hot spot of political violence in Purulia district. Among the four points, one point situated in Amruhasa shows a 90% confidence level, while the other three points such as Balarampur, Makiari, and Dandra show a 95% confidence level as a hot spot of political violence. Among these four points, Amruhasa and Balarampur are located in Balarampur block. The two other hot spots such as Makiari and Dandra are located in Arsha and Jhalda II block, respectively. From the locations of the hot spots, it is quite evident that the western side of the district is more politically vulnerable than the other parts as all the hot spots are located in the blocks of the western and southwestern parts of the district. In the year 2010, with a gap of 5 years, the number of hot spots of political violence went up to 13. Among the 13 hot spots, four hot spots which are located in Supurdi (Jhalda II block), Sikrabad (Arsha block), Jorka, and Amruhasa (Balarampur block) show a 90% confidence level. Amruhasa is a common point for both the years 2005 and 2010. Five hot spots such as Kotshila, Arkali, and Begunkodor (Jhalda II block), Ajodhya (Baghmundi block), and Balarampur (Balarampur block) show a 95% confidence level. The rest of the four hot spots such as Joypur (Joypur block), Makiari, Arsha, and Kadampur (Arsha block) all show a 99% confidence level. All the 13 hot spots of political violence are concentrated among the five blocks namely Joypur, Jhalda II, Arsha, Baghmundi, and Balarampur. The maximum number of hot spots (4 hot spots) is identified in the Jhalda II block, whereas the three hot spots of the Arsha block show a 99% confidence level. From this point of view, Arsha and Jhalda II were the most vulnerable blocks of Purulia in terms of political violence. This year also, the spatial concentration of political violence is in the western part of the district. In the year 2010, four cold spots were identified in Para block which is situated in the northern part of the Purulia district. These cold spots are Deuli, Dubra, Chapari, and Anara. All these points have a 99% level of confidence. In the year of 2011, there was an election of the legislative assembly in West Bengal which acted as a trigger to the rise of political violence in the district of Purulia in the year 2010. Being a rural district, Purulia exhibits a higher number of political violence in comparison to some other districts of West Bengal.

In 2013, the scenario of political violence has changed quite alarmingly. The number of hot spots went up to 27. In the year of 2010, it was only 13 hot spots. The increase is quite high within just a gap of 3 years. All the hot spots are concentrated only within five blocks which are located in the western and southwestern part of the Purulia district. Here, the most significant concentration and increase in hot spots can be identified in the Baghmundi block. In the year 2005, there were no hot spots in this block. In the year 2010, there was only one hot spot at Ajodhya with a 90% confidence level. In the year 2013, it went up to 15 hot spots in the Baghmundi block. A maximum number of hot spots (15 hot spots out of 27 hot spots) were located in one single block of Baghmundi. This particular block had some terrorist activities in the past, and the topography of this particular block is hilly and covered with thick vegetative cover. This kind of legacy along with a natural set-up made this block quite suited for organized political violence. The tribal population and scheduled caste population face a poverty-stricken situation in this block. This type of socio-economic backwardness creates a sense of deprivation among the inhabitants of this block which can be used as a provocation for the political rage against the government by the political goons. Jhalda I block was an addition for the year 2013. This block exhibited five hot spots of political violence in the year 2013 such as Buribir, Ankura, and Bandgram (90% confidence level), Bandgora (95% confidence level), and Bargram (99% confidence level). Jhalda I is a block with a marginal location which has a boundary with Jharkhand state on the western side. The location of Jhalda I and II blocks is quite crucial for crime detection as they share state borders. Political violence in these types of areas with interstate borders is more complicated as the offenders can fly away to the other state making it difficult for the police personnel to locate and arrest the offenders (Banerjee & Ghosh, 2015; Bose, 2021; Ray, 2022). In the case of the Arsha block, the same locations are showing the hot spots of political violence such as Arsha, Sirkabad, Kadampur, and Makiari. In the case of Balarampur, the locations of hot spots prevailed such as Balarampur and Amruhasa. For the year 2013, the cold spots of political violence shifted to Purulia I and Purulia II blocks. The three cold spots were namely Bongabari (Purulia II block) and Simulia and Lagda (Purulia I block).

For the analysis of the latest scenario of political violence in the Purulia district, a hot spot analysis of 2022 has been done. In the year 2022, the number of hot spots of political violence comes down to 12 which is a good sign for the overall political ambience of the district. All the hot spots are concentrated in three blocks, namely, Jhalda I, Baghmundi, and Balarampur. Baghmundi shows the highest concentration of nine hot spots. Among the nine hot spots, four show a 90% level of confidence, and the other five show a 95% confidence level (Raydi, Sopa, Kushaldi, Baghmundi and Sosokhas). Two hot spots are located in Jhalda I block, and one hot spot is located in Balarampur block. The cold spots are located in Purulia I (Lagda), Purulia II (Bongabari), and Bandowan (Kendapara, Udalbani) block. After the passage of almost 10 years after 2013, the political scenario is much more stable for the entire state of West Bengal which is reflected in the case of Purulia district in particular. The situation of Baghmundi block is still vulnerable in terms of political violence in comparison to all the other blocks of Purulia. Jhalda I blocks play a very crucial role in the times of election to date for its bordering location with Jharkhand, and the location of the block is within the famous Jungle Mahal which has a long historical legacy of political turmoil and disturbance. The entire western part of Purulia belongs to Jungle Mahal. The legacy of political vulnerability still persists though in a much lesser tone (Samaddar, 1995; Kamra, 2016; Bhattacharya & Bhattacharya, 2021) (Fig. 5).

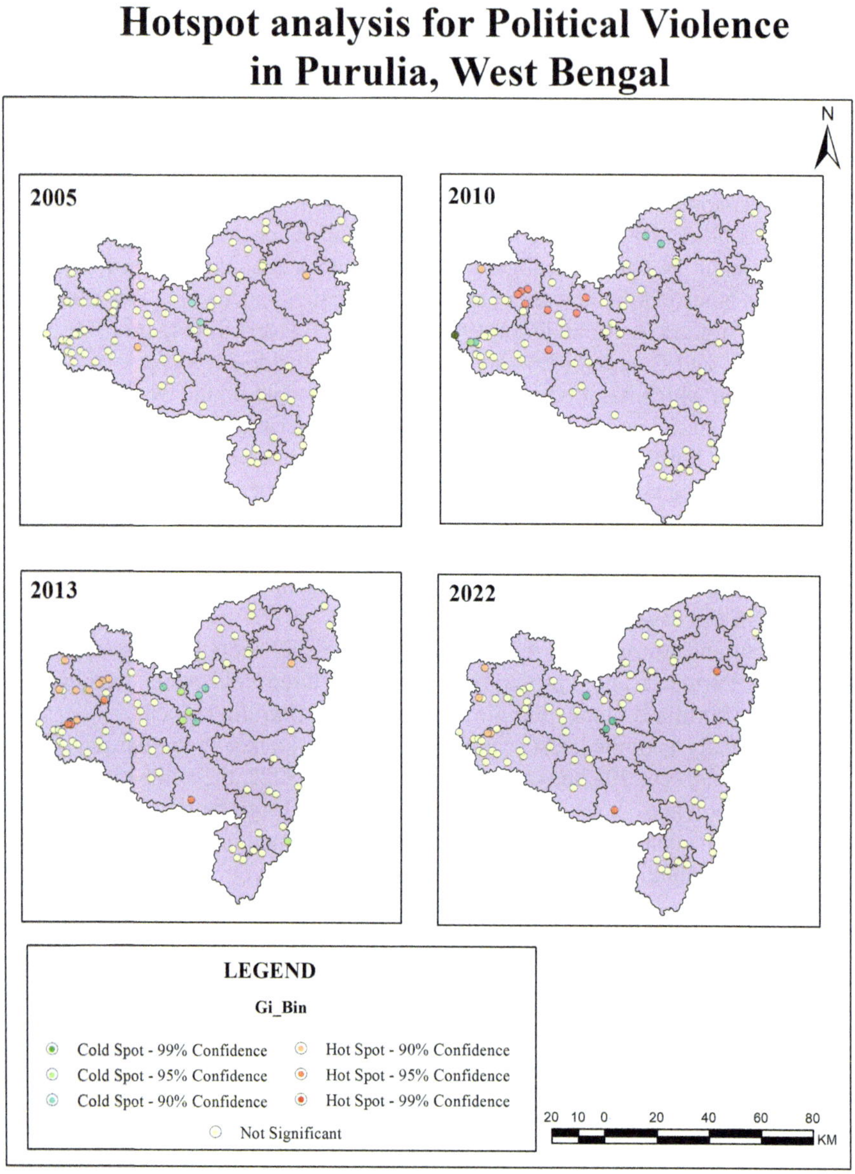

Fig. 5 Hot spot map of political violence in Purulia

4 Pattern of Spatial Distribution of Political Violence

4.1 Spatial Trends of Crime by Kriging

From the hot spot analysis for 4 years (2005, 2010, 2013, and 2022) spread over 17 years, it is evident that all the hot spots have a tendency to be concentrated in the western part of the Purulia district. For further valediction of the spatial distributional pattern of political violence, spatial interpolation has been done using the normal Kriging method for the year 2013. The year 2013 has been chosen because of the cruciality of this particular year for political violence. By this method, the continuous distribution pattern around the sample points has been generated for the cases of political violence. From the spatial interpolation, it is quite evident in the western part of the Purulia district. Highly weighted points are majorly clustered in the western-most blocks of Purulia namely Jhalda I, Jhalda II, Jaypur, and the south-western blocks of Baghmundi and Balarampur. The results of interpolation follow the same trend of the concentration of hot spots. Blocks, having interstate boundaries like Jhalda I, Jhalda II, Jaypur, Baghmundi, and Balaramur, are more vulnerable to political violence. The bordering location makes these blocks full of heterogeneous populations and floating populations which makes these places more crime-prone. Heterogeneity in the nature of the population makes the social dissociation more pronounced which in turn gives rise to social disintegration. Lack of social cohesiveness causes differences in political ideologies and demands from the government (Jackson & Dexter, 2014; Vincent, 2014). These play a crucial role behind any sort of political sensitivity which culminates into political violence. The presence of thick vegetation cover and rugged topography also help the offenders to take shelter in those areas and organize political upsurges. The central location of the blocks like Purulia I and Purulia II helps these places to be politically stable and have cold spots. The bordering location with difficult terrain makes the western side blocks difficult for surveillance. The spatial pattern of political violence unveils the hidden mechanism of social physics. The behaviour of the offenders to organize crimes in a particular area is based on the spatial attributes which can be geospatially quantifiable and explained. Political violence is dependent on mass behaviour (Lacey, 2011). The marginal location of the blocks also determines some typical behavioural patterns of mass. An unaware state of mind regarding the political scenario of the state makes it easy for the offenders to build up violent consensus among the mass people of the mentioned blocks with higher numbers of political violence (Gillooly, 2022) (Fig. 6).

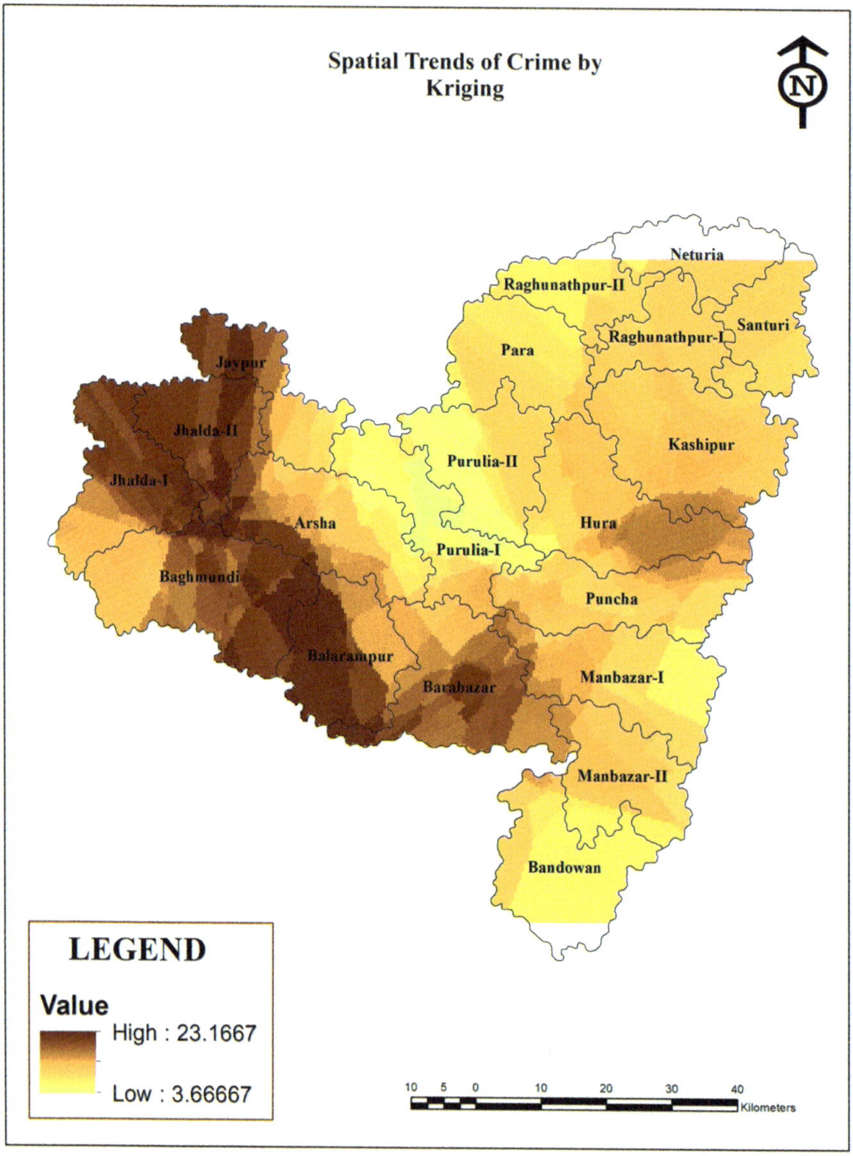

Fig. 6 Spatial trends of crime by Kriging

4.2 Geo-Statistical Analysis of Spatiality of Political Violence

Many times, political violence is an organized form of crime. The political protests often take the form of violence and intolerance. Selection of the places for the gatherings and movements often has a logical thought process behind that. On the basis of the location of the 78 sample crime spots, centroid has been calculated for political violence. These centroids have been calculated for four different years such as 2005, 2010, 2013, and 2022. The centroids give the idea of the most accessible spaces for the occurrence of crimes. The centroids for the years 2005, 2010, and 2022 show the same location in Arsha block, a place near Sirkabad. The location of the centroid is also in the western part of the district because of the higher number of reported cases of political violence from that part of the district. The concentration of the hot spots is in this part of the district. Following the same spatial trends, the centroid or the most accessible point for political violence gets its position in that part. The Arsha block has been highly politically vulnerable throughout these years. Arsha has a location which is not as bordering as Jhalda I, Baghmundi, Jaypur, or Balarampur block, and it is not an interior location like Purulia I and Purulia II block. It has a balance of both which means that Arsha block is distant from the border from all sides and located in the western central location. As the year 2013 has seen a sterling increase in the number of hot spots, as a result, the centroid has shifted to the Baghmundi block. In the year 2013, the Baghmundi block showed a concentration of 15 hot spots. Following that trend, the centroid position has been changed to Baghmundi. It is quite evident from the concentration of the hot spots in 2013 that for that particular year, Baghmundi was the most accessible block for offenders to commit or organize political violence. All the spatial factors like vegetation cover and topography and demographic factors like poverty-stricken population and tribal mass population were favourable for crime. The specific location of the centroid for 2013 was Ajodhya, which is the most difficult space for surveillance due to its hilly topography. These are pull factors in terms of social physics. The statistics of political violence also support the spatial trend of centroid location (Fig. 7).

Calculation of the mean centre of crime gives the idea about the spatial trend of the central tendency of political violence in the Purulia district. For all the years, the shifting of the mean centre is not very remarkable. Again, following the concentration of hot spots, the mean centre is located in the western part of the district. All the mean centres for the years 2005, 2010, 2013, and 2022 are located in the block of Arsha. Arsha block showed a number of concentrations of crime locales of political violence and hot spots also. The shifting of mean centres in the years 2005 and 2010 is almost zero. For the year 2022, the shifting is in the western direction, but that is also a negligible shift. In the year 2013, the shift is towards the southwest in the direction of Baghmundi. It also follows the previously discussed trend when in 2013, the concentrations of hot spots were highly increased in the Baghmumdi block. Along with the central tendency, the spatial nature of dispersion is also measured with the help of location data of political violence in Purulia. A standard distance ellipse (SDE) has been drawn for 4 years. The highest range of dispersion of political violence engulfed the blocks of Hura, Puncha, Manbazar I, and Manbazar II on the eastern side. On the western side, all the blocks like Jhalda I, Jhalda II, Jaypur, Arsha, and Baghmundi were there within the ellipse. In the southern section, Balarampur and Barabazar blocks are incorporated within the ellipse. The spatial expansion and direction have been shifted with the changing years. The highest shift

Fig. 7 Centroid position of crime. (Source: Office Of the Superintendent of Police, Chowk Bazar, Purulia)

in SDE can be seen between the years 2010 and 2013. In the year 2013, the SDE was shifted the most in the western direction, engulfing the block of Jhalda I and the entire block of Baghmundi. Some parts of the SDE even went out of the district border towards the parts of Jharkhand in the west. In the years 2005 and 2010, the SDE always remains inside the district. Even in the year 2022, the SDE again came

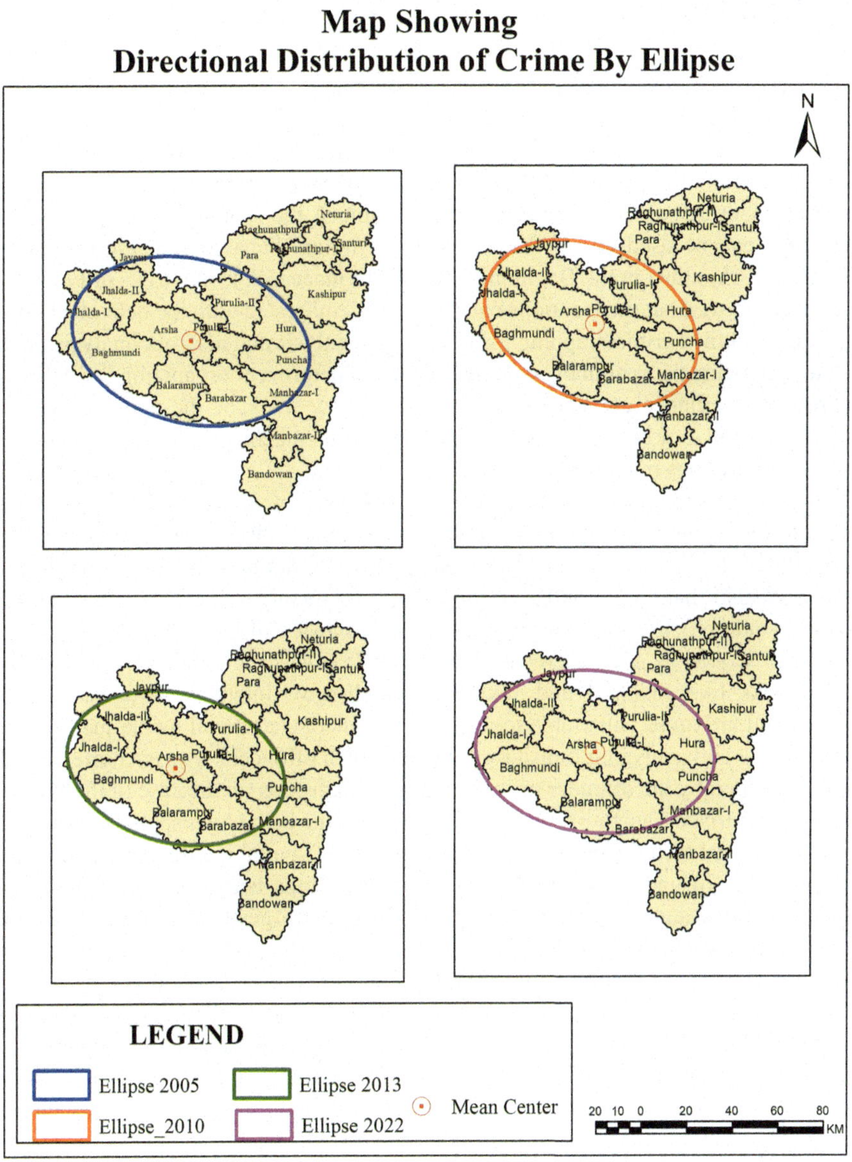

Fig. 8 Directional distribution of crime by ellipse in Purulia. (Source: Office Of the Superintendent of Police, Chowk Bazar, Purulia)

inside the boundaries of the district Purulia. But in the year 2013, it went out of the district boundary as well as the state boundary of West Bengal. It hints towards the interstate connectedness of political violence and the floating nature of offenders who are involved in these criminal activities. The blocks of the northern section of the district are comparatively less prone to political violence. That is why these blocks remained devoid of hot spots and the territory of SDE (Fig. 8).

5 Conclusion

Geospatial techniques play a crucial role in managing organized crimes like political violence by providing law enforcement agencies and policymakers with valuable tools for analysis, monitoring, and intervention. For a district like Purulia, the geospatial remote sensing techniques can be proven very beneficial to crime in the management of political violence. Identification of the spatial pattern of political violence can be used as a reflection of the criminal behaviour of the offenders. Places like Ajodhya in the Baghmundi block are inaccessible due to its hilly terrain and forest cover, but the locations can be easily demarcated with the help of satellite images. Detailed crime mapping using hot spot analysis and network analysis made it easy to define the criminals' movement pattern in terms of organizing the political unrest. Surveillance agencies can have some easy track of the activities of the offenders, and they can allocate the crime prevention machineries over the crime hot spots. Continuous mapping of the historically politically vulnerable zones gives an idea about future criminal activities. The law enforcement bodies can be prepared for the unforeseen and upcoming political turmoil. Political acts of violence are directly connected with mass livelihood. Future prediction can be very helpful in this way also that communities staying in the politically vulnerable areas like the western side blocks of the Purulia district can be aware of the political turmoil and not be provoked by any miscreant intentions. We have seen in the analysis part that the blocks like Jhalda I and Jhalda II which have interstate boundaries with Jharkhand are more vulnerable to political violence. Geospatial techniques help in monitoring and analyzing transit routes by political offenders. The territories which are controlled by the offender groups or antisocial groups (in the case of the Purulia Maoist terrorist organization) can be mapped with the help of geospatial techniques. The geospatial techniques enhance the collection of spatial intelligence, aiding law enforcement in understanding the spatial relationships between criminal entities, their hideouts, and operational centres. Political upsurges can be avoided by the change in mass behaviour in a collective sense. The pattern and nature of human behaviour cannot be entirely predicted by any spatial and statistical modelling. Human behaviour is above any positivist approach which is highly dependent on universal laws. Political violence is highly dependent on human ideologies. If the society can run on such an ideology where the political dimension of the society never goes beyond the ethical limits of humanity, then only political violence can be reduced in a significant way, and it can even be stopped.

References

Adhikari, D., Gazi, K. H., Giri, B. C., Azizzadeh, F., & Mondal, S. P. (2023). Empowerment of women in India as different perspectives based on the AHP-TOPSIS inspired multi-criterion decision making method. *Results in Control and Optimization, 12*, 100271.

Ahen, F., Buabeng, K. O., & Salo-Ahen, O. M. (2023). Market violence through destructive entrepreneurship: Assessing institutional responses to the proliferation of counterfeit traditional and alternative medicines in Ghana. *Heliyon, 9*(3), e13881.

Ahmad, F., Uddin, M. M., & Goparaju, L. (2018). Role of Geospatial technology in Crime Mapping: A case study of Jharkhand state of India. *American Journal of Geographical Research and Reviews, 1*(5), 1–11.

Bochenkova, A., Buonanno, P., & Galletta, S. (2023). Fighting violence against women: The role of female political representation. *Journal of Development Economics, 164*, 103140.

Banerjee, A., & Ghosh, A. (2015). Traversing Jungle Mahal: Field notes on democracy and development in the Santhal regions of Bihar and West Bengal. In *Claiming India from below* (pp. 54–73). Routledge.

Bhattacharya, S., & Bhattacharya, A. (2021). The adivasi resistance in Jungle Mahal: A case study of Chuar Rebellion1. *Journal of Adivasi and Indigenous Studies, 11*(2), 1.

Bohstedt, J. (2016). *The politics of provisions: Food riots, moral economy, and market transition in England* (pp. c. 1550–c. 1850). Routledge.

Boivin, R., & de Melo, S. N. (2019). Comparing global spatial patterns of crime. *Policing: An International Journal, 42*(6), 1097–1106.

Bose, R. (2021). Lalgarh insurgency and political society: How Eastern India erupted in anger against the left front government. In *Civil resistance against the 21st century authoritarianism: Defending the human rights in the global south* (pp. 130–155).

Brathwaite, R., & Park, B. (2023). Deadly influences: Evaluating the relationship between political competition and religious violence. *Political Behavior*, 1–27.

Bruhn, A. L., Wehby, J. H., & Hasselbring, T. S. (2020). Data-based decision making for social behavior: Setting a research agenda. *Journal of Positive Behavior Interventions, 22*(2), 116–126.

Calas, B. (2007). From rigging to violence: Mapping of political regression. In *The general elections in Kenya* (Vol. 167).

Calas, B. (2008). From rigging to violence. Mapping of political regression. *Les Cahiers d'Afrique de l'Est/The East African Review, 38*, 171–192.

Chatterjee, J., & Basu, S. (2017). West Bengal politics at the crossroads. In *Electoral politics in India* (pp. 196–212). Routledge India.

Census of India. (2011). *Government of India.*

Chainey, S., & Tompson, L. (2012). Engagement, empowerment and transparency: Publishing crime statistics using online crime mapping. *Policing: A Journal of Policy and Practice, 6*(3), 228–239.

Chitiga, M., Soropa, G., Dube, T., & Sengera, P. (2023). Spatial interpolation of vertisol physico-chemical properties through ordinary kriging in south-eastern Zimbabwe. *South African Journal of Plant and Soil, 40*(1), 46–57.

Chowdhury, S., Singh, A., Kasemi, N., & Chakrabarty, M. (2022). Decomposing the gap in intimate partner violence between Scheduled Caste and general category women in India: An analysis of NFHS-5 data. *SSM-Population Health, 19*, 101189.

Das, R. (2020). Narratives of the dispossessed and casteless: Politics of land and caste in Rajarhat, West Bengal. *Journal of Contemporary Asia, 50*(5), 806–830.

Dawson, L. (2020). *The politics and perils of space exploration: Who will compete, Who will dominate?* Springer.

Dhattiwala, R. (2017). Mapping the self: challenges of insider research in a riot-affected city and strategies to improve data quality. *Contemporary South Asia, 25*(1), 7–22.

Donner, H. (2004). *The significance of Naxalbari: Accounts of personal involvement and politics in West Bengal* (p. 14). Occ. Papers Series.

Eger, C. (2021). Gender matters: Rethinking violence in tourism. *Annals of Tourism Research, 88*, 103143.

Franklin, K., & Herek, G. M. (2022). Sexual minorities, violence against. In *Encyclopedia of violence, peace, & conflict: Volume 1–4* (Vol. 2, 3rd ed.). Elsevier.

Fuentes, C. M., & Hernández, V. (2013). *Assessing spatial pattern of crime in Ciudad Juárez, Chihuahua, Mexico (2009): The macrolevel, mesolevel and microlevel approaches.*

Ghosh, S. (2022). Political empowerment of women and financial inclusion: Is there a link? *Social Sciences & Humanities Open, 5*(1), 100267.

Gillooly, S. N. (2022). Legacies of political violence and voter behaviour in Colombia. *Journal of Peacebuilding & Development, 17*(1), 58–75.

Gunasekara, R. (2023). *Religious violence against Christians: Disruptions and Distractions-November 2021 to October 2022.*

Hassan, M. (2023). Religion as a Source of Violence: Contending the Narrative of Political Violence Perpetrated in the Name of Islam in Bangladesh. *Journal of Asian Wisdom and Islamic Behavior, 1*(1).

Hipp, J. R. (2016). General theory of spatial crime patterns. *Criminology, 54*(4), 653–679.

Jackson, R., & Dexter, H. (2014). The social construction of organised political violence: An analytical framework. *Civil Wars, 16*(1), 1–23.

Jana, M., & Sar, N. (2016). Modeling of hotspot detection using cluster outlier analysis and Getis-Ord Gi* statistic of educational development in upper-primary level, India. *Modeling Earth Systems and Environment, 2*, 1–10.

Jeppesen, S. (2022). Performance crime and self-surveillant subjects in the capitol riots. In *The Capitol Riots* (pp. 135–157). Routledge.

Joseph, J., Jauhola, M., Arvind, L. S., & Gadavi, S. (2021). Wounded attachments to disaster recovery: Gendered structural violence and everyday life, Indian experiences explored. *International Journal of Disaster Risk Reduction, 59*, 102242.

Joshi, D. K., & McGrath, K. (2015). Political ideology, public policy and human development in India: Explaining the gap between Gujarat and Tamil Nadu. *Journal of Contemporary Asia, 45*(3), 465–489.

Kamra, L. (2016). *The politics of counterinsurgency and state making in modern India*. (Doctoral dissertation. University of Oxford.

Kebede, B. D. (2023). Geographical imagination as portrayed in contemporary Ethiopian creative works that feature the Battle of Adwa. *Social Sciences & Humanities Open, 8*(1), 100446.

Khalko, A. (2015). *Environmental limitations to early colonial expansion in the Jungle Mahals.*

Kim, S. M., & Choi, Y. (2017). Assessing statistically significant heavy-metal concentrations in abandoned mine areas via hot spot analysis of portable XRF data. *International Journal of Environmental Research and Public Health, 14*(6), 654.

Kindynis, T. (2014). Ripping up the map: Criminology and cartography reconsidered. *British Journal of Criminology, 54*(2), 222–243.

Kobayashi, Y., Howell, C., Heinrich, T., & Motta, M. (2022). Investigating how historical legacies of militarized violence can motivate COVID-19 vaccine hesitancy: Evidence from global dyadic survey. *Social Science & Medicine, 311*, 115346.

Kounadi, O. (2015). *Geospatial privacy framework for confidential discrete data with emphasis on spatial crime analysis & visualization*. (Doctoral dissertation. Department of Criminal Justice, Temple University.

Kumar, A. (2013). Impact of West Bengal politics on India–Bangladesh relations. *Strategic Analysis, 37*(3), 338–352.

Lacey, D. (2011). *The role of humiliation in collective political violence*. (Master's thesis. University of Sydney.

Lee, S., & Son, Y. J. (2020). Extended decision field theory with social-learning for long-term decision-making processes in social networks. *Information Sciences, 512*, 1293–1307.

Legg, S. (2020). Political lives at sea: Working and socialising to and from the India Round Table Conference in London, 1930–1932. *Journal of Historical Geography, 68*, 21–32.

Linning, S. J. (2015). Crime seasonality and the micro-spatial patterns of property crime in Vancouver, BC and Ottawa, ON. *Journal of Criminal Justice, 43*(6), 544–555.

Mahadani, S. (2022). Political dilemma in Jangalmahal area: A historical perspective. *IJAR, 8*(8), 18–22.

Mahato, S. (2021). Travel route planning for eco-tourists of Jungle Mahals of West Bengal. *Sambodhi Journal, 46*, 352–361.

Mahato, N. K. (2023). Notion of resilience among the adivasis of Jungle Mahals, Eastern India. In *Indigenous societies in the post-colonial world: Responses and resilience through global perspectives* (pp. 301–313). Springer.
Mahapatra, S. (2021). The Shikar and Hunting: Eradication of Wildlife in Colonial Jungle Mahal. *Global Environment, 14*(2), 335–369.
Makate, M., & Nyamuranga, C. (2023). Changing the mindsets? Education and the intergenerational spread of tolerance for physical violence against women in Zimbabwe. *Economics & Human Biology, 101345.*
Mandel, S. (2024). *Ground control: An argument for the end of human space exploration.* Chicago Review Press.
Manning, P. K. (2008). *The technology of policing: crime mapping, information technology, and the rationality of crime control* (Vol. 4). NYU Press.
Manning, P. K. (2017). Technology's ways: Information technology, crime analysis and the rationalizing of policing. In *Cyberspace Crime* (pp. 491–511). Routledge.
Massey, J. (2019). Police, press and politics: Institutional responses to riots in Bradford 2001 and Manchester 2011. *Internet Journal of Criminology.*
Maiti, P., & Mitra, D. (2023). Ordinary kriging interpolation for indoor 3D REM. *Journal of Ambient Intelligence and Humanized Computing, 14*(10), 13285–13299.
Mukhopadhyay, A., & Mukhopadhyay, A. (1994). Politics in West Bengal. *The Indian Journal of Political Science, 55*(3), 321–334.
Murray, A. T., & Grubesic, T. H. (2013). Exploring spatial patterns of crime using non-hierarchical cluster analysis. In *Crime modeling and mapping using geospatial technologies* (pp. 105–124).
Nath, S., & Chowdhury, S. R. (2019). Mapping polarisation: Four ethnographic cases from West Bengal. *Journal of the Indian Anthropological Society, 54*, 51–64.
Nugroho, H., Utama, N. P., & Surendro, K. (2021). Normalization and outlier removal in class center-based firefly algorithm for missing value imputation. *Journal of Big Data, 8*, 1–18.
Oostveen, D. F. (2022). Material politics, violence, and religion. *Religion, Conflict, and Violence, 58.*
Prathap, B. R. (2022). Geospatial crime analysis and forecasting with machine learning techniques. In *Artificial intelligence and machine learning for EDGE computing* (pp. 87–102). Academic Press.
Piazza, J. A. (2023). Drivers of political violence in the United States. *Journal of Public Policy & Marketing, 42*(1), 11–14.
Piętak, K., Dajda, J., Wysokiński, M., Idzik, M., & Leśniak, Ł. (2016). Geospatial data integration for criminal analysis. In *Man–Machine Interactions 4: 4th International Conference on Man–Machine Interactions, ICMMI 2015 Kocierz Pass, Poland, October 6–9, 2015* (pp. 461–471). Springer.
Pires, B., & Crooks, A. T. (2017). Modeling the emergence of riots: A geosimulation approach. *Computers, Environment and Urban Systems, 61*, 66–80.
Puthillam, A., Kapoor, H., & Karandikar, S. (2021). *Beyond left and right: A scale to measure political ideology in India* (p. 387). Preprint. https://psyarxiv. com/fg.
Ray, S. (2022). Voting for security: Evidence from a Maoist-affected region in India. *Contemporary South Asia, 30*(3), 346–365.
Ratcliffe, J. H. (2005). Detecting spatial movement of intra-region crime patterns over time. *Journal of Quantitative Criminology, 21*, 103–123.
Rikala, J. (2023). *Determining a person's individual-centered home range using GIS–Comparison and evaluation.*
Roy, D. (2013). *Rural politics in India: Political stratification and governance in West Bengal.* Cambridge University Press.
Ru, L., & Zhang, L. (2017, November). A weighted centroid localization algorithm for wireless sensor networks based on weight correction. In *2017 9th international conference on advanced infocomm technology (ICAIT)* (pp. 165–169). IEEE.

Sahbeni, G., & Székely, B. (2022). Spatial modelling of soil salinity using kriging interpolation techniques: A study case in the Great Hungarian Plain. *Eurasian Journal of Soil Science, 11*(2), 102–112.
Samaddar, R. (1995). Territory and people: The disciplining of historical memory. In *Texts of power: Emerging disciplines in Colonial Bengal* (pp. 167–200).
Shafique, I., Zahra, S. A., Farid, T., & Sharif, M. (2017). Role of GIS in crime mapping & analysis. *Sukkur IBA Journal of Computing and Mathematical Sciences, 1*(2), 39–47.
Sinha, S., & Ashwini, S. (2021, May). Rssi based improved weighted centroid localization algorithm in wsn. In *2021 2nd International Conference for Emerging Technology (INCET)* (pp. 1–4). IEEE.
Slater, T. (2011). From 'criminality' to marginality: Rioting against a broken state. *Human Geography, 4*(3), 106–115.
Szołucha, A. (2021). *Space exploration and the imaginaries of living in a climate-changing world.*
Tesar, J. (2020). How do social norms and expectations about others influence individual behavior? A quantum model of self/other-perspective interaction in strategic decision-making. *Foundations of Science, 25*, 135–150.
Thakur, M. (2024). Cyber security threats and countermeasures in digital age. *Journal of Applied Science and Education (JASE), 4*(1), 1–20.
Tiratelli, M. (2018). Reclaiming the everyday: The situational dynamics of the 2011 London riots. *Social Movement Studies, 17*(1), 64–84.
Treadwell, J., Briggs, D., Winlow, S., & Hall, S. (2013). Shopocalypse now: Consumer culture and the English riots of 2011. *British Journal of Criminology, 53*(1), 1–17.
Vincent, M. (2014). Political violence and mass society: A European Civil War? In *The Oxford Handbook of European History* (pp. 1914–1945).
Wika, P. N., & Abdullahi, M. M. (2023). Perspectives on religion and ethnic identity dynamics: Conditions for violent collective actions in Nigeria. *Social Science Research, 9*(1).
Zhang, L., & Gläscher, J. (2020). A brain network supporting social influences in human decision-making. *Science Advances, 6*(34), eabb4159.
Zhao, Z., Zhao, Z., & Zhang, P. (2023). A new method for identifying industrial clustering using the standard deviational ellipse. *Scientific Reports, 13*(1), 578.

Ashit Modak, currently working as a GIS Analyst in Pioneer Geo-Science, completed his post-graduation in Geoinformatics from the Department of Geography, Adamas University in 2024. His area of research interest is the application of GIS-remote sensing in crime detection management and urban climate change issues. He presented research papers at national and international conferences. He has already published a few research papers and book chapters.

Supratim Karmakar, currently working as an Assistant Professor, at the Department of Geography, Adamas University, has completed his Ph.D. from the Department of Geography, Visva Bharati in 2017. Prior to that, Dr. Karmakar served as a Post-graduation Coordinator at Bag Bazar Women's College and a Guest Lecturer at Vidyasagar College. Dr. Karmakar has published 17 research articles in peer-reviewed journals and 11 book chapters. He co-edited a book on Health Geography in 2023. His areas of interest are geography of crime, gender studies, urban ecology, and folk culture studies.

Dr. Kasturi Mukherjee is an Associate Professor and Head, Department of Geography, School of Basic and Applied Sciences, Adamas University, and Visiting Researcher in Stockholm Resilience Centre, Stockholm University. She has received her Ph.D. in Urban Climate, from the University of Calcutta. Dr. Mukherjee has completed a post-graduation degree in Geography with a specialization in Environmental Geography and Post-Graduation Diploma in Geoinformatics. She is a gold medalist for securing first rank in the university in post-graduation, and she has received the best paper award two times in 2010 and 2018. Her research interest is focused on urban climate, sustainable development, and peri-urban development. This year, she has received the prestigious Core Research Grant from the Science and Engineering Research Board (SERB), DST as CoPI.

Framing a Historical Landscape: Role of Tea and Railways in the Creation of 'Colonial Assam' in the Nineteenth and First Half of the Twentieth Century

Srijani Bhattacharjee

1 Introduction

Historical landscape usually refers to a landmass with deep-rooted trees, gardens and an area with old relic patterns and land uses. In the present context, the term is used to denote the province of Assam that was curved into a historically important landscape under the British after attaining its repute as the largest tea-producing territory when indigenous tea plants and local tea consumption practices were discovered in the region. Till then, Assam remained isolated from the rest of India due to its backward transport and communication systems, topography considered to be inaccessible, and disease-prone climate. The province was a huge territory scantily populated after the decline of the Ahom dynasty and Burmese incursions that discouraged British investments in the area. The British considered scarcity of manpower in Assam as an impediment behind colonial industrial and administrative initiatives in the region. Considering these factors, Assam was viewed as a backward territory by the British authorities. The discovery of indigenous tea plants in the terrain by 1823 brought transformations in the colonial mindset. It persuaded the British administrators to commercially explore Assam, develop its communication systems, and establish British administration over the region (Goswami, 1963, pp. 183–186; Hunter, 1879, p. 55).

The isolated position of Assam has a history behind it. The Ahom rulers in the region until the initiation of British rule preserved their language and culture of Tai origin, their political institutions, and societal norms from external intrusions. However, in their efforts to preserve their ethnicity, they faced challenges. The 'noble clans' exercised influence over the kings whose authority was largely decided by them. There was importation of Brahmans from Kanauj who somewhat

S. Bhattacharjee (✉)
Department of History, Amity Institute of Social Sciences, Amity University, Kolkata, West Bengal, India

A. Rai et al. (eds.), *Contemporary Social Physics*, Springer Geography,
https://doi.org/10.1007/978-3-031-77596-3_6

sanskritised the Ahom structure by contributing to its statecraft, diplomacy, revenue organisation, ideology, religion, and military science. The Vaishnavite movements in Assam in the fifteenth and sixteenth centuries could be considered as a reflection of this. The peasants were divided into *Khels* (an official guild system based on occupation or territories given to them) in exchange for an acre of land per householder. The structure thus created was neither entirely a feudal formation nor a total sanskritised organisation. Consequently, there were long drawn series of civil wars with the peasants and newly developing feudal forces on one hand and the Ahom oligarchy on the other (Sharma, 1980, p. 1321). The communities residing on the fringes of the Ahom kingdom also raised their voices for independence. Thus, Assam was engaged in its own crises. Taking advantage of the situation, the Burmese incursions surfaced in. It was during this period that the British assumed the reins of administration in Assam. On one hand, tackling the Burmese attacks was a necessity, while on the other the detection of indigenous tea plants in Assam expedited their access to the region. The Assamese intellectual class perceived the British intervention as the saviour from the situation. As a consequence, they invited British access to the region, and with their support the colonial regime could evict the Burmese from the province, but in the process, the ruling authority of the Ahom dynasty was substituted by the British administration in Assam. The British expanded its governance over the province by framing a colonial bureaucracy mostly composed of immigrant populations from neighbouring East Bengal and other parts of India. Subsequently, immigration to Assam from other regions was encouraged by the government under various pretexts thereby altering the regional demography. The Assamese people perceived this as exclusion (Sharma, 1980, pp. 1321–1322).

The discovery of indigenous tea plants and the commercial importance assigned to Assam hastened the incorporation of the province within the ambit of the British administration. Colonial branches of governance such as the Judicial and Police Departments, the Public Works, Revenue and Agriculture Departments, and the Local Board of Assam came up in the region (Cosh, reprint 1986, p. 124). The British Forest Department was established in 1868 for the reservation and commercial exploitation of forests (Ribbentrop, reprint 1989, 76). The Chief Commissioner of the province directly under the Viceroy in Council was held responsible for the overall control of the province (Assam Gazette, 1886). Subsequently, botanical, zoological, geological, and topographical surveys and Scientific Forestry (German concept of utilisation, exploitation, and preservation of natural resources) applications conducted under colonial initiatives surveyed and exploited the natural and mineral resources of the territory and contributed to revenue generation from the region (Saikia, 2011, pp. 48–50). Such endeavours encouraged further explorations into the resources of Assam to ascertain their commercial possibilities. The British government thought of politically configurating the province by incorporating the entire northeast Indian region except the monarchical units of Tripura and Manipur within the province of Assam. The newly structured territory was thus a huge landmass with diverse topography, geography, flora and fauna, and demography (Figs. 1, 2, and 3).

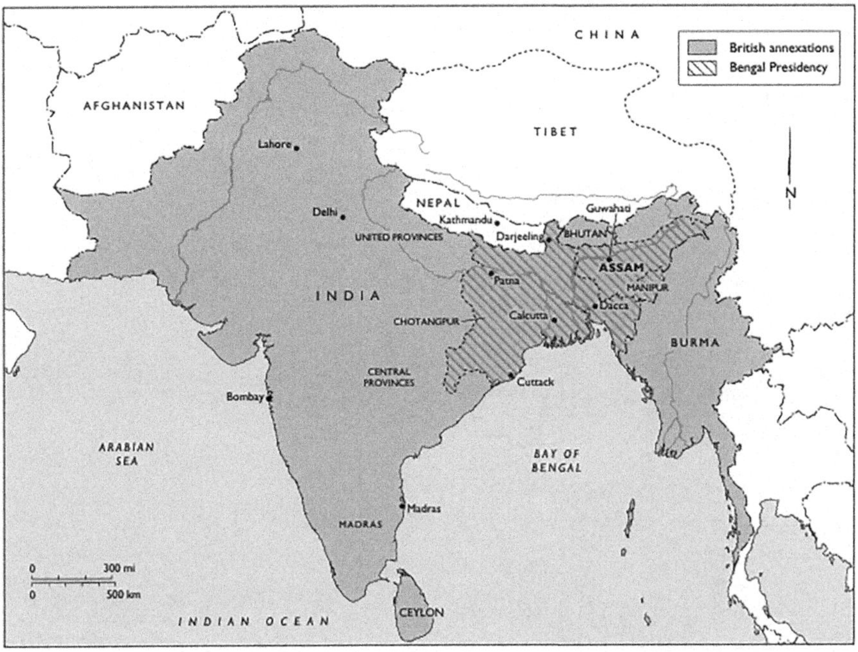

Fig. 1 Location of Assam in British India. (Source: Sharma (2011), p. xvi)

2 Objectives and Methodology

The paper investigates the role of tea and railways in re-acquainting Assam to the rest of India after years of incursions and isolation subsequent to the decline of the Ahom dynasty in the region. It tries to illustrate how a terrain hitherto considered as backward and inaccessible by the British was begun to be perceived as a viable province within India due to the commercial significance and consequent political weightage attached to it after the establishment of these colonial enterprises in the region. The paper tries to understand the associated socio-economic and political developments that took place in British Assam from the perspective of social physics. It has attempted to evaluate the patterns and regularities connected with economic growth in the region and the accompanying commercial importance and political significance attached to them as per the modes of observation undertaken in social physics. The study confines itself to the colonial era in the region till the mid-twentieth century. It has consulted Assam administration reports, forest administration reports, primary and secondary literature on tea and railways, gazetteers, journal articles, unpublished dissertations, and online sources for comprehending these aspects.

The identification of indigenous tea plants in Assam was perceived as a boon by the colonial traders and British administrators in the region. This could be attested from the words of Charles Alexander Bruce, Superintendent of Northern Division

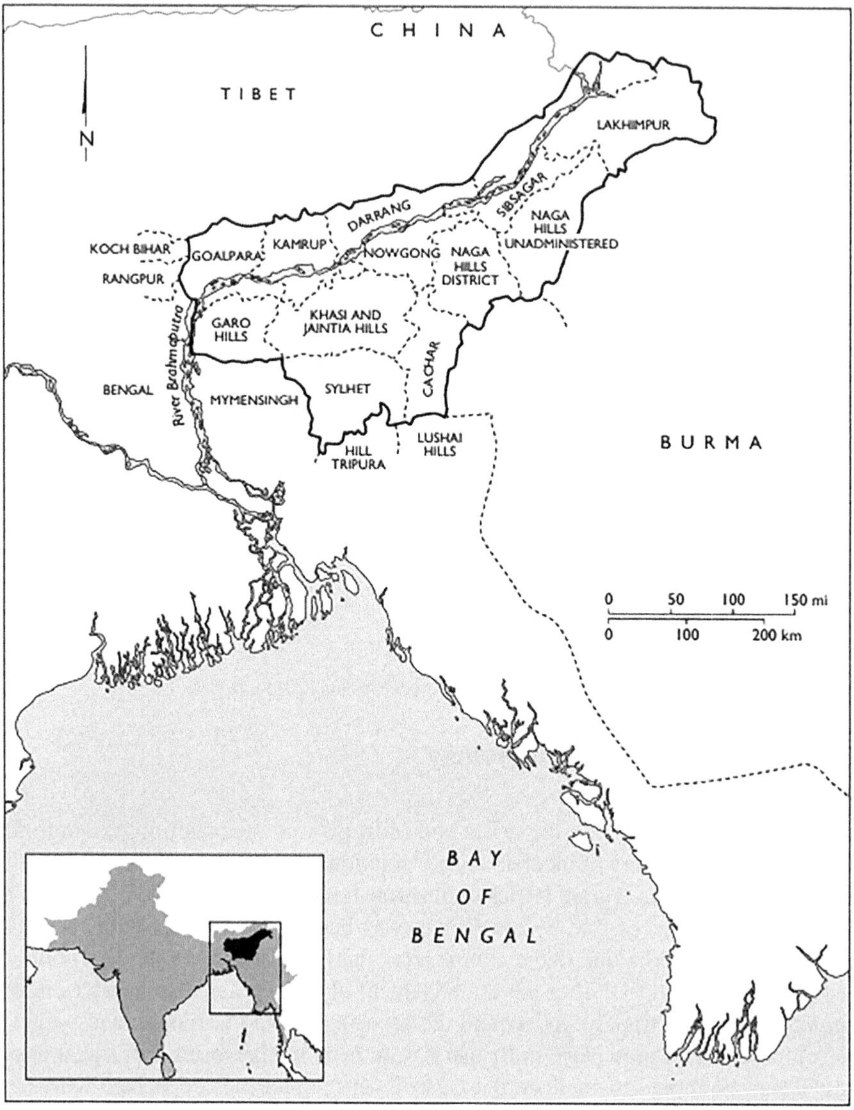

Fig. 2 Map of British Assam in the nineteenth century. (Source: Sharma (2011), p.xvii)

Assam Company who considered it a blessing to his country (Bhuyan, 1949b, pp. 29–30). Maniram Dewan, the Prime Minister of the Ahom King Purander Singha (1818–1819, 1833–1838), and his successors were the first to inform the British about the existence of indigenous tea plants in Assam. Robert Bruce, a Scottish trader and mercenary soldier, communicated the fact to Dr. Nathaniel Wallich, the Secretary to the Tea Committee in 1836, and met Maniram Dewan to place a memorandum on prospects of tea cultivation in the region. Later, when Mr.

Fig. 3 Map of British Assam in the first half of the twentieth century. (Source: Map of Assam Province in 1936, Source data: Survey of India 1:253 k (Perry-Castenada Map Library, Univ of Texas), India 1:1 M, Imperial Gazetteer of India 1:4 M (Digital South Asia Library, Univ of Chicago). https://en.wikipedia.org/wiki/Assam_Province#/media/File:AssamProvince1936_Map.png. Accessed on 18 March 2024)

Charles Alexander Bruce was employed as the Superintendent of the Northern Division Assam Company, he initiated the process of procuring tea plants and seeds from the Sadiya and Muttock areas of the province under an agreement concluded between the Singpho (tribe) chief and Robert Bruce (Griffiths, 1967, pp. 43–44). The denial of the Chinese government in 1833 to renew the agreement to grant the English East India Company the right to trade in tea with China expedited British exploration of indigenous tea plants in the area. In 1834, Lord William Bentinck the Governor General of India, established the 'Tea Committee' under Dr. Nathenial Wallich as the Superintendent who was entrusted with the task of finding out possibilities of tea cultivation in India. The Committee surveyed the soil types and

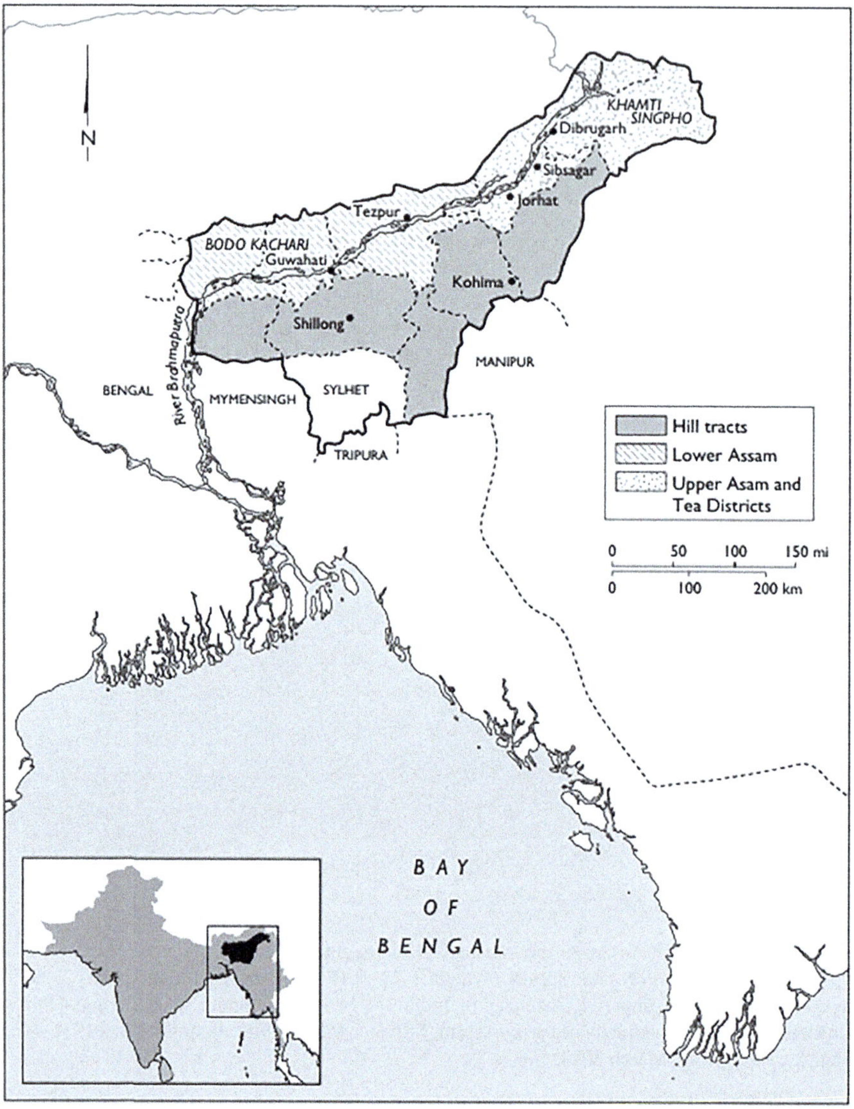

Fig. 4 Tea-producing areas of Assam in the nineteenth century. (Source: Sharma (2011), p. xviii)

climatic conditions of the region and concluded that Assam was conducive for tea plantations. The initial tea plantation that came up in 1835–1836 was formed with the seeds procured from China (Bhuyan, 1949a, b, pp. 29–30). Subsequently the tea types from Assam were found to be identical to the tea varieties from China. John M' Cosh attested this (Cosh, reprint 1986, p. 31) (Fig. 4).

The beginning of geo-political configuration of Assam by the British began with the declaration of unused areas of the region as 'wastelands'. For functional purposes, wastelands are referred to as 'lands not occupied by any owner or allotted to anyone' (Siddique, 1990, p. 6). However, to the British lands that did not accrue revenue to the government were demarcated as wastelands. Dr. Nathaniel Wallich after inspecting the tea-producing areas of Assam suggested ways to Colonel Francis Jenkins the Chief Commissioner of the region in 1833, about productive utilisation of these lands. In his words, 'I must anxiously recommend that all the tracts [referring to wastelands in Assam] ...should be put under systematic course of management in our hands, in order that we may be enabled, with far less cost, and with vast saving in point of time compared to what would be the case if we were to depend exclusively on artificial plantations, to arrive at a satisfactory conclusion as to the capability of Assam tea of yielding a good and merchantable produce' (Siddique, 1990, p. 6). Jenkins in reciprocation stated that the current 'unpromising state of the country' could be restored if European investors were introduced to the scene. He suggested a system of management under which unhindered trade relations between Assam and other regions, development of transport and communication, involvement of the local Assamese in settled landholdings, and imposition of land revenue except areas under wet rice cultivation should be undertaken (Barpujari, 1963, p. 215). Hence, the government framed policies for regulating the wastelands of the region. As Assam was administered as a part of Bengal until then, the unutilised and unoccupied lands of the province passed to the state of Bengal as per Regulation III of 1828. Under the regulation, extensive tracts of lands unoccupied or not owned at the time of the perpetual settlement of the British in the region belonged to the state. The situation began to change when under the orders of Lord Canning in 1861, the first Viceroy of India, local governments in every region were directed to prepare revised rules for the sale of wastelands on favourable conditions (Powell, 1882, p. 60). Henceforth, the British government in Assam took decisions on matters related to wasteland grants in its favour.

The presence of indigenous tea plants and the suitability of Assam for tea cultivation soon converted the wastelands of the territory into commercially lucrative properties that produced wealth. A mad rush developed among the tea companies and the planters for the acquisition of lands in Assam. The government during the period distributed wasteland grants to the tea companies at concessional rates. In this context, A. J Moffat Mills, the Judge of the Sudder Diwani Adalat and Nizamat Adalat, had observed, '.... nothing but absolute rent free tenure of the land will induce people to bring English capital largely into the market'. To him, the aim of the wasteland grants should be immediate returns in terms of government revenue by clearing tracts of forests and promoting immigration (Mills, 1854, p. 17). The Charter Act of 1833 by granting the English East India Company the right to hold lands outside the presidency towns on long-term lease or with freehold rights provided a boost to the situation (Guha, 1988, p. 11). Accordingly, a number of grants like the Assam Waste Lands Rule of 1838, Old Assam Waste Land Rule of 1854, Revised Fee Simple Rule in 1874, and New Lease Rule of 1876 came up. Some rules also put the wasteland grants to auction (Griffiths, 1967, pp. 70–71). In several cases, the tea planters merely occupied lands for cultivating them in future. Lieutenant Colonel Edward Money has described this attitude as 'mad fever' when every man thought owning

Table 1 Average size of tea estates (in acres)

Year	Surma Valley	Brahmaputra Valley	All Assam	All India
1880	226	117	147	71
1885	302	167	210	79
1890	333	231	266	N.A.
1895	346	323	340	80
1900	417	420	420	90
1905	467	465	467	93
1910	480	480	480	128
1915	496	490	493	145
1920	483	477	4801	139
1925	437	453	448	168
1930	437	434	436	167
1935	407	408	408	N.A.
1940	392	386	389	N.A.

Sources: for Surma Valley, Brahmaputra Valley, and All Assam: Report on Tea Culture in Assam relevant issues. For India: Report on Tea Culture in India, relevant issues. Cited by Muhammad Abu B Siddique, Evolution of Land Grants and Labour Policy of Government The Growth of Tea Industry in Assam 1834–1940, pp. 109–110. N.A. (not available)

tea bushes in Assam was to realise riches. Incommensurate to the situation, the government raised the number of wasteland grants in Assam when the figures of the grants increased from 115 in 1861–1862 to 182 in 1862–1863. The numbers grew much bigger in 1867–1868 with 418 (Griffiths, 1967, p. 96) (Table 1).

The expansion and commercial importance attached to tea engaged the indigenous population in different positions within the industry. Maniram Dewan, the initial person to inform the British government in Assam about the presence of indigenous tea plants in the Singpho area of the region, was also the first indigenous tea planter to commercially grow tea in the province and own a tea garden at Cinnamara in Jorhat. He being engaged as the Dewan under the Assam Tea Company was informed about the skills associated with tea cultivation in the province. Maniram Dewan tried to popularise tea consumption in the region by putting forward its benefits among the local populace and established a tea mart at Garhgaon with the purpose. He attempted to provide employment to the tribes in the tea gardens who were thrown out of their work. For example, the Sonowal Kacharis who had given up their gold washing jobs were appointed in tea gardens under his initiatives (Dutta, 2007, pp. 20–31). Yet, owning tea gardens by the indigenous planters was not free from challenges. The privileges associated with wasteland grants were mostly reserved for the British planters and not for the inhabitants. Maniram Dewan did not receive revenue concessions under wasteland grants for his two gardens at Cinnamora and Safrai which were assessed as ordinary rice lands resulting in his payment of high land revenue. The unsupportive attitude of the Europeans was evident when the Assam Company, Margherita Company, and Nakchari Company alleged that Maniram Dewan had incited rebellions among the indigenous people against the British administration in the region. However, here it is also pertinent to mention that there were instances when some relaxations were granted to a few

Assamese planters like Jagannath Barua. Their numbers were nevertheless so meagre that the British hardly considered them as threat. Rather, the small Assamese bourgeoise class thus created was regarded as a potential ally of the British administration (Misra, 1980, p. 1358). Some Assamese cherished aims to establish tea gardens, but their wishes remained unfulfilled because of their lesser purchasing power and limited capital. Their participation in the arena was found as clerks, managers, supervisors, or suppliers who fulfilled the requirements of European tea gardens and agency houses (Dey, 2019, p. 69). Absorbing the tribes in tea plantations was viewed by the British government as a political strategy to administer the hill communities. One type of indigenous engagement in tea cultivation included the employment of tribal chiefs as supervisors of tea estates on behalf of the Europeans and the usage of their manpower as coolies in the gardens. For instance, Ningaroola, the Singpho chief of Ningroo in the Buri Dihing area of Assam, agreed to cultivate a tea estate in the Singpho territory on behalf of the British, provided the government consented to pay the coolies employed by him (Dutta, 2007, pp. 35–36). Such efforts on the part of the colonial government not only allowed them to own tea bushes in the hill areas but also positioned the indigenous communities under British supervision and minimised the problem of labour to a large extent. The Christian missionaries supported such endeavours of the government by motivating the local communities to employ themselves in the tea gardens. The American missionaries operational in Naga Hills considered the engagement of the Nagas in tea gardens not only as a source of manpower but also as an essential way to bring them under colonial control (Barpujari, 1986, pp. 253–254).

The situation started to change from the 1850s onwards and continued in the ensuing years. Attempts to engage the indigenous population in the tea plantation industry failed to fulfil colonial expectations. The British government began viewing the local Assamese as 'lazy' and 'addicted to home grown opium' and hence thought of importing plantation labour from outside the state (Sharma, 2011). Colonial policies were framed to segregate the external coolie population from the local populace. The imported population from outside the state were provided with company tokens that could be converted into commodities in company shops within the tea gardens. They also gathered wood and grazed their cattle in the plantations located adjacent to local villages often resulting in conflicts between the indigenous peasants and plantation workers (Sharma, 1980, p. 1322). Colonial efforts to engage the Naga, Singpho, Khampti, and Kachari communities in the tea plantations soon revealed irregularity on their part as labourers. Endeavours to engage the Chinese also proved unsuccessful. Moreover, Assam during the period was also experiencing a population deficit due to Burmese incursions, civil wars, epidemics, and natural calamities. But the tea plantation industry in the region required labour for plantation works. Under these circumstances, the British government in Assam encouraged large-scale immigration in the region (Nag, 2003, p. 85). The spree of the government to accumulate more lands under its control though primarily for the expansion of tea was further increased when Assam was opened to the immigrant population (Griffiths, 1967, pp. 70–71). The government opined that as the Assamese population were unwilling to work, the tea companies in Assam had to procure labourers from outside the state. From the 1870s to the 1900s, around 700000 to

750000 labour population from Central India and the Chotanagpur plateau were settled in Assam (Sharma, 2009, pp. 1297–1307). The occurrence of famines in Central Provinces in 1896–1898 also stimulated migration of population from those regions towards Assam during the period (Barua, 1980, p. 49). Importation of external population to the province and the creation of export-oriented tea enterprise in Assam produced intimate connections between the state and outside the region centring around Calcutta port and agency houses which soon became the pivot for the accumulation of tea capital, expense and expenditure related to it, employment in the tea industry, management of supplies, and marketing related to tea in the region (Sharma, 2011). Migration continued in the province in the subsequent years with Bengali influx in administrative positions and Muslim immigration as peasants which increased the population of the area sixfold since 1901 when it rose to 3.3 million (Weiner, 1983, p. 283). The following tables illustrate the trend of labour immigration that took place in the tea gardens of Assam from the nineteenth to the mid-twentieth century (Tables 2 and 3).

Table 2 Annual immigration of tea garden labourers 1880–1900

Year	Number of labourer immigration to Assam
1880	15,913
1881	17,116
1882	22,559
1883	32,138
1884	45,511
1885	29,398
1886	30,894
1887	36,463
1888	46,293
1889	55,658
1890	36,080
1891	49,908
1892	56,050
1893	50,675
1894	46,530
1895	72,837
1896	81,115
1897	95,931
1898	49,169
1899	31,908
1900	62,733

Source: Report on Labour Immigration into Assam, relevant issues cited by Muhammad Abu B Siddique, Evolution of Land Grants and Labour Policy of Government The Growth of Tea Industry in Assam 1834–1940, p 141

Table 3 Annual immigration of tea garden labourers 1901–1940

Year	Number of labour immigration to Assam
1901	26,223
1902–1903	26,684
1903–1904	22,162
1904–1905	24,209
1905–1906	31,830
1906–1907	25,617
1907–1908	84,824
1908–1909	60,773
1909–1910	39,332
1910–1911	43,657
1911–1912	58,646
1912–1913	59,873
1913–1914	58,869
1914–1915	63,638
1915–1916	110,376
1916–1917	48,130
1917–1918	19,407
1918–1919	222,171
1919–1920	102,089
1920–1921	25,472
1921–1922	20,961
1922–1923	21,654
1923–1924	41,862
1924–1925	33,727
1925–1926	33,009
1926–1927	45,694
1927–1928	42,845
1928–1929	68,900
1929–1930	59,796
1930–1931	53,519
1931–1932	50,949
1932–1933	39,901
1933–1934	47,960
1934–1935	19,968
1935–1936	23,879
1936–1937	27,842
1937–1938	33,015
1938–1939	30,262
1939–1940	24,777

Source: Reports on Labour Immigration in Assam cited by Muhammad Abu B Siddique, Evolution of Land Grants and Labour Policy of Government The Growth of Tea Industry in Assam 1834–1940, p 161

The wasteland grants given on minimum ceilings to the tea planters made them the largest proprietors of lands in the region. With this, the era of 'Planter's Raj' in Assam was ensued. The tea economy based on external labour or coolie system led to the formation of a structure where the planter was the 'ruler' and the labourer was the 'ruled'. The planter assumed the status of a landlord within the gardens where he was the '*sahib*' of the area, the court of justice, and the provider of livelihood to the people (Hamilton, reprint 2003, pp. 120–121). The acquaintance of the planters with the lands, forests, and natural resources of the region during their exploration of wastelands for tea cultivation placed them in an advantageous position and contributed to furthering their 'Raj' over the local people. They preferred lands covered with trees and forests for tea cultivation as the absence of timbers would generally show that either the place was liable to flood, the soil was sandy, or the rainfall was deficient (Allen, 1905a, pp. 138–139). They were informed about the fauna found in the province as hunting was undertaken by them for both amusement and security purposes (Barker, 1884, p. 90). Exploration of lands in Assam was also undertaken by the tea planters for surveying timber types suitable for charcoal making required in tea factories and for producing tea chests in the sawmills (Brandis, 1879, p. 5). These factors situated the tea planters in Assam in a privileged position not only among the inhabitants of the region but also among the government employees in the area who were nascent to the territory and to the administrative system of the province. For instance, the British Forest Department in Assam initially functioned with information accumulated from the tea planters about local flora and fauna. Even the policies of the department were directed to benefit the tea enterprise in the region (Brandis, 1879, p. 5). The extent of 'Planter's Raj' over Assam could be assessed from the fact that even the Deputy Commissioners in their correspondence made with the Tea Estate Managers referred themselves as 'yours obedient servant' (Gupta, 2001).

The tea planters being the de-facto guardian of lands and forests in the province made endeavours to improve the road and transport systems of the region. Amelioration of road networks within Assam was considered necessary for the conveyance of tea within and outside the province. After the decline of the Ahom dynasty, roads in Assam fell into despair which impaired the movement of tea to the potential markets. Hence, road construction was taken up by the tea companies in the territory. The Assam Tea Company repaired the road between Nazira and Gabruparbut. Considering the commercial possibilities of the province, the British government in the region gave priority to the development of transport and communication. Some roads like the Assam trunk road were built under the joint initiatives of the government and the tea planters. Road construction in Assam received impetus under government road construction projects in association with the Local Boards (Griffiths, 1967, pp. 639 and 641). However, the infrastructural developments undertaken for the expansion of tea mostly remained restricted to the tea sectors only. The areas outside the tea networks of the region stayed underdeveloped. The beginning of railways in Assam is also attributed to the expansion of tea in the region. This aspect is discussed in the next section of the study.

3 The Beginning of Railways

With the expansion of the tea enterprise in Assam, railways came up in the region by the late nineteenth century. This was regarded as necessary for the growing tea industry in the area as transportation of tea to potential markets was a difficult proposition before the establishment of railways in the province. Previously, tea chests were carried in canoes down the small streams flowing into the river Brahmaputra during the rains. In winter, when these streams dried down with a lesser number of bullock carts available, tea chests were carried either by the coolies or by elephants. Even after when tea was brought to the Brahmaputra, it was cumbersome to transport them to Calcutta and other parts of Assam through country boats. A country boat proceeding to Calcutta from Assam went by the river Bhagirathi to Pabna, from there to Goalundo and from Siraj to Goalpara and Gauhati. The transportation process took around two and a half to three months. Later, the Indian General Navigation Steam Company ran steamers in the Ganges from Calcutta to Dibrugarh every 6 weeks. In 1863, the frequency of the services was increased to once a month (Griffiths, 1967, pp. 633–636). Under these circumstances, an urgent requirement was realised to develop a quicker and easier mode of communication for the movement of tea from Assam to larger markets. Shyam Bhadra Medhi has considered the tea industry as responsible for the improvement of communications in Assam. To him, the tea industry in the late nineteenth century was urgently looking for a suitable mode of transport for accessing the prospective markets (Medhi, 1978, pp. 131–132).

Though rail construction began in India in 1853, Assam remained outside the railway networks until the ninth decade of the nineteenth century when the nearest railway connection from Assam was as far as Goalando on the river Padma in South Eastern Bengal and Poradaha in Central Bengal. The line was extended up to Kushtia by 1862 that reached the River Tista by 1879. The line was expanded till Dhubri located on the bank of River Brahmaputra in 1881 (Griffiths, 1967, p. 642). The tea gardens of Dibrugarh and Jorhat sub-divisions were connected by steamer and boat services with the Padma and Tista rivers via the Brahmaputra (Gawthrop, 1951, pp. 9–10). The discovery of coal beds in the Makum area of Upper Assam also encouraged the British to extend railways in the region. The formation of the Assam Railways and Trading Company and the Dibru-Sadiya Railway initiated railway construction by 1881. By 1885, the Jorhat Provincial Railway was opened in the Jorhat sub-division to connect the neighbouring tea gardens to potential markets. In the same year, the Tezpur-Balipara Railway was constructed to connect the river Brahmaputra at Tezpur with the tea gardens around Balipara. These railway systems tied the tea gardens with the adjoining areas and the river ports (Goswami, 1963, pp. 183–186).

In the subsequent years, Assam was connected to other parts of India through a network of railways. In 1892, the Assam Bengal Railways interlinked different parts of the region. In 1903, the hill section from Lumding to Badarpur was accomplished with the establishment of railway communication between Chittagong and

Dibrugarh. In 1902, the Eastern Bengal Railway connected Calcutta with Dhubri. The line was extended up to Amingaon in 1903. By the 1920s onwards, the task of interlinking parts of Assam was more or less accomplished (Goswami, 1963, pp. 183–186). The investments for railway construction in Assam were financed by the ruling government, the Assam government, and the local government. However, paucity of finance still hindered railway construction works during 1882–1900. Hence, the government thought of paying larger attention to branch and feeder lines (Hilaly, 2016, p. 48). Under these circumstances, the tea companies assisted by the local governments came forward to promote tramways and short-distance railways to connect the tea gardens with markets and the ports. (Guha, 1991, pp. 192–193). The tea planters insisted on connecting the major railway networks with short-distance railways as even transporting tea to the main railway sectors appeared difficult for them. Dr. Berry White, a tea planter from Assam, had discussions with Sir Stewart Bayley, the Chief Commissioner of Eastern Bengal and Assam (1911–1912) on the problems experienced by the tea planters due to poor transport and communication in the province (Gawthrop, 1951, p. 11). In 1882, the tea planters from the region collected funds and with the financial assistance received from the provincial governments constructed light tramway projects to connect the gardens and the rivers at Kokilamukh. The line was later converted into the Jorhat-Kokilamukh State Railway. With the support of the tea planters and the Local Board, the Tezpur Balipara Tramway was built in 1892. The tram lines were initially not under the purview of the Railway Act of 1890 but were subsequently included under the act and designated as 'Light Railways' (Hilaly, 2016, p. 48).

The initiation of railways stimulated the economy of Assam remarkably. In 1897, after the Assam Bengal Railways opened up Badarpur and Chittagong, despatching tea to the Chittagong port became easier. In the year, 92,929 *maunds* (a weighing system in India considered equivalent to 37 kilogrammes) of tea were transported from Assam through the Chittagong port and were finally sent to places outside India by sea. There was a 7% increase in tea exports from Chittagong during the period. A substantial amount of tea from Jorhat, Tezpur, Upper Assam, and Cachar was transported for Calcutta port also. The Dibru-Sadiya Railway, Jorhat Provincial Railway, and the Tezpur-Balipara Railway carried tea to the steamers plying to Calcutta (Hilaly, 2001, pp. 188–189). Thus, the development of railways provided a fillip to the tea industry when the production of tea rose tremendously. The tea crop was highly valued with prices rising to more than three and a half million sterling. The total area of plantations amounted at the close of 1912 was 1,238,658 acres of which 29% was under tea. According to the trade statistics of 1912, 15,341 *maunds* of tea seeds were exported from Assam to Calcutta (Sadir, 1913, p. 1028) (Table 4).

Along with this, large-scale labour importation in Assam required in the plantations became convenient with the expansion of railways in the province. Prior to this, labour was usually transported through boats and steamers which often lacked adequate sanitary arrangements resulting in outbreaks of epidemics like cholera among the coolies (Assam Administration Report, 1899). In addition to this, insufficient food supply, lack of cooking arrangements, and want of proper toilet facilities increased the mortality rate among them. Such situations made the government

Table 4 Growth of tea trade from Chittagong Port with expansion of railways

Years	Values (in rupees)
1901–1902	55.59 lakhs
1902–1903	75.01 lakhs
1903–1904	125.10 lakhs
1904–1905	145.10 lakhs
1905–1906	155.79 lakhs

Source: Eastern Bengal and Assam Administration Report 1905–1906, p 151 cited by Sarah Hilaly, pp. 188–189

suspect the prospect of tea plantation in Assam (Chatterjee & Das Gupta, 1981, pp. 1862–1863). Under these circumstances, the conveyance of labourers to the tea gardens in healthy conditions was considered as a priority by the government. Transportation of labour through the railways was thought to be more hygienic and easier. In 1897–1898, the Assam Bengal Railway connecting Chittagong and Silchar was opened for coolie traffic (Assam Administration Report, 1899). Subsequently, the rail route from Goalundo and Eastern Bengal Railway via Chandpur to Karimganj was also opened for coolie transportation. With this, the mortality rate among the labour population reduced considerably (Assam Administration Report, 1901). With decline in coolie mortality, labour population increased in Assam and subsequently labour unions came up in the region that also included the worker classes engaged in Assam Bengal Railway and Assam Steamer Service. The labour unions cooperated with each other in raising their demands (Sharma, 2002, p. 625). However, despite this, scarcity of manual labour continued to hinder the infrastructural development of the province. Railways, tea companies, and government administrative branches operational in the region such as the Forest Department and Public Works Department experienced labour paucity in executing their assignments. Consequently, competition arose among these colonial enterprises for procurement and engagement of labour that estranged their mutual relationship on several occasions (Dutta, 2002, pp. 62–90) (Home, 1896, p. 5). Immigration to Assam was also largely boosted with the expansion of railways in the region. Around 70,000 immigrants from Dacca, Mymensingh, Pabna, and Rangpur were registered to have entered the province between 1905 and 1915 with the expansion of railways (Hilaly, 2016, p. 238).

The distance travelled by railways in Assam increased from 114 miles in 1891 to 715 miles in 1903. The upsurge in distance covered by railways required a stable supply of fuel which was fulfilled by the coal miles available in the region (Baruah, 2011, p. 65). Thus, the expansion of railways provided an impetus to coal mining, production, and trade in the province. The Ledo mines became operational in early 1883. During 1885–1886, 191,701 *maunds* of coal were exported from Assam to Calcutta, Goalundo, and Sirajganj. The Assam Railways and Trading Company took initiatives to open up the Namdang and Tirap mines. Coal beds in Diphu, Longai Hills, Lumding, and Naga Hills were also explored. With these developments, geological surveys in Assam received a boost, and transportation of resources like petroleum became easier. The production at the Digboi refinery rose from

8,82,578 gallons in 1900–1901 to 1,756,759 gallons in 1902–1903. The Dibru-Sadiya Railways and Assam Bengal Railways transported mineral oils from within and outside the province. Production of the Burma Oil Company also increased with the railway expansion (Hilaly, 2001, pp. 172–185).

Timber trade in Assam was enormously promoted by the introduction of railways in the region. Forests which were previously considered inaccessible by the timber traders were opened up for exploitation. B. C Allen mentioned that with the erection of railway lines in the North Cachar Hills district of Assam, traders from Surma Valley could visit the forests in the Nowgong area of the region which were previously inaccessible to them (Allen, 1905a, b, pp. 13–140). The phenomenon was similar in other parts of the region. Now new timber species like *Nahor* (*Mesuaferrea*), *Hollong* (*Dipterocarpuspilosus*), *Hollock* (*Terminaliamyriacarpa*), and *Uriam* (*Bischofiajavanica*) acquired importance as railway sleepers. Previously '*Sal*' (*Shorea Robusta*) was considered as the most suitable timber for manufacturing railway sleepers (Mackarness, 1932, p. 3). The development of railways ensued an increasing demand for the survey of new timber species and an urge to bring them under forest reservation plans of the government. Hence fresh proposals were made to include hitherto unexplored areas like the Langting-Mahur forests in North Cachar Hills, Borail forests in Cachar, and Borpani Valley forests in the Nowgong divisions under government forest reservation plans (Home, 1896, p. 3). Schemes were even framed to reserve wastelands containing economically viable timber species (Cavendish, 1910, p. 2) (Bor, 1941, p. 4). The establishment of tramways could bring timbers from interior forests to the railways which communicated them to prospective markets (Smith & Purukayastha, 1946, pp. 60–61) (Table 5).

4 Markers of 'Development'

In this context, the term 'development' refers to the opening of Assam to the rest of India. The region that was hitherto considered inaccessible was opened up with the establishment of the tea industry and the consequent expansion of railways in the area. With this, other cumulative developments like the exploration of mineral and natural resources, expansion of government infrastructure, and adoption of medical facilities encouraged repopulation of Assam that suffered from population deficit until then. There was the establishment of colonial political infrastructure, expansion of agriculture, and consequent rise in revenue from the province. These appeared as markers of 'development' for the British government in India. The requirements associated with the expansion of the tea industry in the region rendered it necessary for the government to bring more lands under its control. As the Charter of 1833 allowed the English East India Company to hold lands on long-term leases outside the presidency towns and had given freehold rights, land acquisition became easier for the British in India, and the colonial government utilised the opportunity to its optimum (Guha, 1988, p. 11).

Table 5 Statement showing the working of tramway between 1924 and 1925 and 1944 and 1945 in procuring timber logs

	Quantity of logs	Revenue	Maintenance charge	Surplus or deficit
Year	Cubic feet	Rs	Rs	Rs
1924–1925	124,154	82,215	31,436	50,779
1925–1926	261,165	147,277	57,768	89,509
1926–1927	392,336	234,066	101,109	132,957
1927–1928	254,256	299,411	119,448	179,963
1928–1929	368,168	193,643	88,984	104,659
1929–1930	344,962	91,348	154,774	63,426
1930–1931	372,764	58,146	93,595	35,449
1931–1932	199,879	117,644	87,655	29,989
1932–1933	60,378	19,645	31,375	11,730
1933–1934	123,366	45,272	36,661	8611
1934–1935	122,462	41,371	31,282	10,089
1935–1936	116,144	47,365	26,573	20,792
1936–1937	134,163	35,739	24,717	11,022
1937–1938	228,511	36,346	25,726	10,620
1938–1939	234,900	55,712	30,007	25,705
1939–1940	165,971	44,123	36,294	7829
1940–1941	173,460	50,564	35,064	15,500
1941–1942	96,650	32,404	28,924	3480
1942–1943	162,996	42,766	35,027	7759
1943–1944	153,757	31,958	43,009	−11,051
1944–1945	250,752	51,578	100,089	−48,511
Total	4,341,195	1,758,593	1,219,517	539,076

Source: Smith and Purukayastha (1946)

Colonial control over lands as the key to development in Assam was initially suggested by Colonel Jenkins, the Agent to the Governor General in the North East Frontier of Bengal (1831–1864). As the region had abundant unused lands termed as wastelands by the British government, the latter came up with the idea of wasteland grants under which these lands could be brought under governmental control. Jenkins considered this as a better provision for the speedy improvement of Assam (Guha, 1988, pp. 11–12). The government accordingly promulgated policies that made the conversion of indigenous lands into government possession an effortless process. Lord Canning in 1861 published the Free Simple Rules where a provision was included under which outright sale of lands to the planters at extremely low rates could be done (Chatterjee & Das Gupta, 1981, pp. 1863–1868). The wasteland grants granted on long-term leases to the Europeans made retention of lands in Assam a long-term possibility for the British government in the region. These lands were converted into tea plantations which the colonial authorities viewed as a marker of development.

Economic profits produced from the tea industry increased the importance of Assam tremendously among the Europeans and to the ruling government. After the initial issues such as the acquisition of lands, establishment of gardens, management, and recruitment of labour were accomplished, the subsequent years showed a remarkable increase in revenue generation from tea. From 1869 to the next 30 years, the tea sector in Assam made financial profits. The revenue generated from tea in 1881 was £ 37,000,000; in 1891 it was £ 90,000,000; in 1900, the amount rose to £ 141,000,000; and in 1910, the revenue generated was £175,095,069. In the local markets of Lakhimpur and Cachar, the price of tea rose from ₹30 to ₹150 and ₹80 to ₹120 per *maund* (unit of weight equivalent to about 37 kg). There was an area-wise expansion of tea in the region. Areas under tea in Assam increased from 354,716 acres in 1911 to 361,371 acres in 1912. The tea plantations provided employment to the Europeans, imported labour, and also to some local population in the region. Hiring of external labour increased from 107,847 in 1885 to 247,760 in 1900 (Sadir, 1913, pp. 1022–1023). By the 1940s the Assam Valley tea plantations employed about half a million labourers (Behal, 2006, p. 143). Women and children consisted of about one-third of the labour force in 1877 to half of the total workforce by 1930. This was because children and women extended their labour at basic wages. The children received half of the dearness allowance paid to the adults (Chakravarty, 2014, p. 478). Generation of larger revenue with lesser investments was the priority. With the production of higher revenue and the desire to conquer larger markets, the Assam Tea Company was formed in London. Experimental plantations also came up and Assam tea was exported to British markets (Verma, 2017).

With enormous profits earned by the tea companies and huge lands under possession, the tea planters emerged as the highest landlords in Assam. The tea companies, of which they were part, worked under a power structure that operated under two tiers. In the upper tier, the tea companies had their head office in England. In India, the Indian Tea Association supervised the companies within the country. The Indian Tea Association functioned as a lobby for the tea industry, supervised and implemented its policies associated with production, organisation, and labour in the tea sector. In the twentieth century, branches of the association started operating as pressure groups that impacted government policies associated with tea trade in the province. The planters participated in the local committees and decided on matters related to the development of transport and communication and state infrastructure within Assam. In 1909, after the Morley Minto Reforms was passed, the Indian Tea Association with the assistance of the Bengal Chamber of Commerce allocated two seats to the interests of tea in the Council (Behal, 2006, pp. 144–171). With this, the tea planters in Assam started influencing political decision-making in the province. In November 1937, Omeo Kumar Das, the Congress MLA (Member of Legislative Assembly) from the Assam Valley by contesting the tea planters, informed the Secretary of the Assam Legislative Assembly about his desire to move the Assam Tea Garden Labourer's Freedom of Movement Bill. He was also the mode of interaction with the tea authorities including representatives of the Indian Tea Association

Table 6 Ownership of tea estates in Assam 1900–1940

Areas under tea estates by ownership (in acres)			
Year	Under Europeans	Under Indians	Total
1900	327,333	9994	337,327
1905	322,357	16,849	339,206
1930	380,659	53,150	433,809
1935	373,461	62,092	435,553
1940	383,350	55,709	439,059

Source: Report on Tea Culture in India, relevant issues, cited by Muhammad Abu B Siddique, Evolution of Land Grants and Labour Policy of Government The Growth of Tea Industry in Assam 1834–1940, p 109

(Behal, 2006, pp. 144–171). In the sphere of development, some indigenous tea planters contributed significantly (Saikia, 2000). One of them was Bisturam Barooah who made efforts to improve the infrastructure of Jorhat. He established Namghars or prayer houses of congregational worship under the Ekasarana sect of Hinduism in Assam, constructed roads, and dug ponds for the well-being of the town. Another Assamese tea planter from the region named Siva Prasad Barooah also constructed the Vaishnavite Namghar, Bisturam Barooah Hall, Bisturam Barooah Hospital and established Jorhat College under his initiative. Thus, the indigenous tea planters after being associated with the tea industry in Assam took measures to develop the region (Hiwale, 2019, p. 526) (Table 6).

The labour population in the tea gardens had to experience indentured slavery which took a toll on their health that was evident from the high mortality and desertion rates among the labour population in the region. Studies suggest that out of 84,915 labourers brought to Assam between May 1863 and May 1866, only 49,750 survived, and the rest either died or fled away (Chemmalakuzhy, 2016). The labour population also suffered from diseases like kala-azar, beriberi, and malaria. Malaria was so common among the coolies between 1877 and 1920 that around one-tenth of the coolie deaths were caused by it (Saikia, 2014, p. 74). In 1898, an investigation on kala-azar was undertaken by G.M Giles, a surgeon in the Indian Medical Service who was on special duty in Assam. Another survey on kala-azar was conducted by Major Ronald Ross in 1899 who considered it as similar to malaria (Bihar, 2019, p. 634). These investigations undertaken did not make the colonial government adopt better medical facilities in the tea gardens; rather, settlement changes in the coolie lines were thought to be the suitable solution. However, despite the above, a significant beginning in medical education came up in Assam under the initiatives of the tea planters. Medical education in the province started with the contributions of Brigadier Surgeon John Berry who was a tea planter as mentioned earlier. The Berry White Medical School established by him is considered as the pioneer institution for the dissemination of medical education in North East India. Berry White donated a sum of 50,000 rupees for the establishment of a medical school at Dibrugarh which was the industrial hub of Assam at that time. The medical school came up in Dibrugarh in 1900 (Wary, 2020, p. 4).

Transport and communication systems in Assam received an enormous boost with the establishment of tea plantations in the region. The advancements thus produced were viewed as markers of development by the colonial authorities. Transportation of tea from Dibrugarh, Jorhat sub-divisions, and other areas of the region became easier with the coming of railways (Goswami, 1951, pp. 9–10). With the arrival of the railways, the excavation of mineral resources like coal and petroleum received a fillip. The exploitation of forest resources also received an impetus. Along with these, trade and commerce were developed too. Under the initiatives of the tea companies, internal communication within Assam was expanded. The tea companies in collaboration with the district Superintendents constructed roads that connected the gardens with the district towns, commercial centres, and areas from where transportation could be availed. The district administration with the financial support provided by the tea planters could erect bridges in various parts of Assam to expedite the transportation of tea to the commercial centres. In the Cachar district of the region, 15 roads were constructed for the purpose (Bhattacharjee, 1977, pp. 425–426).

5 Conclusion

Thus, Assam was re-opened after years of turmoil subsequent to the decline of the Ahom dynasty and Burmese incursions when indigenous tea plants were found in the region, and the development of railways took place in the territory. The province hitherto considered inaccessible, disease-prone, and backward by the British began to be perceived as a domain with immense commercial potential. The establishment of the tea industry in the region stimulated explorations of natural and mineral resources. The British government after being informed about the commercial possibilities of Assam attached political relevance to the region. The strategic significance of the province was comprehended, and measures to establish governmental infrastructure were undertaken. The initiation of railways in the region that was originally commenced to meet the requirements of the tea industry provided boost to timber trade and excavation of coal and petroleum and carried these resources to the potential markets. Hence, the economy of Assam received an impetus. The internal transportation among the gardens, the district towns and the commercial centres improved the communication systems within the province. There was the emergence of an industrial class (tea planters) among the indigenous population. This class though small in number made initiatives to develop places within Assam by providing infrastructural facilities like hospitals and colleges. The local politicians forwarded the demands of the tea labour unions to the legislature which strengthened the position of Assam within the political map of India. The plantations provided employment to a large section of the population including the Europeans, imported labour, and some local people. These produced changes in the region's demography, land uses, and human settlement patterns. Though the treatment meted out to the labour population in the gardens exhibits the tea industry in Assam as another colonial agency of capitalist commercial exploitation as in other parts of India, yet its contribution in ushering infrastructural advancements to Assam cannot be overlooked. The measures undertaken to fulfil the demands of the tea industry modernised and promoted the progress of the region tremendously. With these developments, the landscape of colonial Assam was created that fulfilled the political and economic aspirations of the British government in India.

References

Allen, B. C. (1905a). *Assam District gazetteers* (Vol. II). Caledonian Steam Printing Works.

Allen, B. C. (1905b). *Assam District gazetteers* (Vol. VI). City Press.

Barker, G. M. (1884). *A planter's life in Assam*. Thacker Sprink and Company.

Barpujari, H. K. (1963). *Assam in the days of the company 1826–1858.*

Barpujari, H. K. (1986). *The American missionaries and North East India 1836–1900*. Spectrum Publications.

Barua, K. (1980, June). 'Foreigners' in Assam and the Assamese middle class. *Social Scientist, 8*(11), 49.

Baruah, D. M. (2011, January). The refinery movement in Assam. *Economic and Political Weekly, XLVI*(1), 65.

Behal, R. P. (2006). Power structure, discipline and labour in Assam tea plantations under colonial rule. *International Review of Social History, 51*(Suppl 14), 143.

Bhattacharjee, J. B. (1977). *Cachar under British rule in North East India*. Radiant Publishers.

Bhuyan, S. K. (1949a). *Anglo-Assamese relations 1771–1826*. Department of Historical and Antiquarian Studies in Assam.

Bhuyan, S. K. (1949b). *Early British relations with Assam*. Assam Government Press.

Bihar, M. K. (2018–2019). Life threatened: Revisiting Assam fever during colonial period. *Indian History Congress Proceedings*, 79th Session, p. 634.

Bor, N. L. (1941). *Working plan for the forests of the Goalpara Division, Western Circle, Assam, 1929–30 to 1938–39*. Assam Government Press.

Brandis, D. (1879). *Suggestions regarding Forest Administration in Assam*. Office of the Superintendent of Government Printing.

Cavendish, F. H. (1910). *Working plan for the Jokai Reserve of the Lakhimpur Division, Eastern Circle, Eastern Bengal and Assam*. Eastern Bengal and Assam Secretariat Printing Office.

Chakravarty, S. (2014). Children employed in the tea plantations of Assam 1880–1930. *Proceedings of Indian History Congress, 75*, 478.

Chatterjee, S., & Das Gupta, R. (1981, November). Tea labour in Assam recruitment and government 1840–80. *Economic and Political Weekly, 16*(44/46), 1862–1863.

Chemmalakuzhy, R. S. (2016). *How Assam's Tea Garden's flourished on abuse and torture of their labourers*. https://www.youthkiawaaz.com/2016/05/slavery-in-assam-tea-gardens-india/. Accessed on 12 July 2023.

Cosh, J. M. (1st published 1837, reprint 1986). *Topography of Assam*. Logos Press.

Dey, A. (2019). Planting 'improvement' tea in British India. *Austrian Journal of Historical Studies, 30*(3), 69.

Dutta, A. K. (2002). *The Final Frontier*. North East Frontier Railway.

Dutta, A. K.. (2007). *The great Assam tea planter martyr Maniram Dewan*.

Gawthrop, W. R. (1951). *The story of the Assam railways and the trading company limited 1881–1951*. Harley Publishing Company Limited.

Goswami, P. C. (1963). *The economic development of Assam*. Asia Publishing House.

Griffiths, P. (1967). *The history of the Indian tea industry*. Weidenfeld and Nicolson.

Guha, A. (1991). *Medieval and early colonial Assam society, polity, economy*. K.P Bagchi and Company.
Guha, A. (1988). *Planter Raj to Swaraj, freedom struggle and electoral politics in Assam (1826–1947)*. People's Publishing House.
Gupta, B. (2001). *History of Cachar tea industry-as I know* (pp. 1901–2001). Surma Valley Branch Indian Tea Association Centenary Souvenir.
Hamilton, A. (1st published 1912 reprint 2003). *In Abor jungles of north East India*. Mittal Publications.
Hilaly, S. (2001). *Development of railways in Assam, and its socio-economic impact 1885–1947*. unpublished PhD Dissertation, Rajiv Gandhi University.
Hilaly, S. (2016, January). Railways and colonialism: The Kachugaon Forest tramway in Assam. *Global Journal of Indian History and Culture, 1*(1).
Hiwale, H. A. (2019). Re-reading the British Indian economy through the Assam tea industry. *Infokara Research, 8*(9), 526.
Home, A. L. (1896). *Progress report of Forest Administration in Assam for 1895–96*. Shillong.
Hunter W. W. (1879). *A statistical account of Assam* (vol. I).
Mackarness, C. G. M. (1932). *Working plan for eleven Forest reserves in the Lakhimpur and Sibsagar Divisions, Eastern Circle Assam, 1931–32 to 1940–41*. Assam Government Press.
Medhi, S. B. (1978). *Transport system and economic development in Assam*. Assam Publication Board.
Mills, A. J. M. (1854). *Report on the province of Assam*. Calcutta Gazette Office.
Misra, T. (1980, 9 August). Assam: A colonial hinterland. *Economic and Political Weekly, 15*(32), 1358.
Nag, S. (2003). Fei-isation of the Nepalis of North East India. In A. C. Sinha & T. B. Subba (Eds.), *The Nepalis in North East India a community in search of Indian identity*. Indus Publishing House.
Powell, B. (1882). *A manual of jurisprudence for Forest Officers: Being a treatise on the forest law and those branches of the General, Civil and Criminal Law which are connected with Forest Administration; with a comparative notice of the Chief Continental Laws*. Superintendent of Government Printing.
Report on the Administration of the Province of Assam for the year 1897–98. (1899). Assam Secretariat Printing Office.
Report on the Administration of the Province of Assam for the year 1899–1900. (1901). Assam Secretariat Printing Office.
Ribbentrop, B.. (reprint 1989). *Forestry in British India*. Indus Publishing House.
Sadir, M. (1913, 10 October). The tea industry in Assam. *Journal of the Royal Society of Arts, 61*(3177), 1028.
Saikia, R. (2000). *Social and economic history of Assam 1853–1951*. Manohar Publishers.
Saikia, A. (2011, 19 March). Imperialism, geology and petroleum: history of oil in Colonial Assam. *Economic and Political Weekly, XLVI*(12), 48–50.
Saikia, A. (2014). Mosquitoes, malaria and malnutrition: The making of Assam tea plantations. In *RCC perspectives, number 3, Asian environment, connections across borders, landscapes and times*.
Sharma, K. M. (1980, 2 August). The Assam question: A historical perspective. *Economic and Political Weekly, 15*(31), 1321–1322.
Sharma, B. K. (2002). Struggle of tea plantation coolies and the Assam Bengal railway strike 1921. *Proceedings of the Indian History Congress, 63*, 625.
Sharma, J. (2009, November). 'Lazy' natives coolie labour, and the Assam tea industry. *Modern Asian Studies, 43*(6), 1297–1307.
Sharma, J. (2011). *Empire's garden and the making of India*. Ranikhet.
Siddique, M. A. B. (1990). *Evolutions of land grants and labour policy of growth the growth of the tea industry in Assam 1834–1940*. South Asian Publishers.
Smith, H. P., & Purukayastha, C. (1946). *A Short History of Assam Forest Service* (pp. 60–61).
The Assam Land and Revenue Regulation 1886 published in the Gazette of India, 13th March 1886 and in the Assam Gazette on 3rd April 1886.

Verma, N. (2017). Tea in the colony, Assam Tea and the making of coolie labour. In *Coolies of Capitalism*. De Gruyter.
Wary, J. (2020, October–December). Introduction of Western medicine in Assam during the colonial period. *Shodh Sanchar Bulletin, 10*(40), 4.
Weiner, M. (1983, June). The political demography of Assam's anti- immigration movement. *Population and Development Review, 19*(2), 283.

Srijani Bhattacharjee works as Assistant Professor in the Department of History, Amity Institute of Social Sciences, Amity University, Kolkata. Her areas of research concentrate on Environmental history of North East India, tribal and indigenous studies of the region, colonial and post-colonial history of North East India.

India-Bangladesh Border Dispute: An Indian Perspective

Subhankar Roy, Gouri Sankar Bhunia, and Pravat Kumar Shit

1 Introduction

According to Van Schendel and De Maaker (2014), border studies have not gained much traction in Asia, despite their significance in the region's highly multi-ethnic and multi-lingual social ecosystems. Borderlands are extremely contentious, brittle, porous, and permeable everywhere, including in South Asia (Majaw, 2021). Due to the border region's dense population and the fact that both sides of the border cultivate their land to the very last inch, conflicts are especially high. The transboundary operations involving Bangladesh and India can be categorized into several principal themes: border killings, fence operations, illegal immigration, border enclaves, and trade across borders.

India confronts many of its neighbors with territorial problems. In the previous 70 years, Bangladesh and Sri Lanka alone have been able to overcome their border concerns. Tensions have frequently arisen over the unmarked border lines with Myanmar, Bhutan, and, most recently, China, Pakistan, and Nepal (Shahriar, 2021). Since British colonialists split Pakistan from India in 1947 and gave both countries sovereignty, the border issue has continued. The land borders between India and Bangladesh are about 4100 km (Leake & Haines, 2017). The 1947 Radcliffe Award as the land border between India and East Pakistan defined this border; however, there were rapid disagreements over key elements (Chatterji, 1999; International, 2019). To resolve the unresolved disputes, following Bangladesh's independence,

S. Roy
Department of Geography, Seacom Skills University, Santiniketan, West Bengal, India

G. S. Bhunia (✉)
Researcher, Paschim Medinipur, West Bengal, India

P. K. Shit
PG Department of Geography, Raja N. L. Khan Women's College (Autonomous), Vidyasagar University, Medinipur, West Bengal, India

A. Rai et al. (eds.), *Contemporary Social Physics*, Springer Geography,
https://doi.org/10.1007/978-3-031-77596-3_7

India and Bangladesh signed the 1974 Land Boundary Agreement ("1974 LBA"). An additional protocol updated the 1974-LBA ("2011 Protocol" and the "Land Boundary Agreement" for LBA of 1974) in 2011. In 1971, after a violent nine-month struggle, Bangladesh earned independence from Pakistan. Boundary conflicts between neighboring territories (whether on land or maritime boundaries) can generate considerable risks and challenges for foreign investors to operate in or invest in the areas at issue (Schultz, 2015).

Bangladesh and India possess historic and cultural links. Bangladesh is the greatest development partner for India (Singh, 2014; Shewly, 2017). Nevertheless, it has one of South Asia's quickest-growing economies and is ideally positioned with India's North-East connectivity. They disagree about the Citizenship Amendment Act of India, and the water problem of Teesta is still to be solved (Ashraf, 2020). Both countries were able to resolve severe border problems successfully. A maritime border controversy was resolved in 2014, and the India-Bangladesh Land Boundary Agreement was confirmed the following year by the Indian Assembly (Kashem & Islam, 2016). Based on the above discussion, the present study is undertaken to discuss cross-border challenges between Bangladesh and India and specifically addresses cross-border interactions between Bangladesh and India in the area of comparative borders.

2 Methods

Data from the study were gathered with a range of online tools such as Scopus, Web of Science (WoS), and Google Scholar database. For the period from 1980 to 2020, Sci Verse Scopus has been used to retrieve literature on border conflicts. The necessary papers were retrieved using nine distinct search situations with varying keyword combinations. Every type of paper, including editorials, published in peer-reviewed journals was considered. Keywords used in the search strategy comprised phrases such as "border conflict," "border dispute," "Trans-Boundary," "Policy& approach," "International relation," "Border management," "Issues and Conflict," "geopolitical strategy," and "Border crisis" in the title or abstract. Both quotation marks and asterisks are used in Scopus and WoS to improve the accuracy and integrity of results. The search approach can be accurate in all these functions, once used. Nevertheless, there are always opportunities for misleading positive outcomes. A search method for the elimination of false-positive documents has therefore been completed by an exclusion step. The elimination process comprised terms like the International Border, Indo-China, and India-Pakistan.

3 Border Dispute and Impact

A total of 65 papers have been collected. Author keyword visualization has demonstrated that the most regularly encountered terms are "border conflict," "Trans-Boundary," "Policy& approach," "Issues and Conflict," and "Border crisis." *Homeland Security*, *Strategic Analysis* and *Int. J. Agril. Res. Innovation & Technology* are the journals most active in this sector. The maximum number of publications was recorded in 2018, followed by 2019 and 2014. Researchers in the Indian area have been involved in around 75% of the retrieved documents, whereas Bangladeshi researchers have made the least contribution (25%).

3.1 Territorial Disputes

India also established the land and sea border with Bangladesh by settling the issue of the Ganga-Brahmaputra delta region, the south of Talpatti Island, and Bangabandhu Island, a small and deserted land sandbar on the Bay of Bengal. The border between India and Bangladesh is the sixth biggest border of land in the world, around 4096.7 km long (Jamwal, 2004). The India-Bangladesh border was regarded by New Delhi as a security problem and tried to secure the Bangladesh border in terms of the porous nature and access to the frontier for activist groups and traffickers (Kashem & Islam, 2016). The borders are 4096 km long in India and Bangladesh. Indo-Bangladesh is bordered on the Indian side by West Bengal (2216.7 km), Assam (263 km), Meghalaya (443 km), Tripura (856 km), and Mizoram (318 km). The whole area includes lowlands, riverbanks, hills, and jungles (Fig. 1). The area is very crowded and cultivated right up to the boundary (Table 1). The river is around 180 km away. The Indo-Bangladesh border is not complete because it is still 6.5 km near Comilla on the Tripura border (Dutta, 2010). There is a disagreement over several areas of the border between the two countries. As a result, unprovoked gunfire near the border persists and sometimes intensifies to attract more administrative attention. In the Indians surrounding the Meghalaya state and Timbli district of Sylhet district in Bangladesh, border clashes took place around the village of Padua, also known as Pyrdiwah (Majaw, 2021). Since 1971, India has been found to have illegally occupied the region. A great deal of attention was given to the disputed area by the Indian military attempting to build a trail in Bangladesh, approximately 300 m wide to Meghalaya from an armed station in Padua. In the three sectors (in the Indian States of West Bengal, Tripura, and Assam, totaling 6.1 km), the Land Boundary Agreement also established the boundary between India and Bangladesh.

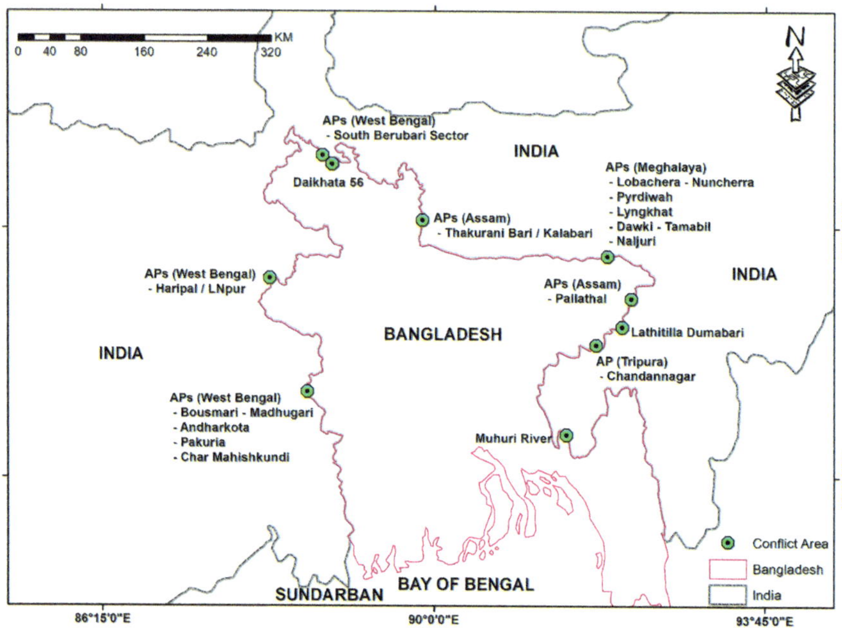

Fig. 1 Conflict zone of India-Bangladesh border

Table 1 The details of border fencing along the Indo-Bangladesh border

State	Total length of international border	International border covered with fence	Balance length (K.M)
West Bengal	2216.7	1638.047	578.653
Assam	263	210.017	52.983
Meghalaya	443	325.773	117.227
Mizoram	318	155.077	162.923
Tripura	856	794.373	61.627
Total	4096.7	3123.287	973.413

Source: Ministry of Home Affairs, Government of India, 2021

3.2 *Enclaves*

An enclave is the divided territory of a sovereign authority within a sovereign state. Built in the eighteenth century in the Indian and Bangladeshi enclaves, war and peace treaties were concluded between the kings in Bengal and Cooch Behar (Mishu, 2018). The north of Bengal has shown essentially in former India to be a gateway to the remainder of Bengal and a border zone between the Hindu, Muslim, and Buddhist realms. Although local tradition tells that the enclaves were the outcome of a chess game between Cooch Behar's Maharaja and a Mughal commander with villages as a wager, in fact, they are a result of a series of peace treaties between Cooch Behar's feudal Kingdom and its Empire inked between 1711 and 1713 (Alam, 2021). The Mughals seized a large range of Cook Behar and were not able to rob any of the Chieftains of Cook Behar in places surrounding the Mughal realm. In 1947, partitioning procedures reduced princely states' independence, including Cooch Behar, to allow them to decide whether to join India or Pakistan. The Cooch Behar Merger Agreement was signed by Cooch Behar in August 1949, one of the last countries to pick their choice in India. The British separated 80 million people and 175,000 square miles in 1947 within a period of 36 days and formed an alliance for nearly 1000 years in different ways. There were several problems disregarded by the fast and ambitious process of dividing the Bengal border by close to 4000 km, including the resolution of access to the enclaves (Cons, 2014). Under the 1950 agreement that allowed government officials access to their side's enclaves, the only step was taken to bring the enclaves together with their mainland (van Schendel, 2002). India granted 110 of 111 enclaves to Bangladesh in the context of the Land Boundary Agreement in 1974 and its 2011 Protocol and received 51 enclaves in the course of July 31 at midnight in 2015 (Table 2). The residents were entitled to remain nationals of the State into which the areas were transferred. The transfer of the encounters, as with unfavorable territories, constitutes legal recognition, de jure, both by India and Bangladesh of the de facto status on the ground.

In 1947, both India and East Pakistan (after Bangladesh's independence) kept enclaves that total around 119 square kilometers within the newly delimited boundaries of each other, after being de-colonized from the British Empire (Roy, 2012). The entire number of enclave dwellers, about 38,000 in Bangladesh and 15,000 in India, was projected to be 53,000 during the exchange. Over time, each country sometimes required full access to their enclaves on the territory of the other but did not want reciprocal access in response. On August 1, 2015, Bangladesh and India legally replaced 162 enclaves by an anomaly that had lasted for centuries and which led to land and population exchange processes starting in the 50's (Banerjee et al., 2017). Consequently, none of the countries have seriously tried to increase governance and build infrastructure in the enclosures in the territory of each other, and in many cases, they have forgotten their population and the victims of bilateral enmity.

3.3 *Cross-border Resource*

The problems with the allocation of water resources between India and Bangladesh may be traced to the time when Bangladesh was still East Pakistan. In particular, 54 rivers, many of which are in origin and flow into the last, are in India and Bangladesh (Kawser & Abdus Samad, 2016). Of course, water is the key resource in all forms of life, and its fair and effective management should therefore become the practice. The Farakka dam—operational until April 1975—was built in 1961 by India to disrupt a part of the dry season and enhance the navigability of the port of Calcutta (Mukherjee, 2011). The development of the port, however, was short-sighted and

Table 2 India's trade balance and export intensity with Bangladesh in the top ten (10) exports

Product label	Trade balance between India and Bangladesh (value in US $)			India's exports to Bangladesh (value in US$)		
	Balance in value in 2018	Balance in value in 2019	Balance in value in 2020	Balance in value in 2018	Balance in value in 2019	Balance in value in 2020
Cotton	2,005,866	1,511,962	1,665,525	2,026,679	1,548,715	1,696,854
Vehicles other than railway or tramway rolling stock and parts and accessories thereof	1,154,078	894,359	585,008	1,159,157	896,841	588,475
Mineral fuels, mineral oils, and products of their distillation, bituminous substances, mineral, etc.	878,858	908,724	855,405	878,921	910,915	855,889
Machinery, mechanical appliances, nuclear reactors, and boilers; parts thereof	640,875	819,499	525,838	647,351	829,404	536,332
Cereals	403,165	45,418	351,693	403,165	45,418	351,693
Electrical machinery and equipment and parts thereof; sound recorders and reproducers, television, etc.	290,883	300,402	258,941	297,641	306,417	264,403
Iron and steel	280,097	333,166	217,999	302,485	359,526	240,225
Plastics and articles thereof	263,116	216,860	226,190	296,902	245,542	236,619
Organic chemicals	244,531	280,986	285,489	246,807	284,014	289,369
Tanning or dyeing extracts, tannins and their derivatives, dyes, pigments, and other coloring	213,583	226,173	219,325	213,689	226,232	219,349

Source: ITC, Trade statistics for international business development, https://www.trademap.org/

took no notice of environmental considerations—changes in the flow of rivers and silt. This required a long-term solution in the form of a barrage. In 1975, India fully energized the dam (Wolf & Newton, 2014). The consequences of lower flows in Bangladesh include surface and groundwater degradation, damaged fisheries, increased salinity, and polluted/declined water sources, which have affected public health (Wolf, 1998). The contradictions in various versions of the EIA prompted unrest between officials and individuals in local communities facing displacement opportunities. According to the official EIA of 2007, this area will comprise reserved forests, unclassified state forests, rocky outcrops, farms, wet rice, gardens and orchards, wastelands, human settlements, and others (Rahaman, 2006). India could begin a new stage of increasing major projects in the North East to implement blocked projects. One of the biggest is Tipaimukh. Even before the FAC's local complaints and rejection, some small buildings began around 2008. There is a chance, though, that India could start to represent Tipaimukh shortly with similar stalled projects, such as Dibang (Thakur, 2020).

In the last decade, under the wide framework of the Indian "Act East" policies of the government, connection agreements with Bangladesh and rapid infrastructure expenditures in North East Region (NER) and its neighbors have decreased the economic insulation of the sub-region. This can assist NER in demonstrating its strengths in agriculture and services together. NER may leverage on increasing demand for fresh, high-quality fruits and vegetables and, in particular, spices produced in a socially responsible and ecological way (Sacks et al., 2015). It can support bamboo, a resource that respects the environment. It can stimulate natural tourism and trade in services like medicine and education. In addition, Bangladeshi firms will have growth opportunities by access to NER markets both for input supplies and for the sale of finished products. They will be provided by the investment in goods and services that may lead to the creation of regional value chains and the prospect of learning more demanding markets.

3.4 *Trade Conflict*

Commercial relations have adversely affected bilateral relations. India has been one of India's greatest business partners because of its geographical proximity to Bangladesh. Bangladesh has urgent concerns about the significant bilateral trade deficit with India and the large volumes of informal imports across the country that escape Bangladeshi import tariffs. The bilateral trade imbalance in Bangladesh with India has risen fast on average by almost 9.5% every year. However, the gap shrank for the first time in the 2005/2006 fiscal year as exportations from Bangladesh increased to $242 million while exports from officially reported exports to India declined to $1.8 billion, down from $2 billion. Due to exports from India of $6.1 billion and imports from Bangladesh of $456 million (US$2.7 billion five years earlier), bilateral trading between India and Bangladesh stood in 2013–2014 for US$6.6 billion. Bilateral commerce between India and Bangladesh has consistently increased in the past decade, and Bangladesh's exports tripled to $1 billion in 2018–2019 in the last decade. India exports in 2019–2020 to Bangladesh amounted to 8.2 billion dollars and imports to 1.26 billion dollars (Table 3). Despite the peaks, the overall commerce in Indo-Bangladesh, particularly imports from India, has

Table 3 Enclave exchange by numbers in 2015

Enclaves	Enclaves number	Area	Residents
Bangladesh	51	7110 acres	14,863 became Indian citizens
India	111	17,161	37,532 became Bangladeshi citizens

Sources: Ministry of External Affairs (MEA), "Exchange of enclaves between India and Bangladesh," (press release, November 20, 2015); MEA, India and Bangladesh Land Boundary Agreement (New Delhi: MEA, n.d.)

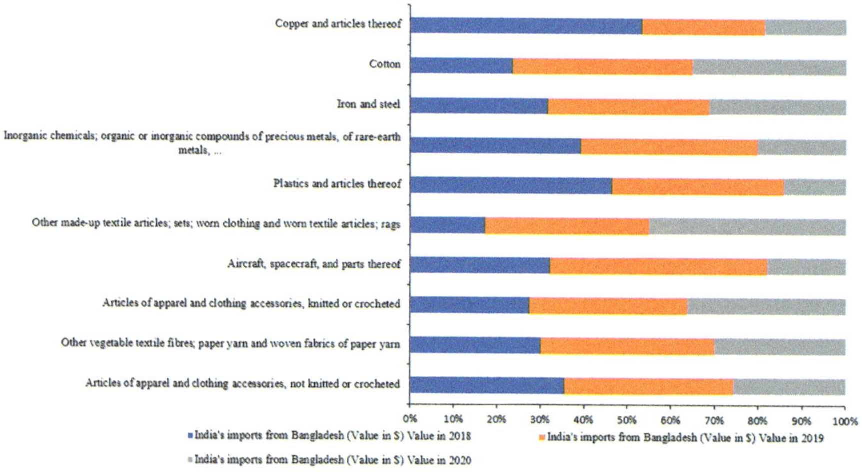

Fig. 2 India's import intensity with Bangladesh in the top ten imports

increased throughout time (Fig. 2). The export receipts for Bangladesh were raised from the Indian market (Table 1). Because India is the second-largest source of Bangladesh's general imports, it accounts for 15.1% of the country's total imports according to the latest figures available from the Bangladesh Bank.

India has adopted the regional power trading system concept, which will allow certain regions of the country to reduce the power deficit through the transfer of surplus power from other regions. The Indian corporations could pool energy in an exchange under the Electricity Act of 2003. To reach the Bhutan and Nepal markets, Bangladesh has formally requested a power corridor (Basher, 2013). India has decided to permit hydroelectricity transmission from Assam to Bihar. Bangladesh imports 1160 MW of electricity now from India. There is a confusing feature in the export of Bangladesh of produced goods by India. Total imports from Bangladesh as a developing country are clearly dominated by production products. However, imports of Bangladesh from India are heavily dominated rather than produced items by primary commodities and food. This, together with our past findings, shows the domination of intra-industry trade in Bangladesh's exports to India also dominated by commodities.

Political disquiet in Bangladesh seems to arise from the significant trade imbalance that India is facing in Bangladesh's exports with several non-trade restrictions.

Although Bangladeshi exports are expanding at a strong pace, there is little hint that this trade imbalance has been reduced. Ninety percent of industrial tariffs are now much lower and much more uniform at 12.5% compared with the previous 50 years. From the SAARC countries, including Bangladesh, the change means Indian domestic markets are extremely competitive for most consumer products, and prices close to international prices are hard to combat, although Indian duties are totally exempt from bilateral or multilateral free trade agreements including those planned in connection with SAFTA.

3.5 *Adverse Possessions*

Unfriendly possessions shall be territories adjoining and controlled by the borders of one country but legally belonging to the neighboring country (e.g., adjoining the Indian border, but legally a part of Bangladesh, under Indian sovereignty). The enclaves of the border between India and Bangladesh were hundreds of years old but continued unsolved after India and East Pakistan became independent. The LBA in 1974 asked for the exchange of these unfavorable possessions. In the 2011 protocol, however, both India and Bangladesh agreed to redraw the global frontier to recognize the status quo. The people who lived on unfavorable lands were recognized to be strongly connected to their land and resistant to being relocated. The 2011 Protocol also acknowledged that in actuality, the territories of unfavorable possession to be allocated by the 1974 LBA had the adverse possession of the country. In this respect, both India and Bangladesh acknowledged this fact through the Protocol of 2011.

In the last 3 years, cows have played an average of 12% in illegal livestock trading from India to Bangladesh. The number of seized animals ranged from 1.5 lakh to 1.75 lakh each year between January 2014 and December 2016. And, in all 3 years, cattle seizures range from 11.7% to 11.9%. The bullock or ox (37.6–40.7%) and calves are maximum convulsions (30.6–43.24%). In the calf category, male calves were 95%. Figure 3 shows the number of seized cattle on India Bangladesh border from 2016 to 2019. According to data supplied by the South Bengal Frontier of the BSF, the border guarding troops took roughly 5449 cattle heads in the year 2020, and that is much lower than the number of cattle heads confiscated in 2019, which stood at 29,720. Senior BSF officials indicated that the decline was due to heightened monitoring along the border. Another notable illustration is the seizure of livestock along the Indo-Bangladesh border when animals from across northern India are moved to Indian borders before being smuggled through the border.

Bangladeshi citizens' death on the border between Bangladesh and India impacted their relationship with Bangladesh and India. The death of those who seek to enter the Indian border illegally from Bangladesh and cross-border shooting and livestock smuggling occur many times every year. The Indian Border Security Force has implemented its contentious "shoot-on-sight" policy to combat smuggling and illegal migration from Bangladesh. According to a study released in Jatiya Sangsad

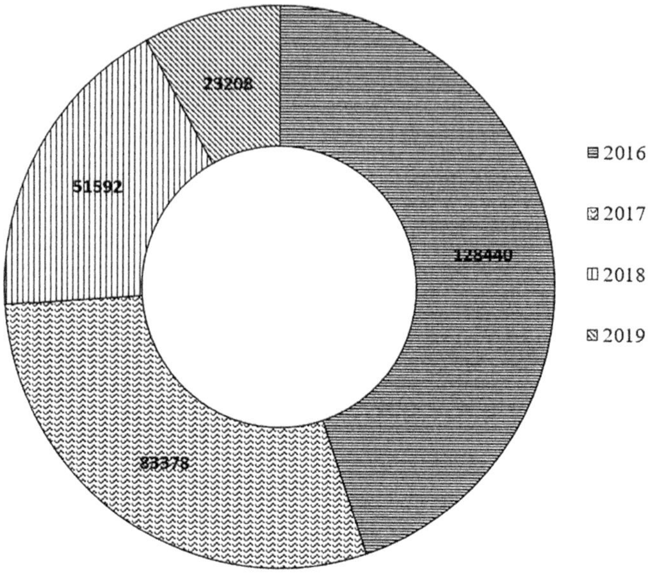

Fig. 3 Number of cattle (cow) has been seized by BSF in Indo-Bangladesh border

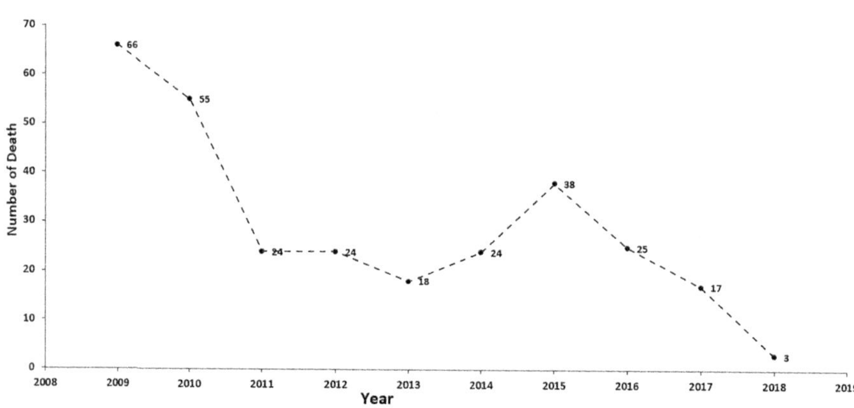

Fig. 4 Number of deaths of Bangladeshi citizens by BSF on the Indo-Bangladesh border for illegal activity

by the Bangladesh Home Minister, about 294 Bangladeshi citizens were murdered by the BSF in 10 years (from 2008 to 2018) (Fig. 4).

The report further indicates that Bangladeshi border guards normally do not help the residents of Bangladesh, whereas Indian paramilitary soldiers intimidate, harass, and torture routinely. The border between the BSF and the Bangladesh Border Guard is also being trafficked by civilians living close to India and Bangladesh cross-border fire. From 2001 to 2010, the Bangladeshi border security forces

Table 4 State-wise details of illegal infiltrators apprehended by Border Security Force along the Indo-Bangladesh border during the last 5 years

State	2016	2017	2018	2019	2020
Assam	6	8	9	3	5
Meghalaya	56	26	26	24	20
Tripura	105	76	96	108	44
West Bengal	1434	797	753	974	886
Total	1601	907	884	1109	955

Source: Ministry of Home Affairs, Government of India, 2021

Table 5 Number of deaths of Bangladesh citizens due to illegal activities from 2000 to 2020. (Source: The Dhaka Tribune, The Daily Observer.; The Daily Star)

Year	Killed	Injured	Abducted	Missing	Rape	Snatching	Push in	Other	Total
2000	31	17	106	0	2	13	0	0	169
2001	84	29	55	0	1	10	0	0	179
2002	94	42	118	30	0	12	0	0	296
2003	27	41	120	7	2	8	0	0	205
2004	72	30	73	0	0	5	0	0	180
2005	88	53	78	14	3	4	0	0	240
2006	155	121	160	32	2	9	0	0	479
2007	118	82	92	9	3	5	198	0	507
2008	61	46	81	0	0	3	20	0	211
2009	98	77	25	13	1	1	90	3	308
2010	74	72	43	2	0	1	5	0	197
2011	31	62	23	0	0	0	0	9	125
2012	38	100	74	1	0	9	0	16	238
2013	29	79	127	0	1	77	41	0	354
2014	35	68	99	2	0	0	0	5	209
2015	44	60	27	1	0	0	0	0	132
2016	29	36	22	0	0	0	0	0	87
2017	25	39	28	0	0	0	0	0	92
2018	11	24	16	0	0	0	0	0	51
2019	41	40	34	0	0	0	0	0	115
2020	51	27	7	0	0	0	0	1	86

murdered roughly 1000 civilians (Tables 4 and 5). According to the Bangladesh Home Minister from 2012 to 2016, around 146 civilians in Bangladesh were killed by BSF and Indian citizens.

3.6 Technological Innovation and Border Management

The border management function asks for collaboration and concerted action within our country by numerous government entities. The purpose is to preserve our borders and to prevent the risk of moving products and persons from India to other nations and vice versa. Border management itself may involve regulating legal and illegal immigration, guaranteeing safe transport, the movement of persons and products authorized, and preventing smuggling, trafficking of people, and infiltrating. It is not restricted to regulating legal and illegal immigration. India's rapid growth today brings a range of opportunities and highlights the necessity for proper management of the borders. However, if these movements are unmanaged, less regulated, or without any oversight, smuggling, trafficking, terror, crime, and illegal migration might increase. To meet the geopolitical, social, and economic issues that we in India confront, our large boundaries are a competent border management system that asks for close interaction between technology and infrastructure.

We believe that the free movement of individuals, goods, and ideas is a necessity for economic growth and globalization, yet the increasing number of frontier risks requires heightened vigilance and stronger border regulations. However, certain agencies use more familiar technology, in particular those that assist them in meeting their data management and operating goals. Thus, while the government has performed many measures, including the building of roads, fencing, flashing, and ground radar installation, technological use, intelligence collecting, and information assimilation from multiple sources are still a challenge to develop the common picture of the organization. The armed border management actors are modernizing their weaponry and equipment. The Indian Army is also modernizing its small arms, such as pistols, carbines, and LMGs, and is looking forward to deploying future weaponry now under development along the border. Uncontrolled ground sensors and up-to-date medical equipment may be implemented, such as ground-penetrating radar systems, unmanned aerial vehicles (UAVs), targeted acquisition devices, corner shots, hand-held thermal sights, and night vision devices (NVDs). The Defense Research and Development Organization (DRDO) has been adding considerable research and development to India's border management capabilities. Technologies in the fields of information, communication, control, and monitoring, the air defense control and reporting system (ADC&RS), and the battlefield management system have been the focus in particular. Moreover, in different pre-emptive and reactive methods, advanced control and control systems can help. DRDO has developed the artillery command and control system for the automatic use of field artillery. In addition to the components in software applications, GIS integration, inputs such as surveillance cameras and sensor types, a dashboard, mobile stations, and modules of high display are the main components of an integrated control and control center. The efficient integration of these components enables the control center to identify intrusion in real time and to become aware of the situation. This can help to react quickly and aid in dispatching an intervention team before a perimeter break occurs.

4 Challenges Ahead

4.1 Farakka Barrage Dispute

In 1950–1951, when Bangladesh was still a part of Pakistan, preliminary plans for the Farakka Barrage in India began. India erected the Farakka dam barrage in 1975, to redirect water to the system of the Bhagirathi-Hoogly River from the Ganges River. The dam drains water from the Ganges-Brahmaputra-Meghna basin, one of the most populated basins in the world. Between India and Bangladesh, there are 54 cross-border rivers (Kawser and Samad, 2016). Given the river's long road in numerous nations, every source of tension between India and Bangladesh puts millions of people on the Ganges and its tributaries at risk of food and water security. The surface water supply of Bangladesh emanates beyond its borders by 94%. Bangladesh's plan to build and operate a dam in China and India is exceedingly fragile. India relies on its military and economic power, in particular by establishing the Farakka Dam, to act unilaterally in water-sharing scenarios.

While salinity in Kolkata is reduced by the dam, river salinity in Bangladesh has increased due to diversion in the Ganges. Reduced flows affect the ecosystem of Bangladesh, specifically the mangrove forest of Sundarbans. Forest deterioration reduced timber production in Bangladesh and cost the country a financial loss. Higher levels of salinity also affect drinking water in Bangladesh. Failure to consume water increases Bangladesh's propensity for venereal infections. Considering the importance of food and water safety and the bad consequences Bangladesh has had as a consequence of the Farakka barrage, relations between India and Bangladesh could worsen in the medium term. With the increasing scarcity of water and population pressure having an additional impact on the availability of water, in the absence of a concrete framework for water sharing, conflicts may increase.

The Ganges water is sensitive to Bangladesh by Farakka barrage, which has damaged its environmental and economic environment. The Ganges basin depends on 37% of the total area and 33% of the people in Bangladesh. Bangladesh has been exposed to the challenge of agriculture, industry, fisheries, shipping, salinity, and ecology in the southwest region because of the diminished flow of the Ganges. Approximately one-third of Bangladesh's entire region is dependent on livelihood in the Ganges basin (Mirza & Hossian, 2000). In this case, water divergence in Farakka is sure to affect the regular path of nature in the attempt to develop a new ecological system. The situation was expected to improve after the treaty; however, many investigations have demonstrated that in some circumstances, things have deteriorated even worse.

4.2 Teesta River Dispute

The river Teesta, a tributary of the river Brahmaputra, rises in the Teesta Kangse glacier and passes through the Sikkim and West Bengal states before arriving in Bangladesh. Since 1947, the Teesta's catchment lands were allocated to India, and it has been a source of conflict. In 1983, an ad hoc agreement was reached after the setting up of the Indian-Bangladesh Joint River Commission, which receives 39% of the water in India and 36% from Bangladesh. After the conclusion of the Ganga Water Treaty in 1996, the Teesta River question took on importance. In 2011, India decided to share 37.5% of Teesta waters while reserving 42.5% of the water between December and March in the lean season. Bangladesh has long been demanding a sharing of the Teesta waters since millions of them survive on the water on the river. In addition, the constant construction of dams along the Teesta in Sikkim led to slight drilling into Bangladesh during the season. The major concern is in the lean season's flow of water. The Teesta issue remains the biggest unresolved disagreement between the two member countries after the conclusion of the LBA and the resolution of the maritime border. Whether or not it comes from security cooperation, trade improvement, or the Indian connectivity initiatives to the east, including its own North East, a timely settlement is certainly necessary to preserve the progress made in bilateral relations and to prevent Bangladesh from being fully subsumed into the Chinese sphere of influence. The agreement is certainly necessary.

4.3 China's Influence

Now, the continuous confrontation between China and India can echo in South Asia's geopolitical dimensions, leading to new connections within the region. Bangladesh is an interesting situation, in contrast to the conviction that China-India is a rival to other neighbors. The strategic location, population, markets, and production capabilities of Bangladesh could seize the chance of the current war. It also has to be welcomed in China—both to assure commercial gains via Bangladesh's ports and to control India as well. China is also welcome to continue its connection with Bangladesh. In 1976, Beijing started its Dhaka diplomatic relations. As a process of comprehensive commercial collaboration, economic cooperation, and technical exchange, China-Bangladesh relations began.

The expanding diplomatic links between Bangladesh and China could directly challenge India and damage the relationship between Bangladesh and India. Since the border with India (with the Indo-China situation much more controversial in nature) is continuing, both China and Bangladesh can scare India and exacerbate existing conflict by moving toward closer militarism and diplomacy with China. In addition, in 2005, China took over India as the top trading partner for Bangladesh and challenged India's leadership in South Asia.

On the other side, to build and sustain its hegemony in South Asia, China strengthened its relations with the Southern Asian countries. The overall aim of Bangladesh has been to reduce Dhaka's dependence on India and establish new ways of cooperating with China and Southeast Asia. This policy of Bangladesh has been well connected. In 2015, China substituted India as Bangladesh's main trading partner. Bangladesh has received help from Peking, including 27 agreements inked during the visit to Bangladesh of President Xi Jinping in 2016 as a participant of

China's Belt and Road Initiative (BRI). The investment amounted to over 24 billion dollars. It is safe to say that Bangladesh is tilting toward China. Beijing has recently granted duty-free access, effective on July 1, to 97% of Bangladeshi imports. In addition, Bangladesh recently approached China for a $64 billion infrastructure fund. This request was made in connection with the Working Group on Investment Cooperation with China set up last year. With its current dispute, both India and China will endeavor to expand in Bangladesh, but so far, China has provided Bangladesh with the best card. While certain Indian critics banned the offer of China for Bangladesh as "charity." In addition, to protect regional exclusivity that would ensure Bangladesh's commitment to regional peace, the government should further improve economic and defense links among its Southern Asian neighbors.

5 Way Forward

The foundations for India-Bangladesh relations were strengthened for the first 50 years. The future offers itself plenty of chances to help both countries reach a new level of bilateral relations more than before, if they are more mature and pragmatic in the leadership of the two neighbors, taking into account the regional aspirations and nuances of both countries. In all these countries, the relations between India and Bangladesh have been maintained by shared destinies due to the common geography and values based on the sacrifice of the two sides who battled to defend those principles. The umbilical cord of friendship and trust between India and Bangladesh will soon become a major driver for global progress and stability in both countries' immediate and extended districts. A major force for the protection of physical and digital interconnection has been established through cooperation between both administrations.

Among the key concerns discussed during the consultation were immigration, water sharing, frontier demarcation, minorities, trade, and transit. The importance of North East India also has been emphasized by most social scientists in the bilateral relations between the two countries, given that there is a greater awareness on both sides that trade, and connectivity growth in North-eastern India is in the best interest of both countries. Although illegal migration and water-sharing conflicts have plagued the two countries' bilateral relations for the longest period, the social scientist concluded that recent trends in migration, sharing of water, and demarcation at borders are significant. Both governments have worked actively to make the two countries more connected. There are several unused roads and tracks re-established, inland rivers restored, and connecting bridges erected. Honorable Prime Minister of India recently opened the "Maitri Setu" bridge in Tripura, emphasizing the creation of "HIRAs," an information mesh between both countries, including roads, trains, and airways. Better connectivity can leverage the latent commercial opportunities between the two countries. Enhanced connection to Bangladesh will prevent the Chinese influence of the north-eastern states of India. It also will lead to enhanced governance, business, and relationships with people.

Regular interactions between young entrepreneurs and civil society centered on fields including cultural, musical, sporting, filmmaking, and sharing in sustainable development, development of human capital, and equitable growth for women and men, among other things, must be promoted. Taking advantage of technology and social media, such conversations can be a blend of physical and online engagements. Steps can be taken to enable greater interaction between communities living at the frontiers, border, and mainstream and between persons at the common borders. The increase and facilitation of a mechanism of exchange through the creation of a common entertainment zone at the border with a curated calendar of events can go far and wide in spurring interest on the "other" side of tourism at select border locations, for example, in Akhaura in Tripura and Dawsky in Meghalaya.

Bangladesh has great tourist potential, with its rich cultural and natural history. Popular places for Indians and international travelers might also include beaches such as Cox Basar, Ratergul Swamp, Sylhet Tea Gardens and the Rabindranath Tagore heritage, the Shaktipeetah Temples, and the Chittagong Buddhists. The offer of North-eastern India combinations can be an important way to entice mainstream Indian tourists to Bangladesh and to provide value to the interest of the younger crowd in visiting India in Bangladesh. A new security paradigm at shared borders must be worked together. A paradigm that allows borders to be not just thick lines that delimit national borders, but work as "connecting areas" for inclusive growth and wealth addresses constant problems on both sides—questions such as illegal immigration and trafficking.

6 Conclusion

The future of India's East Policy Act, stability and North-eastern development, economic integration into the broader region, people flowing across borders, and peace and security depend on links to Bangladesh. With Bangladesh's 50th anniversary in 2021 and India's 75th anniversary in 2022, we should think about how the "third space" between the two nations could be strengthened. The area where confidence and affection breathe can supply the ecosystems for the next half-century to nurture growth and stability.

Smart border management is aimed at modernizing the existing border management of our country through the improvement of border crossing procedure quality and efficiency. It also aims to help India and neighboring nations manage growing travel patterns without increasing the number of border guards or patrol teams needed. Smart border management aims mainly at promoting movement between India and other countries in a safe environment while helping to combat terrorism, drug trafficking, smuggling, illegal activities, and other major crimes. Intelligent border management is a stronger and more efficient and transformative border management solution that emphasizes enhanced border security controls, intelligent intelligence information systems, and more planned and sophisticated prevention and action on all threats using the latest technologies.

With every innovation in technology and research and development capability, the productive potential of organizations has expanded. These developments are also expanding the sources of dangers to its interruption and destruction. New approaches for automation, analysis, and communication devices are introducing new vulnerabilities to police, security, and defense companies. New technologies such as infrared, thermal camera combinations, surveillance drones, the Internet, smart handheld devices, data analysis, artificial intelligence, and cloud computing have become more important for border management agencies in the sphere of surveillance and security. These innovations must be developed and adopted by the border's management authorities, and at the same time, capacity building and management of change should be focused.

Research Grant Not applicable.

Conflict of Interest No conflict of interest.

References

Alam, R. (2021). *From stateless people to citizens: The reformulation of territory and identity in India-Bangladesh border enclaves*. Electronic Thesis and Dissertation Repository. 7666. Available at: https://ir.lib.uwo.ca/etd/7666

Ashraf, T. (2020). *Bangladesh and India: The domestic and international repercussions of asymmetric and securitized hydro-politics*. FIU Electronic Theses and Dissertations. 4430. https://digitalcommons.fiu.edu/etd/4430

Banerjee, S., Guha, A., & Chaudhury, A. B. R. (2017). *The 2015 India-Bangladesh land boundary agreement: Identifying constraints and exploring possibilities in Cooch Behar* (pp. 1–46). ORF Occasional Paper # 117.

Basher, A. (2013). *Indo-Bangla trade: Composition, trends and way forward*. Commonwealth Secretariat. Available at: http://ris.org.in/

Chatterji, J. (1999). The fashioning of a frontier: The radcliffe line and Bengal's Border landscape, 1947–52. *Modern Asian Studies, 33*(1), 185–242.

Cons, J. (2014). Impasse and opportunity: Reframing postcolonial territory at the India-Bangladesh border. *South Asia Multidisciplinary Academic Journal*. http://journals.openedition.org/samaj/3791; https://doi.org/10.4000/samaj.3791.

Dutta, P. (2010). *India-Bangladesh relations. Issues, problems and recent developments*. Institute of Peace and Conflict Studies.

International, C. (2019). *Bordering on happiness: An assessment of socioeconomic impacts of Bangladesh-India border haats*. Retrieved from CUTS International, Jaipur, India. https://cuts-crc.org/pdf/bordering-on-happiness-an-assessment-of-socio-economic-impacts-of-Bangladesh-India-border-haats.pdf. Accessed on 12 February 2021.

Jamwal, N. S. (2004). Border management: Dilemma of guarding the India-Bangladesh border. *Strategic Analysis, 28*(1), 5–36.

Kashem, A., & Islam, S. (2016). Narendra Modi's Bangladesh policy and India–Bangladesh relations. *India Quarterly, 72*(3), 250–267.

Kawser, M. A., & Abdus Samad, M. (2016). Political history of Farakka Barrage and its effects on environment in Bangladesh. *Bandung: Journal of the Global South, 3*(1), 1–14. https://doi.org/10.1186/s40728-015-0027-5

Leake, E., & Haines, D. (2017). Lines of (In) convenience: Sovereignty and border-making in Postcolonial South Asia, 1947–1965. *Journal of Asian Studies, 76*(4), 963–985. ISSN 0021-9118.
Majaw, B. (2021). Indo-Bangladesh borderland issues in Meghalaya. *South Asia Research, 41*(1), 100–118. https://doi.org/10.1177/0262728020966100
Mirza, M., & Hossian, M. (2000). Adverse effects on agriculture in the Ganges basin in Bangladesh. In M. Q. Mirza (Ed.), *The Ganges water diversion: Environmental effects and implications*. Kluwer Academic Publishers. Return to ref 2000 in article.
Mishu, H. M. (2018). *The politics of International Law and India-Bangladesh land border management: A critical theory approach*. Nottingham Trent University.
Mukherjee, J. (2011). The Farakka Barrage: A review from the Indo-Bangladeshi perspective after 36 years. *Proceedings of the Indian History Congress, 72*(2), 1054–1063. https://www.jstor.org/stable/44145717
Rahaman, M. M. (2006). The Ganges water conflict: A comparative analysis of 1977 agreement and 1996 treaty. *International Water Law Project, 1*(2), 195–208.
Roy, H. (2012). *Partitioned lives: Migrants, refugees, citizens in India and Pakistan, 1947–65*. History Faculty Publications. Paper 21. http://ecommons.udayton.edu/hst_fac_pub/21
Sacks, R., Yi, S. S., & Nonas, C. (2015). Increasing access to fruits and vegetables: Perspectives from the New York City experience. *Am J Public Health, 105*(5), e29–e37. https://doi.org/10.2105/AJPH.2015.302587. Epub 2015 Mar 19. PMID: 25790427; PMCID: PMC4386516.
Schultz, K. A. (2015). Borders, conflict, and trade. *Annual Review of Political Science, 18*, 125–145.
Shahriar, S. (2021). Bangladesh-India border issues: A critical review. *Geoforum, 124*, 257–260.
Shewly, H. J. (2017). Life in de facto statelessness in enclaves in India and Bangladesh. *Singapore Journal of Tropical Geography, 38*(1), 108–122. https://doi.org/10.1111/sjtg.12181
Singh, N. (2014). India and development partnership: special reference with Bangladesh in 21st century. International relations conference on India and development partnerships in Asia and Africa: Towards a new paradigm (IRC-2013). *Procedia - Social and Behavioral Sciences, 157*(2014), 137–142.
Thakur, J. (2020). *India-Bangladesh trans-boundary river management: Understanding the Tipaimukh Dam controversy*.', ORF Issue Brief No. 334. Observer Research Foundation.
van Schendel, W. (2002). Stateless in South Asia: The making of the India-Bangladesh enclaves. *The Journal of Asian Studies, 61*(1), 115–147.
Van Schendel, W., & De Maaker, E. (2014). Asian borderlands: Introducing their permeability, strategic uses and meanings. *Journal of Borderlands Studies, 29*(1), 3–9. https://doi.org/10.1080/08865655.2014.892689
Wolf, A.T., & Newton, J.T. (2014). *Case study of transboundary dispute resolution: The Ganges River controversy*.
Wolf, A. T. (1998). Conflict and cooperation along international waterways. *Water Policy, 1*(1998), 251–265.

Subhankar Roy has been dedicated to the field of education for over a decade. Since 2013, he has been serving as an Assistant Professor in Geography at Jalangi B.Ed. College. His role involves teaching, mentoring students, and contributing to the academic community through research and scholarly activities. His experience and dedication have made him a respected figure in the academic circles of Geography.

Dr. Gouri Sankar Bhunia has more than sixteen years of experience in geospatial technology and involved in several Government and Private Organizations in India. He received his Ph.D. from the Department of Geography, University of Calcutta, India in 2015. He also worked as a visiting Faculty in Private University, West Bengal, India. His Ph.D. dissertation work focused on environmental control measures of infectious disease using Geospatial technology. Currently, he is involved in Smart city and urban planning project. He has received Senior Research Fellowship Award from Indian Council of Medical Research. His research interests include environmental modelling, Health GIS, risk assessment, natural resources mapping and modelling, data mining and information retrieval using Geospatial technology. He has published Ten books with International Publishing Group. He is currently the editor of the GIScience and Geo-environmental Modelling (GGM) Book Series, Springer-Nature.

Dr. Pravat Kumar Shit Assistant Professor has been working in the Postgraduate Department of Geography, Raja N. L. Khan Women's College (Autonomous), Gope Palace, Midnapore, West Bengal, India. He received his Ph.D. in Geography from Vidyasagar University (India) in 2013, M.Sc. in Geography and Environment Management from Vidyasagar University in 2005 and PG Diploma in Remote Sensing & GIS from Sambalpur University in 2015. His main fields of research are soil erosion spatial modelling, Non-timber forest and rural livelihood, Forest Biomass and Carbon Stocks, water resources and natural resources mapping and environmental modelling and published more than 45 international and national research articles in various renowned journals. His research work has been funded by the University Grants Commission (UGC), Govt. of India and Department of Science and Technology (DST), Govt. of West Bengal. Dr. Shit is an associate editor and on the editorial boards' member of three international journal in geography, and earth environment science. He has served as a reviewer for many international journals. He published four (04) books. Also he published two books in Springer (https://www.springer.com/gp/book/9783030232429 and https://www.springer.com/gp/book/9783030016791).

Effects of COVID-19 on Migration in Murshidabad District of West Bengal

Koushik Kumar Das, Ajay Debnath, and Bapan Mondal

1 Introduction

People who migrate have an impact on a place's social cohesiveness, identity development, and general behavior. The socio-economic condition has been greatly impacted by migration for a very long time, especially in rural India. Typically, people move from less prosperous areas to wealthier ones because of the economy. The flow of people throughout this process somewhat erases the regional economic imbalance. Migration alters social networks, demography, and cultural dynamics inside a nation; it also has a significant effect on West Bengal's social physics and land use patterns in migrant places. Over time, the district of Murshidabad has seen a significant number of labor movements. Spatial patterns within cities and regions are also impacted by migration. This may lead to the development of ethnic communities and adjustments to urban movement patterns. In general, migration is a complicated phenomenon that interacts with many different facets of social physics, such as economic modeling, spatial dynamics, network analysis, and cultural evolution. Numerous factors, including economic opportunities, a lack of employment options in the area, and the desire for better living conditions, can be linked to the labor movement in the Murshidabad district. The major destinations for labor migration from the Murshidabad district are cities like Kolkata, Delhi, Mumbai, and Chennai because they offer a wide range of employment opportunities across different industries. In addition to being dispersed throughout the nation, migrant laborers

K. K. Das (✉)
Department of Geography, Nathulal Das B.Ed. College, Murshidabad, West Bengal, India

A. Debnath
Department of Geography, Sripat Singh College, Murshidabad, West Bengal, India

B. Mondal
Department of Geography, Asutosh College, Kolkata, West Bengal, India

A. Rai et al. (eds.), *Contemporary Social Physics*, Springer Geography,
https://doi.org/10.1007/978-3-031-77596-3_8

from Murshidabad have also been found in many other nations, particularly in Central Asia, the Arab world, South Africa, Singapore, Hong Kong, Malaysia, and East Asia. The Burj Khalifa, the highest structure in the world, was constructed by masons from Beldanga C.D. block, a village in the Murshidabad district (Das, 2020). Approximately one-fourth of the population of Murshidabad district, which has a current population of 72 lakh, currently makes their living outside of the area and sends their excess income to their families (Das, 2020). COVID-19 has made migrant laborers' lives even more precarious. Migrant workers have been the most severely affected by the pandemic because they have been stranded in their destination countries or forced to leave, are unemployed, and lack access to social assistance. All people are imprisoned in their rooms during a pandemic. The same things take place in both our state and nation.

2 Study Area

Our study area is located in the central part of West Bengal which is Murshidabad. Murshidabad district of West Bengal is selected as the study area for this present study (Fig. 1). Murshidabad is positioned roughly between latitudes 23° 43′ N and 44° 52′ N and longitudes 87° 49′ E and 88° 44′ E. The overall size of the area is 5316.11 sq. km. (District Census Handbook, 2011). Murshidabad district comprises 26 C.D. blocks and seven statutory towns. There are a total of 2166 villages, 254 gram panchayats, and 65 census towns in this district. This district is bordered on the north by the Malda district, on the south by the East Bardhaman and Nadia districts, on the east by Bangladesh, and on the west by the Jharkhand state and Birbhum district. Murshidabad district is located along the Bhagirathi River's western bank. This is an agriculturally based land area full of usefulness. But the present situation is quite different from the past situation. So people are interested in migration and have left their home place for many reasons, like labor workers, jobs, education, and treatments. Mainly, the maximum number of migrations happens from Lalgola, Bhagwangola-I, Bhagwangola-ii, Suti-I, Suti-ii, Beldanga, Sagardighi, Samserganj, Nabagram C.D. blocks, etc. of Murshidabad district.

3 The Major Objectives of Our Study

I. To comprehend the socio-economic standing of the local population in Murshidabad district.
II. To know the proper occupational status of the population and social physics of Murshidabad district.
III. To know the effect of COVID-19 on the migration and social status of Murshidabad district.

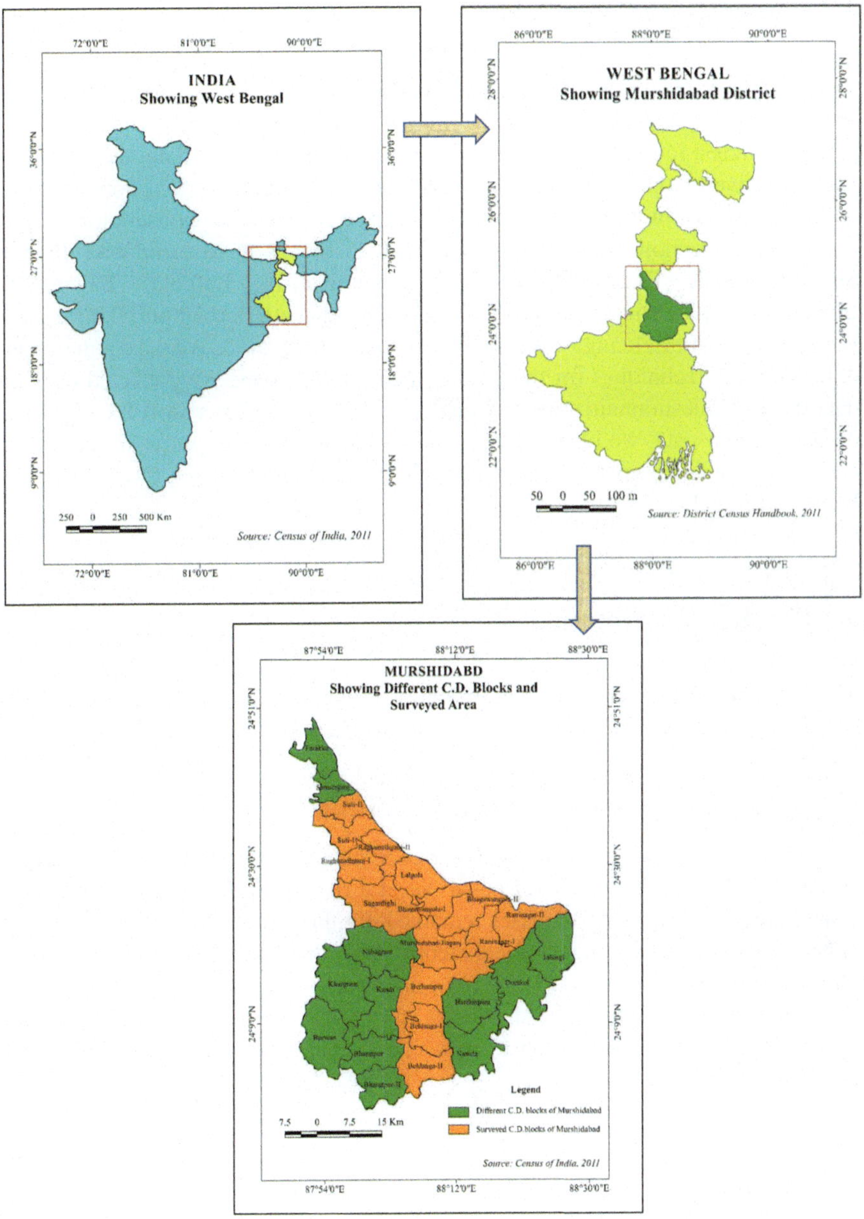

Fig. 1 Geographical location of the study area

4 Materials and Methods

This study is based on qualitative and quantitative data. There are mainly two types of data sources: secondary and primary data. Secondary data were collected from various articles, journals, different types of handbooks, census data, etc., and primary data were collected from household surveys through random conversations with local people in the surveyed block. We have selected 380 households for primary data collection from 14 blocks out of 26 blocks in Murshidabad district. Suti-II, Suti-I, Raghunathganj-I, Raghunathganj-II, Lalgola, Sagardighi, Bhagawangola-I, Bhagawangola-II, Murshidabad-Jiaganj, Berhampur, Raninagar-I, Raninagar-II, and Beldanga-I are the blocks that we surveyed. We used simple random sampling techniques for the primary household survey. We collected primary data through questionnaires and created data tables using Microsoft Excel. Based on the primary data, we identified the major objectives of the study. We have analyzed the data and prepared appropriate tables and diagrams using QGIS software, Microsoft Excel, and Microsoft Word.

5 Physical Environment

Murshidabad district has abundant groundwater and surface water. The district is traversed by the main two rivers among the surface water bodies.

- Ganga-Padma River channels and its distributaries.
- Mayuraskhi River channel and its distributaries.

Bhagirathi, Jalangi, and its distributaries Sialmari and Bhairab are the primary rivers of the Ganga-Padma River basin. Several rivers run through Murshidabad's "Bagri" region. These distributaries remained dry the rest of the year but conveyed water during the monsoon. Historical records show that boating was once permitted on these distributaries, although the river is currently shallow due to siltation. The Bhairab River has taken on the shape of a saucer year after due to the silt from the Padma River. Paddy cultivation is now carried out in the river bed throughout the winter.

5.1 Climate

This is a tropical region. Here have many seasonal variations of rainfall and temperature. The land at night is 18 °C winter it becomes 28 °C. In the day it is 42 °C and at night it is 27 °C average temperature belongs between in the summer season is 33 °C. The year is divided broadly into three seasons (Fig. 2):

Summer (pre-monsoon): starts in March and continues till middle of the June.

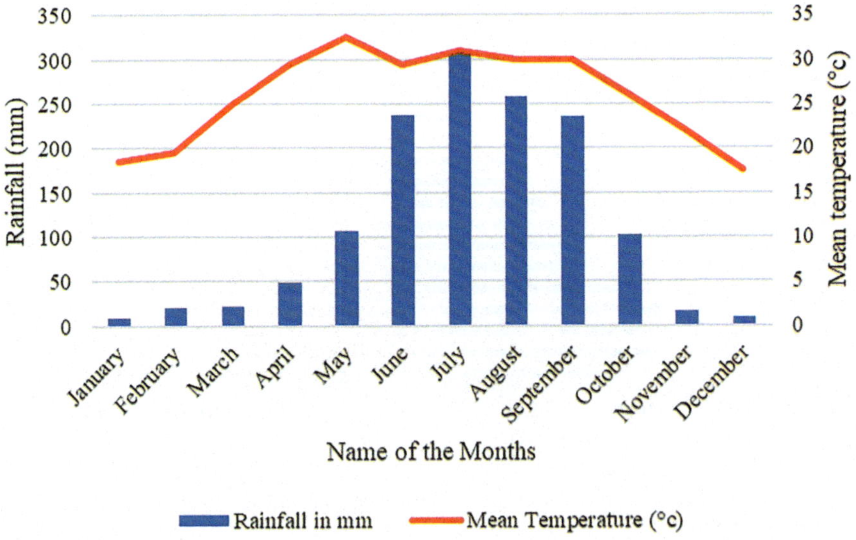

Fig. 2 Monthly temperature and rainfall distribution of Murshidabad, 2014. (Source: District Statistical Handbook, 2014)

Rainy (monsoon): from the middle of June to October.

Winter (post-monsoon): begins in November and continues till the beginning of March.

The season is hot in summer, and the winter season remains dry. Maximum temperature is recorded in the month of June, whereas January is the coldest month. Rainfall occurs in three phases, i.e., monsoon, summer, and winter months, but most of the rainfall (Fig. 2) happens in the monsoon time due to south-west monsoon rainfall. In the winter season, there is much less rainfall than summer season in this region. Monsoon season mainly shows from June to September. The average rainfall is mainly 150–200 cm throughout the year.

6 Socio-Economic Status

6.1 Demography

Murshidabad, one of West Bengal's most densely populated districts, has the third-highest decadal population increase rate, according to the 2011 Census of India. With a total area of 5341 square kilometers, the district has a male population percentage of 51.06% and a female population percentage of 48.94%. There are 1334 persons living in the district per square km (Fig. 3).

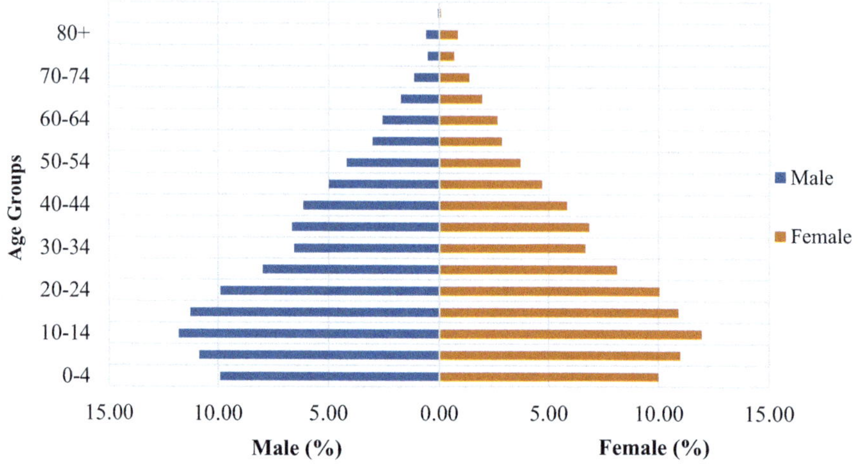

Fig. 3 Distribution of population by age-sex group in Murshidabad district. (Source: District Census Handbook, 2011)

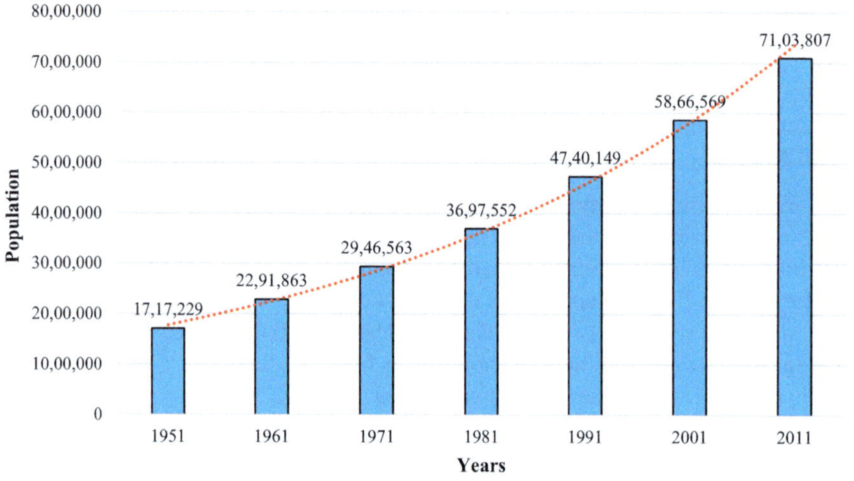

Fig. 4 Population growth of Murshidabad district according to the year. (Source: District Census Handbook, 1951–2011)

6.2 *Population Growth of Murshidabad District*

The population growth of Murshidabad district is represented by this bar graph. The increasing trend of the bars clearly shows that the population of Murshidabad district is increasing day by day. According to this diagram (Fig. 4), the population of Murshidabad district was 1,717,229 in the year 1951. And, according to the 2011 Census, the population of Murshidabad district is 7,103,807. In the 60 years from

1951 to 2011, Murshidabad's population increased by 5,386,508. Population growth from 1951 to 1961 was 80,066. That means, over 20 years, Murshidabad's population has grown by 12%. Population growth was 14.53%, with an increase of 96,289 more people between 1961 and 1981. The population growth rate between 1971 and 1991 was the highest in Murshidabad district as shown by the graph, reaching 291,608 people. It means around 44% population increased in this duration. The population growth rate was 12.65% between 1981 and 2001. This indicates a declining rate of population. And again, the growth of the population rate started to increase between 1991 and 2011. The rate of population growth was 16.72%. As the population grew, so did the problem of unemployment and the number of migrant workers.

6.3 *Work Participation Rate*

Work participation rate is an important factor in understanding the occupational status of an area. According to the Indian Census, a worker is anyone who performs any kind of gainful activity, regardless of the payment received. The work participation rate, which is calculated as the ratio of total employees (main + marginal) to the total population, shows the proportion of the people in a region that is employed. The work participation rate for the rural and urban areas of the Murshidabad district is shown in Fig. 5. The diagram shows that from 2001 to the most recent 2011 Census, the district's overall work participation rate was often lower than the state's. Between the 2001 and 2011 censuses, the district's rural work participation rate continued to rise. From 1,718,547 in 2001 to 2,007,805 in 2011, the work participation rate for all workers (main and marginal) increased, and from the 2001 Census to the 2011 Census, it grew in rural areas. Between the 2001 and 2011 censuses, the

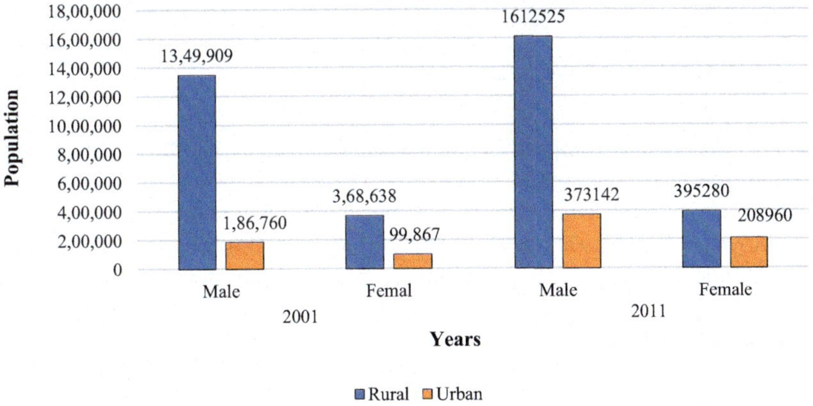

Fig. 5 Work participation rate in rural and urban areas (2001–2011). (Source: District Census Handbook, 2001–2011)

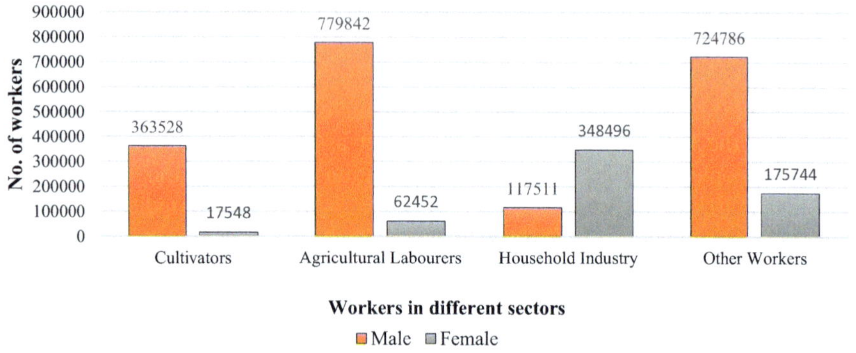

Fig. 6 Work participation rate in different sectors. (Source: District Census Handbook, 2011)

district's urban work participation rate continued to rise. The work participation rate in total workers (main and marginal) increased from 286,627 in 2001 to 582,102 in 2011, and then, from the 2001 Census to the 2011 Census, it follows an increasing trend in urban areas.

6.4 Work Participation Rate in Different Sectors

Four different worker types—cultivators, agricultural laborers, household industry workers, and other workers—are taken into consideration in this study. Figure 6 shows that 381,076 work as cultivators where 363,528 are male and 17,548 are female workers. A total of 842,294 are agricultural laborers where 779,842 are male and 62,452 are female workers. A total of 466,007 are workers in the household industry where 117,511 and 348,496 are male and female workers, respectively.

7 COVID-19 and Migration

The socio-economic composition of the surveyed areas in Murshidabad district refers to the characteristics of population composition, work composition, caste composition, religion composition, income, family status, education, and so on. Here, all these variables are described briefly with the illustration. And, finally, the socio-economic composition found in these areas is in developing condition.

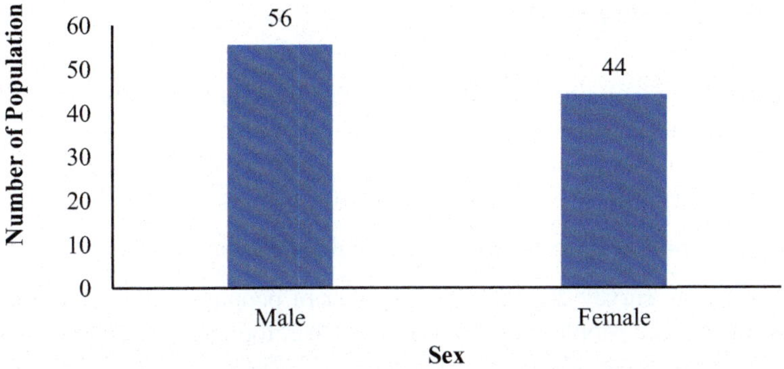

Fig. 7 Percentage distribution of male and female population of the surveyed areas of Murshidabad district, 2023. (Source: Primary Data, 2023)

7.1 *Population Composition*

From Fig. 7, it has been shown that the total populations of the surveyed households are 1592 persons, in which the male population has been registered at 55.65% and the female population has recorded at 44.35%. It has been established that the surveyed C.D. blocks have an unbalanced distribution of male and female inhabitants and a high sex ratio. A total of 796 females were detected in the surveyed C.D. blocks, which have a calculated sex ratio of 1000 males.

7.2 *Dependency Ratio*

A statistic that can be used to determine how many people-including children and senior citizens are dependent on wage earners is the dependency ratio. It is also considered to be the overall or youth dependency ratio.

Dependency Ratio Total population of age 0–14 and population of >60 years/ population of the working age group (15–59 years) × 100.

According to survey data, the total population of the 0–14 age group of the surveyed area of Murshidabad district is 214. The total population of people aged >60 years is 74, and the population of the working age group 15–59 is 1304. Therefore, the dependency ratio in the surveyed area of Murshidabad district is 22.08%.

7.3 Work Participation Rate

The rate of labor participation is determined by dividing the number of workers as a whole by the total population. The working age group is defined as 15–64 years of age. This indicator is divided into age groups, which are defined by a percentage of each age group.

Work Participation Rate (Total worker/total population) × 100.

According to survey data, the total number of people is 1592, where the total number of engaged people (migration) is 480. Therefore, the work participation rate of the surveyed area of Murshidabad district is 30.15%.

7.4 Occupation Category of Migrant Worker

People are migrated from their motherland to other places for different purposes or work. In this study area, people go from their place to another place for different work or job purposes like cooking, masonry, and rickshaw pulling. Over 182 persons are labor workers, and this is followed by masons (82), hotel workers (32), factory laborers (64), etc. The lowest person is engaged with the post of a hawker, business, serviceman, conductor, etc. (Fig. 8).

7.5 Time Duration of Migration

Different types of migrant workers work in different places with different time durations. The total no. of migrated persons is 480. Over 49% of workers are migrating from their home place to their current place 5–10 years ago. Most people migrated during this period. During the period 0 to 5 years, 23.75% of workers are migrating from their native place as follows: 22% during the period 10–20 years and 3.75% and 2.22% between the ages of 20 and 30 and 30 and 40, respectively. This different time duration happened mostly because of the different age groups of people migrating and also the economic development perspective (Fig. 9).

7.6 Migration Status

Migration means when people go somewhere from their birthplace and stay for some time in this place. Due to different work purposes, international and national migration is happening in this study area. In Fig. 10, we can see that over 460 people

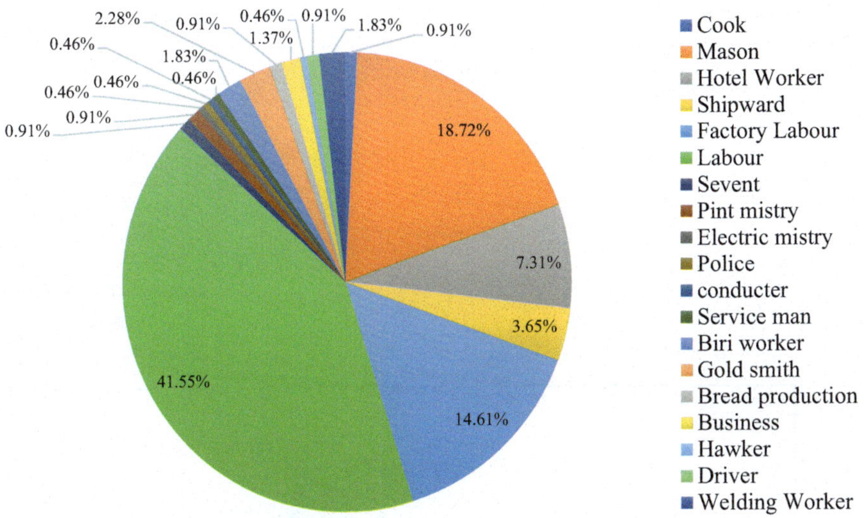

Fig. 8 Occupation category of surveyed migrant worker. (Source: Primary Data, 2023)

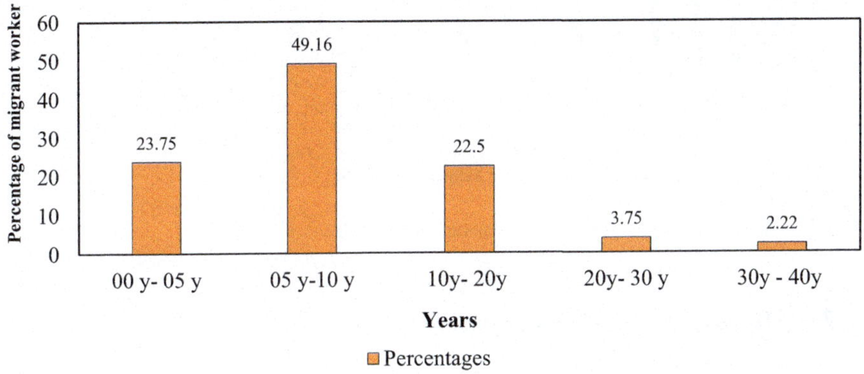

Fig. 9 Duration of migration of the surveyed areas of Murshidabad district, 2023. (Source: Primary Data, 2023)

are going under national migration for their work and over 20 people are going to international migration workplaces like Saudi Arab, Nepal, and Dubai. According to those who have migrated abroad, their job opportunities are much better than in India, and the level of income is higher, so they migrate abroad. Again, it is also seen that a large number of migrant workers go outside West Bengal in search of work in different states. Migrant workers in other states get higher wages as compared to West Bengal so they go to other states with great difficulty to earn their living.

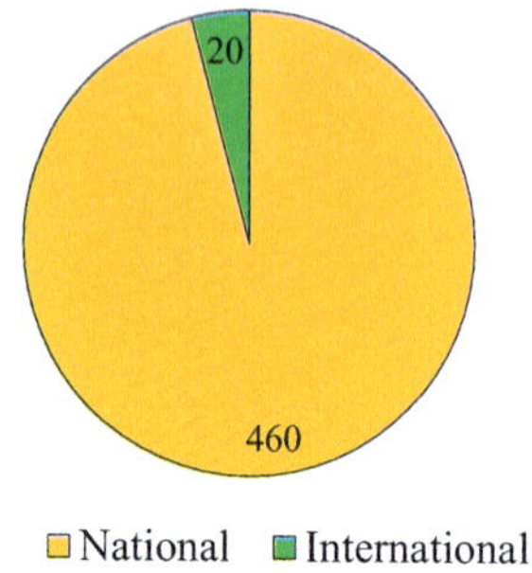

Fig. 10 Migration status of the surveyed areas of Murshidabad district, 2023. (Source: Primary Data, 2023)

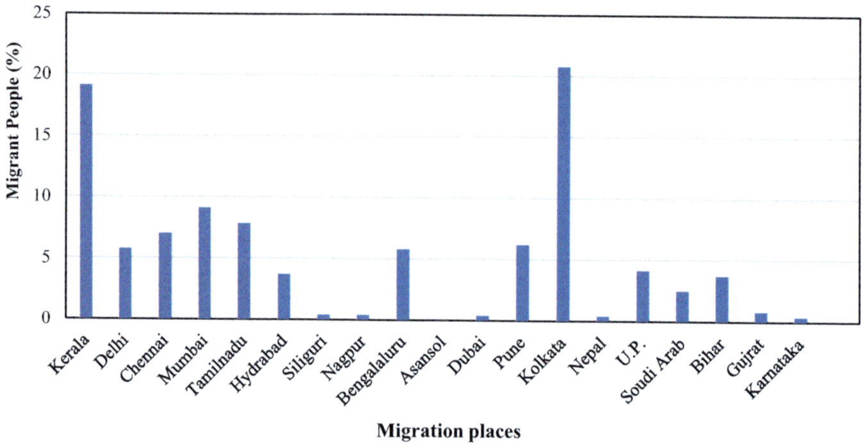

Fig. 11 Surveyed migrant people's workplace. (Source: Primary Data, 2023)

7.7 *Migrant People's Workplaces*

Migrant people's workplaces mean where the migrant people migrate themself for their work. In this diagram (Fig. 11), we can see that people migrate to different states or countries for their employment. People of this particular district migrate from there to Kolkata, Kerala, Karnataka, Chennai, Mumbai, Delhi, and Hyderabad, and some people also go outside of our county as international migrants. Over 20.83% and 19.16% of people are going to Kolkata and Kerala, respectively. Then, most are going to Mumbai, Chennai, and Tamil Nādu followed by Bengaluru, Delhi, Gujrat, and Pune. Many people are going outside the international border like Dubai (0.44%), Nepal (4.1%), and Saudi Arab (2.5) for their work. People rarely go to Asansol for work.

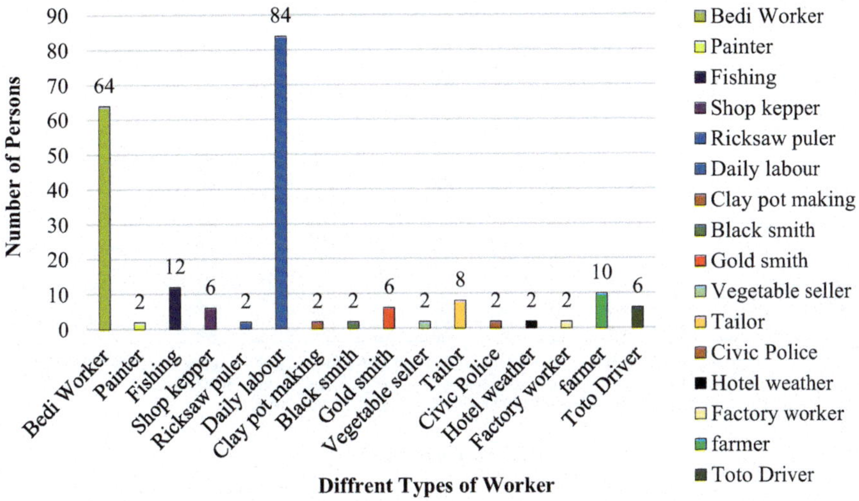

Fig. 12 Non-migrant family member's income source in the surveyed areas. (Source: Primary data, 2023)

7.8 *Non-migrant Family Member's Income Source*

In our study area, Murshidabad district, we have surveyed around 380 households. After analysis of primary data, we have found that 480 persons are migrant workers among the total population of 1592. The migrant workers have been working in different places in West Bengal, different states of India, and different places of the World. And we have also noticed that 212 people worked as non-migrant workers at their local place, they made another source of income, and they earned the money to contribute the family, for the family's well-being.

In this diagram (Fig. 12), it has been clearly shown that the non-migrant workers worked in category types of work. A total of 64 people worked as beedi workers, and 84 people worked as daily laborers. Additionally, the following persons worked in these different types of work: painters 2, fishermen 12, shopkeepers 6, rickshaw pullers 2, clay pot makers 2, blacksmiths 2, goldsmiths 6, vegetable sellers 2, tailors 8, civic police 2, hotel workers 2, factory workers 2, farmers 10, and toto drivers 6.

7.9 *Monthly Income of the Migrant Worker*

Murshidabad district is one of the most important districts based on migrant workers. From the Murshidabad district, thousands of migrant workers, day by day, spread in the different states of India as well as the different countries of the world. The maximum number of migrant workers migrated just because of better job opportunities. And they have earned a good amount of money in a month.

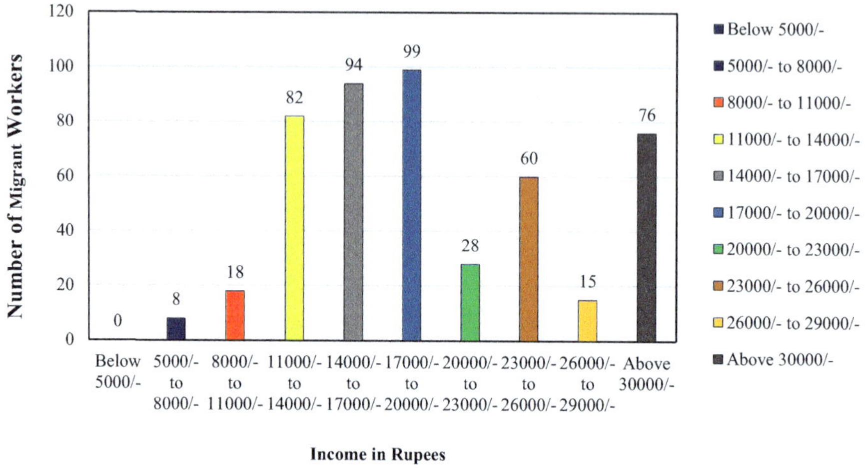

Fig. 13 Monthly income from surveyed migrant workers. (Source: Primary data, 2023)

We have surveyed around 380 households, and 480 persons have been migrant workers among the 1592 total surveyed population. The migrant workers worked in different places at a different time duration. Maximum migrant workers worked as a laborer (Raj Mistry), factory laborers, masons, etc., and they earned a good amount of money in a month.

The monthly income of the migrant workers has been divided into ten categories accordingly. In Fig. 13, it has been clearly shown that no migrant workers have to belong to the below 5000 monthly income categories. Over 99 migrant workers belong to 17,000 to 20,000, and 76 persons of migrant workers belong to the above 30,000 monthly income categories. We also noticed that the highest and second highest migrant workers belong to the 17,000 to 20,000 and 14,000 to 17,000 monthly income categories.

Monthly Income of the Family Members

From a household's point of view, monthly income plays an important role in any household. Because it has worked as a backbone of the family. In our study area, Murshidabad district, we have surveyed around 380 households. The maximum number of surveyed households was migrant workers' households. Around 480 persons were migrant workers, and 212 persons of workers have been found who have been working in their local places.

The monthly income of the family has been divided into ten categories accordingly. In Fig. 14, it represents the total income of the family. After primary data analysis, we have found that 17 households belong to the below 5000 monthly income categories, and according to the primary data survey, 27 households belong

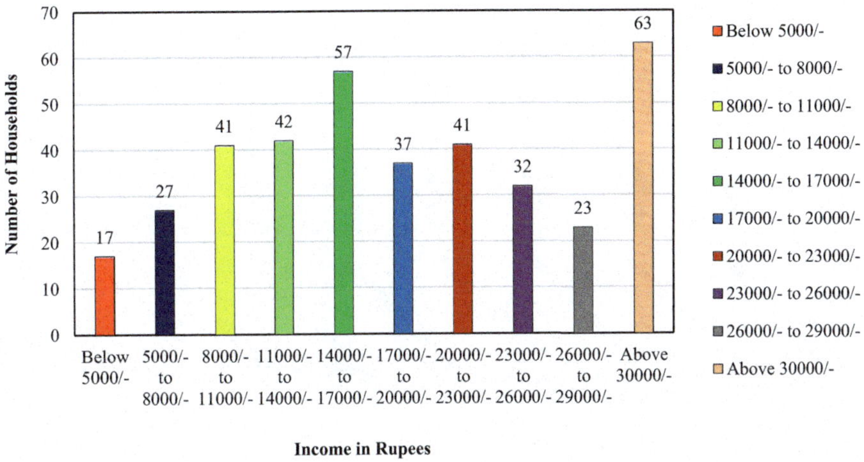

Fig. 14 Monthly income of the surveyed family members. (Source: Primary data, 2023)

to the 5000 to 8000, and 23 households belong to 26,000 to 29,000 categories, respectively. And the highest number of households 63 among the 380 surveyed households belong to the highest category of the monthly income which was above 30,000.

7.10 Role of Remittance in Migrant Family Welfare

Family welfare means the well-being of every indicator of human development, like child education, food quality, use of modern amenities, and living conditions. In our study area, Murshidabad district has a surplus labor class district. We have surveyed around 380 households. According to our surveyed primary data, we noticed that 430 people have been migrant workers among the 1592 surveyed total population. The maximum number of families' welfare has depended on the remittance of the migrant workers.

Figure 15 clearly shows that 198 respondents responded that the development of a boy child's education is moderate, 140 persons responded that a boy child's education is low, and few people responded that the development of a child's education is very high. Respectively, a maximum number of respondents responded to other indicators of family welfare like girl child education, use of modern amenities, food quality, living conditions, and health care facilities. Development rating has been moderate, and around 50 persons in each category responded that the family welfare rating is high, while few people have responded that the indicators of family welfare rating are very high.

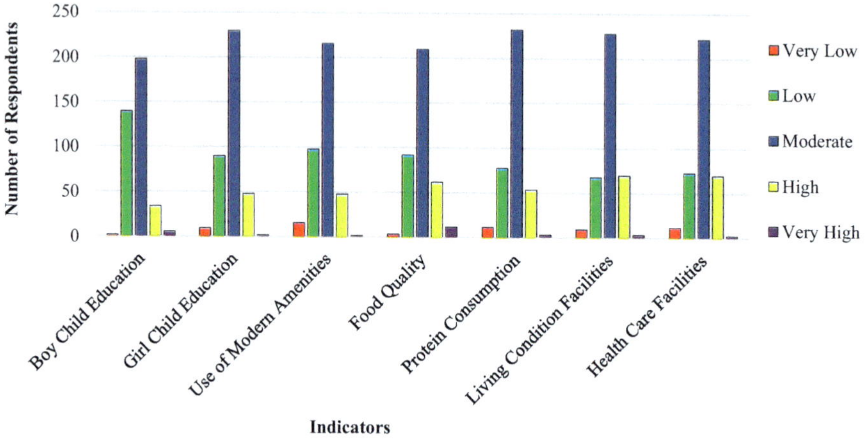

Fig. 15 Development rating of family welfare of the surveyed areas of Murshidabad district, 2023. (Source: Primary data, 2023)

7.11 *Utilization of Remittance*

Remittance is the money that is sent by migrant workers to their families. The migrant worker's families have used the money in different sectors for their family well-being. We conducted a survey of about 380 houses in our study region and discovered that, out of the 1592 people living there, 480 are migrant workers. The migrant workers are employed in various locations around West Bengal, various Indian states, and many countries worldwide. The migrant family utilized their remittance in different types of sectors, like education, health, food, and clothing (Fig. 16).

7.12 *Changes of Occupation After COVID-19 Effect*

Figure 17 shows the changes in occupation after COVID-19. Lots of people lost their occupations due to the COVID-19 effect. Therefore, migrant people are joining different occupations like toto drivers, daily laborers, beedi workers, vegetable sellers, farmers, painters, and other occupations. Therefore, many migrated laborers have changed their occupations due to the effects of COVID-19.

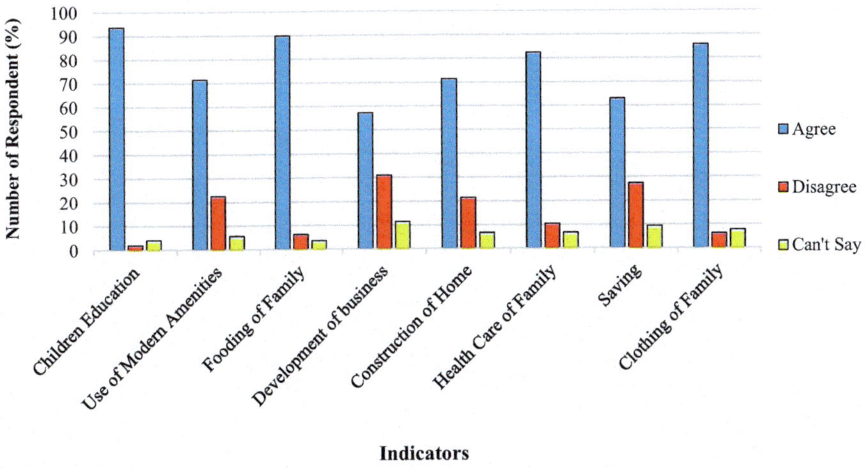

Fig. 16 Utilization of remittance of the surveyed areas of Murshidabad district, 2023. (Source: Primary data, 2023)

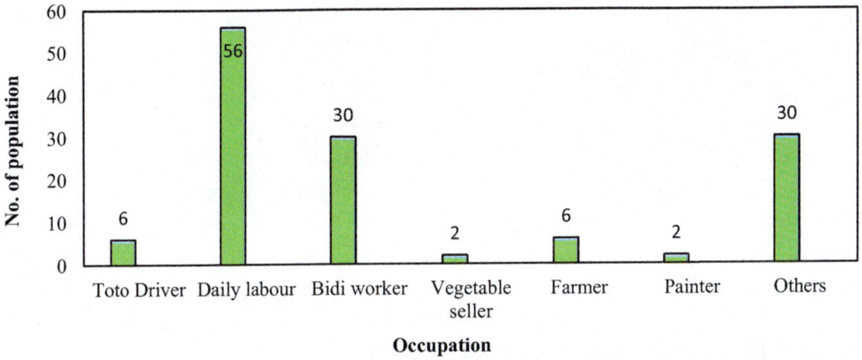

Fig. 17 Changes of occupation after COVID-19 effect of the surveyed areas of Murshidabad district, 2023. (Soirce: Source: Primary Data, 2023)

7.13 *Trends of Migration of Workers*

A survey of the houses of migrant workers in different blocks of Murshidabad revealed that, after the COVID-19 pandemic, their financial situation worsened. Financial hardship has forced them to go to other states or countries to earn money. There are more migrant workers than ever before due to the loss of income locally. The diagram (Fig. 18) given makes it easy to understand how fast the number of migrant workers has grown in the last three years. The growing number of migrant workers in the past few years gives a preliminary idea of how much the number of migrant workers may increase in the future.

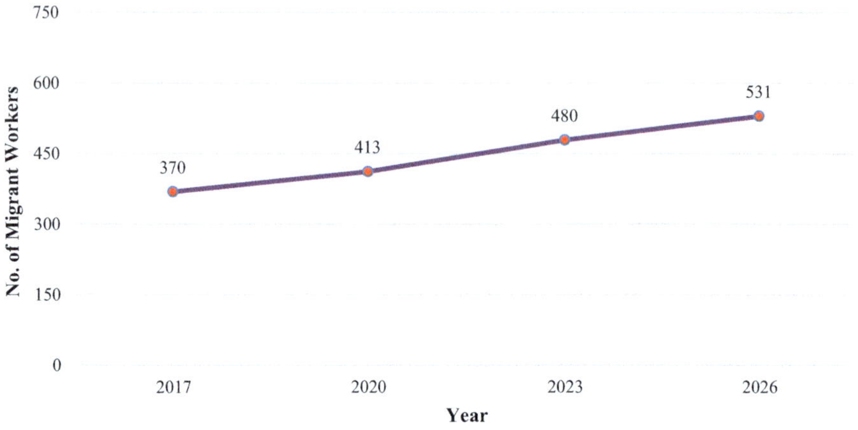

Fig. 18 Future trend of the migrant workers in the surveyed areas of Murshidabad district. (Source: Primary Data, 2023)

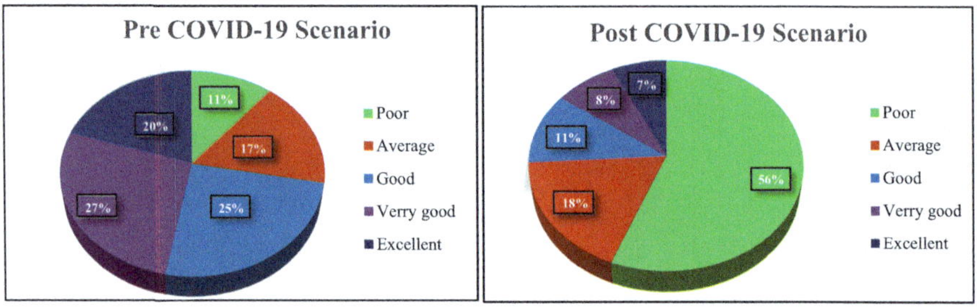

Fig. 19 Impact of COVID-19 on economic scenarios in the surveyed areas of Murshidabad. (Source: Primary Data, 2023)

7.14 Impact of COVID-19 on Economic Scenarios

Migration modifies labor markets, consumer trends, investment flows, employment dynamics, and economic inequality, hence impacting economic interactions. In our study area, the district of Murshidabad, we surveyed around 380 households. After analyzing primary data, we noticed that the economic status of the district was good before COVID-19. But after the arrival of COVID-19, that situation has changed and deteriorated. The diagram (Fig. 19) nicely illustrates the economic environment before and after the arrival of COVID-19. According to interviews with locals, this deadly disease has affected them more economically than physically. As the number of working people in poor families is less, they are forced to work outside as migrant workers against their will. The workers who were previously employed in the same occupation now have a much lower income but a comparatively higher expenditure. They have now moved from above the poverty line to below the poverty line.

8 Why People Migrate from Murshidabad

Migration in geography refers to the movement of people from one place to another. It occurs when the perceived interaction of push and pull factors overcomes the friction of moving. Human migration involves people moving from one place to another with the intention of settling, either permanently or temporarily, at a new location. This movement often takes place over long distances and can involve moving from one country to another.

There are several reasons that influence people's decisions to migrate from Murshidabad. People usually migrate on the basis of the push and pull factors. Push factors are basically the lack of job opportunities, lack of health facilities, lack of education facilities, poverty, high crime rate, etc. And the pull factors are likely better job opportunities, a good climate, political stabilization, a wealth of education and health facilities, etc.

After the COVID-19 effect, the migration rate in the Murshidabad district has most effectively increased. During COVID-19, many small workers, farmers, and other workers have totally perished.

9 Effect of COVID-19 on Social Physics

(i) Everything abruptly comes to a stop, except for necessities like food and water. Surviving in such bad conditions was the biggest challenge for them during this period.

(ii) Large numbers of migrant workers lost their jobs, and many factories faced closure as a result of the huge financial crisis.

(iii) Almost all forms of public transportation stopped. Due to the lack of public transport, many people walked or cycled hundreds of km to get home. This long journey resulted in many infections, and many lost their lives due to a lack of adequate treatment after returning home. A large number of migrant workers were unable to go back home.

(iv) Migrant workers returning from their workplaces were forced to refrain from attending social gatherings and cultural events to avoid the spread of this infection.

(v) Many migrant workers lost their shelter, family members, financial security, etc.

(vi) Many migrant workers returned to their places of residence and engaged in new occupations, but it was very difficult to support their families with the money earned in that occupation.

(vii) Workers from poor families fall below the poverty line in this situation. They became dependent on government aid.

(viii) When migrant workers return to their workplaces after the end of the COVID-19 period and join their old occupations, most of them find that their wages have been reduced from before.

10 Suggestions

(i) If settlers can improve their accommodation, they will be able to improve their conditions.
(ii) They need to improve their economic status so that they can improve their basic needs.
(iii) Create awareness among the migrant workers about the next generation's future.
(iv) They should be aware of the education and health status of their children.
(v) Authority should assure the migrant workers so that they can start their businesses when they return to poverty.
(vi) The authority can grant them silent loans.
(vii) These workers may also have priority in opening small and medium enterprises.
(viii) The government should establish a pension fund and a contingency fund for them.
(ix) And initially create attention to migrant workers for their health, education, and also profitable development.

11 Conclusion

An effort has been made in this research paper to describe migrant workers' risk, vulnerability, and level of life satisfaction during the COVID-19 period. An unprecedented population of Murshidabad travels from state to state or abroad in search of work and hopes to earn two meals a day. Higher pay, job availability, and better working conditions are said to be the main drivers of migration in Murshidabad. A more vulnerable group of people are migrant laborers who are unskilled or semi-skilled. Their remittances sometimes make it impossible for their family to thrive and to make them happy. Even so, they continue to try hard to shoulder the burden of their family. However, the unexpected worldwide lockdown and COVID-19 resulted in the overnight loss of hundreds of thousands of jobs for migrant workers, who then took to the streets. Everything was shut down during this lockdown; there were no supplies or facilities. As a result, the entire economy suffered losses, but it was excessive in the unorganized sector. This led to rising poverty and a decline in the labor force. A key priority across these dimensions is migrant workers who have been the backbone of the Indian economy and are now some of the hardest hits (Das, 2020). Their lives and way of life have suffered as a result of the lockdown. According to the responders, the migrant workers who were stuck elsewhere did not have any basic survival food, and there was no money to drive back home. According to several media reports, they faced mental and emotional stress. Many have walked or cycled hundreds of km to return home. They lacked jobs, shelter, and hope. The Indian and West Bengal governments have taken several actions, but they have not been properly implemented. Many migrant workers changed their livelihood into other unskilled work, intentionally or unintentionally, during and after the COVID-19 lockdown. This coronavirus is a dark chapter in the lives of migrant workers and their families.

References

Das, A. R. (2020). Socio-economic problem and challenges of migrant Workers in Murshidabad District: A study in the context of Covid-19. *Journal of People's History and Culture, 6*(1), 140–145.

District Census Handbook, Murshidabad, 2011—Accessed on 11th May, 2023. https://censusindia.gov.in/nada/index.php/catalog/1341

District Statistical Handbook, Murshidabad, 2014—Accessed on 27th June, 2023. http://wbpspm.gov.in/publications/District%20Statistical%20Handbook

Koushik Kumar Das possesses a great academic background and is an accomplished individual. In 2020, he earned his Bachelor of Science in Geography from Jiaganj Murshidabad's Sripat Singh College. He passed his Master's in Geography from Kalyani University with specialization in Environmental Science in 2022. Currently, he is pursuing a Bachelor of Education (B.Ed.) degree from Nathulal Das B.Ed. College in Murshidabad. Currently, he is continuing his research on various topics in social sciences. His manuscripts on various topics have already been published in many famous journals of West Bengal and the world.

Dr. Ajay Debnath is an assistant professor in the Department of Geography at Sripat Singh College, Jiaganj, Murshidabad. He has completed his graduation and master's from Presidency College at the University of Calcutta. He completed his Ph.D. degree from the University of Calcutta in 2018. He is currently doing research on the social welfare of Murshidabad district.

Bapan Mondal is an accomplished individual with a strong academic background. He completed his Bachelor of Science in Geography in 2020 from Sripat Singh College, Jiaganj Murshidabad. He furthered his education by earning a Master of Science in Geography from Asutosh College at the University of Calcutta, Kolkata. Currently, he is pursuing a Bachelor of Education (B.Ed.) degree from Netaji Subhas Chandra Bose P.T.T.I. in Murshidabad, with a passion for teaching and education.

Part III
Social Learning and Environmental Management

Access to Maternal and Child Healthcare in India: Mapping of Regional and Social Variation

Vanishree Joseph, Prasad N. S. R., N. V. Madhuri, and Kesava Rao Pyla

1 Introduction

Ensuring sufficient maternal and child health services is a fundamental human right and a crucial indicator of a country's healthcare system. In India, despite significant efforts to improve healthcare services, substantial disparities persist in accessing these essential services, particularly among different regions and social groups. Understanding and addressing the regional and social variation in the availability of maternal and child healthcare are essential for confirming equitable healthcare delivery and improving health outcomes for women and children throughout the country.

The aim of this research article is to conduct an in-depth analysis of the mapping of regional and social variation in maternal and child healthcare accessibility in India. By synthesizing existing literature, this study intends to shed light on the factors contributing to disparities in accessing healthcare services, identify key challenges and barriers, and propose strategies for achieving more equitable healthcare provision. Maternal and child healthcare remains a pressing concern in India. Despite significant progress over the years, the nation still grapples with elevated rates of maternal and infant mortality, indicating a need for further improvements in healthcare access and quality.

V. Joseph · N. V. Madhuri
Centre for Gender Studies and Development at the National Institute of Rural Development and Panchayati Raj (CGSD) NIRDPR, Hyderabad, India
e-mail: vanishree.nird@gov.in

P. N. S. R. (✉) · K. R. Pyla
Centre for Geo-Informatics for Rural Development at the National Institute of Rural Development and Panchayati Raj (CGARD) NIRDPR, Hyderabad, India
e-mail: nsrprasad.nird@gov.in; kesava.nird@gov.in

A. Rai et al. (eds.), *Contemporary Social Physics*, Springer Geography,
https://doi.org/10.1007/978-3-031-77596-3_9

Moreover, India's vast geographical and cultural diversity adds complexity to the challenge of ensuring equitable access to healthcare services across different regions and social groups. Variations in different regions significantly influence the nearness of maternal and child healthcare facilities. Healthcare infrastructure, availability of skilled healthcare providers, and utilization of services vary significantly between states and geographical areas. Factors such as economic development, urban-rural divide, and governance mechanisms contribute to these regional disparities. Understanding the specific challenges faced by different regions is vital for designing targeted interventions and healthcare policies to address their unique healthcare needs.

Furthermore, social variation is a key determinant of healthcare access in India. Factors such as caste, religion, income, education, and gender significantly influence individuals' ability to gain entry and make use of services. Marginalized populations, including tribal communities, Scheduled Castes, and economically disadvantaged groups, face significant barriers to accessing quality healthcare services due to societal discrimination, limited resources, and inadequate awareness.

Recognizing and overcoming obstacles in reaching maternal and child healthcare services are vital to enhancing health outcomes and minimising inequalities. Gupta et al. (2014) examined factors related to antenatal care visits in Tanzania, stressing the significance of recognizing and tackling challenges in accessing maternal healthcare, including the declining utilization rates observed over time. Financial constraints, cultural beliefs and practices, distance to healthcare facilities, and inadequate infrastructure are among the key obstacles faced by individuals seeking healthcare services. Manna and Basu (2023) conducted a study mapping social and regional disparities in access to maternal healthcare in India, highlighting the importance of addressing these inequities to ensure improved maternal and child health outcomes. Moreover, social norms, gender discrimination, and healthcare provider attitudes can perpetuate inequalities, further hindering equitable access to care.

Through this chapter, we aim to synthesize the existing literature on pregnant women, lactating mothers, and children's healthcare in India, mapping regional and social variation. By critically analyzing the research findings, we will highlight the persistent disparities in healthcare access, identify the underlying determinants, and propose evidence-based strategies to address these challenges. It is imperative to develop comprehensive and targeted interventions, strengthen healthcare infrastructure, promote awareness, and ensure policy implementation to improve accessibility to healthcare services for pregnant women, new mothers, and children across regions and social groups in India.

1.1 Objectives

In light of the aforementioned context, this research aims to meticulously evaluate the accessibility and distribution of maternal and child health (MCH) care services across India, with a particular focus on identifying the underlying regional and social disparities. The objectives of this study are as follows:

- To assess the prevalence of key MCH indicators, including antenatal check-ups, institutional deliveries, underweight children, complete immunization coverage, and infant mortality rate (IMR) across different states and union territories of India.
- To analyze the socioeconomic and demographic variables that contribute to disparities in MCH outcomes, thereby identifying the most vulnerable groups within the population.
- To evaluate the coverage gap in MCH services based on social stratification, thereby highlighting the extent of disparity in access to healthcare services across different regions and communities.
- To propose targeted interventions based on the findings, aimed at reducing the inequalities in access to MCH care, contributing to the advancement towards achieving the Sustainable Development Goals (SDGs) pertaining to maternal and child health in India.

By setting these aims and objectives, the study seeks to provide a comprehensive understanding of the current state of MCH care access in India and to offer actionable insights for policymakers to address the critical gaps and challenges identified.

1.2 Review of Literature

This review critically examines the current landscape of maternal and child health (MCH) care in India, focusing on disparities and interventions. Despite India's enhanced economic growth, health inequities in MCH outcomes persist, with significant disparities observed across different socioeconomic groups and states. Research by Gopika et al. (2022) highlighted the alarmingly low rates of cancer screening participation in India, underscoring the need for improved public awareness and screening programs. Similarly, Girotra et al. (2023) discussed the underutilization of antenatal care services, revealing that a substantial portion of mothers do not complete the recommended number of antenatal visits, exacerbating the risk of low birth weight and adverse health outcomes.

Studies emphasize the critical role of socioeconomic factors, with findings indicating that the poorest and rural populations suffer the most in terms of MCH outcomes. Bango and Ghosh (2022) addressed social and regional disparities in MCH service utilization, noting that disadvantaged social groups face limited accessibility and availability of complete healthcare services, which hinders the achievement of "Health for All". This is further compounded by the insights from Jana et al.

(2023), who illustrate the relationship between low birth weight and infant mortality, stressing the importance of targeted interventions to mitigate these risks. Mishra et al. (2023) conducted an explorative qualitative study in Odisha, revealing that while community health workers provided significant support, transportation limitations and fear of infection led to a reluctance to seek care, impacting antenatal and postnatal services negatively. Similarly, Manna and Basu (2023) highlighted the financial and logistical challenges faced by families in remote rural regions of West Bengal, leading to increased out-of-pocket expenses and reliance on private healthcare facilities for childbirth during the pandemic.

Interventions aimed at bridging these disparities often encounter challenges due to misaligned governance structures, as observed by Etiaba et al. (2023) in their examination of maternal, neonatal, and child health policy processes in Nigeria. Their findings suggest that collaborative efforts across government levels are essential yet frequently hampered by systemic issues. Uneke et al. (2019) advocated for equity-focused interventions under the reproductive, maternal, neonatal, child and adolescent (RMNCH+A) strategy, emphasizing the need for programs that address human resource shortages in rural areas and the economic and social empowerment of women. Addressing these disparities requires a multifaceted approach that involves improving public awareness, expanding access to quality care, and fostering intergovernmental collaboration.

1.3 Study Design

The research paper employed a cross-sectional study approach, utilizing data gathered from the NFHS 5 survey. This methodology enables the analysis of various maternal and child health indicators along with their corresponding factors.

1.4 Methodology

In this research article, a descriptive methodology was adopted to examine the data obtained from the National Family Health Survey 5 (NFHS 5). NFHS 5 is a nationwide survey conducted in India, capturing data on diverse health indicators, including those related to maternal and child health. The survey employs a multi-stage sampling design, ensuring the inclusion of diverse regions and social groups across the country.

To conduct the analysis, the researchers obtained the relevant data from the NFHS 5 dataset, which includes information on maternal and child health outcomes, access to healthcare services, demographic characteristics, and socioeconomic factors. The study also incorporated a literature review to contextualize the NFHS 5 findings within the existing body of knowledge on maternal and child healthcare in India. The synthesis of the NFHS 5 data with insights gleaned from the literature review was guided by a thematic analysis framework. This approach enabled the identification of recurring themes and patterns related to healthcare access and disparities, facilitating a nuanced understanding of the complex interplay between socioeconomic determinants, demographic factors, and health outcomes. Key themes were categorized and analyzed to highlight regional and social variations in healthcare accessibility and to draw connections between these variations and the broader health indicators captured in the NFHS 5 dataset.

1.5 Selection of Criteria

The criteria for selecting specific maternal and child health (MCH) indicators in this study were based on their relevance to assessing the overall health status of mothers and children, as well as their ability to reflect disparities in access to healthcare services across different regions and social groups in India. The selected indicators, including antenatal check-ups, institutional deliveries, underweight children, and complete immunization coverage are widely recognized by health professionals and international organizations such as the World Health Organization (WHO) as key measures of MCH care effectiveness. These indicators were chosen for their potential to provide insights into the quality of healthcare services, the success of public health interventions, and areas where targeted efforts are necessary to improve health outcomes. The selection was also influenced by the availability of comprehensive and reliable data within the NFHS 5 dataset, ensuring that our analysis was grounded in robust evidence.

2 Data Analysis

2.1 Antenatal Check-Up

"Child marriage in India poses significant barriers to accessing maternal and child health care services, contributing to adverse outcomes for young women and their children" (Raj et al. 2009). Figure 1 illustrates notable differences in the percentage of mothers who received a minimum of four antenatal care visits across various states and districts in India.

Certain districts exhibit elevated percentages, signifying that a significant portion of expectant mothers in those regions received consistent antenatal care. Notably, districts such as South Goa, Dadra and Nagar Haveli, and Ernakulam have percentages exceeding 90%, implying that the majority of mothers in these districts availed themselves of antenatal care services.

On the other hand, there are districts with lower percentages, indicating a relatively lower utilization of antenatal care services. For instance, districts like Kaimur

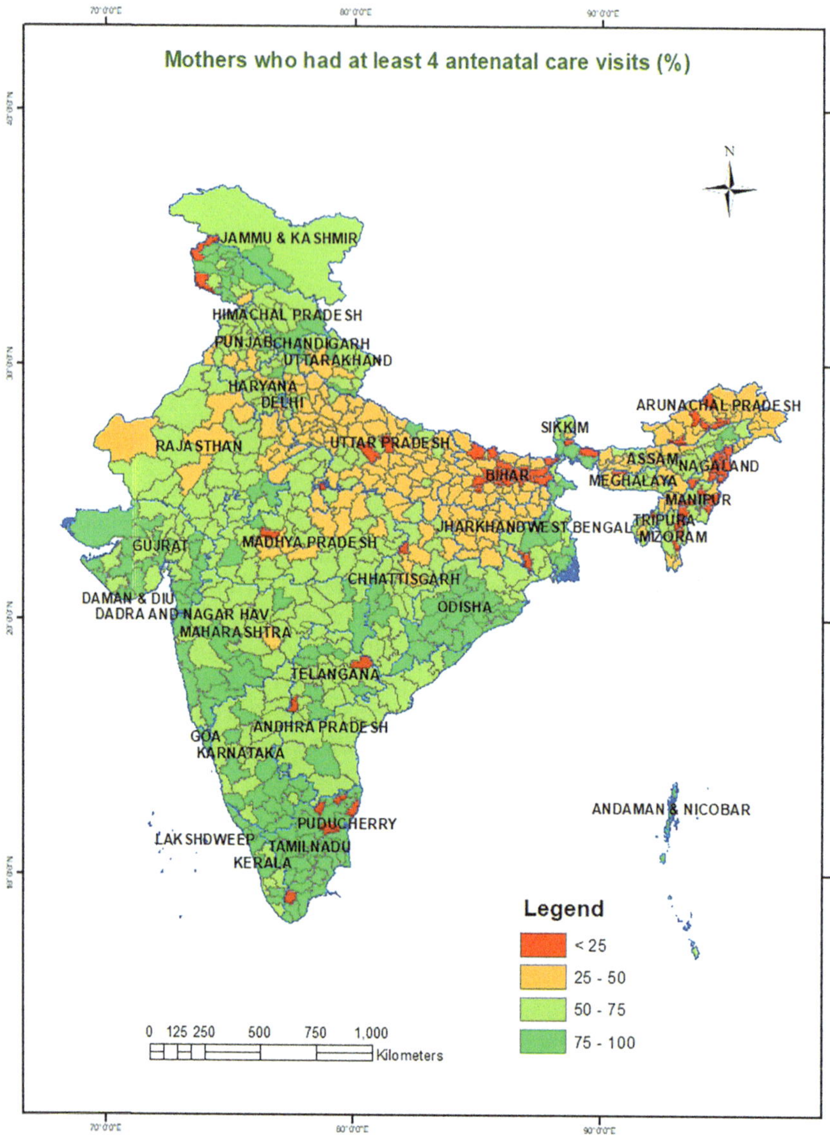

Fig. 1 Mothers who had at least four antenatal care visits

(Bhabua) in Bihar, Dhubri in Assam, and Bastar in Chhattisgarh have percentages below 40%, suggesting that a significant number of mothers in these districts did not obtain the advised quantity of antenatal care appointments.

It is important to note that several factors can influence these percentages, including availability and accessibility of healthcare services, awareness and education levels of the population, cultural practices, and socioeconomic factors. Here are some key observations from the map:

Variation Across States There is significant variation in the percentage of mothers receiving adequate antenatal care across different states. For example, in Dadra and Nagar Haveli, all districts have high percentages ranging from 70.5% to 90.9%. But states like Jharkhand and Bihar have several districts with low percentages, indicating a need for improvement in antenatal care access and utilization.

Regional Differences Within states, there are variations in antenatal care utilization across districts. For instance, in Andhra Pradesh, the YSR district has the highest percentage (82.8%) of mothers who require a number of antenatal care visits, while East Godhavari (51%) has the lowest. Similarly, in Maharashtra, some districts like Gadchiroli (86.8%) and Kolhapur (81.8%) have high percentages, while others like Nandurbar (57.3%) and Mumbai (57.1%) have relatively lower percentages.

Urban-Rural Divide The table does not explicitly indicate whether the districts are rural or urban, but it is likely that districts with major cities or urban centres tend to have higher percentages of mothers receiving adequate antenatal care. For example, in Gujarat, districts like Ahmedabad (77.8%), Surat (93.4%), and Rajkot (93.5%) have relatively higher percentages, which may be attributed to better healthcare facilities and accessibility in urban areas.

State-Wide Patterns Some states exhibit consistent patterns across their districts. For instance, in Kerala, most districts have high percentages ranging from 65.7% to 94.2%, indicating a generally strong antenatal care system throughout the state. Similarly, in Himachal Pradesh, the majority of districts have percentages above 55%, suggesting good overall access to antenatal care services.

Overall, the map highlights the disparities in antenatal care utilization across different regions of India. It emphasizes the need for targeted interventions and improvements in healthcare infrastructure, particularly in districts with lower percentages, to ensure that all pregnant women receive proper care during their pregnancies.

2.2 *Mothers Consumed Iron and Folic Acid Tablets*

The table provided represents the proportion of mothers who adhered to iron folic acid supplementation for a duration of 180 days or longer throughout their pregnancy in different states and districts of India (Fig. 2).

Here are some key observations and interpretations based on the data:

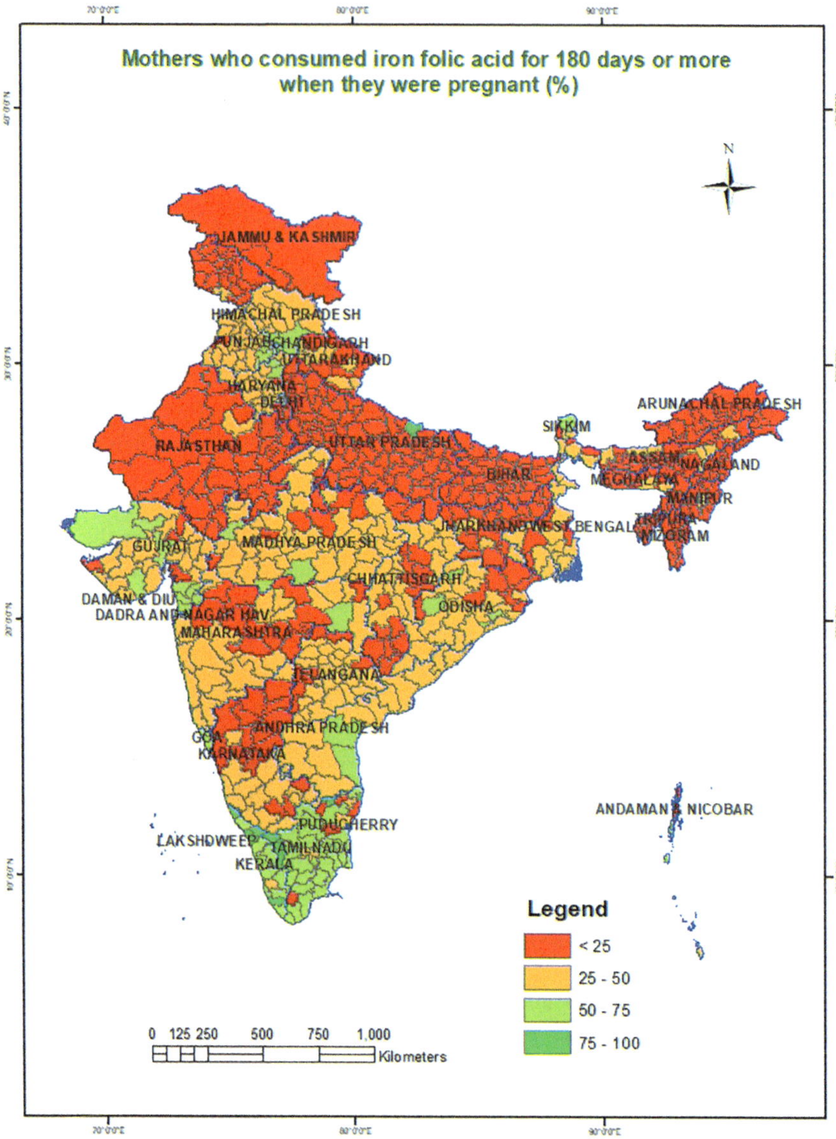

Fig. 2 Mothers who consumed iron folic acid for 180 days or more

Variation Among States There is considerable variation in the consumption of iron folic acid across different states. In Andaman and Nicobar Islands, the percentages range from 24.1% to 61.9%, while in Arunachal Pradesh, the percentages range from 1.2% to 17.6%.

High and Low Performers Some districts stand out as having relatively high percentages of mothers consuming iron folic acid, such as South Goa in Goa (59.1%), Kasaragod in Kerala (83.7%), and Dadra and Nagar Haveli (51.2%), whereas districts such as Leh (Ladakh) in Ladakh (2.6%), Burhanpur in Madhya Pradesh (60.7%), and Aurangabad in Maharashtra (6.9%) have very less consumption.

Regional Patterns There are regional patterns in the data. For instance, districts in Kerala generally have higher percentages, indicating better adherence to consuming iron folic acid during pregnancy. Similarly, districts in Gujarat, particularly Navasari (67%) and Valsad (59.1%), also show higher percentages. However, districts in Jammu and Kashmir, Jharkhand, and Bihar tend to have lower percentages overall.

Intra-State Variation There is significant variation within states as well. For example, in Andhra Pradesh, the percentages range from 32.3% to 61.9%, indicating differences in awareness and access to iron folic acid across districts within the same state.

Health Implications The intake of iron folic acid during pregnancy is vital for the well-being of both the mother and the child. Iron folic acid helps prevent anaemia, supports fetal development, and reduces the risk of birth defects. The data suggests that efforts should be made to improve the overall percentage of mothers consuming iron folic acid, particularly in districts with lower percentages.

Need for Intervention The data emphasizes the necessity for focused interventions and awareness initiatives to advocate for the significance of iron folic acid intake during pregnancy, particularly in regions and states where the percentages are lower. Improving access to healthcare facilities, antenatal care, and maternal education can contribute to increasing these percentages.

2.3 *Breastfeeding Within One Hour*

The data shows notable differences in the rates of early breastfeeding initiation among districts and states, highlighting varying levels of success in promoting and implementing this practice across different regions. At the district level, Prakasham district in Andhra Pradesh stands out with the highest percentage of children (73.8%) being breastfed within one hour of birth. This suggests that efforts in the Prakasham district have been successful in creating awareness and providing support to mothers for early breastfeeding.

On the other hand, Kishanganj district in Bihar has the lowest percentage of children (16.9%) being breastfed within one hour of birth. This indicates a need for targeted interventions and improved healthcare practices in Kishanganj to encourage and enable early breastfeeding (Fig. 3).

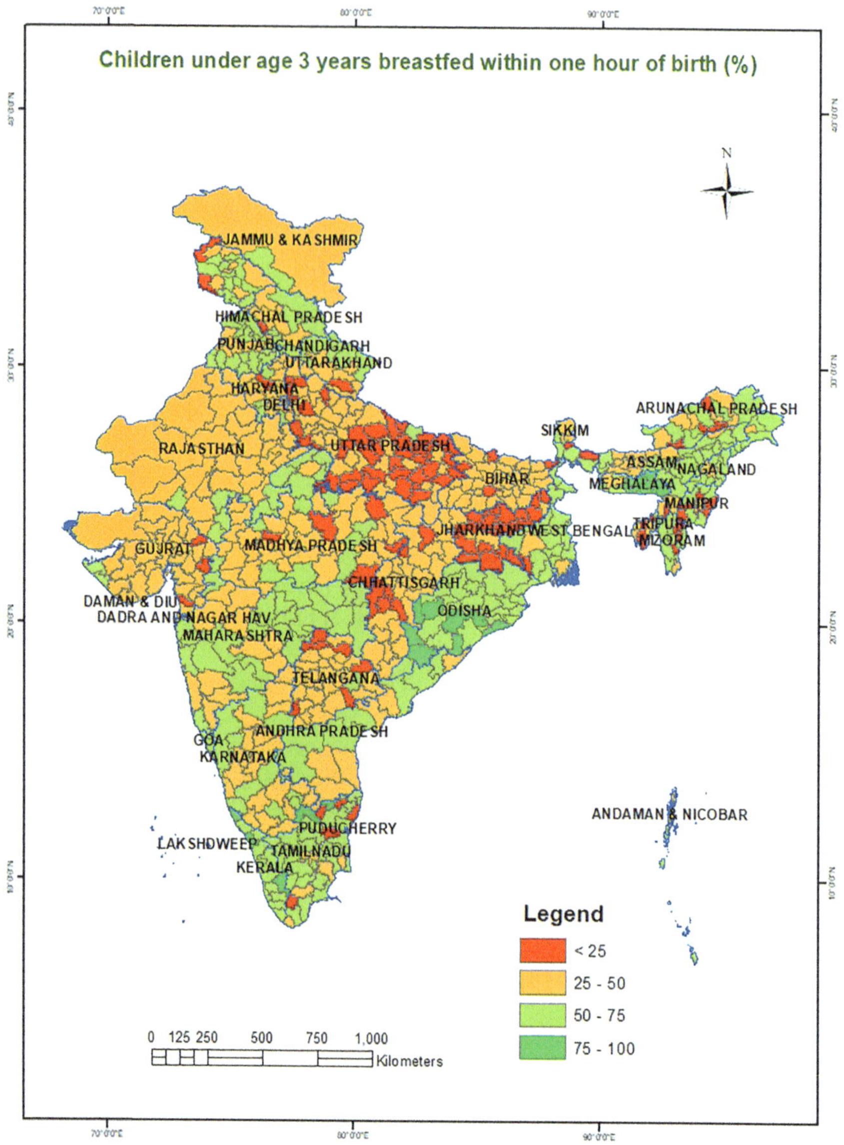

Fig. 3 Percentage of children breastfed within the first hour of birth

Andaman and Nicobar Islands exhibit the highest overall percentage of early breastfeeding initiation. Nicobar district within the Andaman and Nicobar Islands records the highest rate of 55.4%, indicating a positive trend towards promoting early breastfeeding in this region.

Conversely, Bihar reports the lowest percentage of early breastfeeding initiation at the state level. Kishanganj district within Bihar specifically demonstrates the lowest rate of 16.9%. This highlights the importance of implementing effective strategies and initiatives in Bihar to improve early breastfeeding practices and enhance child health outcomes.

These findings underscore the need for targeted interventions, education, and support for mothers across districts and states in India to encourage early initiation of breastfeeding. Early breastfeeding plays a vital role in providing essential nutrients, antibodies, and bonding between mother and child, contributing to the overall health and development of infants.

2.4 *Stunted Children*

The map displays information regarding the proportion of children under 5 years old who experience stunted growth (height-for-age) in various states and districts of India. Stunting is a condition that reflects chronic malnutrition and inadequate growth in children (Fig. 4).

Here are some observations from the map:

Variation Across States The stunting percentages vary significantly across different states. For example, in Karnataka, Yadgir district has a high stunting rate of 57.6%, while Chamarajanagar district has a relatively lower rate of 32.2%. Similarly, in Bihar, Sheikhpura district has the highest stunting rate of 53.6%, while Patna district has a lower rate of 34.5%.

Variation Within States There is also considerable variation within states. For instance, in Andhra Pradesh, Kurnool district has a significantly higher stunting rate of 50.5% compared to other districts like Srikakulam (19.7%) and East Godavari (23.1%).

Regional Patterns Some regions within states consistently show higher stunting rates. For example, districts like Sheikhpura (Bihar), Pashchimi Singhbhum (Jharkhand), Yadgir (Karnataka), and Dima Hasao (Assam) consistently have high stunting rates compared to other districts within the same state.

Urban-Rural Divide The data includes both urban and rural districts. It is worth noting that certain urban districts, such as Mumbai (Maharashtra), also have stunting rates above the national average, indicating that the issue of stunting is not limited to rural areas.

States with Lower Stunting Rates Some states have relatively lower stunting rates across most districts. For example, Kerala, Goa, Himachal Pradesh, and parts of northeastern states like Arunachal Pradesh show lower stunting rates compared to the national average.

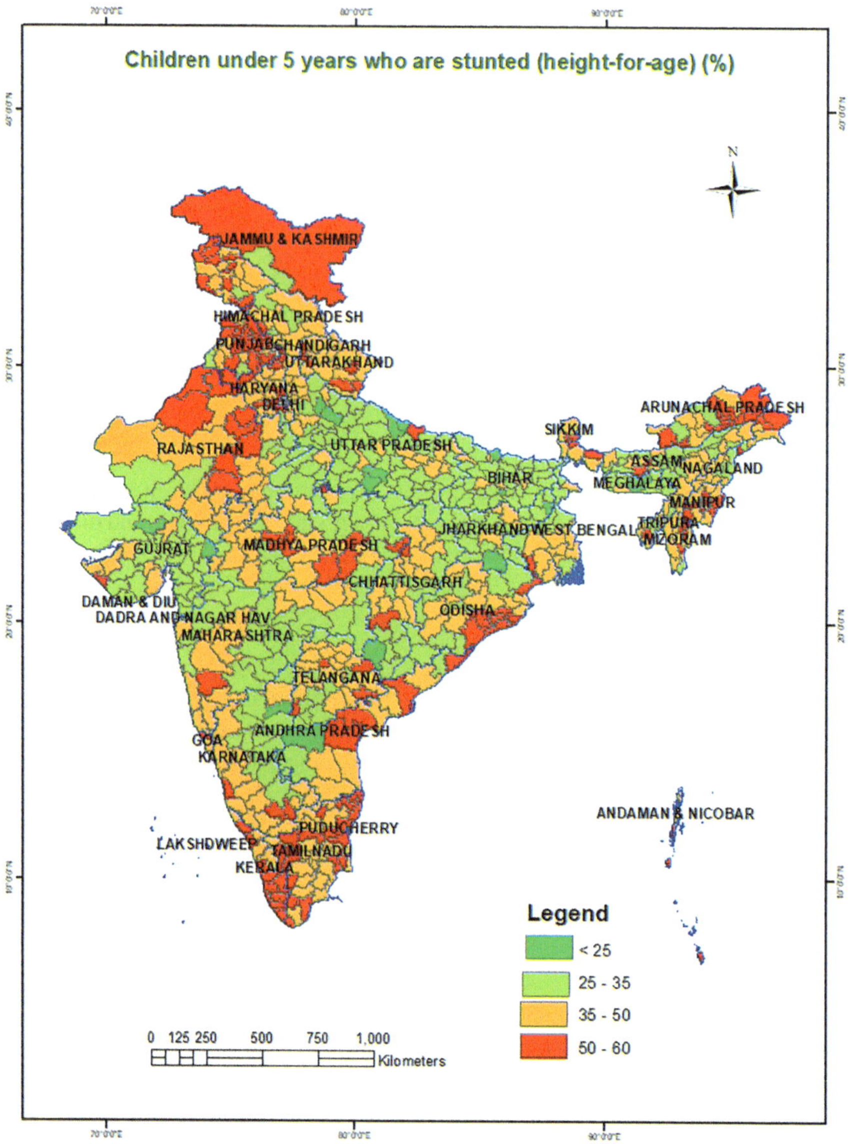

Fig. 4 Percentage of stunted children under 5 years old

Overall, the table highlights the regional disparities in stunting rates among children under 5 years old in India. It underscores the need for targeted interventions and focused efforts to address malnutrition and promote healthy growth and development among children, particularly in high-stunting areas.

2.5 Wasted Children

Figure 5 illustrates information about the occurrence of undernourished children under the age of 5 (weight-for-height) in various states and districts of India. "Wasted" refers to a condition of acute malnutrition where children have low weight in relation to their height, indicating a significant health risk.

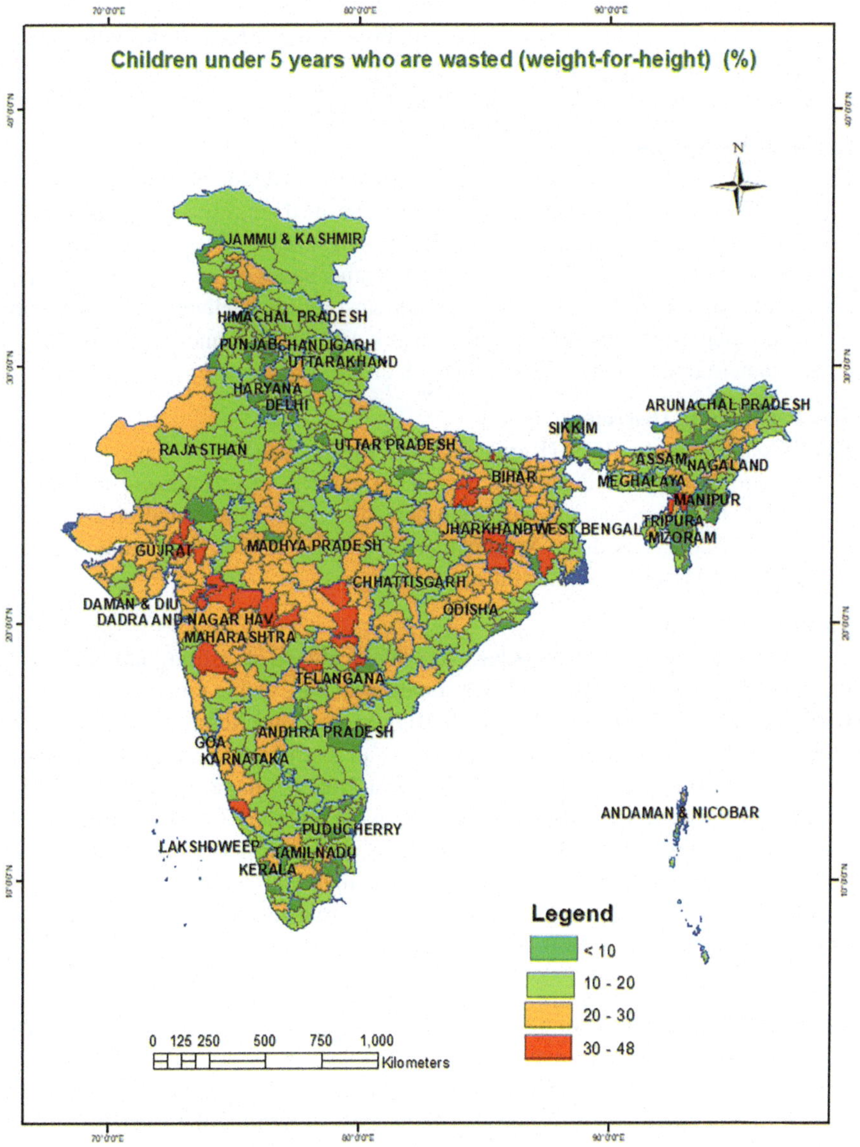

Fig. 5 Percentage of wasted children under 5 years old

Here are some key observations from the data:

Variation in Wasting Prevalence The prevalence of wasted children varies across different states and districts, indicating regional disparities in child nutrition.

2.6 *Institutional Births*

To analyze the data and provide some variations, we can look at different aspects such as regional patterns, top-performing districts, and overall trends. Here are some variations based on the data (Fig. 6):

Regional Variations
The Andaman and Nicobar Islands, Andhra Pradesh, and Goa have the highest percentage of institutional births across all districts in their respective states. Some districts in Arunachal Pradesh, Assam, Bihar, and Jharkhand have relatively lower percentages of institutional births compared to other states. "The Chiranjeevi Yojana program in Gujarat, India, aimed at promoting institutional deliveries, had a positive impact on increasing access to maternal health care services and improving neonatal and maternal outcomes" (Mohanan et al. 2013).

Top-Performing Districts
South Goa, Bangalore Rural (Karnataka), and Kasaragod (Kerala) have a 100% rate of institutional births, indicating a high level of access to healthcare facilities in these districts. Other districts with high percentages of institutional births include North and Middle Andaman (Andaman and Nicobar Islands), Srikakulam (Andhra Pradesh), and Mumbai (Maharashtra).

Variations by States
In Andhra Pradesh, most districts have a high percentage of institutional births, with the majority above 95%. Bihar shows some variations, with a range of percentages from 54.6% in Kishanganj to 93.2% in Munger. Goa and Kerala have consistently high percentages of institutional births across all districts, indicating strong healthcare infrastructure and awareness. In Madhya Pradesh, there is a mix of districts with varying percentages, with Mandla having the lowest (87.6%) and Bhopal having one of the highest (98.3%). "Mistreatment during childbirth, including verbal and physical abuse, is associated with adverse maternal health complications and may deter women from seeking necessary maternal healthcare services in Uttar Pradesh, India" (Raj et al. 2017).

Overall Trends
Generally, there is a positive trend of high percentages of institutional births across the country, indicating improved access to healthcare facilities and awareness about the importance of institutional deliveries. Some districts in states like Bihar and Jharkhand still have relatively lower percentages, suggesting the need for targeted

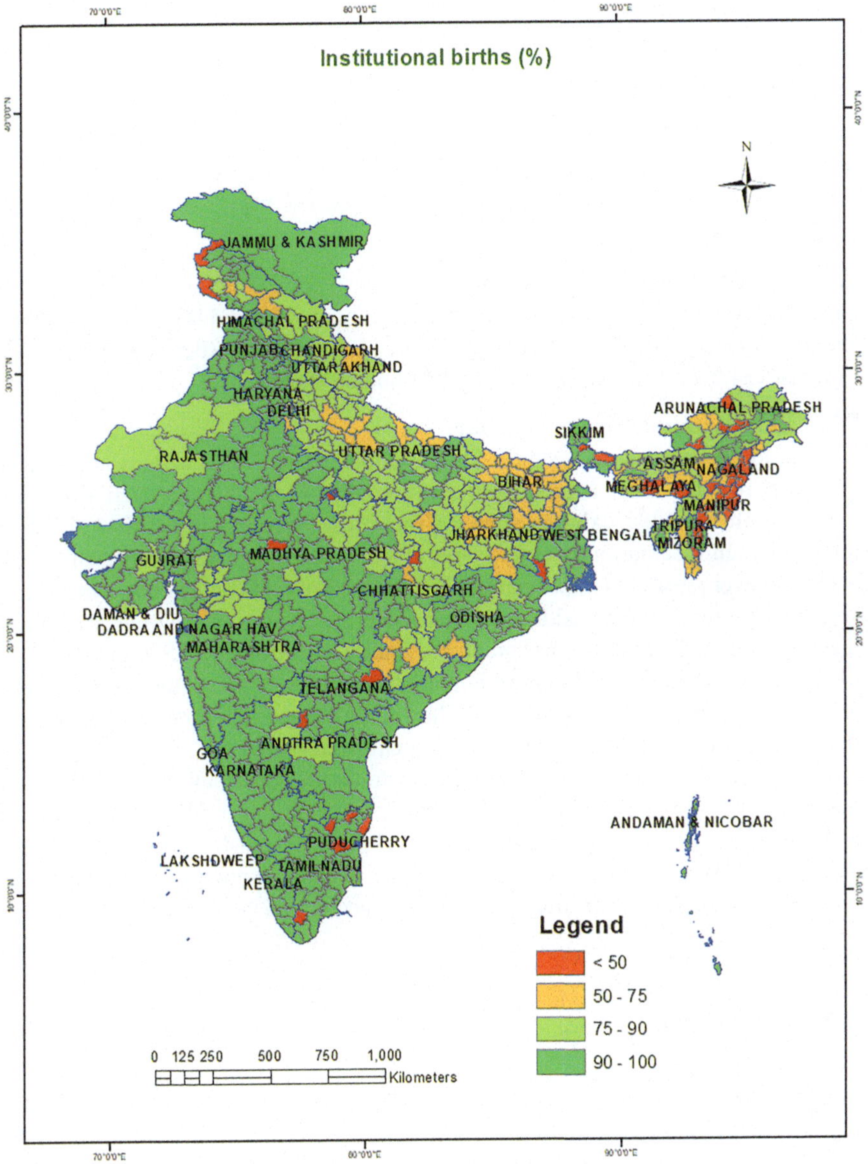

Fig. 6 The institutional births

interventions to improve healthcare infrastructure and promote institutional deliveries in these areas. These variations provide insights into the regional disparities and the progress made in promoting institutional births across different states and districts in India.

2.7 Immunization

To examine the differences in vaccination rates among states and districts, let us analyze the given data. We can observe the following patterns:

States with High Vaccination Rates
Dadra and Nagar Haveli (100%). Karnataka: Hassan (96.8%), Tumkur (97.7%), and Mysore (97.2%).

States with Low Vaccination Rates
Uttar Pradesh: Shravasti (20.9%), Siddharthnagar (25.7%), and Balrampur (30.6%). Bihar: Supaul (73.6%), Madhepura (72%), and Sheohar (62.9%). Jharkhand: Deoghar (54.5%), Garhwa (77.2%), and Sahibganj (66%).

Disparities Within Districts of a State
Uttar Pradesh demonstrates significant variations in vaccination rates across districts, where some districts report extremely low rates (e.g., Shravasti 20.9%) compared to others. Bihar exhibits similar disparities, with certain districts showing relatively high vaccination rates (e.g., Arwal 84.2%, Aurangabad 89.7%), while others have lower rates. It is important to note that the provided data is not comprehensive, and there might be additional disparities and variations within states and districts. Below the map is a bar graph presenting data from NFHS 5 and NFHS 4, likely representing two different time points. Each state's percentage of fully vaccinated children aged 12–23 months is shown, allowing for a comparison over time to assess improvements or declines in vaccination coverage.

The map, along with the bar graph, illustrates the distribution of vaccination coverage among young children across India, highlighting both high-performing and underperforming regions. Such visualization is crucial for public health officials to identify areas where vaccination efforts need to be intensified and to evaluate the progress made over time in improving vaccination coverage (Fig. 7).

2.8 Underweight

Figure 8 presents data concerning the incidence of underweight children under the age of 5 in different states and districts of India. The table presents the name of each state, followed by the name of the district within that state, and the corresponding percentage of children who are underweight based on weight-for-age measurements.

The prevalence of underweight children varies across different states and districts. Some districts have a relatively high percentage of underweight children, indicating a greater level of malnutrition in those areas. Certain states show a wider variation in underweight percentages among their districts, suggesting disparities in child health within those states. It is crucial to note that the data solely offers details about underweight children and does not encompass other indicators of child health or nutritional status. Overall, the data highlights the need for targeted interventions

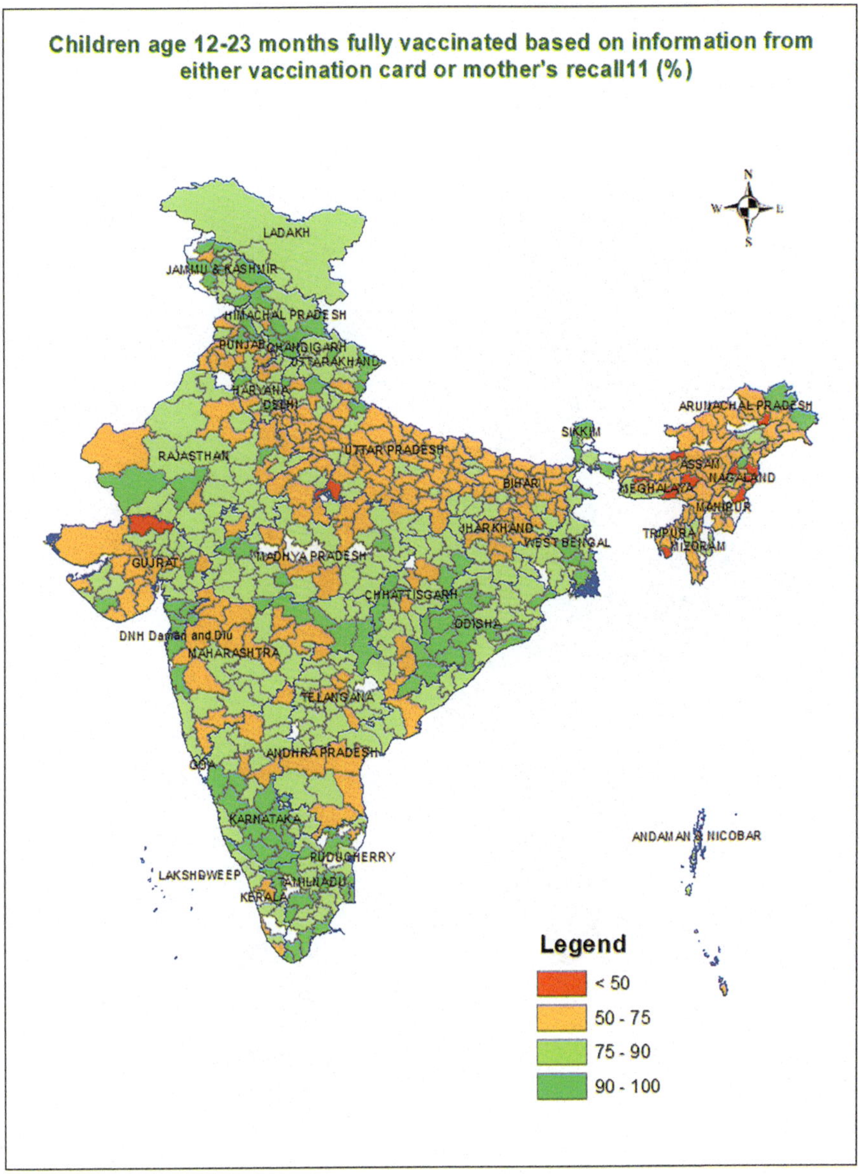

Fig. 7 Children aged 12–23 fully vaccinated

and policies to address the issue of child undernutrition, particularly in districts with higher prevalence rates.

The data highlights the variation in the prevalence of underweight children across different states and districts in India. Some districts have a high percentage of underweight children, indicating a higher prevalence of malnutrition in those areas. For example, districts like Kurung Kumey in Arunachal Pradesh, Katihar in

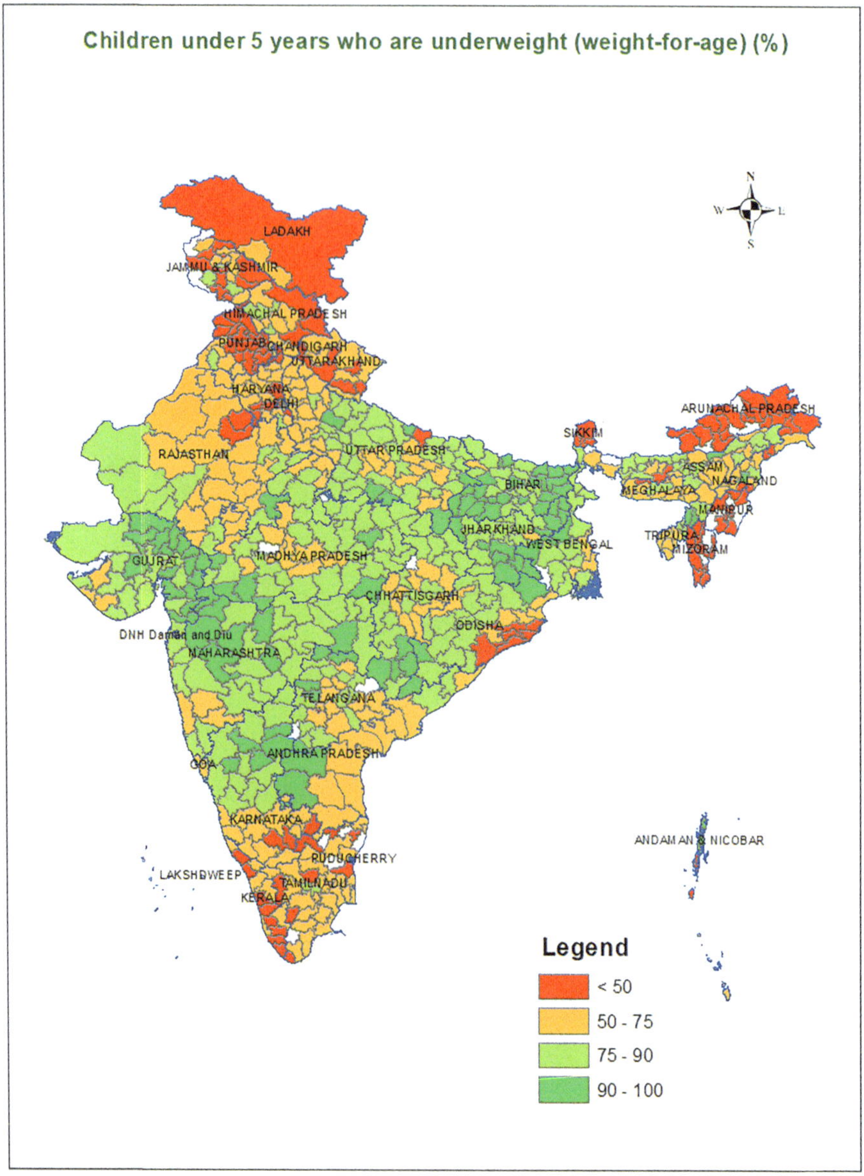

Fig. 8 Underweight children under the age of 5

Bihar, Pashchimi Singhbhum in Jharkhand, and Daman in Dadra and Nagar Haveli have relatively high percentages of underweight children.

On the other hand, some districts have a lower percentage of underweight children, suggesting better nutritional conditions. For instance, districts like Hamirpur in Himachal Pradesh, Alappuzha in Kerala, and Leh(Ladakh) in Ladakh show relatively lower percentages of underweight children. The data also reveals variations within states. For example, in Andhra Pradesh, districts like Kurnool and Ananthapur have higher percentages of underweight children compared to districts like South Andaman and Sri Potti Sriramulu.

3 State-Level Analysis

This section highlights the state-level variation in the maternal and child health indicators. It is imperative to highlight that state-level analysis plays a critical role in formulating broader policies, allocating resources effectively, and enabling comparative analysis between states. Such analysis not only identifies overarching trends and issues transcending district boundaries—crucial for state and central agencies—but also simplifies data interpretation for a wider range of stakeholders, including policymakers and the public. By aggregating district data at the state level, the analysis integrates with other socioeconomic and demographic data, offering a comprehensive view of health determinants within states. Hence, state-level analysis complements the granular insights from district-level data, providing a multifaceted understanding essential for strategic public health planning and improvement.

3.1 Institutional Delivery

Figure 9 provides information on institutional births as a percentage in different states of India that take place in healthcare facilities such as hospitals, clinics, or maternity centres. The states with the lowest percentages of institutional births in NFHS 5 (the most recent National Family Health Survey) are Nagaland (45.7%) and Meghalaya (58.1%). However, both states have shown an improvement in institutional births compared to NFHS 4 (the previous survey).

The states with the highest percentages of institutional births in NFHS 5 are Kerala (99.8%), Goa (99.7%), and Puducherry (99.6%). These states have nearly universal coverage of institutional births. Overall, there has been an increase in institutional births across most states between NFHS 4 and NFHS 5, indicating an improvement in access to healthcare facilities during childbirth. Some states that have shown significant improvements in institutional births include Andhra Pradesh (from 91.5% to 96.5%), Dadra and Nagar Haveli and Daman and Diu (from 88.5% to 96.5%), and Karnataka (from 94% to 97%).

States like Kerala, Sikkim, Lakshadweep, and Puducherry have consistently high percentages of institutional births in both NFHS 4 and NFHS 5. This trend is important as institutional births are associated with better maternal and neonatal outcomes due to access to skilled healthcare professionals and emergency obstetric care.

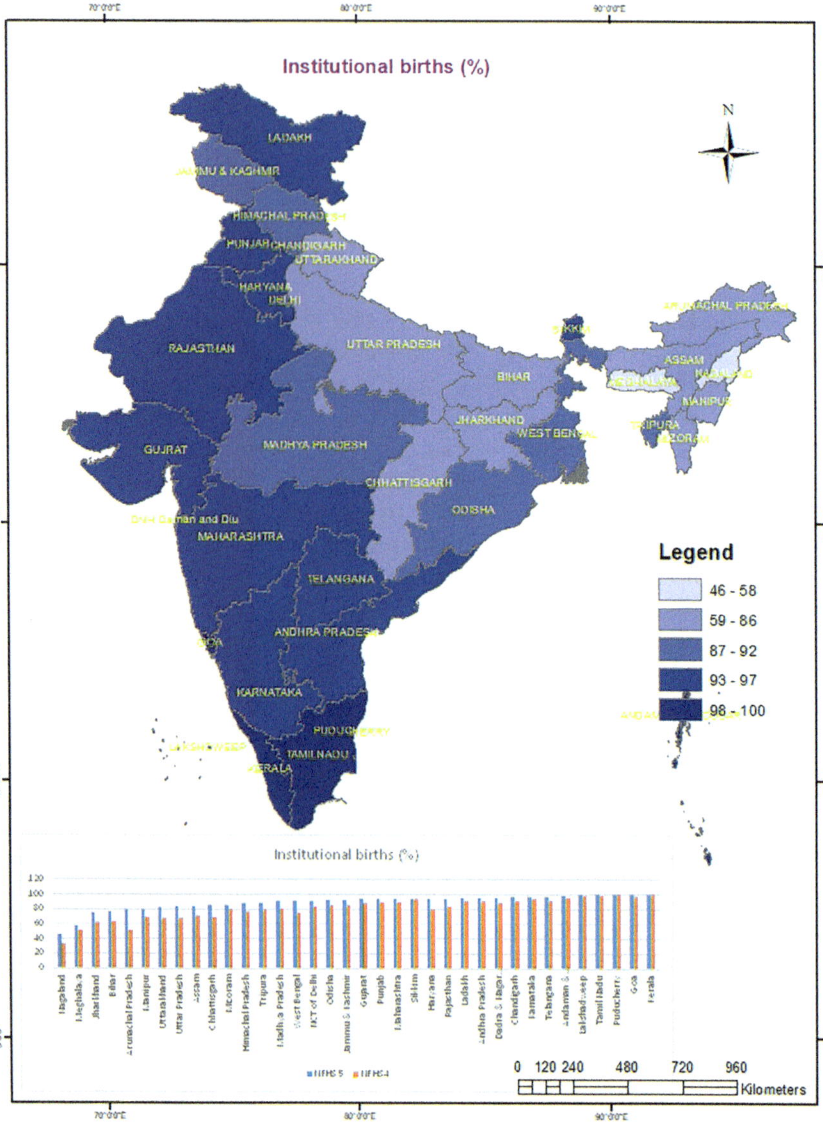

Fig. 9 Institutional births

3.2 *Breastfed*

Figure 10 offers data about the proportion of children below the age of 3 years who are breastfed within one hour of birth across various states in India. Early breastfeeding within one hour of birth is deemed essential for the welfare of the child and the mother. The states with the lowest percentages of children breastfed within one hour

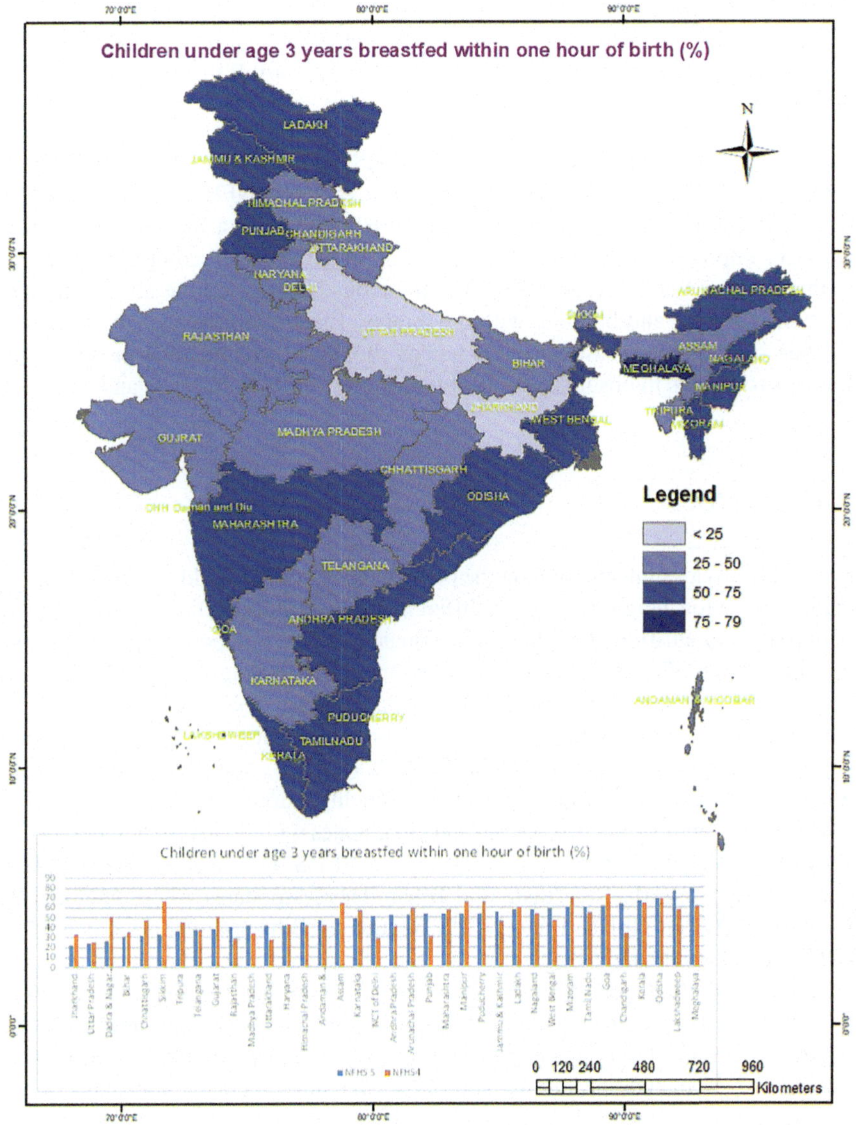

Fig. 10 Breastfed children under the age of 3

of birth in NFHS 5 (the most recent National Family Health Survey) are Meghalaya (78.8%) and Lakshadweep (76.3%). However, it is important to note that both states have shown improvements compared to NFHS 4 (the previous survey).

The states with the highest percentages of children breastfed within one hour of birth in NFHS 5 are Odisha (68.5%), Kerala (66.7%), and Chandigarh (63.7%). Overall, there has been a decrease in the percentage of children breastfed within one hour of birth in several states between NFHS 4 and NFHS 5. States like Goa, Mizoram, Manipur, Puducherry, and Jammu and Kashmir have consistently high percentages of children breastfed within one hour of birth in both NFHS 4 and NFHS 5.

Some states that have shown a significant decline in breastfeeding within one hour of birth include Rajasthan (40.7% in NFHS 5 compared to 28.4% in NFHS 4), Uttar Pradesh (23.9% in NFHS 5 compared to 25.2% in NFHS 4), and Jharkhand (21.5% in NFHS 5 compared to 33.1% in NFHS 4). The data suggests that there is room for improvement in promoting early initiation of breastfeeding across several states in India. Early breastfeeding has numerous benefits for the child's immune system, nutrition, and bonding with the mother. Efforts to increase awareness and provide support for breastfeeding practices should be strengthened in states with lower percentages to ensure better health outcomes for both mothers and children.

3.3 Stunted

Figure 11 provides information on the percentage of stunted children under 5 meaning they have low height-for-age, in different states of India. Stunting is an indicator of chronic malnutrition and reflects inadequate nutrition and poor health conditions during early childhood. The states with the lowest percentages of stunted children are Puducherry (20%), Sikkim (22.3%), and Andaman and Nicobar Islands (22.5%).

The states with the highest percentages of stunted children in NFHS 5 are Bihar (42.9%), Meghalaya (46.5%), Uttar Pradesh (39.7%), and Jharkhand (39.6%). Some states have shown improvements in reducing stunting compared to NFHS 4 (the previous survey), such as Kerala and Goa. States with relatively higher percentages of stunted children in both NFHS 4 and NFHS 5 include Madhya Pradesh, Gujarat, Chhattisgarh, Rajasthan, and Bihar. There are states where stunting has increased between NFHS 4 and NFHS 5, such as Bihar (42.9%), Uttar Pradesh (39.7%), and Jharkhand (39.6%).

The data highlights the persistent challenge of childhood stunting in many states of India. Efforts to address malnutrition and improve the health and nutritional status of children need to be intensified, particularly in states with high stunting rates. Strategies may include promoting appropriate feeding practices, improving access to nutritious food, enhancing healthcare services, and implementing interventions targeting maternal and child health.

3.4 Iron Folic Acid (IFA)

Figure 12 provides information on mothers who had iron folic acid (IFA) supplements for 180 and during their pregnancy. Iron folic acid supplementation is crucial during pregnancy to prevent maternal anaemia and ensure proper fetal development.

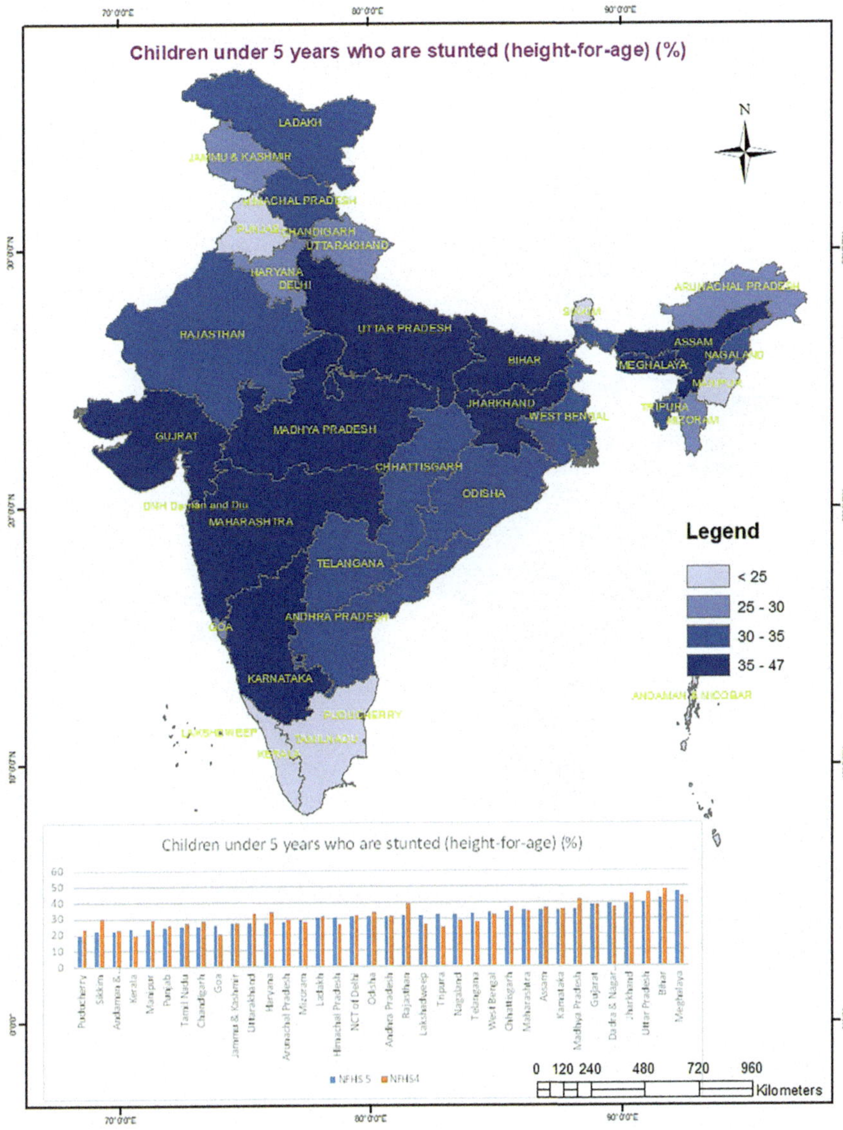

Fig. 11 Percentage of stunted children

The states with the lowest percentages of mothers consuming IFA for 180 days or more in NFHS 5 (the most recent National Family Health Survey) are Nagaland (4.1%), Ladakh (7.3%), and Arunachal Pradesh (8.6%). The states with the highest percentages of mothers consuming IFA for 180 days or more in NFHS 5 are Kerala (67%), Goa (65%), and Puducherry (64.9%). Some states have shown improvements in IFA consumption compared to NFHS 4, such as Tamil Nadu (63.1%) and

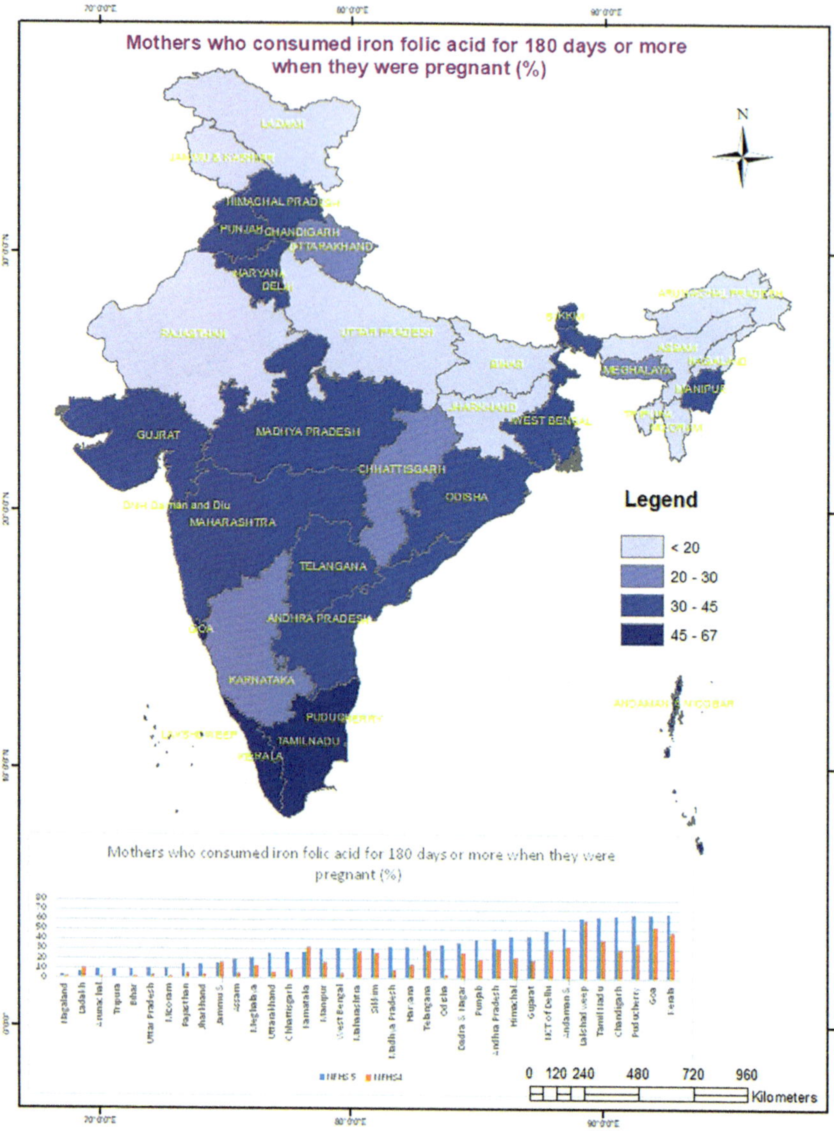

Fig. 12 Consumption of iron and folic acid

Chandigarh (64.5%). States with relatively higher percentages of mothers consuming IFA for 180 days or more in both NFHS 4 and NFHS 5 include Kerala, Goa, Puducherry, Tamil Nadu, and Lakshadweep.

There are states where the percentage of mothers consuming IFA for 180 days or more has decreased between NFHS 4 and NFHS 5, such as Uttar Pradesh, Rajasthan, and Maharashtra. The data highlights the importance of improving IFA supplementation during pregnancy, particularly in states with low percentages. Ensuring access to and awareness about the benefits of IFA supplements, as well as strengthening healthcare systems to provide adequate antenatal care, are crucial in promoting the health and well-being of both mothers and their babies.

3.5 *Wasted*

Figure 13 provides information on the percentage of children with wasted growth, which is a measure of acute malnutrition characterized by low weight-for-height. Wasting indicates a recent and severe decline in a child's nutritional status and is linked to a higher likelihood of illness and death.

The states with the lowest percentages of wasted children in NFHS 5 are Chandigarh (8.4%), Mizoram (9.8%), and Manipur (9.9%). The states with the highest percentages of wasted children in NFHS 5 are Gujarat (25.1%), Maharashtra (25.6%), and Bihar (22.9%). Some states have shown improvements in reducing wasted growth compared to NFHS 4 such as Jammu and Kashmir and Ladakh. States with relatively higher percentages of wasted children in both NFHS 4 and NFHS 5 include Bihar, Jharkhand, Madhya Pradesh, Chhattisgarh, and Daman and Diu. There are states where the prevalence of wasted growth has increased between NFHS 4 and NFHS 5, such as Karnataka and Gujarat.

The need for continued efforts to address child malnutrition, particularly wasted growth, in several states is shown from the data. Strategies to improve nutrition and healthcare services, promote breastfeeding and appropriate complementary feeding practices, and ensure access to nutritious food are crucial in combating acute malnutrition for the overall well-being of children.

3.6 *Immunization*

Figure 14 presents the proportion of fully vaccinated children aged 12–23 months. This is collected based on information available from vaccination cards or from the mother's recall. Full vaccination refers to the completion of the recommended vaccination schedule for children, which includes various vaccines to protect against diseases.

Dadra and Nagar Haveli and Daman and Diu have the highest percentage of fully vaccinated children in NFHS 5, with 94.9% coverage. This represents a significant improvement compared to NFHS 4, where the coverage was only 50.5%. Other states with high percentages of fully vaccinated children in NFHS 5 include Odisha (90.5%), Himachal Pradesh (89.3%), Tamil Nadu (89.2%), Ladakh (88.2%), and West Bengal (87.8%).

The states with the lowest percentages of fully vaccinated children in NFHS 5 are Meghalaya (63.8%), Nagaland (57.9%), and Arunachal Pradesh (64.9%).

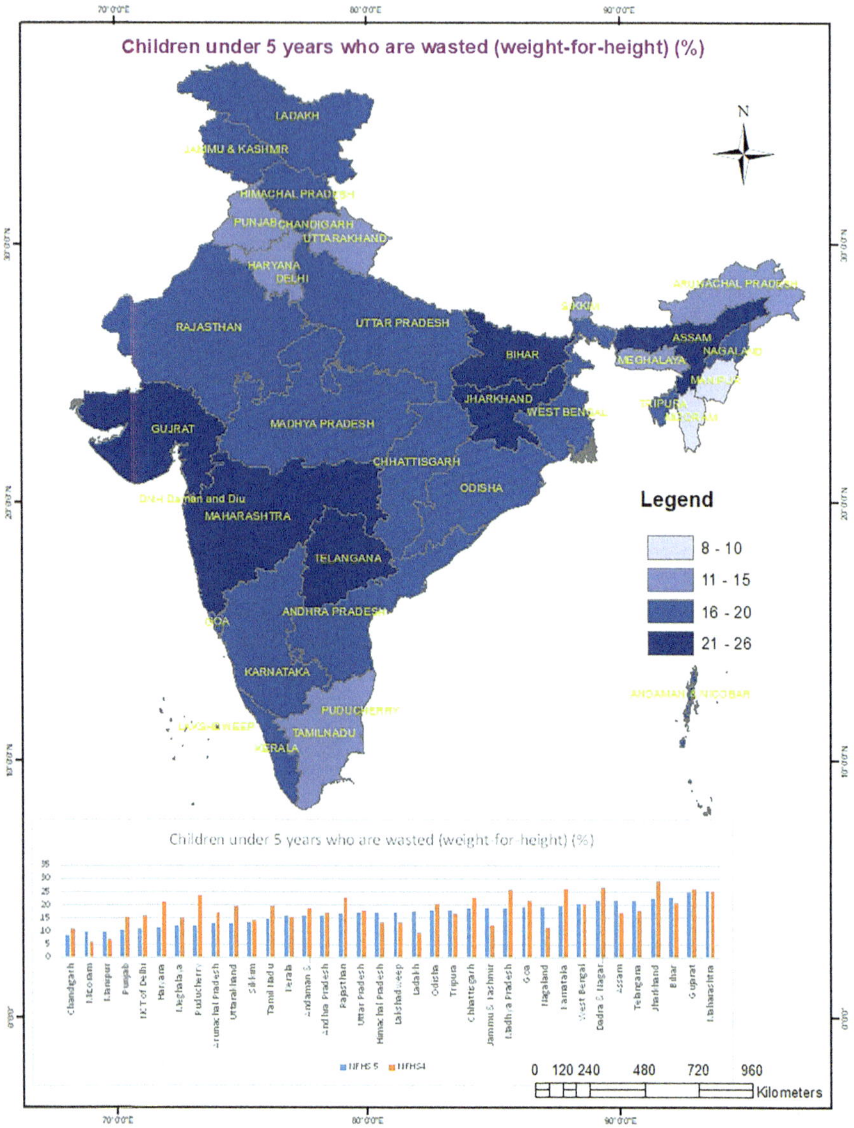

Fig. 13 Percentage of children with wasted growth

Several states have shown improvements in fully vaccinated coverage compared to NFHS 4, such as Gujarat (76.3%), Maharashtra (73.5%), and Punjab (76.2%).

However, there are still some states where the coverage of full vaccination remains relatively low in both NFHS 4 and NFHS 5; states like Bihar, Uttar Pradesh, Madhya Pradesh, and Rajasthan are some examples. In general, the data indicates advancements in enhancing vaccination among children aged 12–23 months.

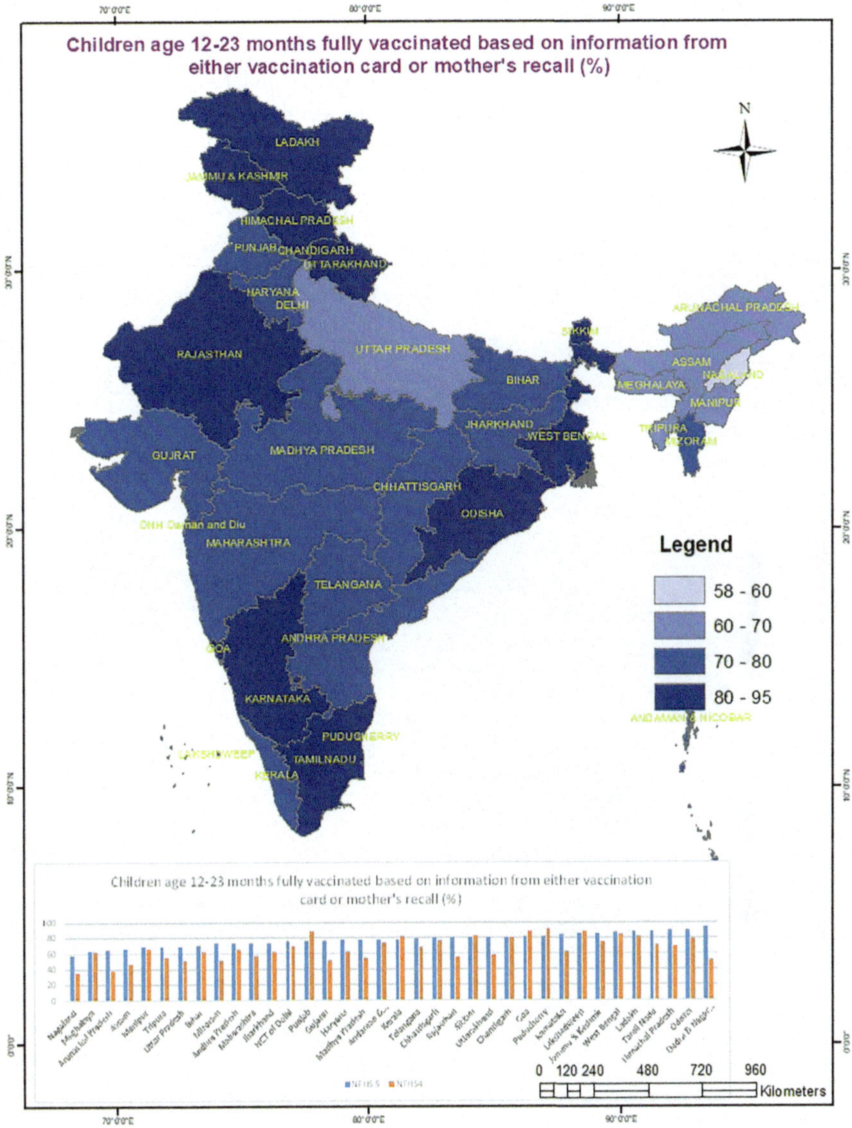

Fig. 14 Fully vaccinated children aged 12–23 months

However, there is still a need to focus on increasing vaccination rates, particularly in states with lower coverage, to ensure that more children receive the recommended vaccines and are protected against vaccine-preventable diseases. Efforts should include strengthening immunization programs, enhancing awareness and education about the importance of vaccination, and ensuring accessible and affordable healthcare services for all communities.

3.7 *Underweight*

Figure 15 provides information on the percentage of underweight children under 5 years based on weight-for-age indicators. Underweight is an indicator of malnutrition and can be indicative of inadequate nutrition and overall poor health among children.

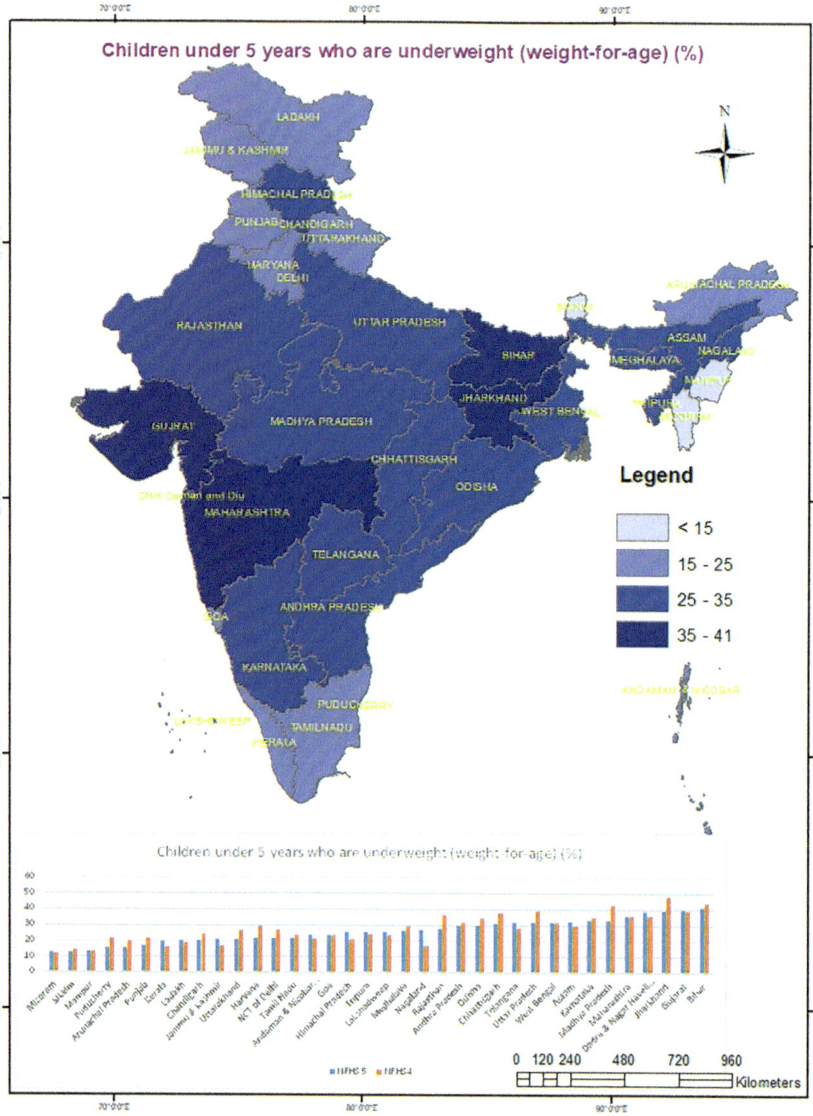

Fig. 15 Underweight children under 5 years

Mizoram has the lowest percentage of underweight children in NFHS 5, with a rate of 12.7%. However, it shows only a slight improvement compared to NFHS 4, where the rate was 12%. Other states with relatively low percentages of underweight children in NFHS 5 include Sikkim (13.1%), Manipur (13.3%), Puducherry (15.3%), and Arunachal Pradesh (15.4%).

On the other hand, Bihar has the highest percentage of underweight children in both NFHS 4 and NFHS 5, with a rate of 41% in NFHS 5 and 43.9% in NFHS 4. Other states with high percentages of underweight children in NFHS 5 include Jharkhand (39.4%), Gujarat (39.7%), Madhya Pradesh (33%), Uttar Pradesh (32.1%), and Chhattisgarh (31.3%). Some states have shown improvements in reducing underweight rates compared to NFHS 4, such as Mizoram, Sikkim, Manipur, Punjab, Kerala, Jammu and Kashmir, and Nagaland. However, there are still states where the underweight rates have increased or remained high, including Jharkhand, Bihar, Madhya Pradesh, Chhattisgarh, and Uttar Pradesh.

The data highlights the persisting issue of underweight children in certain states, indicating a need for targeted interventions and programs to address malnutrition and improve the overall nutritional status of children. Efforts need to be directed towards enhancing the availability of nutritious food, encouraging breastfeeding, improving services, and implementing effective nutrition programs. These initiatives are crucial in addressing malnutrition and its negative impact on child growth and development.

3.8 ANC

Figure 16 displays data regarding the proportion of mothers who had a minimum of four antenatal visits during their pregnancy. These visits are essential for safeguarding the health and welfare of both the mother and foetus, as they involve regular check-ups, screenings, and guidance from healthcare professionals. Below the map, there is a bar graph comparing the data from two different sources or surveys: NFHS 5 and NFHS 4. The bars represent each state's percentage of mothers who had at least four antenatal visits. This provides a temporal comparison, potentially indicating whether there has been an improvement or decline in antenatal care visits over time between the two surveys.

Goa has the highest percentage of antenatal visits in NFHS 4 as well as NFHS 5, with a rate of 93% in NFHS 5 and 89% in NFHS 4. Other states with high percentages in NFHS 5 include Tamil Nadu (89.9%), Lakshadweep (88.3%), Puducherry (86.9%), and Dadra and Nagar Haveli and Daman and Diu (86.2%).

Some states have shown improvements in antenatal care utilization compared to NFHS 4, such as Gujarat, NCT of Delhi, Odisha, Ladakh, Kerala, Chandigarh, Manipur, Jammu and Kashmir, and Andaman and Nicobar Islands. However, the utilization of antenatal care is relatively lower in Nagaland, Bihar, Arunachal Pradesh, Jharkhand, Uttar Pradesh, and Assam. These states should focus on

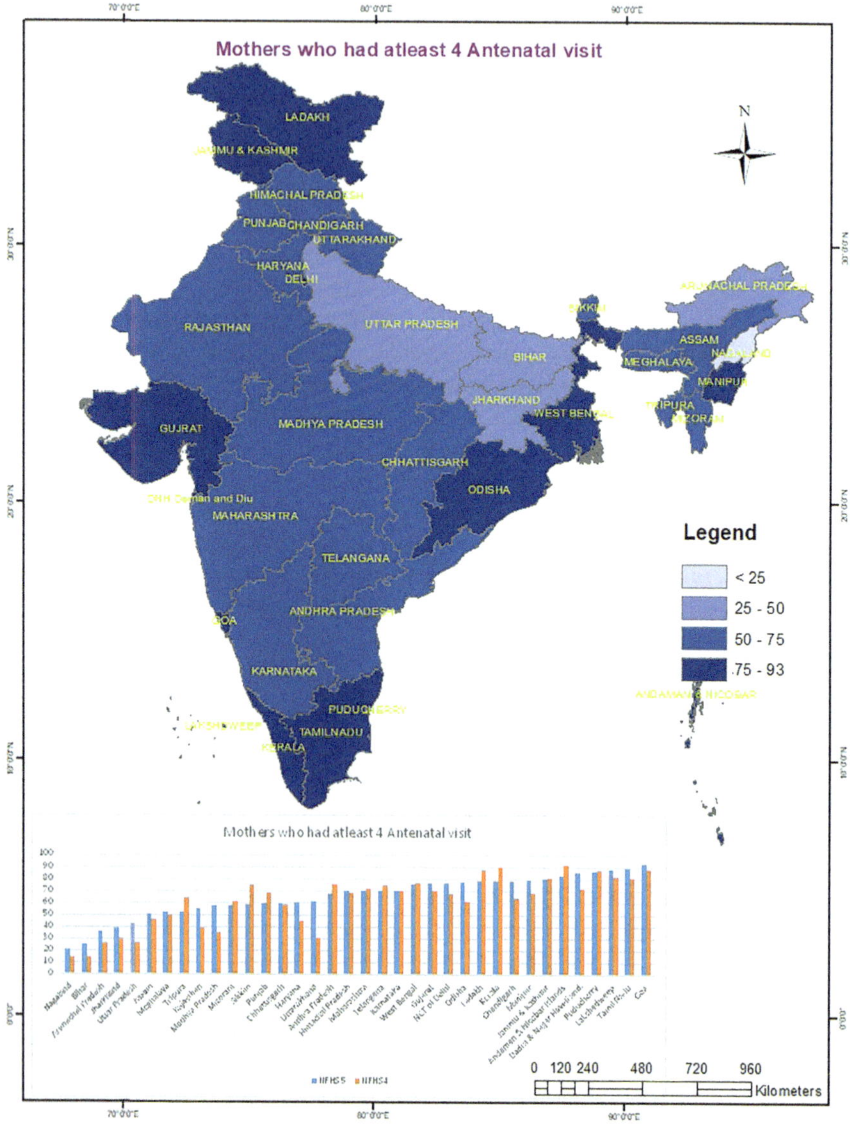

Fig. 16 Percentage of mothers who had a minimum of four ANC visits

increasing awareness and access to antenatal care services to ensure the well-being of both mothers and infants.

It is worth noting that some states that have relatively high percentages of antenatal care utilization still show room for improvement, such as Goa, Tamil Nadu, and Lakshadweep. The data highlights the importance of promoting and improving antenatal care services across all states to ensure that pregnant women receive the necessary healthcare support and interventions for a healthy pregnancy and childbirth. It is crucial to enhance awareness, access, and quality of antenatal care services, particularly in states with lower utilization rates, to improve maternal and child health outcomes. The map and the accompanying bar graph together provide a clear visual representation of the variation in maternal healthcare access across India. By comparing the data from two surveys, one can evaluate the progress or lack thereof in different regions.

3.9 Social Variation

Evidence from NFHS 5 on social variation indicates the existence of disparities based on various social factors. Srivastava et al. (2014) emphasized the role of factors such as education, socioeconomic status, and awareness in determining access to healthcare services. Karmaker and Behera (2022) examined the effect of social determinants of health in rural West Bengal, India. Their findings emphasized the significance of factors like caste, religion, income, and gender in shaping access to maternal healthcare services.

Education

"Socioeconomic factors, including educational attainment and income levels, play a crucial role in determining the choice of home delivery over facility-based delivery, leading to disparities in access to skilled maternal and child health care services" (Sreeramareddy et al. 2006). Compared to women with lesser education levels, those with higher education are likely to have better access to maternity and child healthcare services. For instance, women with higher levels of education are more likely to give birth in an institution and receive prenatal care from a qualified practitioner. There is a significant correlation between maternal education and improved child health outcomes, as seen by the higher likelihood of important immunizations received by children of educated moms.

Antenatal Care

Women with no education had a lower proportion (63%) of obtaining four prenatal care visits at a minimum compared to women with 12 or more years of education (90%). Skilled Birth Attendance: Women with greater levels of education had a much larger proportion of skilled health staff attending their births. For example, among women with 12 years or more of education, 92% of deliveries were attended by qualified health staff, while 67% of births were attended by women with no education.

Wealth

Household wealth is a substantial factor in access to healthcare services. Women from wealthier households have better access to all care services and institutional deliveries. Wealthier households also demonstrate higher rates of institutional delivery, indicating that financial resources play a role in seeking healthcare from formal healthcare facilities.

Institutional Delivery: Merely 63% of women in the lowest quintile of wealth gave birth at a medical institution, whereas 94% of women in the highest quintile of wealth did so. Postnatal Care: Among women in the greatest wealth quintile, 80% received postnatal care within two days after birth, compared to 48% of women in the poorest wealth quintile.

Caste and Tribe

Disparities exist among different castes and tribes in accessing maternal and child healthcare. Scheduled Tribe populations face greater challenges in accessing essential services compared to other social groups. Prenatal care coverage, institutional delivery rates, and postnatal care utilization are lower among Scheduled Caste and Scheduled Tribe populations, indicating the need for targeted interventions to address these disparities. "Efforts to improve maternal health services in rural areas of Maharashtra, India, have shown mixed results in terms of equity, with disparities persisting in access to quality care among different social groups" (Dettrick et al. 2013).

Antenatal Care: The percentage of women who received a minimum of four antenatal care visits varied across different social groups. It was highest among women from "Other Backward Classes" (OBC) at 80%, followed by Scheduled Castes (75%) and Scheduled Tribes (65%). Institutional Delivery: The percentage of institutional deliveries was 86% among women from OBC, 83% among Scheduled Castes, and 77% among Scheduled Tribes.

Religion

Access to maternity and pediatric healthcare varies throughout religious communities. For example, compared to other religious groups, Muslims use antenatal care services at a lower rate. These discrepancies could be caused by variations in the way people seek healthcare as well as cultural aspects pertaining to preferences and behaviours around birthing.

Antenatal Care: Religion affiliation affected the percentage of women who had a checkup at least four times before delivery. Christian women (86%), Hindu women (77%), and Muslim women (73%) had the greatest rates. Postnatal Care: Of Muslim, Hindu, and Christian women, 63%, 68%, and 79%, respectively, of women received postnatal care within two days after giving birth.

Gender

NFHS 5 data highlights gender disparities in accessing maternity and pediatric healthcare. Women often face obstacles in decision-making regarding their own healthcare and that of their children. This includes limited autonomy in choosing healthcare providers and delayed decision-making for seeking timely care. Gender inequities also impact the utilization of healthcare services for children, with male children more likely to receive healthcare compared to female children.

These findings emphasize the need for targeted interventions and policy measures to work on eradicating social disparities in healthcare access. By considering these social determinants of health, healthcare programs and policies can be designed to promote equity and improve health outcomes for vulnerable populations. “The autonomy of women in household decision-making is closely linked to their ability to access maternal and child health care services in Nepal. Empowering women and promoting gender equality are crucial for improving access to healthcare” (Acharya et al. 2010).

Full Immunization Coverage: Boys had a higher percentage (62%) of full immunization coverage compared to girls (58%), indicating a gender disparity in accessing essential vaccinations. These data points highlight the social variations in accessing maternal and child healthcare services based on education, wealth, caste, tribe, religion, and gender. They show the differences that occur and stress how critical it is to address these socioeconomic determinants of health in order to guarantee that all Indian population groups have fair access to healthcare services.

Regional Disparities

Significant regional differences exist in the availability of maternity and child healthcare among various states and geographical areas, as indicated by the NFHS 5 statistics. It is clear from examining metrics like institutional delivery rates, prenatal care coverage, and postnatal care utilization that some states have significantly improved access while others are still falling behind. This highlights the importance of targeted interventions and tailored healthcare strategies to address the unique challenges faced by each region.

Urban-Rural Divide

The data from NFHS 5 allows for an analysis of the urban-rural divide in access to healthcare services for mothers and children. It shows that urban areas generally have better access to healthcare services compared to rural areas. This disparity can be attributed to factors such as better healthcare infrastructure, availability of skilled healthcare providers, and proximity to healthcare facilities in urban settings. Addressing this divide requires efforts to strengthen healthcare infrastructure in rural areas and ensure the availability of quality healthcare services in remote and underserved regions. “Vulnerable slum communities in Mumbai, India, face significant inequalities in accessing maternity care, leading to adverse newborn outcomes. Factors such as poverty, lack of education, and limited availability of healthcare services contribute to these disparities" (More et al. 2009).

Socioeconomic Factors

The NFHS 5 data sheds light on how socioeconomic variables affect a mother's and child's ability to receive healthcare. It demonstrates that women who belong to marginalized social groups have lower levels of education and that those who are economically disadvantaged encounter more obstacles when trying to get healthcare services. In order to increase access and support equitable healthcare delivery, it is critical to address socioeconomic determinants of health like poverty, education, and social prejudice. These discrepancies underscore this need.

4 Limitations of the Research

While this study provides important insights into the disparities in maternal and child healthcare access in India, it is not without its limitations. Firstly, the reliance on NFHS 5 data, though extensive, means that our analysis is constrained by the survey's design, data collection methods, and the variables included. Secondly, the descriptive nature of the study limits our ability to establish causality between socioeconomic and demographic factors and health outcomes. Furthermore, regional variations within states and union territories may not be fully captured, potentially overlooking localized disparities. Another limitation is the potential for reporting bias in survey responses, which could affect the accuracy of the data on healthcare access and utilization.

4.1 Future Scope

Given the limitations identified, future research should consider employing longitudinal designs to track changes in maternal and child health indicators over time and to better understand the causal relationships between policy interventions and health outcomes. Additionally, qualitative studies exploring the experiences of healthcare providers and recipients could provide deeper insights into the barriers to accessing MCH services and the effectiveness of current healthcare delivery models. Investigating the impact of emerging healthcare technologies and digital health initiatives on improving access to MCH services in remote and underserved areas also represents a promising area for future research. Finally, more granular analysis at the district or community level could help in identifying and addressing hyper-localized disparities in healthcare access, informing targeted interventions that are tailored to the specific needs of different populations.

5 Conclusion

The findings from this paper provide substantial evidence regarding the regional and social variation in access to healthcare services for pregnant women, mothers, and children in India. The data shed light on the persistent disparities that exist across different states and social groups, emphasizing the urgent need for targeted interventions and policy measures to ensure equitable access to healthcare services. Factors such as caste, religion, income, education, and gender significantly shape individuals' ability to access and utilize healthcare services. Marginalized populations, including tribal communities and Scheduled Castes, are particularly vulnerable, facing multiple barriers to accessing quality maternal and child healthcare. Focusing on social determinants of health, such as poverty, discrimination, and limited education, is important for achieving equitable healthcare access and reducing disparities.

The evidence from NFHS 5 highlights the urgent need for comprehensive and targeted interventions to address regional and social disparities. Policymakers, healthcare practitioners, and researchers must collaborate to develop evidence-based strategies that improve healthcare infrastructure, strengthen healthcare systems, enhance awareness, and ensure the provision of quality healthcare services. Efforts should focus on addressing social determinants of health, empowering marginalized populations, and promoting gender equality to achieve equitable access and improved health outcomes for women and children across India. With concerted and sustained efforts, we can overcome the regional and social disparities in access to healthcare services for mothers and children and make significant strides towards achieving better health outcomes for all in India.

References

Acharya, D. R., Bell, J. S., Simkhada, P., Van Teijlingen, E. R., & Regmi, P. R. (2010). Women's autonomy in household decision-making: A demographic study in Nepal. *Reproductive Health, 7*(1), 1–12.

Bango, M., & Ghosh, S. (2022). Social and regional disparities in utilization of maternal and child healthcare services in India: A study of the post-National Health Mission Period. *Frontiers in Pediatrics, 10.*

Dettrick, Z., Firth, S., & Jimenez Soto, E. (2013). Do strategies to improve quality of maternal and child health care in lower and middle income countries lead to improved outcomes? A review of the evidence. *PLoS One, 8*(12), e83070.

Etiaba, E., Eboreime, E., Dalglish, S. L., & Lehmann, U. (2023). Aspirations and realities of inter-governmental collaboration in national- level interventions: Insights from maternal, neonatal and child health policy processes in Nigeria, 2009–2019. BMJ. *Globalization and Health, 8.*

Girotra, S., Malik, M., Roy, S., & Basu, S. (2023). Utilization and determinants of adequate quality antenatal care services in India: Evidence from the National Family Health Survey (NFHS-5) (2019–2021). *BMC Pregnancy and Childbirth, 23.*

Gopika, M. G., Prabhu, P., & Thulaseedharan, J. (2022). Status of cancer screening in India: An alarm signal from the National Family Health Survey (NFHS-5). *Journal of Family Medicine and Primary Care, 11*, 7303–7307.

Gupta, S., Yamada, G., Mpembeni, R., Frumence, G., Callaghan-Koru, J. A., Stevenson, R., et al. (2014). Factors associated with four or more antenatal care visits and its decline among pregnant women in Tanzania between 1999 and 2010. *PLoS One, 9*(7), e101893.

Jana, A., Saha, U., Reshmi, R., & Muhammad, T. (2023). *Relationship between low birth weight and infant mortality: Evidence from National Family Health Survey 2019–21* (p. 81). Archives of Public Health.

Karmakar, R., & Behera, B. (2022). Geographical inequalities in access to public healthcare facilities in rural West Bengal: Exploring major health system challenges. *Journal of Rural Development, 41*(1), 50–63.

Manna, S., & Basu, S. (2023). It cost us all of our savings to deliver our baby: A qualitative study to explore barriers and facilitators of maternal and child health service access and utilization in a remote rural region in India during the COVID-19 pandemic. *Cureus, 15*.

Mishra, B., Kanungo, S., Panda, S., Patel, K., Swain, S., Dwivedy, S., Karna, S., Bhuyan, D., Som, M., Marta, B., Bhattacharya, D., Kshatri, J., Pati, S., & Palo, S. K. (2023). Access to maternal and child health services during the COVID-19 pandemic: An explorative qualitative study in Odisha, India. *Indian Journal of Community Medicine: Official Publication of Indian Association of Preventive & Social Medicine, 48*(3), 459–464.

Mohanan, M., Bauhoff, S., La Forgia, G., Babiarz, K. S., Singh, K., & Miller, G. (2013). Effect of Chiranjeevi Yojana on institutional deliveries and neonatal and maternal outcomes in Gujarat, India: A difference-in-differences analysis. *Bulletin of the World Health Organization, 92*, 187–194.

More, N. S., Bapat, U., Das, S., Barnett, S., Costello, A., Fernandez, A., & Osrin, D. (2009). Inequalities in maternity care and newborn outcomes: One-year surveillance of births in vulnerable slum communities in Mumbai. *International Journal for Equity in Health, 8*, 1–11.

Raj, A., Saggurti, N., Balaiah, D., & Silverman, J. G. (2009). Prevalence of child marriage and its effect on fertility and fertility-control outcomes of young women in India: A cross-sectional, observational study. *The Lancet, 373*(9678), 1883–1889.

Raj, A., Dey, A., Boyce, S., Seth, A., Bora, S., Chandurkar, D., et al. (2017). Associations between mistreatment by a provider during childbirth and maternal health complications in Uttar Pradesh, India. *Maternal and Child Health Journal, 21*, 1821–1833.

Sreeramareddy, C. T., Joshi, H. S., Sreekumaran, B. V., Giri, S., & Chuni, N. (2006). Home delivery and newborn care practices among urban women in western Nepal: A questionnaire survey. *BMC Pregnancy and Childbirth, 6*(1), 1–10.

Srivastava, A., Mahmood, S. E., Mishra, P., & Shrotriya, V. P. (2014). Correlates of maternal health care utilization in rohilkhand region, India. *Annals of Medical and Health Sciences Research, 4*(3), 417–425.

Uneke, C., Sombié, I., Uro-Chukwu, H., & Johnson, E. (2019). Developing equity-focused interventions for maternal and child health in Nigeria: An evidence synthesis for policy, based on equitable impact sensitive tool (EQUIST). *The Pan African Medical Journal, 34*.

Dr. Vanishree Joseph is the head of the Centre for Gender Studies and Development at the National Institute of Rural Development and Panchayati Raj (NIRD&PR) in Hyderabad. With a strong academic foundation, she holds two Master's degrees in Development Administration and Public Administration, along with a PhD in Gender and Governance. Dr. Joseph has over ten years of experience in teaching and research, focusing on areas such as gender and development, human development, and decentralization and governance.

Dr. Prasad N. S. R. is working as an Assistant Professor at the Centre for Geo-informatic Application in Rural Development at NIRDPR, Hyderabad. He has a Ph.D. in Earth Sciences from Osmania University and Masters in Remote Sensing and GIS from JNTU, Hyderabad. He has more than a 20 years' experience in the field of Application of Geo-informatics in Rural Development. His area of interest is spatial planning for village development, natural resource management, rural infrastructure development, and Geo-ICT applications. He has multiple national and international publications to his name.

Dr. N. V. Madhuri is an esteemed Associate Professor at the University of Hyderabad, specializing in Sociology and Social Anthropology. With a rich educational background, including MAs in Sociology and Social Anthropology, as well as a PhD in Sociology, she brings over 17 years of experience in training, teaching, and research. Dr. Madhuri's areas of expertise include the promotion of micro-enterprises, rural women-based livelihoods, and gender mainstreaming in rural development.

Dr. Kesava Rao Pyla is a prominent figure in the field of geo-informatics, currently serving as the head of the Centre for Geo-Informatics Applications in Rural Development at the National Institute of Rural Development and Panchayati Raj (NIRD&PR) in Hyderabad. With a robust academic background that includes a Bachelor of Engineering in Civil, a Master of Technology in Remote Sensing, and a Ph.D. in Remote Sensing, Dr. Pyla brings over 24 years of experience in the application of geo-informatics to rural development.

Strategic Reference for Adopting Sustainable Agricultural Practices Through Social Learning in Indian Agriculture for Environmental Management

Suman Chatterjee and Ishita Tudu

1 Introduction

Social learning entails individuals acquiring knowledge through the act of observing and mimicking the conduct of their peers. It occurs through social interactions and can happen both intentionally and unintentionally. Social learning theory suggests that individuals can learn new behaviors and skills by watching others perform them and receiving feedback on their own performance (Davis & Luthans, 1980; Ladd & Mize, 1983). The foundation of this theory lies in the concept that behavior is influenced by both environmental factors and the outcomes of individual actions. Social learning can take place in a variety of settings, including schools, workplaces, and social networks. Examples of social learning include observing a teacher demonstrate a new math problem, learning a new cooking technique from a friend, or copying the behavior of a successful businessperson (McCarthy et al., 2011).

Environmental management in agricultural systems involves the use of practices and techniques that minimize negative impacts on the environment while maximizing agricultural productivity and profitability (Schwoob et al., 2018). Implementing various key strategies is essential for effective environmental management in agriculture. One approach involves conservation tillage, which reduces soil disruption during planting and addresses erosion by retaining crop residues atop the soil. Integrated pest management (IPM) is another crucial strategy that combines biological, chemical, and cultural methods to control pests, aiming to decrease reliance on synthetic pesticides. Nutrient management involves optimizing fertilizer and manure use to minimize nutrient runoff and greenhouse gas emissions. Efficient water management, including practices like irrigation scheduling, water quality monitoring, and erosion control, helps conserve water resources and prevent

S. Chatterjee (✉) · I. Tudu
Department of Geography, School of Basic and Applied Sciences, Adamas University, Kolkata, West Bengal, India

A. Rai et al. (eds.), *Contemporary Social Physics*, Springer Geography,
https://doi.org/10.1007/978-3-031-77596-3_10

pollution. Crop rotation and diversification entail growing different crops sequentially to reduce disease and pest risks, enhance soil fertility, and promote biodiversity. By incorporating trees into farming practices, agroforestry aims to offer various advantages including preserving soil, capturing carbon, and enhancing crop productivity. Soil conservation practices like contour farming, terracing, and cover crops are employed to reduce soil erosion and maintain soil health. Through the adoption of these strategies, farmers can play a vital role in mitigating the adverse environmental impacts of agriculture, ensuring the long-term sustainability of their farms (Rayhan Shaheb et al., 2022).

Water resource management in agricultural systems entails implementing practices and strategies to foster the efficient and sustainable utilization of water resources in agricultural production. Key strategies encompass optimizing irrigation scheduling, minimizing water losses through evaporation and runoff, and employing efficient irrigation systems such as drip irrigation under irrigation management (Delgado et al., 2020). Soil conservation involves adopting practices like conservation tillage, cover cropping, and soil amendments to enhance soil structure and water-holding capacity, thereby diminishing the water needed for crop production (Olawuyi, 2018). Crop selection plays a crucial role by choosing crops well suited to local conditions and requiring less water, ultimately reducing overall water consumption. Water reuse, achieved through treating and recycling wastewater from agricultural activities, aids in decreasing the demand for freshwater resources. Implementing water conservation technologies, including rainwater harvesting, water storage, and water-efficient equipment, contributes to reducing water consumption in agriculture (Dinar & Mody, 2004). Policies promoting efficient water resource use, such as pricing structures reflecting the true cost of water and incentives for conservation, encourage farmers to adopt water-saving practices. Regular monitoring of water resources facilitates the identification of high water use areas or potential water stress, allowing timely interventions for effective water resource management. Ensuring the long-term sustainability of agricultural systems is contingent upon efficient water resource use, sustaining agricultural production over time. The adoption of new sustainable techniques by farmers necessitates a combination of education, technical assistance, and financial incentives. Strategies for promoting the adoption of sustainable practices include providing education and outreach programs, offering technical assistance and support, and providing financial incentives such as grants or tax breaks to offset adoption costs. Establishing demonstration farms, creating peer networks or farmer-led organizations, and ensuring access to markets valuing sustainable practices contribute to increasing awareness and fostering the adoption of sustainable practices. Supporting research and development efforts focusing on sustainable agriculture identifies new and innovative techniques for adoption by farmers. In summary, promoting the adoption of sustainable practices among farmers requires a comprehensive approach addressing adoption barriers and providing necessary education, technical assistance, and financial incentives to support the transition to more sustainable farming practices (Piñeiro et al., 2020; Sulaiman & Misnan, 2022).

There are some studies which have been conducted in a similar context. The study by Tran et al. (2020) investigated social learning's impact on farmers' adaptive capacity to local flooding in the Vietnamese Mekong Delta, facing climate change and water management complexities. It highlights the link between social learning and adaptive capacity, emphasizing the role of rural communities in innovative practices. Gender disparities, age, education, and disparities across areas and household groups are identified. The study concludes by urging local adaptive capacity promotion and formal learning for marginalized groups, acknowledging key contributors and sponsorship from ALA/AusAID at the Australian National University. In their study across three communes—Phu Thanh B, Phu Xuan, and Thoi Hung—Bookwalter et al. (2006) utilized a stratified sampling approach to ensure a representative sample. They designated households as the primary unit of analysis, recognizing their direct exposure to flood impacts and active involvement in adaptation efforts. Survey administration targeted male household heads, as they were deemed optimal representatives capable of articulating the household's perspectives. According to Bookwalter et al. (2006), the responses of household heads serve as reliable proxies for the entire household, thereby enhancing the validity of the data collected. Consequently, their responses provide valid proxy data for the entire household. Hoffman's (2009) dissertation delved into the dynamics of social learning, decision-making, and the adoption of practices within California's viticulture sector. The research offers valuable insights into sustainable agricultural approaches within the viticulture domain. Another investigation by Morgan (2011) centered on employing the community of practice (CoP) framework to decipher the social learning mechanisms among practitioners of organic farming. This examination assesses the applicability of the CoP framework by conducting case studies on social learning platforms. The analysis employs three fundamental aspects of the CoP model—collaborative involvement, collective endeavors, and shared skill sets—to scrutinize the gathered data. The results indicate that organic farmers are more inclined to engage in social learning with peers who exhibit congruent views regarding agricultural business practices, farming methodologies, and the principles of organic agriculture. The study emphasizes the fluid and self-organizing nature of CoPs within unstructured and dispersed communities and suggests that extension approaches can benefit from viewing farmers as de facto members of CoPs, thus contributing to a better understanding of how these communities form and develop. Shinodi's (2021) study investigated the challenges and opportunities in the adoption of conservation agriculture (CA) technologies in India. Despite the potential soil protection benefits, the adoption rate of CA among Indian farmers remains low, with factors such as the "social learning model" and the "progressive farmer strategy" identified as major contributors to this trend. The study highlights the economic and ecological benefits demonstrated by CA technologies in various cropping systems, emphasizing their potential to improve farming practices in India. It underscores the knowledge-intensive nature of CA, emphasizing the importance of generating awareness among rural farmers to facilitate successful adoption. The heterogeneity of farmlands in India necessitates more field demonstrations tailored to specific farming needs for effective implementation. The study also delves into references providing a comprehensive overview of research related to CA adoption, explores practices in rainfed uplands of India, and discusses the diffusion of innovations, economic implications, and the need to tailor CA to small farmers in developing countries. Additionally, it addresses issues of desertification, land degradation, and the role of innovation diffusion and social influence in agriculture, drawing on relevant studies and theories. Overall, Shinodi's study provided a multifaceted examination of the status, challenges, and potential of conservation agriculture technologies in the Indian agricultural context.

2 Issues Related to the Scarcity of Irrigation Water

In India, the arid and semi-arid regions face significant challenges with the scarcity of water for irrigation purposes. Here are a few challenges linked to the scarcity of irrigation water in India. In general, the issues tied to the lack of irrigation water in India are intricate and diverse, necessitating various interventions and policy measures for resolution. These solutions may include more efficient water use, better infrastructure and management of irrigation systems, increased investment in water storage facilities, and better coordination and conflict resolution mechanisms for managing water resources (Fishman et al., 2021; Horan et al., 2021).

3 Risk of Bioaccumulation from Contaminated Irrigation Water

Contaminated irrigation water can pose a significant risk of bioaccumulation, particularly of toxic chemicals and heavy metals. Bioaccumulation refers to the slow build-up of various substances, like pesticides or heavy metals, within the body tissues of living creatures as time progresses. This process, if left unchecked, may result in detrimental impacts on the well-being of both humans and animals. When plants are irrigated with polluted water, they can take in the pollutants through their root systems and store them in their tissues. This phenomenon, referred to as phytoremediation, allows the contaminants to be passed on to animals and humans that ingest the polluted plants, potentially posing health hazards. Bioaccumulation can have serious consequences for human health, including developmental abnormalities, cancer, and other chronic diseases. It can also have ecological impacts, such as the decline of populations of wildlife species. To mitigate the risks of bioaccumulation from contaminated irrigation water, it is essential to monitor the quality of irrigation water and to take measures to reduce the use of contaminated water. This may involve implementing water treatment technologies, such as filtration or reverse osmosis, to remove contaminants from irrigation water before it is used, or using alternative sources of water, such as rainwater harvesting, to reduce dependence on contaminated water sources. Additionally, farmers can adopt best management practices to minimize the use of pesticides and other chemicals in agriculture and use alternative methods, such as organic farming, to reduce the risk of bioaccumulation (Chalakh et al., 2021; Pande et al., 2022).

4 Mitigating Surface Water and Soil Contamination with Phytoremediation and Riparian Buffers in Agricultural Systems

Phytoremediation and riparian buffers are two Green Infrastructure-related techniques that can be used to mitigate the impact of agricultural practices on the environment. Phytoremediation employs plants to eliminate pollutants like heavy metals, pesticides, and various harmful substances from both soil and water sources. Phytoremediation can be particularly effective for treating contaminated soils and groundwater, and it has been used successfully to remediate contaminated agricultural lands (Dushenkov et al., 1997). Certain plant species have the ability to absorb and accumulate contaminants from the soil and water into their tissues, which can then be harvested and disposed of safely. Phytoremediation is a cost-effective and environmentally friendly technique for remediating contaminated sites (McIntyre, 2003).

Riparian buffer zones bordering streams, rivers, and various water sources play a crucial role in shielding these environments against both pollution and erosion. Riparian buffers can be designed to reduce the impacts of agricultural practices by filtering runoff from fields, absorbing nutrients and sediment, and providing habitat for wildlife (Millhollon et al., 2009). This can help to reduce the amount of pollutants entering water bodies and protect them from the effects of agricultural practices. In agricultural systems, phytoremediation and riparian buffer techniques can be used together to reduce the impact of agricultural practices on the environment. For example, riparian buffers can be planted with species that are effective at phytoremediation, such as willows and poplars, to enhance their ability to remove contaminants from runoff and groundwater. This can help to reduce the amount of pollutants entering water bodies, protecting them from the effects of agricultural practices and improving the health of the ecosystem (Jordahl et al., 2003).

In this chapter, we delve into the application of social learning to construct primarily two strategic references:

(i) A strategy reference aimed at mitigating surface and irrigation water contamination and reducing the risk from pollution load bioavailability.
(ii) A strategy reference focused on the restoration of degraded soil and mitigating soil contamination.

Our objective is to develop strategy references for the adoption of innovative agricultural practices, including phytoremediation and riparian buffers, through social learning. This approach aims to address surface and irrigation water contamination, reduce the risk associated with pollution load bioavailability, and contribute to the restoration of degraded soil while mitigating soil contamination.

To successfully craft these strategy references, the initial step involves identifying stakeholders responsible for water supply, pollution, and land degradation. Understanding the local needs and challenges faced by farmers in their current agricultural practices is essential. Emphasis is placed on identifying the interests of stakeholders, particularly farmers. Finally, armed with this information, we can create strategy references to facilitate the effective adoption of new agricultural practices. To illustrate these concepts, we present a case study of the Lower Damodar River Basin (Fig. 1).

5 Case Study of Lower Damodar River Basin

The Lower Damodar River Basin in India, primarily an agricultural hub, confronts a myriad of challenges intricately linked to its farming practices. These challenges encompass acute water scarcity during the dry season, adversely impacting irrigation and crop production. Additionally, unsustainable agricultural practices, such as monoculture and excessive use of chemical inputs, have led to soil degradation, diminished fertility, and declining crop yields. Small landholdings, inadequate infrastructure, and prohibitive mechanization costs further impede the adoption of modern agricultural technologies. Pest and disease outbreaks pose a constant threat, causing significant crop losses. Furthermore, farmers grapple with limited access to markets and financial resources, hindering their ability to sell produce and invest in their farms. Compounding these challenges is the region's vulnerability to climate change, manifested in erratic rainfall patterns, rising temperatures, and extreme weather events.

Addressing this multifaceted agricultural landscape necessitates a holistic approach that considers the interconnected social, economic, and environmental dimensions of farming. The deployment of social learning networks and capacity building initiatives emerges as pivotal tools in enhancing farmers' knowledge and skills. By fostering collaborative learning and knowledge-sharing, these approaches aim to empower the agricultural community to adopt sustainable practices tailored to the specific challenges faced in the Lower Damodar River Basin. The comprehensive goals of the case study involve understanding local needs, identifying stakeholder interests, and developing context-specific solutions. Tailored to the nuanced circumstances of the region, the study endeavors to formulate effective and feasible strategies, promoting resilience and sustainability in agricultural practices. An additional focus lies on establishing a social learning network, fostering a community-driven platform for the exchange of experiences and best practices, ultimately contributing to a more interconnected and informed agricultural community in the study area. The study contributes to social physics by providing a deeper

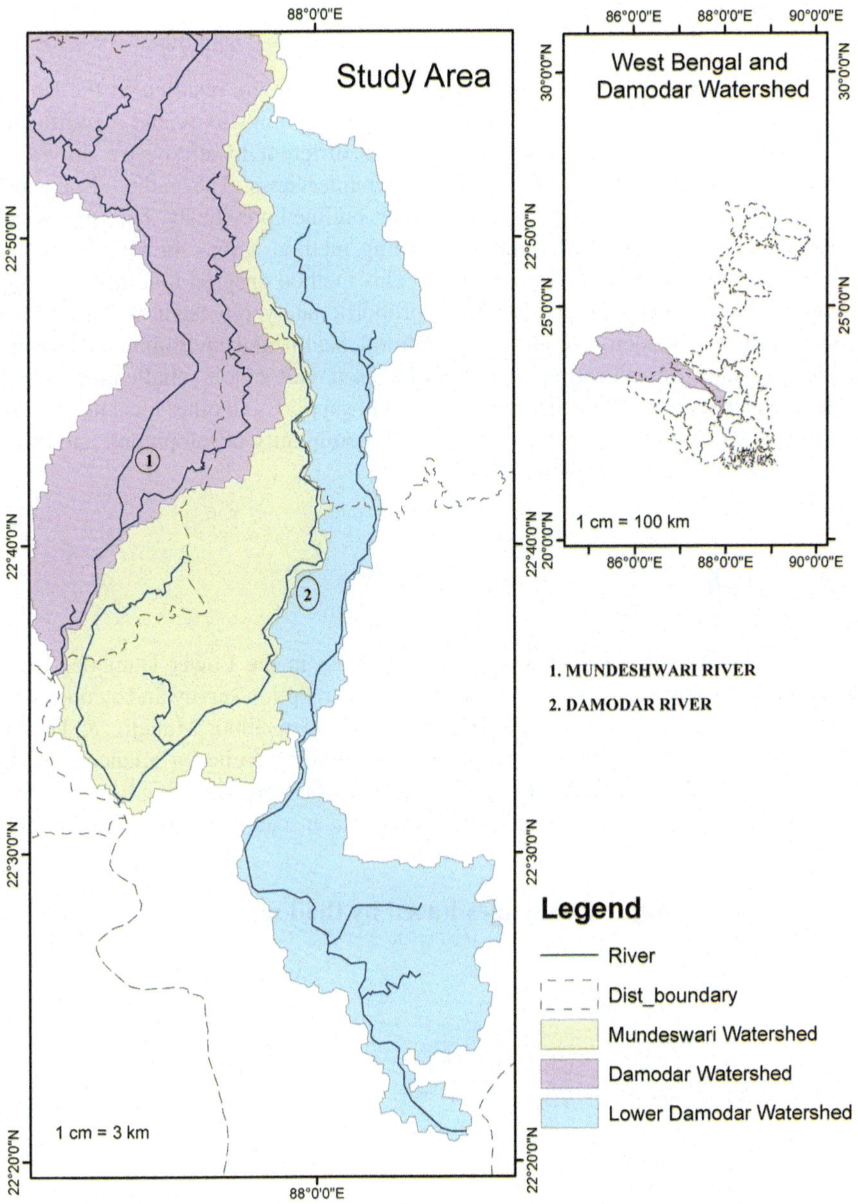

Fig. 1 The lower damodar river basin

understanding of the spatial dimensions of agricultural decision-making and the factors that influence the adoption of sustainable practices. This knowledge can be used to inform policy decisions, improve agricultural practices, and promote more resilient and sustainable agricultural systems.

5.1 Methods

To understand the perspectives and preferences of farmers residing in the Lower Damodar Basin, we conducted two surveys: a perception survey and a contingent valuation survey. Targeting 60 farmers across different locations, we utilized a structured questionnaire through door-to-door interviews. The gathered data has been carefully analyzed to fulfill the objectives outlined previously. The survey was conducted using a stratified random sampling method across 44 villages in the Burdwan, Hooghly, and Howrah districts. This method ensured that different segments of the farmer's population were proportionately represented. The survey focused on various socio-economic, agricultural, health, and environmental aspects, collecting data on education, landholding, crop cultivation, challenges, health issues, water sources, soil concerns, government support, and other local needs. The collected data aims to inform policy-making, community development, and interventions to address the challenges identified.

5.2 Result

In the recent survey conducted among 60 farmers in the Lower Damodar River Basin, key insights were gleaned through both a perception survey and a contingent valuation method (CVM) survey (Alberini & Cooper, 2000; Mutaqin & Usami, 2019; Venkatachalam, 2004). The perception survey unveiled a nuanced understanding of local needs and challenges faced by the farmers, shedding light on prevalent agricultural practices (Fig. 2). The issue of water scarcity has become a notable

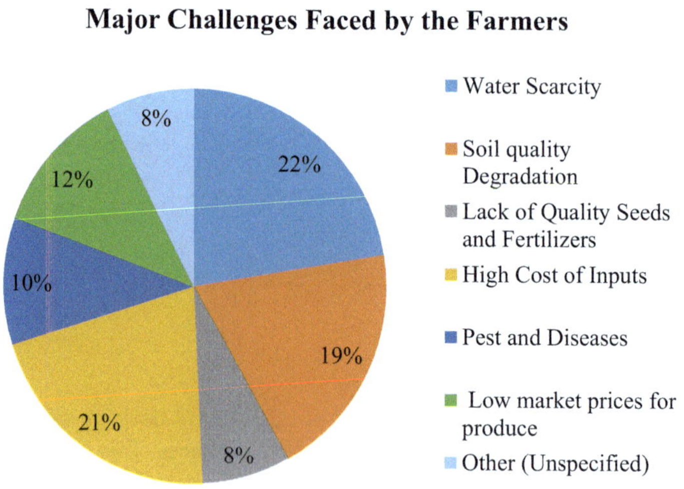

Fig. 2 Major challenges faced by the farmers in the Lower Damodar Basin. (Source: field survey)

worry, with a considerable portion of survey participants (22.39%) emphasizing its effects on their agricultural pursuits. Additionally, issues such as soil quality degradation (19.40%), the high cost of inputs (20.90%), and the lack of quality seeds and fertilizers (7.46%) were identified as pressing challenges. Stakeholder interests were also discerned, with farmers expressing a keen interest in subsidies for inputs (26.47%), improved irrigation facilities (20.59%), and financial support for purchasing modern farming equipment (23.53%). The farmers' educational backgrounds varied, with distribution across categories such as secondary, higher secondary, eight pass, graduate, and illiterate. The major crops cultivated encompassed a diverse array, including large field crops, fruits, vegetables, and medicinal and aromatic crops.

The CVM survey delved into the farmers' willingness to embrace change and adopt new practices. A notable 68% expressed a readiness to cultivate different plants, showcasing receptiveness to diversifying their agricultural portfolio (Fig. 3). Furthermore, a substantial 78% were open to incorporating novel techniques like precision agriculture or organic farming. The analysis of Likert scale responses indicated a positive inclination toward cultivating different plants if provided with incentives and economic benefits. Over 53% either agreed or strongly agreed, portraying a potential for increased adoption (Fig. 4). Moreover, the majority (76.75%) demonstrated a favorable stance toward adopting new farming practices for environmental management, emphasizing the significance of economic incentives in steering such changes. These findings suggest a promising prospect for promoting sustainable agricultural practices in the Lower Damodar River Basin, underscoring the need for further research to validate and refine these preliminary insights. The farmers' collective openness to change presents an opportunity to enhance both livelihoods and environmental sustainability in the region.

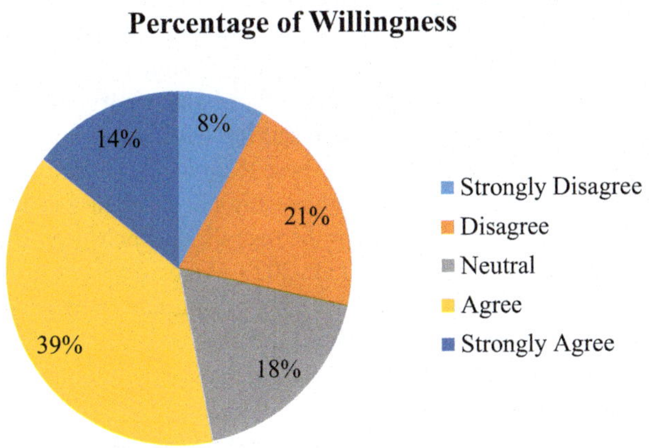

Fig. 3 Willingness to cultivate different plants other than your main crops if given incentives and if there are economic benefits. (Source: field survey)

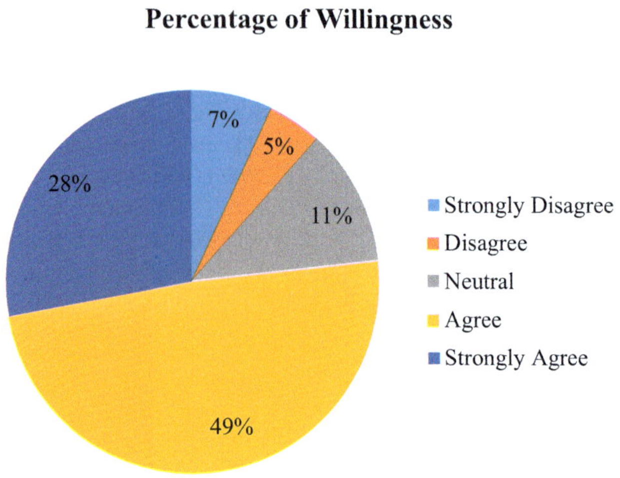

Fig. 4 Willingness to adopt new farming practices for the sake of environmental management if given incentives and if there are economic benefits. (Source: field survey)

6 Discussion

Farmers in India wield substantial influence in environmental management, grappling with challenges to active participation. Their pivotal role encompasses adopting sustainable practices such as conservation agriculture and organic farming, engaging in government programs, participating in community-based initiatives, accessing information and resources, and advocating for policy change. The backdrop of this responsibility is the prevalent issue of irrigation water scarcity, particularly afflicting arid and semi-arid regions, bringing with it a host of challenges including water scarcity itself and the risk of bioaccumulation from contaminated water (Arumugam & Manida, 2023). To combat these challenges, the text proposes the implementation of two crucial techniques: phytoremediation and riparian buffers. Phytoremediation employs plants to remove pollutants from soil and water, providing an eco-conscious remedy. On the other hand, riparian buffer zones comprise vegetation strips along water bodies, serving as a natural filtration system for runoff and a means to absorb pollutants. These techniques are integral to mitigating surface water and soil contamination, addressing not only the immediate environmental concerns but also contributing to the overall well-being of agricultural ecosystems (Anbumozhi et al., 2005; Shi et al., 2023; Wu et al., 2023). Recognizing the significance of social learning in fostering the adoption of phytoremediation and riparian buffer techniques among farmers, the text underscores its effectiveness. Social learning manifests through various channels, including farmer-to-farmer knowledge sharing, peer-to-peer learning networks, capacity building and training programs, participatory research, demonstration projects, and policy support. This collaborative approach is not only informative but also facilitates experiential learning, allowing farmers to adapt these techniques to their specific contexts (Davis &

Luthans, 1980; Hermans et al., 2023; Leta et al., 2018). To formalize and streamline the integration of phytoremediation and riparian buffer techniques into Indian agriculture, the text proposes a comprehensive strategy reference framework. This framework comprises key components such as the assessment of local needs and challenges, the development of context-specific solutions, capacity building and training, farmer-to-farmer knowledge sharing, research and development, policy support, and monitoring and evaluation. By addressing these aspects systematically, the proposed framework aims to create a robust foundation for the successful implementation and promotion of these sustainable agricultural practices. The operationalization of social learning in Indian agriculture involves a step-by-step process, commencing with the assessment of local needs and challenges. This initial step is followed by the establishment of social learning networks, training, and capacity building initiatives. Collaborative learning and experimentation, knowledge sharing, and dissemination then play a pivotal role in spreading awareness and understanding. Scaling up and institutionalization complete the process, ensuring that these practices become ingrained in the agricultural fabric of India.

This framework outlines a comprehensive strategy for implementing and promoting phytoremediation and riparian buffers in Indian agriculture through social learning. The Lower Damodar River Basin (LDBR) case study serves as a crucial reference point, providing valuable insights for tailoring interventions to specific contexts.

Strategy Framework:

Phase 1: Contextual Assessment and Stakeholder Engagement

1. Local needs and challenges: Conduct an in-depth assessment, drawing insights from the LDBR case study. This includes understanding water scarcity, soil degradation, and other agricultural concerns. Which we have already conducted through this study.
2. Stakeholder identification: Identify and engage key stakeholders, including farmers, authorities, and environmental organizations. Leverage LDBR findings to understand their roles and interests (Bezerra et al., 2023).
3. Farmer engagement and knowledge sharing: Create social learning networks based on the LDBR model. Facilitate farmer-to-farmer knowledge sharing and peer-to-peer learning.

Phase 2: Capacity Building and Participatory Research

1. Capacity building programs: Design programs addressing LDBR challenges, focusing on water scarcity, soil degradation, and adoption of phytoremediation and riparian buffers. Use LDBR insights for content and format (Meena et al., 2014).
2. Participatory research and demonstration projects: Implement participatory research, drawing inspiration from the LDBR approach. Collaborate with farmers to co-create demonstration projects showcasing context-specific solutions.

Phase 3: Policy Advocacy and Development of Context-Specific Solutions

1. Policy support advocacy: Leverage LDBR findings to advocate for supportive policies at local and national levels. Emphasize positive impacts and the need for widespread adoption (Agricultural Policy Monitoring and Evaluation 2023, 2023).
2. Development of context-specific solutions: Use the LDBR case study as a reference to develop customized solutions for other regions. Consider local agricultural practices, climate conditions, and socio-economic factors.

Phase 4: Research, Monitoring, Evaluation, and Scaling-Up

1. Research and development: Continuously improve phytoremediation and riparian buffer techniques. Stay responsive to advancements and scientific insights, incorporating LDBR learnings (Henriksson et al., 2021).
2. Monitoring and evaluation: Apply the LDBR monitoring framework to assess the impact of interventions and ensure continuous improvement.
3. Scaling-up and institutionalization: Based on LDBR successes, scale up initiatives to other regions. Integrate social learning practices into local agricultural frameworks (Brown et al., 2022).

Phase 5: Case Study Integration and Learning

1. Case study integration and learning: Integrate key LDBR lessons into the strategy framework. Utilize the case study as a model and promote knowledge transfer to other communities.
2. Addressing local challenges: Prioritize strategies that directly address LDBR-identified challenges, tailoring interventions to region-specific issues.
3. Farmer preferences and economic incentives: Align the strategy with farmer preferences, especially regarding economic incentives. Develop structures that resonate with diverse regions.
4. Community-driven platform: Build upon the LDBR community–driven platform. Foster exchange of experiences and best practices, promoting a more interconnected and informed agricultural community.

7 Conclusion

The study of the Lower Damodar River Basin (LDBR) reveals a multifaceted agricultural landscape facing challenges like acute water scarcity, soil degradation, and limited resources. Social learning networks and capacity building initiatives emerge as crucial tools to empower the agricultural community and foster sustainable practices. The case study, delving into local needs and challenges, identifies water scarcity, soil degradation, and high input costs as pressing concerns. Farmers express interest in subsidies, improved irrigation facilities, and financial support for modern equipment, showcasing the region's agricultural diversity. The contingent valuation method (CVM) survey indicates farmers' openness to change, with a significant percentage willing to diversify crops and adopt novel techniques when provided with incentives. This positive inclination, coupled with a favorable stance toward new practices for environmental management, suggests a promising avenue for sustainable agriculture in the Lower Damodar River Basin. While these findings are optimistic, further research is essential to validate and refine these insights. The collective willingness to embrace change among farmers presents an opportunity to enhance livelihoods and environmental sustainability. The study underscores the need for ongoing efforts to support and amplify these positive trends, establishing a foundation for resilient and sustainable agricultural practices in the Lower Damodar River Basin. By emphasizing the LDBR case study, this framework provides a replicable and impactful model for promoting sustainable agricultural practices. The framework can be adapted to diverse regions across India, ensuring a context-specific and effective implementation of social learning for agricultural sustainability. The study contributes to social physics by providing empirical data on the spatial distribution of challenges and willingness to adopt new practices among farmers. It also highlights the importance of social learning in promoting sustainable agriculture. The proposed framework for implementing phytoremediation and riparian buffers offers a roadmap for spatial planning in agriculture, contributing to a deeper understanding of the spatial dimensions of agricultural decision-making and the factors that influence the adoption of sustainable practices.

References

Agricultural Policy Monitoring and Evaluation 2023. (2023). OECD. https://doi.org/10.1787/b14de474-en/1.

Alberini, A., & Cooper, J. (2000). *Applications of the contingent valuation method in developing countries*. FAO Economic and Development Paper 146.

Anbumozhi, V., Radhakrishnan, J., & Yamaji, E. (2005). Impact of riparian buffer zones on water quality and associated management considerations. *Ecological Engineering, 24*(5), 517–523. https://doi.org/10.1016/j.ecoleng.2004.01.007

Arumugam, U., & Manida, M. (2023). Sustainable farming Management in India. *Shanlax International Journal of Management, 11*(1), 54–60. https://doi.org/10.34293/management.v11i1.6367

Bezerra, M. O., Vollmer, D., Souter, N. J., Shaad, K., Hauck, S., Marques, M. C., Mtshali, S., Acero, N., Zhang, Y., & Mendoza, E. (2023). Stakeholder engagement and knowledge co-production for better watershed management with the freshwater health index. *Current Research in Environmental Sustainability, 5*, 100206. https://doi.org/10.1016/j.crsust.2022.100206

Bookwalter, J. T., Fuller, B. S., & Dalenberg, D. R. (2006). Do household heads speak for the household? A research note. *Social Indicators Research, 79*(3), 405–419. https://doi.org/10.1007/s11205-005-4925-9

Brown, B., Chaudhary, Karki, Timsina, & Sharma. (2022). *How should we scale agricultural practices? Inputs to the development of an assessment framework to compare scaling mechanisms. October*. https://doi.org/10.13140/RG.2.2.14571.41760.

Chalakh, S., Desai, P., & Rajput, D. (2021). Impact of pesticide awareness module (pam) in improving the knowledge of farmers community about pesticide poisoning: A pre- and post-intervention study in Wardha district in Maharashtra, India. *Journal of Pharmaceutical Research International, 160*, 160–166. https://doi.org/10.9734/jpri/2021/v33i38B32111

Davis, T. R. V., & Luthans, F. (1980). A social learning approach to organizational behavior. Academy of Management Review *Doi:10.5465/Amr.1980.4288758, 5*(2), 281–290. https://doi.org/10.5465/AMR.1980.4288758.

Delgado, A. R. S., Ventura, S. D., & Ferreira, P. M. (2020). Agricultural optimization with limited resources using duality. *Pesquisa E Ensino Em Ciências Exatas E Da Natureza, 4*. https://doi.org/10.29215/PECEN.V4I0.1477

Dinar, A., & Mody, J. (2004). Irrigation water management policies: Allocation and pricing principles and implementation experience. *Natural Resources Forum, 28*(2), 112–122. https://doi.org/10.1111/j.1477-8947.2004.00078.x

Dushenkov, S., Kapulnik, Y., Blaylock, M., Sorochisky, B., Raskin, I., & Ensley, B. (1997). *Phytoremediation: A novel approach to an old problem* (pp. 563–572). https://doi.org/10.1016/S0166-1116(97)80071-4

Fishman, R., Gine, X., & Jacoby, H. G. (2021). *Efficient irrigation and water conservation: Evidence from South India*. The World Bank. https://doi.org/10.1596/1813-9450-9713

Henriksson, P. J. G., Troell, M., Banks, L. K., Belton, B., Beveridge, M. C. M., Klinger, D. H., Pelletier, N., Phillips, M. J., & Tran, N. (2021). Interventions for improving the productivity and environmental performance of global aquaculture for future food security. *One Earth, 4*(9), 1220–1232. https://doi.org/10.1016/j.oneear.2021.08.009

Hermans, T. D. G., Smith, H. E., Whitfield, S., Sallu, S. M., Recha, J., Dougill, A. J., Thierfelder, C., Gama, M., Bunderson, W. T., Museka, R., Doggart, N., & Meshack, C. (2023). Role of the interaction space in shaping innovation for sustainable agriculture: Empirical insights from African case studies. *Journal of Rural Studies, 100*, 103012. https://doi.org/10.1016/j.jrurstud.2023.103012

Hoffman, M. (2009). Extending sustainable agriculture: Social learning, decision-making, and practice adoption in California viticulture.

Horan, R., Wable, P. S., Srinivasan, V., Baron, H. E., Keller, V. J. D., Garg, K. K., Rickards, N., Simpson, M., Houghton-Carr, H. A., & Rees, H. G. (2021). Modelling small-scale storage interventions in semi-arid India at the basin scale. *Sustainability, 13*(11), 6129. https://doi.org/10.3390/su13116129

Jordahl, J. L., Madison, M. F., Smesrud, J. K., Emond, H. M., & Motte, M. Q. (2003). Waste management using trees: Wastewater, leachate, and groundwater irrigation. In *Phytoremediation* (pp. 717–751). Wiley. https://doi.org/10.1002/047127304X.ch23

Ladd, G. W., & Mize, J. (1983). A cognitive-social learning model of social-skill training. *Psychological Review, 90*(2), 127–157. https://doi.org/10.1037/0033-295X.90.2.127

Leta, G., Stellmacher, T., Kelboro, G., Van Assche, K., & Hornidge, A.-K. (2018). Social learning in smallholder agriculture: The struggle against systemic inequalities. *Journal of Workplace Learning, 30*(6), 469–487. https://doi.org/10.1108/JWL-12-2017-0115

McCarthy, D. D. P., Crandall, D. D., Whitelaw, G. S., General, Z., & Tsuji, L. J. S. (2011). A critical systems approach to social learning: Building adaptive capacity in social, ecological, epistemological (SEE) systems. *Ecology and Society, 16*(3), art18. https://doi.org/10.5751/ES-04255-160318

McIntyre, T. (2003). *Phytoremediation of heavy metals from soils* (pp. 97–123). https://doi.org/10.1007/3-540-45991-X_4

Meena, M. S., Singh, K. M., Meena, H. M., Kumar, M., Meena, M. L., & Dupare, B. U. (2014). Assessing impact of capacity building program. In M. S. Meena, K. M. Singh, & B. P. Bhatt (Eds.), *Water management in agriculture* (pp. 141–148). Jaya Publishing House. https://doi.org/10.13140/2.1.1585.7605

Millhollon, E. P., Rodrigue, P. B., Rabb, J. L., Martin, D. F., Anderson, R. A., & Dans, D. R. (2009). Designing a constructed wetland for the detention of agricultural runoff for water quality improvement. *Journal of Environmental Quality, 38*(6), 2458–2467. https://doi.org/10.2134/jeq2008.0526

Morgan, S. L. (2011). Social learning among organic farmers and the application of the communities of practice framework. *Journal of Agricultural Education and Extension, 17*(1), 99–112. https://doi.org/10.1080/1389224X.2011.536362

Mutaqin, D. J., & Usami, K. (2019). Smallholder farmers' willingness to pay for agricultural production cost insurance in rural West Java, Indonesia: A contingent valuation method (CVM) approach. *Risks, 7*(2), 69. https://doi.org/10.3390/risks7020069

Olawuyi, S. O. (2018). Farmers' preference for soil and water conservation practices in Nigeria: Analytic hierarchic process approach. *Journal of Economics and Behavioral Studies, 10*(4(J)), 68–80. https://doi.org/10.22610/JEBS.V10I4(J).2408

Pande, A., Mun, B.-G., Methela, N. J., Rahim, W., Lee, D.-S., Lee, G.-M., Hong, J. K., Hussain, A., Loake, G., & Yun, B.-W. (2022). Heavy metal toxicity in plants and the potential NO-releasing novel techniques as the impending mitigation alternatives. *Frontiers in Plant Science, 13*. https://doi.org/10.3389/fpls.2022.1019647

Piñeiro, V., Arias, J., Dürr, J., Elverdin, P., Ibáñez, A. M., Kinengyere, A., Opazo, C. M., Owoo, N., Page, J. R., Prager, S. D., & Torero, M. (2020). A scoping review on incentives for adoption of sustainable agricultural practices and their outcomes. *Nature Sustainability, 3*(10), 809–820. https://doi.org/10.1038/s41893-020-00617-y

Rayhan Shaheb, M., Sarker, A., Shearer, A., & S. (2022). Precision agriculture for sustainable soil and crop management. In *Soil science—emerging technologies, global perspectives and applications*. IntechOpen. https://doi.org/10.5772/intechopen.101759

Schwoob, M.-H., Timmer, P., Andersson, M., & Treyer, S. (2018). *Agricultural transformation pathways toward the SDGs* (pp. 417–436). https://doi.org/10.1142/9789813278356_0012

Shi, Q., Kaur, P., & Gan, J. (2023). Harnessing the potential of phytoremediation for mitigating the risk of emerging contaminants. *Current Opinion in Environmental Science & Health, 32*, 100448. https://doi.org/10.1016/j.coesh.2023.100448

Shinogi, K. C., Srivastava, S., Rashmi, I., Jayaraman, S., Sinha, N. K., Gurav, P. P., & Kamble, A. L. (2021). Socioeconomic challenges and prospects in the adoption of conservation agriculture practices in India. *Conservation Agriculture: A Sustainable Approach for Soil Health and Food Security: Conservation Agriculture for Sustainable Agriculture*, 611–621. https://doi.org/10.1007/978-981-16-0827-8_29/COVER

Sulaiman, N. K., & Misnan, S. H. (2022). Environmental sustainability through agriculture: Perspectives of extension agents on adoption of sustainable practices. *IOP Conference Series: Earth and Environmental Science, 1082*(1), 012024. https://doi.org/10.1088/1755-1315/1082/1/012024

Tran, T. A., James, H., & Nhan, D. K. (2020). Effects of social learning on rural farmers' adaptive capacity: Empirical insights from the Vietnamese Mekong Delta. *Society & Natural Resources, 33*(9), 1053–1072. https://doi.org/10.1080/08941920.2019.1693677

Venkatachalam, L. (2004). The contingent valuation method: A review. *Environmental Impact Assessment Review.* https://doi.org/10.1016/S0195-9255(03)00138-0

Wu, S., Bashir, M. A., Raza, Q.-U.-A., Rehim, A., Geng, Y., & Cao, L. (2023). Application of riparian buffer zone in agricultural non-point source pollution control—A review. *Frontiers in Sustainable Food Systems, 7*. https://doi.org/10.3389/fsufs.2023.985870

Dr. Suman Chatterjee is an Assistant Professor in the Department of Geography at Adamas University. He specializes in the application of Remote Sensing and GIS in environmental management and biogeography. Dr. Chatterjee completed his PhD under the UGC fellowship, focusing on geospatial technologies and their role in sustainable environmental solutions. With over three years of teaching and research experience, he is dedicated to advancing knowledge in his field and fostering innovation through his academic and professional endeavors.

Ishita Tudu is a 2023 graduate of the M.Sc. program in Geoinformatics at Adamas University. Her research interests lie at the intersection of social and environmental geography, with a focus on leveraging geospatial technologies to address critical societal and environmental challenges. Ishita is passionate about using GIS and remote sensing tools to analyze and solve complex issues related to human-environment interactions, sustainability, and resource management.

Role of Anthropogenic Activities on the Riverbank Instability: A Geospatial Analysis on Part of the Bhagirathi-Hugli River, India

Sanat Das and Rajib Sarkar

1 Introduction

Anthropogenic activities have an overarching impact on the environment. The human population is increasing rapidly over the last few decades. According to United Nations estimates, the world population will increase to eight billion by 2030 (United Nations, 2015). With a larger population count, stress on the environment is also increasing. The impact of population growth and related anthropogenic activities are also impacting the riparian system through various mechanisms and feedback loops. Historically, it can be found that most of the large settlements have evolved around a hydraulic source, be it riparian a coastal settlement. River-based settlements may alter the natural characteristics of rivers through different activities in and around the river. These activities influence the physiochemical properties of water as well as bank stability.

River courses are dynamic. These courses change from the sources to regulate themselves with the topography (Bandyopadhyay & De, 2017). Alluvial channels, which are morphological, hydraulic, and sedimentological components, are extremely dynamic to changing environmental phenomena or disturbances. Upstream degradation, downstream accumulation, channel enlargement or narrowing, and changes in the amount and type of sediment loading influence channel responses (Simon & Downs, 1995). Channel responses due to altered environmental settings can lead to different landscape features in the downstream stage, such as meanders, cut-offs, and abandoned channels (Das et al., 2014; Pal et al., 2016). Riverbank erosion and migration are linked with each other. The Bhagirathi-Hugli River has also shown a tendency of migration and cut-off from the historical period (Laha, 2015; Mitra, 2015).

S. Das · R. Sarkar (✉)
Department of Geography, Adamas University, Barasat, West Bengal, India
e-mail: rajib.sarkar@adamasuniversity.ac.in

A. Rai et al. (eds.), *Contemporary Social Physics*, Springer Geography,
https://doi.org/10.1007/978-3-031-77596-3_11

Erosion and deposition processes are part of river migration (Pal et al., 2016; Thakur et al., 2012). The erosion of banks is closely related to near-bank velocities caused by hydraulic actions (Odgaard, 1987; Hasegawa, 1989). The hydraulic actions are often controlled by different human-induced activities. The erosion of the river sometimes occurs in agricultural areas due to changes in the water discharge in the river. It mainly occurs by water flowing over or through flooding or even due to runoff of water from various sources.

The Bhagirathi-Hugli River has large fluctuations in flow depth and velocity over several years due to changes in seasonal rainfall. Usually, the maximum rates of bank withdrawal are related to the bending crest and high shear stresses that occur in contrast to the outward meander bends (Knighton, 1998). The channel is enforced to maintain a static condition by urban growth and embankment construction on riverbanks (Rudra, 2010; Pal et al., 2016). The running water of the river erodes and ejects directly onto the outer bank at meander. The water stacks stress the skin of the curvature due to centrifugal force. During a corkscrew motion, the bottom current determines and is thrown into the mid-channel and inner bank. Therefore, sedimentation occurs within the inner bank, creating a gentle slip-off slope (Morisawa, 1985). Erosion and accumulation of the river cause a change in the surface area.

Although bank erosion and migration are natural processes, typically, anthropogenic activities significantly impact the rate of hydro morphological modification (Chakraborty & Choudhury, 2009; Billi & Rinaldi, 1997; Isik et al., 2008; Debnath et al., 2015). Different anthropogenic activities such as soil, sand cutting on the stream banks, sand lifting from the stream bed, infrastructure building, cut-offs, embankments, bridge construction on the river, and land use modifications affect the river's morphology and natural dynamics (Yamani et al., 2011). Human activities are changing bank stability (Duru, 2017; Kang et al., 2004; Dwivedi et al., 2016; Chatterjee, 2013). Due to urbanization, it mostly occurs adjacent to rural areas (Voogt & Oke, 2003). Normalized Difference Vegetation Index (NDVI), Normalized Difference Built-up Index (NDBI), and Normalized Difference Water Index (NDWI) are the main indicators of urban environmental studies. The negative correlation among them is significant (Liu et al., 2015; Carlson et al., 1994) for this study. Recently, Mohammed-Ali et al. (2020a) and Mohammed-Ali et al. (2020b) evaluated the riverbank stability characteristics using a combined BSTEM model with the HEC-RAS model.

Therefore, this study attempts to assess how much the riverbank has changed over time. The temporal frame for this study is taken from 1972/1973 to 2017, and the study attempts to analyze the impact of human activities on the riverbank stability of the Bhagirathi-Hugli River in West Bengal, India. For this study, riverbank erosion, deposition, valley length, river width, river basin, population growth, and various anthropogenic doings such as uneven brickfield progress, land cover changes, and land use change impacts such as NDVI, NDBI, and MNDWI were discussed.

The study is also important as it will help understand the anthropogenic footprint's role on the river ecosystem and associated social dynamics. Much literature has pointed out the direct and pseudo role of river dynamics and following social change due to displacement, migration, and realignment of personal relations. The application of geoinformation technology is meshed with social data to highlight the areas of enhanced river erosion and zones of altered social relations.

The study also focuses on the causative variables for bank instability. Another aim of the study is to unearth the impact of anthropogenic factors on bank failure using different biophysical components along the river. This study reveals that there is a significant impact of anthropogenic activities around the erosional zones of the rivers.

2 Material and Method

2.1 Study Area Description

The chosen path of the Bhagirathi-Hugli River usually forms the district boundary between Purba Bardhaman and Nadia districts. According to the 2001 Census, this watershed is located in the center of the Bengal Delta. A segment of the Bhagirathi-Hugli River, a Ganga watercourse branch characterized by monsoonal seasonality, forms the study region (Pal et al., 2016). The Bhagirathi-Hugli River's northern portion is situated between the Ketugram-I block and the Kaligangj block in Nadia districts, and its southern portion is situated between the Kalna-II block and the Santipur block in Nadia districts within the study area. The study area is regarded as one of the most dynamic sections of the lower Ganges plain due to its rapid plan-form changes of the river (Rudra, 2010; Bandyopadhyay et al., 2014; Chatterjee, 2013). The latitudinal and longitudinal extension of the study area lies between 23° 12′ 08.20″ N and 23° 46′ 11.88″ N and 88°05′ 05.44″ E and 88°29′ 43.50″ E (Fig. 1). The Bhagirathi-Hugli River is located between Ghasuriadanga and Medgachhi in West Bengal and is approximately 122.4 km long within the study area. The river divides the study area into two parts. The Nadia district is located on the eastern bank of the river, while the Purba Bardhaman district is situated on the western bank of the river. Hugli is the common name given to the Bhagirathi starting from the downstream Jalangi River confluence.

Ganga-Hugli River system has formed an alluvial deposit in this study region. The study area usually contains entisols, also known as fluvaquents and ustifluvent soils. The surface is made up of rivers, sand, and fine silt, which in the shallower areas combine to form clay. The Bhagirathi-Hugli River basin has a maximum elevation of 33 m above sea level in the north and northwest of the study area and a minimum elevation of 3 m above sea level in the south and southeast along the river course. These data are gathered by the Shuttle Radar Topography Mission (SRTM DEM). The highest amount of rainfall occurs in June and July, with an average of 1455 mm each year. Not all of the rainfall occurs at the same time. The districts do not receive the same amount of rainfall; there is more rainfall in the northern and southern regions than in the inner part. According to Das and Sarkar (2019), the average maximum temperature is 36 °C, and the average lowest temperature is 8 °C.

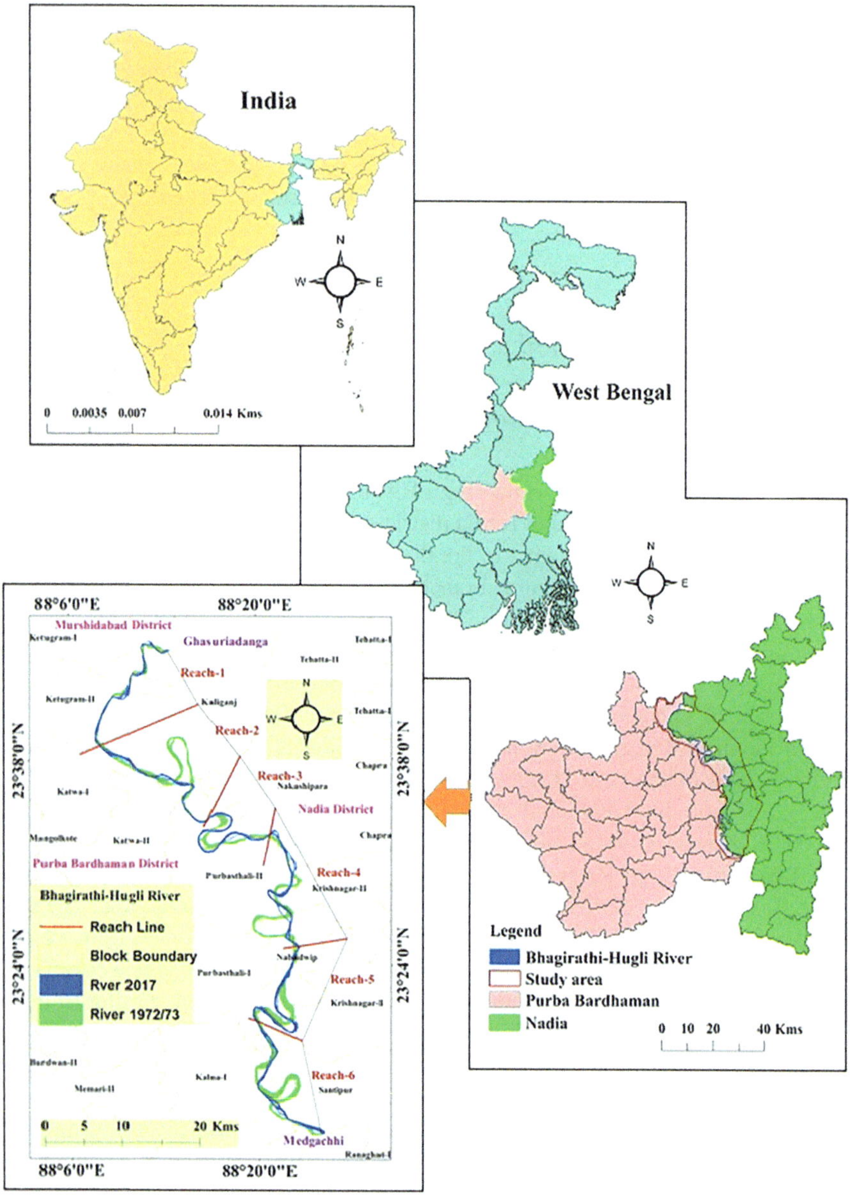

Fig. 1 Location map of the study area

Table 1 Data used in this study

Reference system (path and row)	Satellite/sensor	Date of image	Spatial resolution/ scale	Map No./No. of band used	Data source
	SOI Topographical Map	1972–1973	1:50,000	79A/1, 79A/2,79A/6, 79A/7,79A/8, 79B/5	SOI
WRS_138 & 44	Landsat_5/TM	01.04.1989	30 m	1, 2, 3, 4	USGS
WRS_138 & 44	Landsat_7/ ETM+	26.04.2001	30 m	1, 2, 3, 4	USGS
WRS_138 & 44	Landsat_5/TM	11.04.2010	30 m	1, 2, 3, 4	USGS
WRS_138 & 44	Landsat_8/ OLI_TIRS	14.04.2017	30 m	2, 3, 4, 5	USGS
	SRTM DEM	11.02.2000	90 m		USGS

SOI Survey of India, *USGS* United States Geological Survey

2.2 Study Area and Data Sources

In this investigation, data from the Landsat TM and ETM+ satellites were employed. Table 1 shows that those were acquired from the USGS Erath Explorer website. Furthermore, Google Maps and Survey of India topographical maps at a scale of 1:50,000 were put to use to create the study area boundary map, and census handbooks were used to gather population data. The location of bank erosion has been determined through the utilization of topographical maps and an extensive field survey.

2.3 Methodology

Georeferencing and Digital Processing of Satellite Data

The data that is used in this study needs to be geo-referenced before commencing the analysis. The Semi-automatic Classification Plugin (SCP) in QGIS 2.18.14 has been used to apply the dark object subtraction (DOS) method of atmospheric correction to all of the images. Because of its great outcomes and ease of use, DOS is the method of choice over other approaches. Toposheets from 1972/1973 and Landsat imagery from 1989, 2001, 2010, and 2017 were used to assess riverbank vulnerability. Toposheets (79A/1, 2, 6, 7, 8, 79B/5) and Landsat imagery were geo-referenced using ground control points (GCPs) considering the UTM (WGS_1984) projection system. The geo-referenced images from the same year have been mosaiced together. The Landsat images are subject to geometric corrections and are accustomed to delineating the river course using on-screen digitizing. The Bhagirathi-Huglistretch from Ghasuriadanga to Medgachhi, which comprises approximately 122.4 km (from satellite image 2017), has been divided into six reaches. The river width was measured along the 96 cross-sectional lines between the bank lines. The valley length represents the mid-line valley length, and the river area represents the river's surface area. Software such as ArcGIS 10.5 and Erdas Imagine 2014 were used to analyze and present the data.

Estimation of Erosion and Deposition

Riverbank lines from two different years—1972 and 1989, 1989 and 2001, 2001 and 2010, 2010 and 2017, and 1972/1973 and 2017—have been prepared to calculate the surface area of erosion and deposition. The arc length for each left and the right bank was calculated using the GIS software for all of the above years. Throughout the 45-year study period, polygon areas with shifting shorelines were used to estimate the areas subject to erosion and deposition using appropriate GIS tools. The rate of erosion was measured by the amount of surface area lost over time. The erosion rate is calculated using the following formula:

$$R_e = A_e / T_p \tag{1}$$

where R_e = erosion rate in km^2/year, A_e = amount of the lost area (k^2), and T_p = time compared for two different years.

The rate of deposition is also measured by the amount of surface area gained over a specified time. The following formula calculates the rate of deposition:

$$R_d = A_d / T_p \tag{2}$$

where R_d = rate of deposition in km^2/year, A_d = amount of the area gain (k^2), T_p = time compared to two different years.

Inverse Distance Weighting (IDW) Model

As a precise interpolator, IDW allows the maximum and lowest values of the interpolated surface to occur only at sample points. It depends on the selection of a power value (p) and the neighborhood search approach. IDW implies that local variation is at the root of the phenomena, which can be identified by defining an appropriate search neighborhood. This method has been suggested for creating a zonation map of riverbank erosion that takes into account the overall erosion of every polygon in the study areas. To illustrate the spatial distribution of the eroded zones and the highly erodible areas of the channel bank, the buffer (1.5 km wide from the channel) and IDW have been utilized. This method in this study has been used to estimate the riverbank erosional values of an unknown point from known points weighted by an inverse function of the distance. Interpolation was determined using the following equation (Veda Samhitha and Geetha Srikanth 2017):

$$x_i = \frac{\sum_{i=1}^{n} \frac{Vi}{d_i^p}}{\sum_{i=1}^{n} \frac{1}{d_i^p}} \tag{3}$$

Where X_i represents an unknown value, n indicates the number of sample stations, *Vi* represents the data value (here, the amount of eroded land), d indicates the distance, and p represents the power of the coefficient. This was implemented using the ArcGIS software environment.

Kernel Density Model

Kernel density estimation is a nonparametric approach for calculating the possibility density function of an arbitrary variable. This model was applied to estimate the spatial distribution of brickfields based on the principle of a brickfield-concentrated map between core areas (kernel) and localities. It computes the area in km^2 from point features using a kernel function to fit an easily tightened surface to each point. Kernel functions are used to represent point information markers on a continuous density surface (Krisp & Špatenková, 2010; Langford & Unwin, 1994). Furthermore, the mapping of point density by kernel functions—which enables the polling radius (bandwidth) of the kernel to control the degree of smoothing—is examined (Bowman & Azzalini, 1997). The following formula is used to get the density at the location (x, y):

$$\rho = \frac{1}{(r)^2} \sum_{i=1}^{n} \left[\frac{3}{\pi}^{*} W_i \left(1 - \left(\frac{d_i}{r} \right)^2 \right)^2 \right] \tag{4}$$

where ρ denotes the density at (x, y) coordinates, r represents the radius of search (bandwidth), i indicates $1\ldots, n$ input points (only points that fall within the search radius of the (x, y) location are considered), *Wi* denotes the weight of point i (if no weight field is supplied, the weight for all points will be 1), and *di* is the distance between point i and the locations (x, y) (distance must be smaller than the radius of the search). This technique is also typically applied by the GIS software.

Population Density

The rapid increase in population, along with the riverside effect, river dynamics, as well as natural resources and the environment. It is vital to understand how the population has increased in this region, as well as their activities. Census data from

1981 to 2011 was used to examine the population growth of neighboring blocks within the research region. The population density formula is shown below:

$$P_d = N / A \tag{5}$$

In this equation, P_d is the population density, N is the total population, and A is the area (k^2) covered by the population. The population data was prepared and finally mapped with the help of MS Excel in 2016.

LULC Classification and Various Indexes

The main datasets in this study consist of different types of Landsat images acquired in April 1989, 2001, 2010, and 2017 (Table 1). These satellite images were used to generate land use and land cover maps. The supervised categorization of land use and land cover (LULC) has been carried out utilizing the maximum likelihood classifier (MLS) scheme with decision rules in Erdas Imagine 2014 and ArcMap 10.5 application. Measurements of land use changes and their consequences on streambank stability have been assessed over the past 30 years since this technique is thought to be appropriate for identifying riverbanks. Five groups of land use and cover have been categorized in this study: vegetation, built-up areas, agricultural land, agricultural fallow land, and water bodies. The goal of this inquiry is to calculate the alterations in land use that have taken place over time.

The NDVI, a numerical indicator constructed from NIR and R bands, stands for Normalized Difference Vegetation Index. The NDVI is commonly used to assess and monitor the health and abundance of vegetation. Comparing the NIR and R bands' reflectance provides valuable information about plant growth, biomass, and overall vegetation density. This index is sensitive to vigor and high-quality vegetation. Equation (6) was used to calculate the NDVI (Macarof & Statescu, 2017; Luo & Li, 2014):

$$\mathrm{NDVI} = \left(NIR - \mathrm{R}\right) / \left(NIR + \mathrm{R}\right) \tag{6}$$

The Normalized Difference Vegetation Index (NDVI) is a numerical value that ranges from −1 to +1. Its estimation helps in identifying different types of land cover. An NDVI value ranging from −1 to 0 signifies water bodies, while −0.1 to 0.1 signifies barren rocks, sand, or snow. An NDVI value between 0.2 and 0.5 characterizes shrubs and grassland or senescing crops, and 0.6 to 1.0 symbolizes dense vegetation or tropical rainforest (Kshetri, 2018).

The NDBI Normalized Difference Built-up Index calculates the difference between the SWIR and NIR reflectance values, which helps in distinguishing built-up areas from other land cover types. Zha et al. (2003) introduced a threshold-based approach to classify pixels as built-up or non-built-up based on the NDBI values. Zha et al. (2003) created this index by employing a certain technique (7):

$$\text{NDBI} = (\text{SWIR} - NIR) / (\text{SWIR} + NIR) \tag{7}$$

Correspondingly, The NDBI value ranges from −1 to +1 in accordance. Higher numbers indicate areas that are populated, whereas a negative estimate indicates the formation of water bodies.

The Normalized Difference Water Index (NDWI) is applied to explore the body of water constructed from the NIR and SWIR bands. The more desirable water records operating with the NDWI are frequently blended with noise from the built-up area, and the separated water records are overestimated. The Modified Normalized Difference Water Index (MNDWI) is a suitable method for detecting and separating water data. It is advantageous because it can reduce and eliminate built-up noise. This index uses green (G) and SWIR bands. The modified NDWI was developed by Xu (2006) by calculating the following method (8):

$$\text{MNDWI} = (G - \text{SWIR}) / (G + \text{SWIR}) \tag{8}$$

Similarly, the estimation of MNDWI lies between −1 and +1. For the most part, the estimation of this index above 0.5 represents water bodies. Vegetation has a lot of little qualities that separate vegetation from water bodies without any problems. The built-up areas with positive values remain between 0 and 0.2. For these investigations, a 1.5 km buffer from the channel has been used within the study area.

Correlation Between NDVI, NDBI, and MNDWI

This study examines the correlation between different types of land cover and land use. For this analysis, the relationships between NDVI and NDBI and between MNDWI and NDBI were studied using a linear regression model and Pearson's correlation. This has been implemented using the MS Office 2016 and SPSS Statistic 25 software package.

3 Results and Discussion

3.1 Dynamic Nature of the Riverbank of the Bhagirathi-Hugli River

Spatiotemporal Variation of Bank Erosion and Deposition of the Bhagirathi-Hugli River

For a detailed investigation and analysis of the spatiotemporal variation of bank erosion and deposition, the chosen section of the Bhagirathi-Hugli River has been split into six reaches (Fig. 2). From 1972/1973 to 2017, the area affected by erosion and

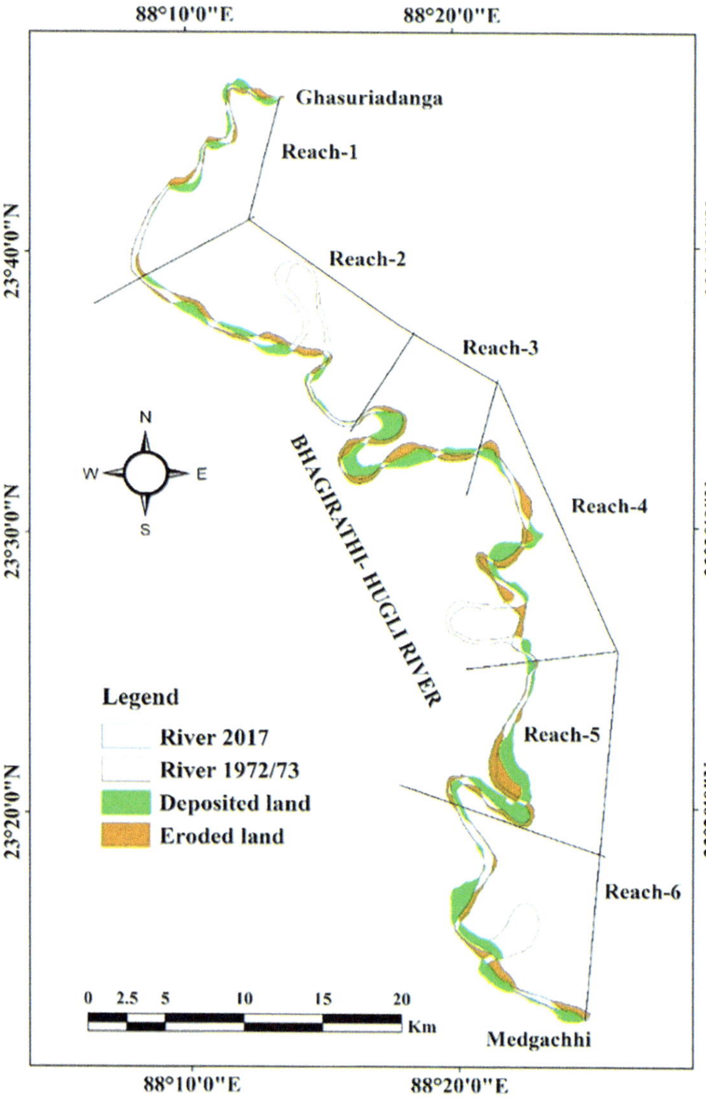

Fig. 2 Riverbank erosion and deposition in various reaches from 1972/1973 to 2017

depositional activities was quantified as the change in area on a reach-wise basis (Table 2). The result shows that the erosion and deposit areas varied in different parts of the riverbanks. Total eroded land and deposited land in this area were approximately equivalent during the study periods. However, eroded land on the left bank (22.45 km^2) was exceptionally high compared to the right bank (14.17 km^2), while the land deposited on the left bank (17.70 km^2) was low compared to the right bank (19.10 km^2) from 1972/1973 to 2017. Table 3 indicates that erosion occurred

Table 2 Area of erosion and deposition along the Bhagirathi-Hugli River from 1972/1973 to 2017

Reach name	Spatial location	Erosion (km^2)			Deposition (km^2)		
		Right bank	Left bank	Total	Right bank	Left bank	Total
Reach 1	Ghasuriadanga to Ballabhpara	1.16	1.95	3.11	1.73	1.27	3.00
Reach 2	Ballabhpara to Patuli	1.98	2.20	4.17	2.14	1.66	3.80
Reach 3	Patuli to Tamaghat-Kashiadanga	3.04	5.26	8.30	4.12	3.73	7.85
Reach 4	Tamaghat-Kashiadanga to Nabadwip	2.53	4.84	7.37	2.57	2.69	5.26
Reach 5	Nabadwip to Nashatpur	4.68	2.87	7.55	3.37	6.29	9.66
Reach 6	Nashatpur to Medgachhi	0.78	5.34	6.12	5.17	2.07	7.24
Total		14.17	22.45	36.62	19.10	17.70	36.80

Table 3 Rate of erosion and deposition (km^2/year) in the right and left banks of the Bhagirathi-Hugli River from 1972/1973 to 2017

Reach	Right bank		Left bank	
	Erosion	Deposition	Erosion	Deposition
Reach 1	0.026	0.038	0.043	0.028
Reach 2	0.044	0.047	0.049	0.037
Reach 3	0.068	0.092	0.117	0.083
Reach 4	0.056	0.057	0.108	0.060
Reach 5	0.104	0.075	0.064	0.140
Reach 6	0.017	0.115	0.119	0.046
Average rate	0.052	0.071	0.083	0.066

at a higher rate on the left bank compared to the right bank, while deposition was lower on the left bank than on the right bank. This indicates that the Bhagirathi-Hugli River is moving toward the left side of the riverbank (Pal et al., 2016). As shown in Fig. 3, one can see that the erosion and deposition by reach fluctuated significantly from 1972/1973 to 2017.

During the study period, the left bank of this river was more significantly eroded than the right bank, according to this study. On the left bank, the greatest amount of erosion (5.34 km^2) occurred in reach 6 between Nashatpur and Medgachhi at Tewarimath, Chaudhuripara, Char Malatipur of Santipur, and Udayganj of Kalna-I and then Patuli to Tamaghata-Kashiadanga in reach 3 at the southern and western parts of Udaychandpur of Nakashipara block, Jhaudanga, and Gangapur of Purbasthali-II block. Reach 6 between Nabadwip and Nashatpur in Godkhali and Godkhali Char of Nabadwip, Jaluidanga to Nashatpur of the Purbasthali-I block, had the highest erosion of the riverbank (4.68 km^2) on the right bank. This particular point indicates that there is heterogeneity in the intensity of erosion between the banks. As anthropogenic activities are more intense on the right bank of the river, a concrete embankment has been constructed to stabilize it. Most brick kiln industries are also situated on the right flanks of the river. Most of their raw material is obtained

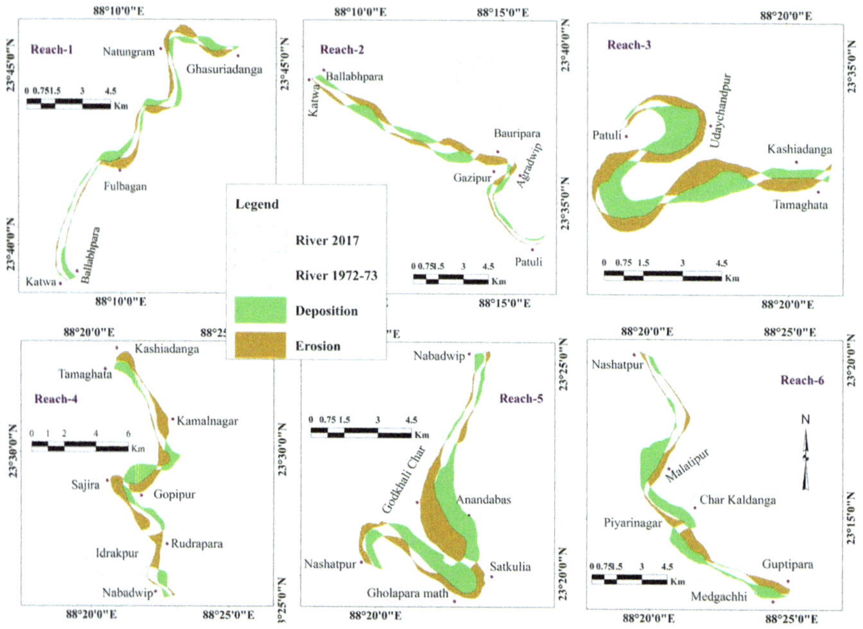

Fig. 3 Bank erosion and deposition of the Bhagirathi-Hugli River in different reaches (1972/1973–2017)

Table 4 Spatiotemporal variation of rate of bank erosion of the Bhagirathi-Hugli River from the different periods (1972/1973–2017)

Reach name	Spatial location	1972/1973–1989	1989–2001	2001–2010	2010–2017
Reach 1	Ghasuriadanga to Ballabhpara	0.10	0.10	0.11	0.10
Reach 2	Ballabhpara to Patuli	0.32	0.25	0.34	0.12
Reach 3	Patuli to Tamaghat-Kashiadanga	0.27	0.25	0.26	0.08
Reach 4	Tamaghat-Kashiadanga to Nabadwip	0.45	0.34	0.27	0.16
Reach 5	Nabadwip to Nashatpur	0.19	0.37	0.15	0.11
Reach 6	Nashatpur to Medgachhi	0.33	0.20	0.16	0.13
Total		1.67	1.50	1.29	0.71

from the left bank. So left banks lose valuable soil due to these activities. Moreover, the unconsolidated left bank is more affected by the kinetic energy of water than the embankment right bank. The study also demonstrates how the Bhagirathi-Hugli River's riverbank erosion rate varies over time and in different locations. From 1972/1973 to 2017, Table 4 illustrates that the rate of bank erosion fluctuated greatest in reach 4 and least in reach 1. Revetments and development along riverbanks lowered the rate of bank erosion at different times between 1972/1973 and 2017 (Debnath et al., 2015).

Riverbank-Eroded Zones

The IDW method was used to create the bank-eroded zones, which took into account the total erosion of each polygonal area in the study areas. The bank vulnerability graphs were created for the period 1972/1973 to 2017 along the river channel of the chosen study area (Fig. 4).

The dark brown areas represent high bank erosion on this zonation map, while the dark green shaded areas represent comparatively lower erosion. It was also observed that reaches 3 and 5 were the most vulnerable portions compared to the other reaches of 6, 4, 2, and 1 in the context of bank erosion. In reach 3, the maximum erosion occurred on the left riverbank at Udaychandpur, Jhaudanga, and Gangapur and on the right riverbank near Narayanpur, Dampal, Singirbag, and Noapara, Bholadanga, Dapharpota, Tamaghata, and Patuli. Maximum right bank erosion occurred near Godkhali, Godkhali Char, and Jaluidanga in reach 5, while Satkulia, Gholapara, and Kishoriganj were on the left bank of this river.

A field survey was conducted from November 2018 to April 2020 to evaluate the erosional state of the highly damaged segments 3 and 5 (Fig. 4). Residents (locals) provided information such as detailed field photography, damaged area coordinates, erosion mark identification, and an incident severity survey, which were used to determine the extent of erosion. The areas most affected were located within the high-priority zones where natural and anthropogenic activities dominate, inferring that the calculated results are valid. In these locations, riverbank erosion was caused by the kinetic energy of water during the monsoon season, uncovered vegetation land, agricultural activities, soil and mud collection from riverbanks for brick kilns, and other related activities (Yamani et al., 2011). James and Marcus (2006) also focused on the intervention of anthropogenic activities on riverbank instability.

Variation of the Physiological Properties of the Bhagirathi-Hugli River

The physiological characteristics of this river within the study area fluctuated at different times from 1972/1973 to 2017. These have greatly reduced the computational time since 1972/1973 (Table 5). The finding of the change in channel planform is also identified by the different authors using remote sensing data (Ghosh et al., 2020).

Due to erosion and deposition, the river altered its areal extension and width. The destruction of riverbank vegetation and the removal of sand from the streambed are the two main causes of riverbank erosion. Where agricultural and unvegetated lands predominate, riverbanks are severely eroded. The eastern part is dominated by agricultural and fallow lands. Over time, the size of the river and its banks can change due to erosion and deposition. Riverbank erosion is mainly caused by the removal of vegetation along the riverbank and sand from the stream bed. Riverbanks are heavily eroded where agricultural and unvegetated lands predominate. The eastern part is dominated by agricultural and fallow lands. The Bhagirathi-Hugli River's western (right bank) portion is more inhabited and urbanized than its eastern portion. These factors cause the river's left bank to erode more quickly than its right. The eastern part is dominated by agricultural and fallow lands. Due to the greater

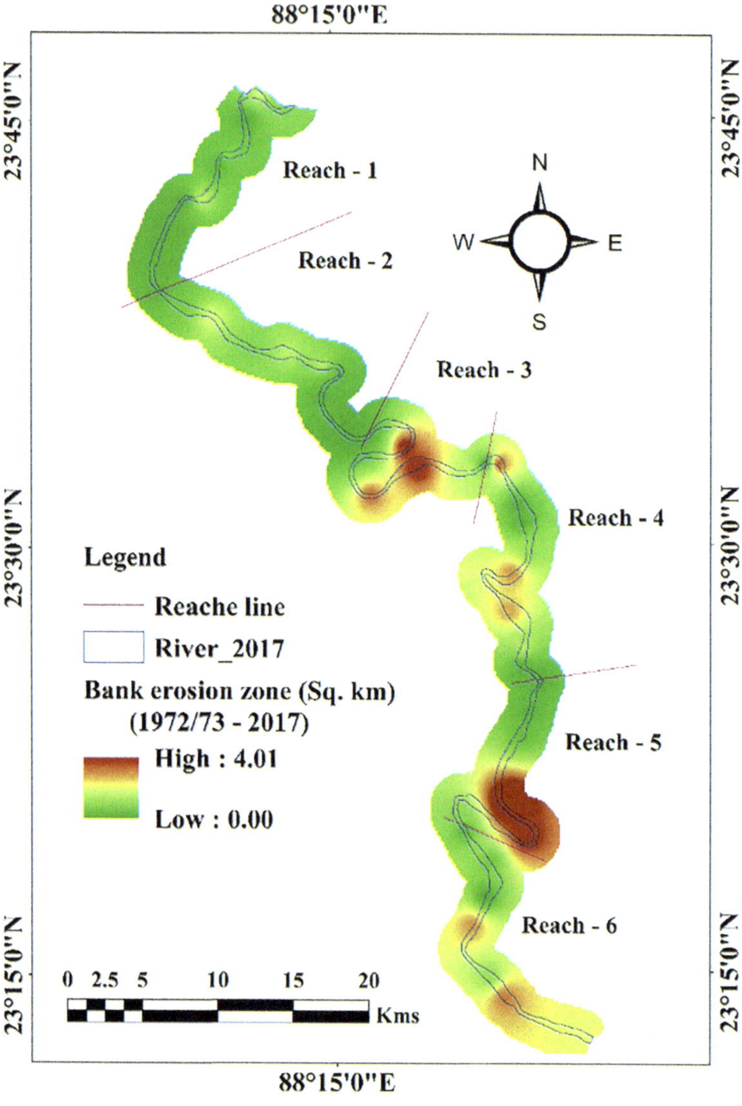

Fig. 4 Riverbank-eroded zoning map from 1972/1973 to 2017

Table 5 Variations of physiographic properties of the Bhagirathi-Hugli River at different times

Year	1972–1973	1989	2001	2010	2017
Valley length (km)	137.85	140.62	123.96	124.21	123.93
Average width (*m*)	546.65	499.57	478.71	487.58	466.61
Area (km^2)	61.82	58.00	48.55	50.92	47.19

flow rate, the left bank of the river is more susceptible to erosion than the right bank. The construction of urban areas and embankments forces the river channel to remain static (Rudra, 2010; Pal et al., 2016). As per Rudra (2008), the depth of the Bhagirathi-Hugli River is decreasing day by day. Therefore, the deposit occurs within the inner bank, creating a gentle slip-off slope (Morisawa, 1985). Human activities can cause an increase in the rate of hydromorphological changes (Chakraborty & Choudhury, 2009; Billi & Rinaldi, 1997; Isik et al., 2008; Debnath et al., 2015). Various human activities like soil, sand cutting on the stream banks, sand lifting from the stream bed, infrastructure building, cut-offs, embankments, bridge construction on the river, and land use modification affect the river dynamics and morphology (Yamani et al., 2011).

3.2 Anthropogenic Impacts on Riverbanks of the Bhagirathi-Hugli River

Riverbank erosion is a natural phenomenon that becomes catastrophic when the riparian buffers are not well maintained and human settlements are constructed too close to the eroding riverbanks. Moreover, human activities that take place along the river stretch disrupt the river's natural balance and expedite the rate of bank erosion. The morphology and dynamics of rivers are changed by the distinctive nature of human activities. Anthropogenic activities are more forcible in changing aspects of the river than natural events like floods and droughts (Yamani et al., 2011). James and Marcus (2006) also focused on the intervention of anthropogenic activities on riverbank instability, etc. Another study on the Kaveri River by Uday Kumar and Jayakumar (2020) has also identified that anthropogenic activities cause alteration of river regimes and thus trigger instability in the river hydraulics vis-à-vis the banks of the river, though natural factors such as soil homogeneity and flow angle to banks, as identified by scholars in their study of the river Brahmaputra (Kotoky et al., 2005), also influence bank stability. The significant anthropogenic impacts on the Bhagirathi-Hugli riverbanks were marked during the field survey. This river is heavily affected by the growth of the population. There were various types of anthropogenic activity, like reckless development of brickfields, land use changes, and bridge construction on the river.

3.3 Population Growth Along the Riverbanks

The population growth of this analysis space has been essential for the past four decades, leading to increasing build-up and industrial (mainly brick industry) activity, as well as alternative human activities along the stream channel (Das, 2020). To

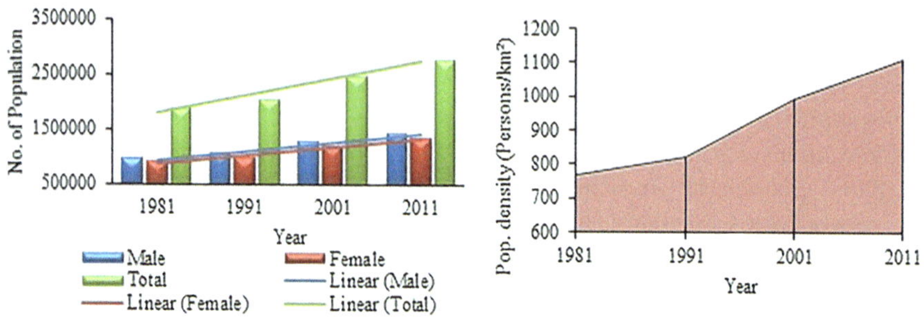

Fig. 5 Growth of population along the Bhagirathi-Hugli riverbanks within the study area from 1981 to 2011

understand the population trend along the Bhagirathi-Hugli riverbanks from 1981 to 2011, the population of adjacent blocks of the river is taken. From the graph (Fig. 5), it can be seen that both the population and the population density have grown since 1981 in this area. The population density in 2011 was 1114 persons/km^2, compared to 768 persons/km^2 in 1981. This continuous increase in the population leads to excessive pressure on the environment (Bandyopadhyay & De, 2017). Most of these people depend on the water of the river. Besides this, population growth accelerates the need for irrational construction, industrialization, and urbanization along the river. For these reasons, various anthropogenic activities like cutting sand and soil from riverbanks and unplanned construction along the river are detrimental to riverbank stability. The instability of the riverbank is amplified by population growth and land use change near the river, as noted by Bandyopadhyay and De (2017).

3.4 Reckless Development of Brick Industries

Increasing Brick Industries Along Riverbanks

Within the study area, there were 101 brickfields on both sides of the Bhagirathi-Hugli River. Between 1950 and 1999, only 23 brickfields existed, but 78 additional brickfields were developed from 2000 to 2016 (Fig. 6). This study found that out of the total of 101 brickfields, 98 were situated within 1.5 km on both sides of the riverbank. Among these, 69 were found to be located within a distance of 0.5 km. The majority of the brickfields saw a rapid increase in growth after 2000, according to Das (2020).

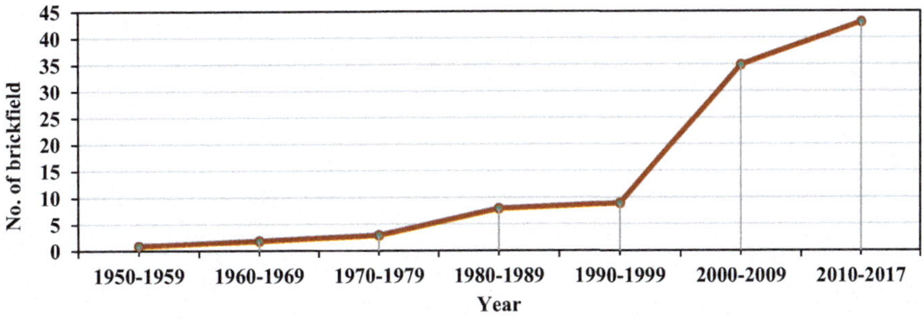

Fig. 6 Reckless developments of brick industries along the Bhagirathi-Hugli riverbanks. (Data source: field survey from December 2016 to 2017)

Brickfield Concentrated Zones Along the Riverbanks

Figure 7 shows the concentrated brickfield zones located along the Bhagirathi-Hugli riverbanks. This map has been prepared using the kernel density estimation method for picturing the spatial distribution of brickfields. Kernel density estimates are classified into three types; they are low (6.17–24.30 km^2), medium (24.30–42.42 km^2), and high (42.42–60.55 km^2) kernel sizes.

On this map, the yellow regions represent high density, while the red areas represent medium density, and the brown areas indicate the low density of the brickfield. There are high-density concentrated zones in Gajipur to Kalikapur and Nibhuji to Nandagram, which are located within 0.5 km of the river channel. These zones exist due to the availability of raw materials, the low cost of land beside the river, and the high demand for bricks in the surrounding areas of the study location.

According to an empirical survey, it has been estimated that the 101 brickfields in the research region collectively use around 944,632.30 m^3 of mud. On average, each brickfield uses 9352.80 m^3 of mud per year as raw material for making bricks. All this mud is collected from the riverbanks and riverbeds by brickmakers (Das & Sarkar, 2019). During the field surveys, it was observed that the sands and silts were being collected from the riverbanks near the Udaichandrapur ghat of the Nakashipara block, the Balir Char of the Nabadwip block, the Char Malatipur of the Santipur block, and the river bed near the Nabadwip Bridge in the Nabadwip block in an unscientific and greedy manner in the Nadia district.

The problem of soil removal from riverbanks is a serious issue that is often linked with brickfields. It has been observed that most of the brickfields are situated on the high banks of rivers, and their owners excavate soil (Shankar of Krishnagar II, Gholapara Math of Nabadwip, Medgachhi of Kalna II block) or mud (Malatipur of Santipur block) from along the riverbanks to collect silt during floods. This has led to increased erosion rates of the riverbanks, as seen in Fig. 8 which displays the soil, sand, and mud collected from different sites along the Bhagirathi-Hugli riverbanks.

As a result of the gradual migration of the river, the tremendous current on concave riverbanks made the existing holes even bigger. Additionally, soil and sand cut from these banks further contributed to the erosion of riverbanks (López, 2004). All of these factors combined lead to the acceleration of various hydro-geomorphological

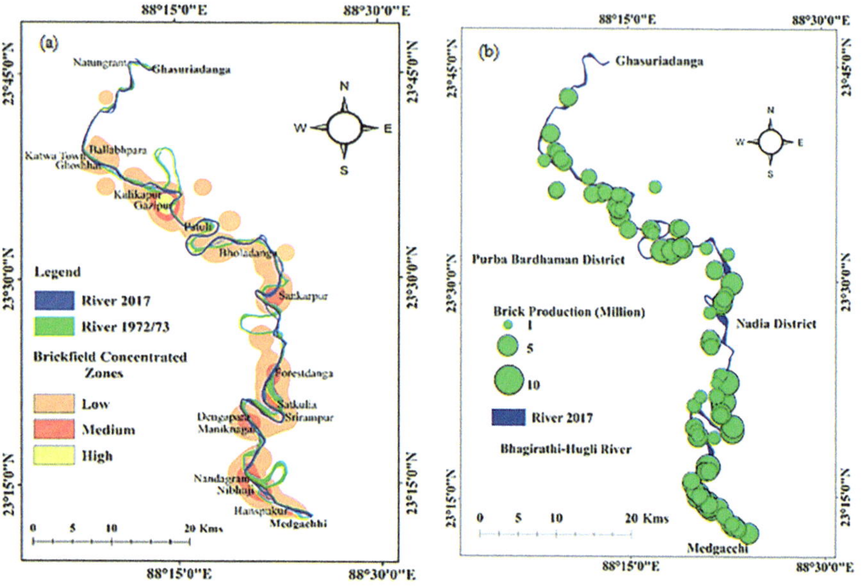

Fig. 7 (**a**) Brickfield concentrated zones and (**b**) brick production (million) on the Bhagirathi-Hugli riverbanks

changes within the river system. It has been discovered that continuous cutting of riverbanks leads to the destruction of vegetation on the land, which in turn increases riverbank erosion and causes river shifting (see Fig. 9). The development of industries, particularly brick kilns along the river, also harms the health of the riverbank. The present study has identified that the location of brick industries along the Bhagirathi-Hugli River is a major cause of riverbank instability. Bandyopadhyay and De (2017) also identified a similar scenario for the role of brick kiln industries along the Haora River in Tripura, India. Other studies also identified that the aquatic environment degrades with the location of brick kiln industries (Dey & Dey, 2015).

3.5 Land Use Alteration Alongside the Bhagirathi-Hugli Riverbanks

The Spatial Pattern of Land Cover and Land Use

Land cover modifications are well associated with socioeconomic conditions. It tends to alter severely over time in rural and urban areas. Land cover and land use are significantly altered by human activities, such as urbanization and industrialization. These activities have led to rapid economic development which, in turn, has caused changes in the land cover along the Bhagirathi-Hugli River. These changes can be monitored and identified through the use of remotely sensed images. As

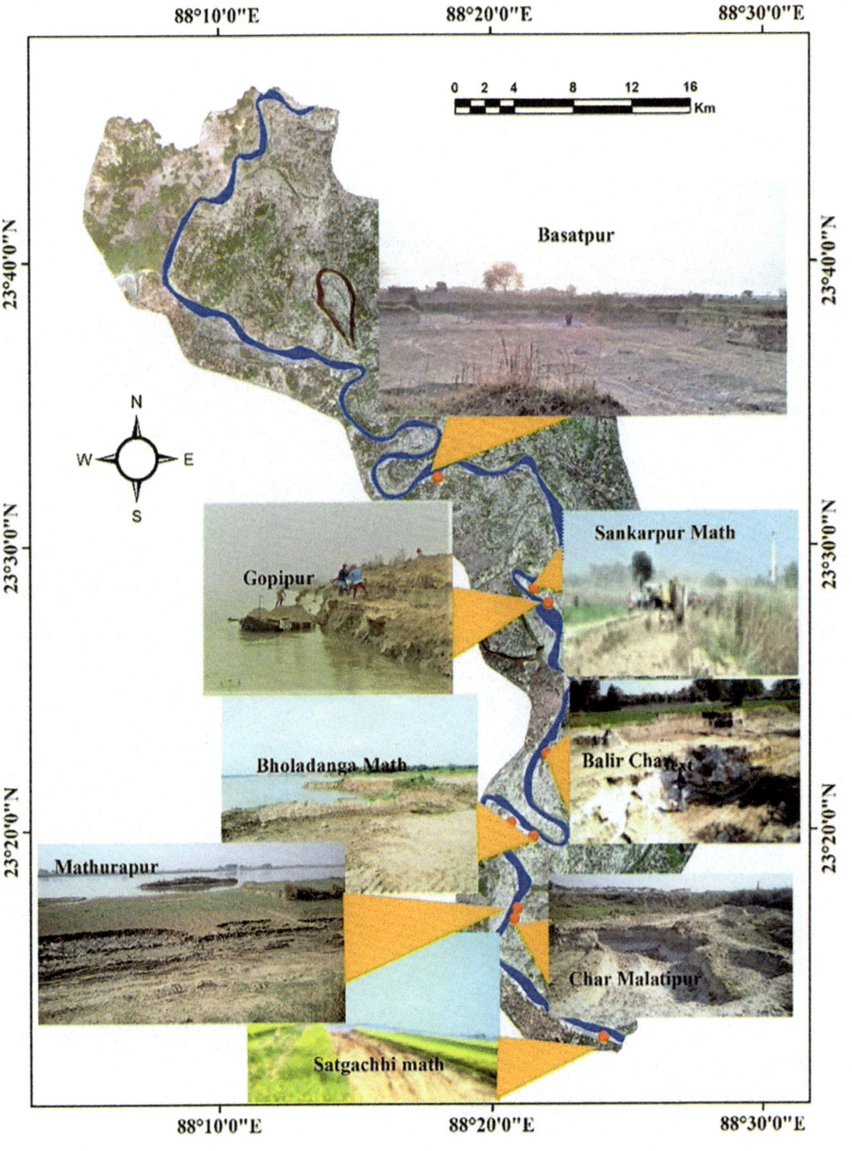

Fig. 8 Field photographs showing the cutting soil, sand, and mud from different sites of the Bhagirathi-Hugli riverbanks

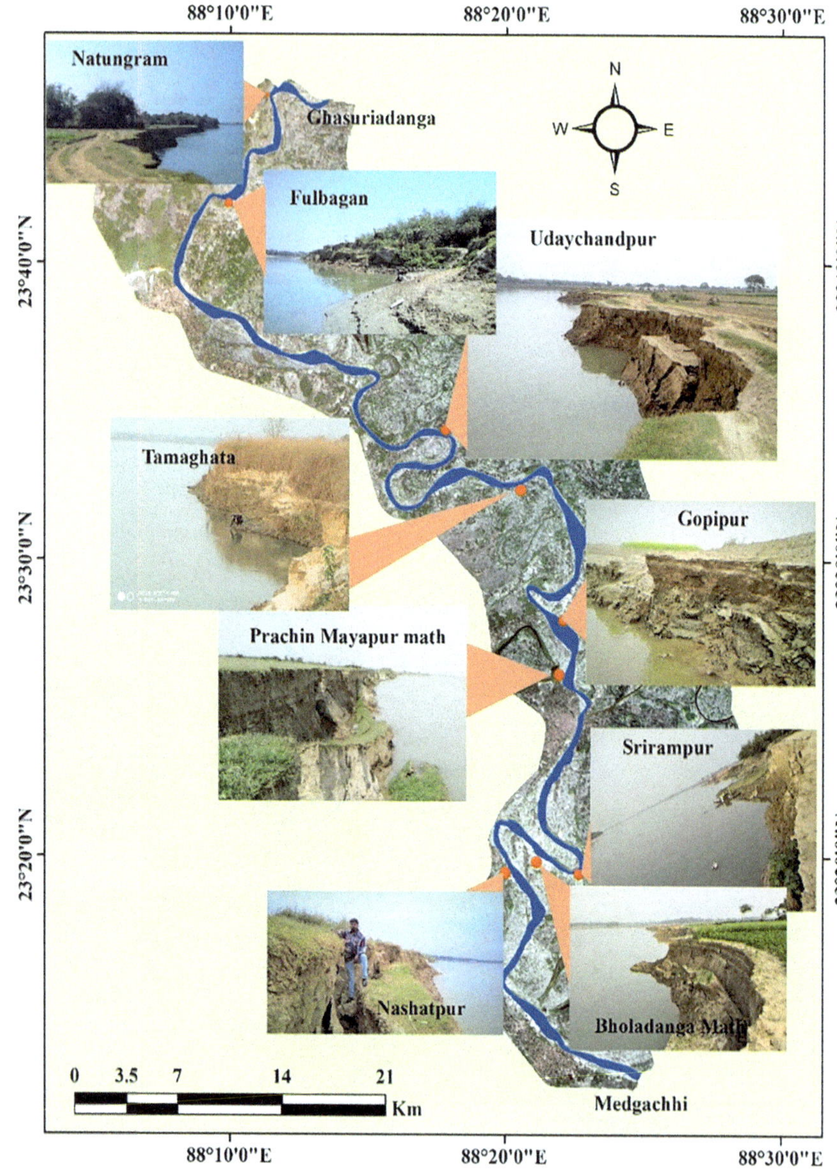

Fig. 9 Field photographs showing the riverbank failure and erosion of different sites of the Bhagirathi-Hugli River

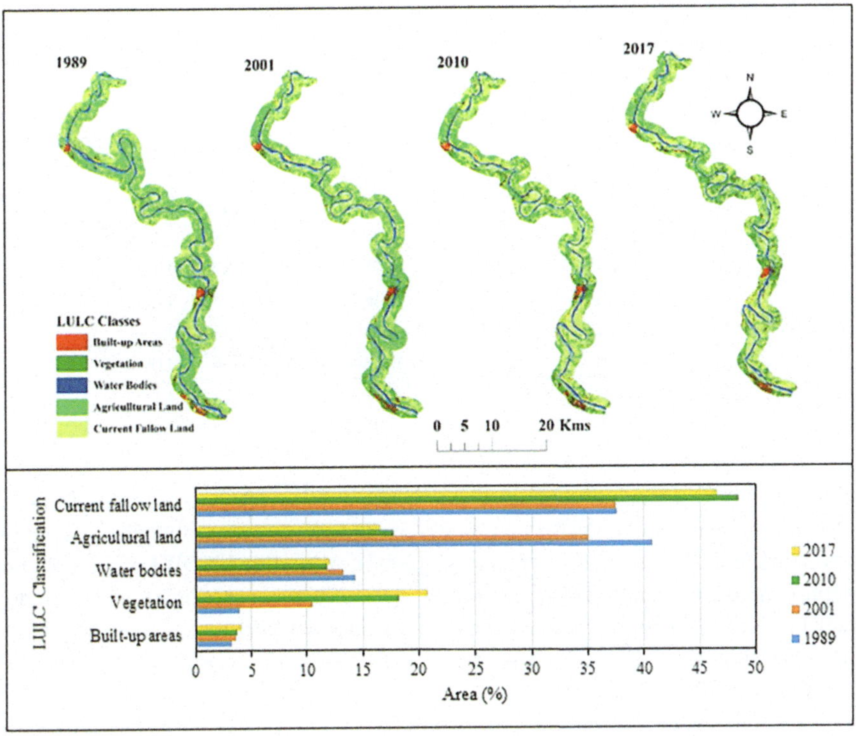

Fig. 10 Land use change along the Bhagirathi-Hugli riverbanks between 1989 and 2017

shown in Fig. 10, there has been a gradual increase in built-up areas and a decrease in water bodies from 1989 to 2017. Over the last 30 years, changes in agricultural land use and water bodies such as ponds, as well as industrial practices along riverbanks, have disrupted the natural dynamics of channels. These disruptions have led to an extensive change in the rate of riverbank erosion. Satellite images reveal that human activities in the buffer zone have increased due to the construction of infrastructure, settlements, and industrial settings. This development was enhanced by the displacement of river areas, which ranged from 61.82 to 47.19 km^2 in 1989 and 2017, respectively. Das and Sarkar (2019) also observed a similar result in their study of the Bhagirathi-Hugli River basin. The authors noted that developmental activities associated with human progression cause a change in riverbank strength.

NDVI, NDBI, and MNDWI

To assess the impact of land cover on the riverbanks of the Bhagirathi-Hugli River, NDVI, NDBI, and MNDWI were evaluated. They are the three main impact factors that are being investigated for their spatial influence (Luo & Li, 2014). The NDVI is

Table 6 Statistical data for NDVI, NDBI, and MNDWI

Index	Year	Minimum	Maximum	Variations	Mean	Standard deviation
NDVI	1989	−0.422	0.647	1.069	0.159	0.199
	2001	−0.484	0.379	0.864	−0.102	0.160
	2010	−0.275	0.584	0.860	0.176	0.174
	2017	−0.062	0.442	0.504	0.193	0.102
NDBI	1989	−0.727	0.371	1.098	0.040	0.171
	2001	−0.273	0.397	0.669	0.138	0.106
	2010	−0.408	0.346	0.754	0.089	0.129
	2017	−0.342	0.122	0.464	−0.083	0.080
MNDWI	1989	−0.516	0.854	1.370	−0.166	0.293
	2001	−0.023	0.011	0.034	−0.006	0.006
	2010	0.368	0.537	0.169	−0.114	0.177
	2017	−0.447	0.541	0.987	−0.135	0.212

a useful tool for mapping vegetation due to the unique reflectance curve of plants. On the other hand, the NDBI is commonly used to identify areas that are covered by buildings and other structures (Zha et al., 2003). A modified NDWI was created to depict uncovered water features and more desirable water information. This index is better for improving and extracting water data for a water area controlled by built-up areas than the NDWI because it reduces or even eliminates constructed land noise (Xu, 2006).

Table 6 presents the minimum, maximum, variation, mean, and standard deviation of NDVI, NDBI, and MNDWI for the years 1989, 2001, 2010, and 2017 within the study area. The variation and standard deviation of the NDVI, NDBI, and MNDWI of the year 1989 were higher than the others. The lowest variation and standard deviation of NDVI and NDBI were shown in 2017, and MNDWI was depicted in 2001. The highest mean values of NDVI, NDBI, and MNDWI were 0.193 (2017), 0.138 (2001), and −0.006 (2001). All of this indicates that different values of all indices changed and varied significantly from 1989 to 2017. The NDVI, NDBI, and MNDWI maps are presented in Figs. 11, 12, and 13.

Relationship Between NDVI, NDBI, and MNDWI

The statistical analysis shows that there is a significant correlation between NDVI and NDBI, as well as between MNDWI and NDBI, at a significance level of p >0.01. The relationship between NDVI and NDBI for different years (1989–2017) reveals that NDBI has a negative correlation with NDVI. The correlation coefficient values for the years 1989, 2001, 2010, and 2017 are −0.35, −0.60, −0.91, and −0.79, respectively. The association between NDVI and NDBI has shifted from significantly moderate to strongly negative. NDVI can be used to illustrate the expansion and development of built-up land. Malik et al. (2019) conducted a study to examine the correlation between LST, NDBI, and NDVI using Landsat 8 data in the Kandaihimmat watershed of Hoshangabad, India. The linear correlation of

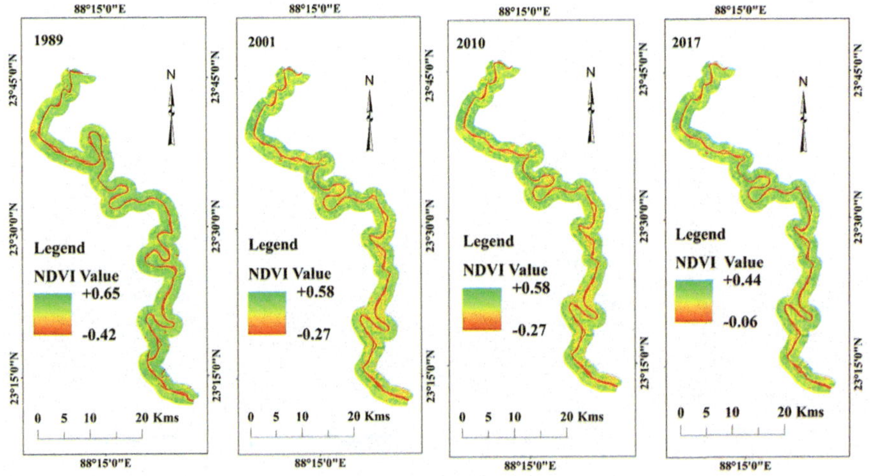

Fig. 11 NDVI maps for the years 1989, 2001, 2010, and 2017

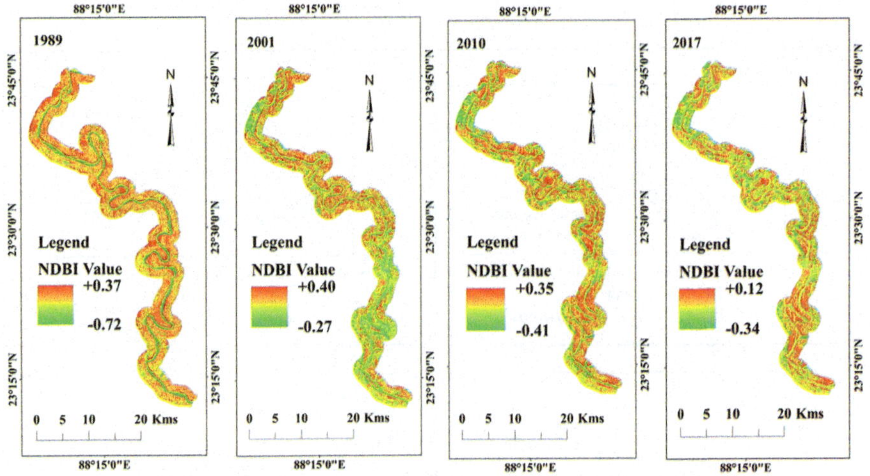

Fig. 12 NDBI maps for the years 1989, 2001, 2010, and 2017

NDVI vs. NDBI is presented in the scatter plot (Fig. 14). From the coefficient of determination (R^2), it can be found that the linear model can explain more than 50% of the variability in the dataset of four different years.

Taking into account the relationships between MNDWI and NDBI in different years from 1989 to 2017, it was found that NDBI was negatively correlated with MNDWI in 1989, 2001, 2010, and 2017 with r values of −0.67, −0.71, −0.56, and −0.59, respectively. The linear correlation between MNDWI and NDBI is presented in the scatter plot (Fig. 15). The coefficient of determination (R^2) explains the

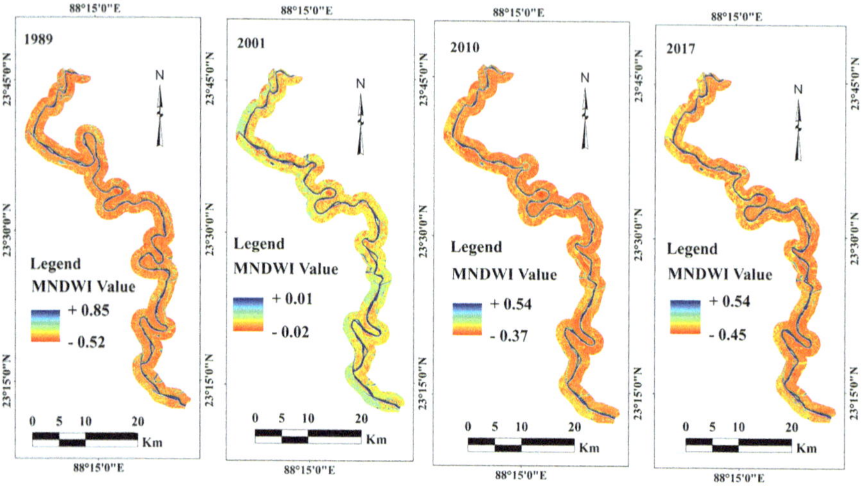

Fig. 13 MNDWI maps for the years 1989, 2001, 2010, and 2017

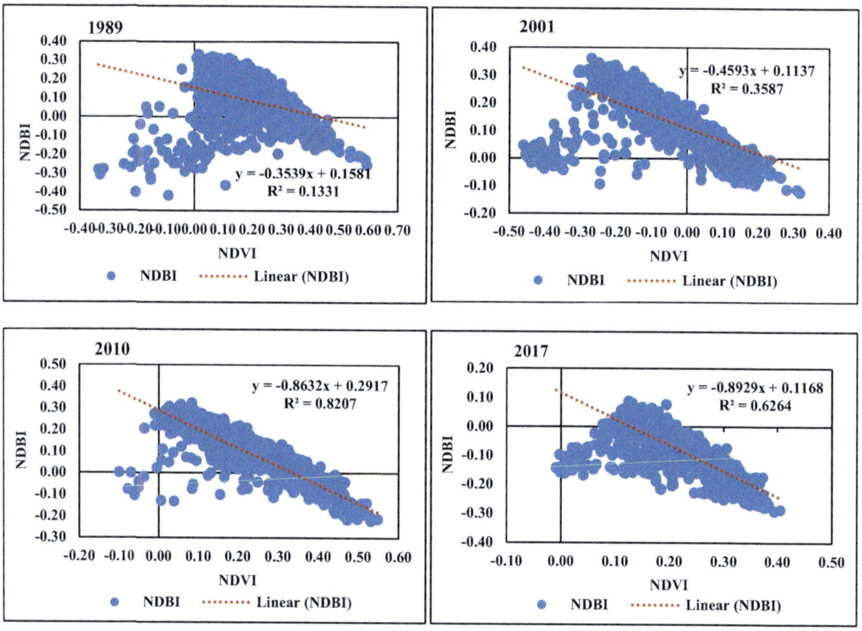

Fig. 14 Relationship between NDVI and NDBI in the years 1989, 2001, 2010, and 2017

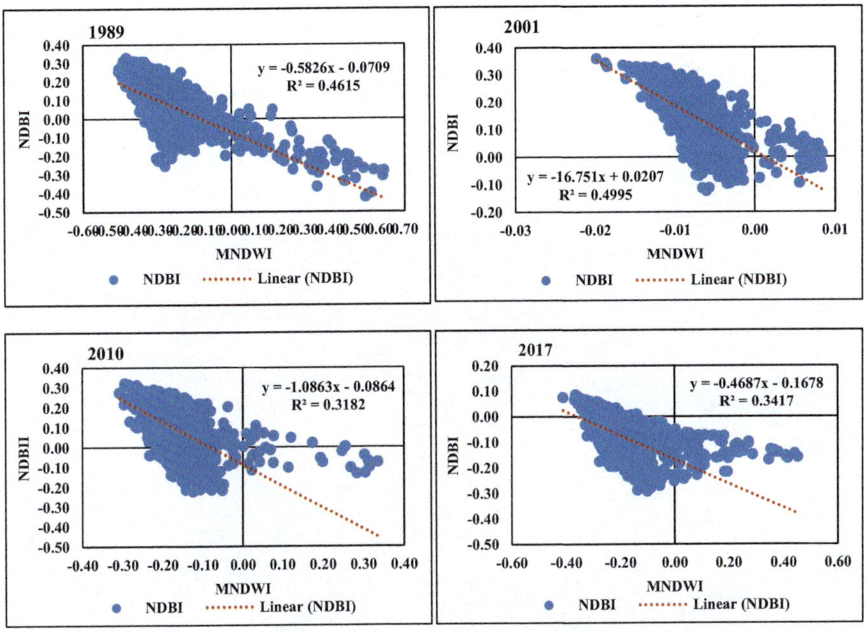

Fig. 15 Relationship between MNDWI and NDBI in the years 1989, 2001, 2010, and 2017

moderate rate of variability explained by the linear model in the dataset of four different years. NDBI, NDVI, and NDWI were used to analyze urban heat islands and land cover changes in a study by Chen et al. (2006).

This study investigates the relationships between the NDVI and NDBI and between MNDWI and NDBI in the investigation region. If the NDBI value is high, the NDVI value is low in urban or building regions. Urban zones have minimal measures of uncovered land and agricultural and other vegetation-secured regions. The massive settlements and urban zones influence the reflection and assimilation of vitality from the sun. For low NDBI, farmland had a higher NDVI value than other vegetation-secured zones with low-density building areas (Sanecharoen et al., 2019). It is additionally noted that the negative MNDWI value represents the high NDBI value of this zone. The MNDWI is increasingly blending into the perfection of water with many developed regions or built-up areas.

Negative correlations between NDVI, NDBI, and MNDWI were found during study periods in these areas. It indicated that anthropogenic activities were dominant in this study area. The problems of soil erosion and bank erosion are severe in some areas because of the rapid growth of brick kilns and urban areas in recent decades (Khatri & Tyagi, 2015; Ho & Hui, 2001). Changes in land use types from agriculture to development and industrial practice disrupt the nature of channel dynamics. Yamani et al. (2011) also observed similar impacts in their study on the Atrak River. Figure 16 shows the different types of land use on the banks of the Bhagirathi-Hugli River.

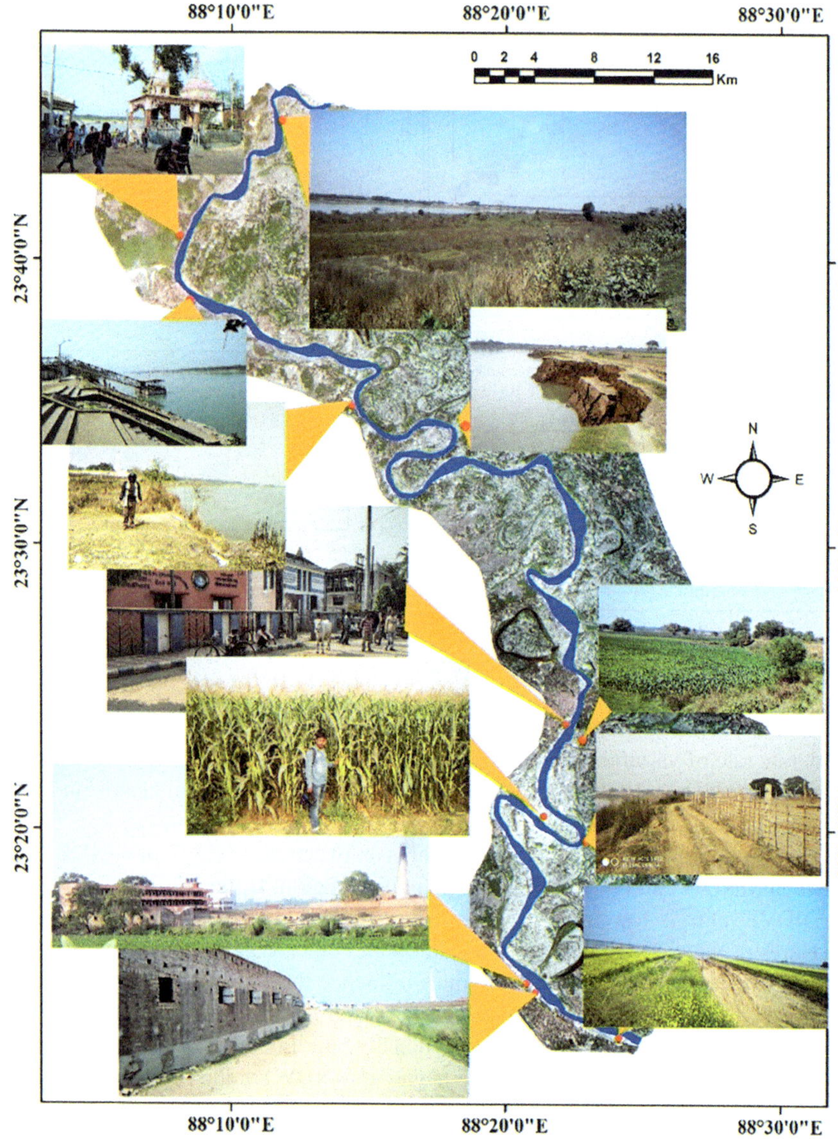

Fig. 16 Field photograph showing the different types of land use on the riverbanks of the Bhagirathi-Hugli River

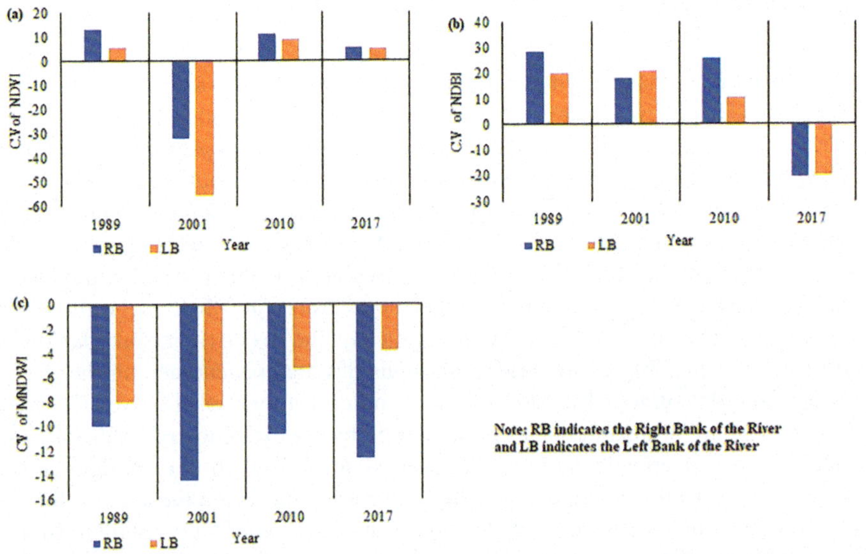

Fig. 17 Year-wise differences of coefficient of variation (CV) of (**a**) NDVI, (**b**) NDBI, and (**c**) MNDWI between riverbanks

Zonal statistical functions have been utilized to precisely measure the vegetation cover, developmental pressure or urbanization, and water status of the riverbank. Multi-temporal datasets of NDVI, NDBI, and MNDWI were extracted to assess the stability of riverbanks. A comparison of the NDVI between the right and left banks shows greater variability on the right bank during different periods. The variability between banks has been measured based on the coefficient of variation of the NDVI values in and around the banks. On the right bank, the NDVI values varied more than on the left bank. A similar trend is also prominent for the other two variables, namely, NDBI and MNDWI. The presence of these variables is indicated by a decrease in the Normalized Difference Vegetation Index (NDVI), Normalized Difference Built-up Index (NDBI), and Modified Normalized Difference Water Index (MNDWI). Figure 17 displays the year-wise variability of NDVI, NDBI, and MNDWI variables on both banks of the Bhagirathi-Hugli River.

The authors noted that developmental activities associated with human progression cause a change in riverbank strength. Applications of the remote sensing data product in association with auxiliary data can be used to identify the role of humans in riverbank stability. In addition to this, bridge construction on the stream, barriers in the river for fishing, bank re-revelation, and construction along riverbanks are also affecting the stability of river dynamics. The construction of physical infrastructures, such as bridges, is also considered in the study. A study conducted on the Atrak River in Iran focused on the effect of human activities on riverbank stability (Yamani et al., 2011). Two other studies on the Muhuri River, Tripura, and the Gumti River, Tripura by Debnath et al. (2015) and Ahmed et al. (2015), respectively, also observed a similar impact.

4 Conclusions

The impact of human activity on the bank stability of a particular section of the Bhagirathi-Hugli River is investigated through many factors, such as the vegetation index, built-up characteristics, population density, and the expansion of industries along the riverbanks. This study demonstrates that different human activities have varying effects on the dynamics of the Bhagirathi-Hugli riverbanks at different periods and places. Anthropogenic activities of all kinds have an impact on bank stability. According to the IDW model, the most risky areas in this research region were reaches 3 and 5. The left bank eroded at a faster rate than the right bank. Anthropogenic activities are more concentrated on the river's right bank than on its left side. The majority of the brick kiln industries are concentrated on the river's right bank. The survey also revealed the significant growth of the riverside brick klin. The nature of the riverbanks is influenced by the extraction of sand, soil, and mud from the riverbanks and beds, as well as the construction of bridges in the streams. The kinetic energy of water has a greater influence on the unconsolidated left bank than on the concreted right bank. A comparison of NDVI and NDBI and MNDWI and NDBI reveals a moderate to strong inverse relationship, indicating that human activities are common in this research region. The main causes of the change in the structure of the region and the erosion of the bank of the study area are hydrological changes, the growth of brick kilns, land use changes, and agricultural patterns, especially the intensification of rice cultivation in the floodplains on the river. Along with urban locations, stabilized riverbanks, primarily bank protection and embankment sections, were seen. The study can also be helpful to administrators in making better decisions by looking into different derivatives of hydraulic actions and social fragmentation concerning the study region. The study can also act as a pointer for sustainable regional planning by taking into consideration myriad aspects of social relations developed out of anthropogenic-induced river regime alteration.

This research, like many others, has significant limitations. Some of these, such as the lack of high-quality or expensive satellite data and the lack of accurate records in government agencies, hamper the study. Further research can be analyzed including the use of a 2D morphodynamic model, the Riverbank Stability Index (RSI), and the Bank Stability and Toe Erosion Model (BSTEM). It is also important to pay attention to the process-based explanations for the influential characteristics that regulate river alteration. This, however, is outside the scope of this research. This will be addressed in future studies. At the same time, the government should develop a strategy to ensure the stability of the riverbank. Local administration should also be vigilant to stop these illegal activities on both sides of this river.

Acknowledgments The authors would like to thank the Department of Geography for their assistance during the study period. The authors also thank Sanjay Das for his assistance with the field survey for this work.

Authors' Contributions Sanat Das prepared the original draft: conceptualization, methodology, formal analysis, inquiry, and writing; Sanat Das did the review and editing, and Dr. Rajib Sarkar did the supervision.

Statements and Declarations

Declaration of Interest
The authors declare that they have no conflict of interest.
Ethics Approval and Consent to Participate
Not applicable.
Consent for Publication
Not applicable.
Conflict of Interest
There is no conflict of interest.
Funding
This research did not receive specific grants from funding agencies in the public, commercial, or non-profit sectors.

References

Ahmed, I., Debnath, J., & Das (Pan), N. (2015). Impact of the river on human life: A case study on the Gumti River, Tripura. *Journal of Radix International Educational and Research Consortium, 4*(9), 1–13.

Bandyopadhyay, S., & De, S. K. (2017). *Human interference on river health-A study on the Haora River, Tripura*. Springer International Publishing. https://link.springer.com/book/10.1007/978-3-319-41018-0

Bandyopadhyay, S., Kar, N. S., Das, S., & Sen, J. (2014). River systems and water resources of West Bengal: A review. *Geological Society of India Special Publication, 3*, 63–84. https://www.researchgate.net/publication/275833658

Billi, P., & Rinaldi, M. (1997). Human impact on sediment yield and channel dynamics in the Arno River basin (central Italy); Human impact on erosion and sedimentation (Proceedings of Rabat Symposium S6). *IAHS Publ, 245*, 301–311.

Bowman, A. W., & Azzalini, A. (1997). *Applied smoothing techniques for data analysis: The kernel approach with S-plus illustrations*. Oxford University Press.

Carlson, T. N., Gillies, R. R., & Perry, E. M. (1994). A method to make use of thermal infrared temperature and NDVI measurements to infer surface soil water content and fractional vegetation cover. *Remote Sensing Reviews, 9*, 161–173. https://doi.org/10.1080/02757259409532220

Chakraborty, D., & Choudhury, D. (2009). Investigation of the behavior of tailings earthen dam under seismic conditions. *American Journal of Engineering and Applied Sciences, 2*(3), 559–564. https://doi.org/10.3844/ajeassp.2009.559.564

Chatterjee, M. (2013). An enquiry into the evolution and impact of human interference on the Churni River of Nadia district, West Bengal. *International Journal of Current Research, 5*(5), 1088–1092. http://www.journalcra.com

Chen, X. L., Zhao, H. M., Li, P. X., & Yin, Z. Y. (2006). Remote sensing image-based analysis of the relationship between urban heat island and land use/cover changes. *Remote Sensing of Environment, 104*(2), 133–146. https://doi.org/10.1016/j.rse.2005.11.016

Das, S. (2020). Impact of brick industries on geo-environment alongside the Bhagirathi-Hugli River, West Bengal, India using statistical techniques and RS and GIS applications. *International Journal of Science and Research, 9*(8), 1266–1277. https://doi.org/10.21275/SR20710122337

Das, S., & Sarkar, R. (2019). Predicting the land use and land cover change using Markov model: A catchment level analysis of the Bhagirathi-Hugli River. *Spatial Information Research, 27*(4), 439–452. https://doi.org/10.1007/s41324-019-00251-7

Das, S., Adak, K., & Samanta, K. (2014). Hydrodynamic changes of river course of part of Bhagirathi – Hugli in Nadia district. A Geoinformatics appraisal. *International Journal of Geomatics and Geosciences, 5*(2), 284–299.

Debnath, J., Ahmed, I., & Das (Pan), N. (2015). Impact of anthropogenic activities on channel characteristics: A case study of Muhuri River, Tripura, North-East India. *Archives of Applied Science Research, 7*(7), 27–36. http://scholarsresearchlibrary.com/archive.html

Dey, S., & Dey, M. (2015). Deterioration and degradation of aquatic systems due to brick kiln industries – A study in Cachar District, Assam. *Current World Environment, 10*(2), 467–472. https://doi.org/10.12944/cwe.10.2.10

Duru, U. (2017). The role of human activities in streambank stability: Lower Sakarya River (NW Turkey). *Journal of Geosciences and Geomatics, 5*(3), 130–135. https://doi.org/10.15551/prgs.2017.44

Dwivedi, A. C., Mayank, P., & Tiwari, A. (2016). The river as transformed by human activities: The rise of the invader potential of Cyprinus carpio and Oreochromis niloticus from the Yamuna River, India. *Journal of Earth Science & Climate Change, 7*(7), 1–6. https://doi.org/10.4172/21577617.1000361

Ghosh, A., Roy, M. B., & Roy, P. K. (2020). Estimation and prediction of the oscillation pattern of meandering geometry in a sub-catchment basin of Bhagirathi-Hooghly river, West Bengal, India. *SN Applied Sciences, 2*(9), 1–24. https://doi.org/10.1007/s42452-020-03275-z

Hasegawa, K. (1989). Universal bank erosion coefficient for meandering rivers. *Journal of Hydraulic Engineering, 115*, 744–765.

Ho, K. C., & Hui, K. C. C. (2001). Chemical contamination of the East River (Dongjiang) and its implication on sustainable development in the Pearl River Delta. *Environment International, 26*, 303–308.

Isik, S., Dogan, E., Kalin, L., Sasal, M., & Agiralioglu, N. (2008). Effects of anthropogenic activities on the Lower Sakarya River. *Catena, 75*, 172–181. https://doi.org/10.1016/j.catena.2008.06.001

James, L. A., & Marcus, W. A. (2006). The human role in changing fluvial systems: Retrospect, inventory and prospect. *Geomorphology, 79*(3–4), 152–171. https://doi.org/10.1016/j.geomorph.2006.06.017

Kang, S., Su, X., Tong, L., Shi, P., Yang, Z., Abe, Y., Du, T., Shen, Q., & Zhang, J. (2004). The impacts of human activities on the water–land environment of the Shiyang River basin, an arid region in northwest China. *Hydrological Sciences Journal, 49*(3), 413–427. https://doi.org/10.1623/hysj.49.3.413.54347

Khatri, N., & Tyagi, S. (2015). Influences of natural and anthropogenic factors on surface and groundwater quality in rural and urban areas. *Frontiers in Life Science, 8*(1), 23–39. https://doi.org/10.1080/21553769.2014.933716

Knighton, D. (1998). *Fluvial forms and processes – A new perspective*. Arnold. http://www.hoddereducation.com

Kotoky, P., Bezbaruah, D., Baruah, J., & Sarma, J. N. (2005). Nature of bank erosion along the Brahmaputra river channel, Assam, India. *Current Science, 88*(4), 634–640.

Krisp, J. M., & Špatenková, O. (2010). Kernel density estimations for visual analysis of emergency response data. In *Geographic information and cartography for risk and crisis management. Lecture notes in geoinformation and cartography*. https://doi.org/10.1007/978-3-642-03442-8_27

Kshetri, T. (2018). *NDVI, NDBI & NDWI Calculation using Landsat 7, 8*. https://www.linkedin.com/pulse/ndvi-ndbi-ndwi-calculation-using-landsat-7-8-tek-bahadur-kshetri

Laha, C. (2015). Oscillation of meandering Bhagirathi on the alluvial flood plain of Bengal Basin, India; as controlled by the palaeogeomorphic architecture. *International Journal of Geomatics and Geosciences, 5*(4), 564–572.

Langford, M., & Unwin, D. J. (1994). Generating and mapping population density surfaces within a geographical information system. *The Cartographic Journal, 31*, 21–26.

Liu, K., Su, H., Zhang, L., Yang, H., Zhang, R., & Li, X. (2015). Analysis of the urban heat island effect in Shijiazhuang, China using satellite and airborne data. *Remote Sensing, 7*, 4804–4833.

López, S. J. L. (2004). Channel response to gravel mining activities in mountain rivers. *Journal of Mountain Science, 1*, 264–269. https://doi.org/10.1007/BF02919330

Luo, X., & Li, W. (2014). Scale effect analysis of the relationships between urban heat island and impact factors: Case study in Chongqing, *8*(1). https://doi.org/10.1117/1.JRS.8.084995

Macarof, P., & Statescu, F. (2017). Comparison of NDBI and NDVI as indicators of surface urban heat island effect in Landsat 8 imagery: A case study of Iasi. *De Gruyter, 11*(2), 141–150. https://doi.org/10.1515/pesd-2017-0032

Malik, M. S., Mishra, S. N., & Shukla, J. P. (2019). Relationship of LST, NDBI and NDVI using Landsat-8 data in Kandaihimmat Watershed, Hoshangabad, India. *Indian Journal Of Geo Marine Sciences, 48*(1), 25–31.

Mitra, S. (2015). Shifting courses of Ganga River, it causes and resultant hazards of Manikchak block, Malda district, West Bengal. *International Journal of Humanities & Social Science Studies, II*(I), 343–352.

Mohammed-Ali, W., Mendoza, C., & Holmes, R. R. (2020a). Riverbank stability assessment during hydro-peak flow events: The lower Osage River case (Missouri, USA). *International Journal of River Basin Management*. https://doi.org/10.1080/15715124.2020.1738446

Mohammed-Ali, W., Mendoza, C., & Holmes Jr, R. R. (2020b). Influence of hydropower outflow characteristics on riverbank stability: Case of the Lower Osage River (Missouri, USA). *Hydrological Sciences Journal*. https://doi.org/10.1080/02626667.2020.1772974

Morisawa, M. (1985). *Rivers: Form and process*. Longman Higher Education.

Odgaard, J. (1987). Streambank erosion along two rivers in Iowa. *Water Resources Research, 23*, 125–136.

Pal, R., Biswas, S. S., Pramanik, M. K., & Mondal, B. (2016). Bank vulnerability and avulsion modelling of Bhagirathi-Hugli river between Ajay and Jalangi confluences in lower Ganga Plain, India. *Modeling Earth Systems and Environment, 2*, 65. https://doi.org/10.1007/s40808-016-0125-7

Rudra, K. (2008). *Banglar Nadikatha (in Bengali)*. Sahitya Samsad.

Rudra, K. (2010). *Banglar Nadikatha (in Bengali)*. Sahitya Samsad.

Sanecharoen, W., Nakhapakorn, K., Mutchimwong, A., Jirakajohnkool, S., & Onchang, R. (2019). Assessment of urban heat island patterns in Bangkok Metropolitan Area using time-series of LANDSAT thermal infrared data. *Environment and Natural Resources Journal, 17*(4), 87–102. https://doi.org/10.32526/ennrj.17.4.2019.34

Simon, A., & Downs, P. W. (1995). An interdisciplinary approach to evaluation of potential instability in alluvial channels. *Geomorphology, 12*, 215–232.

Thakur, P. K., Laha, C., & Aggarwal, S. P. (2012). River bank erosion hazard study of river Ganga, upstream of Farakka barrage using remote sensing and GIS. *Natural Hazards, 61*, 967–987. https://doi.org/10.1007/s11069-011-9944-z

Uday Kumar, A., & Jayakumar, K. V. (2020). Hydrological alterations due to anthropogenic activities in Krishna River Basin, India. *Ecological Indicators, 108*. https://doi.org/10.1016/j.ecolind.2019.105663

United Nations, Department of Economic and Social Affairs, Population Division. (2015). Population 2030: Demographic challenges and opportunities for sustainable development planning (ST/ESA/SER.A/389).

Veda Samhitha, S., & Geetha Srikanth, P. (2017). Prediction of rainfall using inverse distance weighting method and artificial neural networks in Ponnaiyar River Basin. *Indian Journal of Scientific Research, 15*(2), 144–148.

Voogt, J. A., & Oke, T. R. (2003). Thermal remote sensing of urban climates. *Remote Sensing of Environment, 86*, 370–384.

Xu, H. (2006). Modification of normalised difference water index (NDWI) to enhance open water features in remotely sensed imagery. *International Journal of Remote Sensing, 27*(14), 3025–3033. https://doi.org/10.1080/01431160600589179

Yamani, M., Goorabi, A., & Dowlati, J. (2011). The effect of human activities on riverbank stability (Case Study). *American Journal of Environmental Sciences, 7*(3), 244–247.

Zha, Y., Gao, J., & Ni, S. (2003). Use of normalised difference built-up index in automatically mapping urban areas from TM imagery. *International Journal of Remote Sensing, 24*(3), 583–594. https://doi.org/10.1080/01431160210144570

Dr. Sanat Das has been serving as an assistant teacher at Bablari Ramsundar High School (H.S.) since 2008. He completed his honours and master's degrees in Geography from Krishnagar Government College under Kalyani University, West Bengal, in 2004 and 2006, respectively. In 2012, he qualified for the National Eligibility Test (NET) for a lectureship in Geography, conducted by the University Grants Commission (UGC). He has been awarded Ph.D. from Adamas University, Kolkata, with a research title on the "Impacts of Anthropogenic Activities Along the Bhagirathi-Hugli River Between Nadia and Purba Bardhaman Districts, West Bengal." In addition to his teaching career, Sanat Das has experience as a guest lecturer at Dwijendralal College, Krishnagar. He has published several research papers in prestigious journals, including Springer Nature and Elsevier, and has served as a reviewer for various journals. His work has been recognized with the Young Researcher Award 2021 from the Institute of Scholars.

Dr. Rajib Sarkar is an esteemed academic in the Department of Geography at Adamas University. With a Ph.D. in Geography, Dr. Sarkar's research interests span a wide range of subjects including geomorphology, environmental studies, urban geography, and climate change. His work frequently explores the intersection of geographical phenomena and societal impacts, significantly contributing to sustainable development and disaster management research. Dr. Sarkar has authored numerous publications in reputed national and international journals, showcasing his in-depth knowledge and innovative approaches to contemporary geographical issues. He is actively engaged in field research and has led several interdisciplinary projects, which have garnered attention for their practical applications in policy formulation and regional development planning. At Adamas University, Dr. Sarkar is a dedicated educator deeply committed to fostering academic excellence among his students. His teaching methods emphasize critical thinking and real-world applications, encouraging students to develop a comprehensive understanding of geographical concepts. He is also an active participant in academic conferences and workshops, contributing to the growth of geography as a discipline. Dr. Sarkar's academic leadership and research continue to inspire and shape future generations of geographers.

Paddy Yield Modelling Using Selected Physical and Socio-economic Parameters: A Synergy Between Geospatial and Machine Learning Approaches

Simran Kumari Sah and Abhisek Santra

1 Introduction

Nearly half the world relies on rice, and by 2050, its production needs to surge by 60% to meet demand, due to a projected population growth of 9.8 billion (Dugan, 2015; Godfray et al., 2010). With the rapidly growing trend of the global population and limited resource availability for agriculture, there is an increasing need to produce more food, especially rice, which is widely consumed worldwide. This situation leads to higher cultivation costs and a rise in poverty rates. India is the world's second-largest rice producer, accounting for 42% of its food grains and 45% of its cereal output (Nikitha et al., 2018). The United Nation's 2030 Agenda for SDGs places significant emphasis on addressing future demand, ensuring food security, eliminating hunger, and promoting sustainable agricultural practices to meet these challenges head-on (Fernandez-Beltran et al., 2021).

In the course of time, a series of technological changes have occurred specifically in the monitoring and yield assessment aspects. Geoinformatics has become a popular, cost-effective method for monitoring crop health and guiding timely harvest decisions (Zhang et al., 2022; Zhao et al., 2022). These methods have successfully mapped rice-growing areas and shown potential in estimating yields (Islam et al., 2023). According to Yoshida (1981), rice yield refers to the weight of paddy grains with 14% moisture, including the husk, per unit area. This traditional field-based approach sometimes caused the under- and/or over-estimation of yield (Sud et al., 2016). Geospatial data and methods have been extensively applied in past research to estimate various types of crop yield (dela Torre et al., 2021; Dong &

S. K. Sah · A. Santra (✉)
Department of Geography, School of Basic and Applied Sciences, Adamas University, Kolkata, West Bengal, India
e-mail: simrankumari.sah@stu.adamasuniversity.ac.in; abhisek1.santra@adamasuniversity.ac.in

A. Rai et al. (eds.), *Contemporary Social Physics*, Springer Geography,
https://doi.org/10.1007/978-3-031-77596-3_12

Xiao, 2016; Jin et al., 2018; Kasampalis et al., 2018; Kuenzer & Knauer, 2013; Mosleh et al., 2015). The combination of remotely sensed data and crop simulation models is increasingly utilized for crop yield estimation due to satellite sensor capability to gather extensive coverage with frequent revisits. Enhanced prediction is attainable by incorporating controlling factors influencing crop growth and yields into models. This process combines weather, soil traits, and farming practices with remote sensing data, such as slope and vegetation indices, for accurate crop modelling and yield prediction (Dorigo et al., 2007; Wu et al., 2021).

Yield models concentrate on the temporal behaviour of one or multiple variables, connecting remotely sensed data with biophysical parameters. For instance, the Leaf Area Index (LAI) from satellite images indicates leaf coverage and is a key predictor in many yield estimation models (Li et al., 2017). Accurate mapping of rice areas is crucial due to region-specific yield traits (dela Torre et al., 2021). Besides LAI, common spectral indices also encompass the Normalized Difference Vegetation Index (NDVI), Enhanced Vegetation Index (EVI), and Soil Adjusted Vegetation Index (SAVI) (Di et al., 1994; Palakuru & Yarrakula, 2019; Rahman et al., 2009, 2012).

Panigrahy et al. (1997) studied rice acreage estimation in the Haora and Hugli districts of West Bengal, India, using temporal backscattering signatures under flooded conditions. European Remote Sensing Satellite-1 (ERS-1) synthetic aperture radar (SAR) data was used in this regard. In their research, Li et al. (2003) demonstrated the significance of Radarsat ScanSAR data as a valuable resource for microwave remote sensing in monitoring and estimating paddy yield in tropical and sub-tropical regions, particularly in Guangdong, China. They formulated a technique to address issues arising from topography, farming practices, uneven paddy distribution, and varied rice-cropping patterns. The suggested model links multi-temporal SAR data's backscatter coefficients with rice biomass, providing an efficient method for broad-scale rice yield estimation, challenging traditional approaches.

In their study, Iizumi et al. (2009) utilized the Bayesian method known as Markov Chain Monte Carlo (MCMC) to optimize region-specific parameters and evaluate the uncertainty associated with yield estimation in a large-scale paddy crop model. The effectiveness of the MCMC technique was evident in optimizing various parameters within a complex model. The Bayesian approach proved valuable in quantifying parameter uncertainty when assessing yield estimation uncertainty in crop modelling. Additionally, sensitivity analysis revealed heightened uncertainty in yield estimation under future warming conditions compared to historical climates. Gandhi et al. (2016) and Charoen-Ung and Mittrapiyanuruk (2018) utilized machine learning to estimate crop yield at district and plot levels, respectively.

Khanal et al. (2018) applied high-resolution images and five machine learning approaches for estimating crop yields. The study combined multi-sensor datasets along with field-measured soil characteristics. Filippi et al. (2019) developed a machine learning technique that predicted crop yield using comprehensive data from multiple farms. Ranjan and Parida (2019) conducted a study in Jharkhand employing Sentinel 1A and 2B datasets to map paddy acreage and estimated crop yield during the rainy season. They employed random forest (RF) for area modelling and devised a linear regression for yield prediction. Joshua et al. (2021) investigated machine learning methods, including support vector regressor (SVR), general regression neural network (GRNN), radial basis function neural networks (RBFNNs), and back propagation neural networks (BPNNs), to refine paddy yield forecasts in Tamil Nadu, India. Mia et al. (2023) applied a multi-model deep learning technique on unmanned aerial vehicle (UAV)-based multispectral images and weather data to estimate paddy yield. Islam et al. (2023) similarly integrated remote sensing and meteorological data and applied a machine learning approach to estimate rapid paddy yield. Iniyan et al. (2023) applied different regression models taking different physical factors, e.g., precipitation and humidity, to estimate rice yield in Maharashtra. Noorunnahar et al. (2023) applied an XGBoost machine learning model to predict paddy production in Bangladesh. Lingwal et al. (2024) applied artificial intelligence and combined an RF algorithm with a neural network to predict paddy yield. However, all these studies highlight two-fold aspects, i.e., firstly, these are based on data-intensive machine learning models, and secondly, many of them are aspatial in nature.

Apart from the physical factors, several socio-economic factors are also responsible for modelling paddy yield (Siagian et al., 2019). Shanmugan and Baria (2019) applied the Labour Productivity Index to estimate agricultural labour productivity in different dimensions. They commented that the changes in wage rates, technology adoption, on-the-job learning, experience, and education can significantly influence labour productivity in agriculture amid shifting migration patterns and fluctuations in agricultural growth rates. All the above-mentioned research applied various data sources containing optical and microwave images both separately and jointly to predict crop yield with special reference to paddy crops. On the other hand, very few studies have been reported related to the importance of socio-economic factors in paddy yield. Those studies are mostly based on statistical analysis without considering geospatial techniques. No such studies have yet been reported combining physical and socio-economic factors to estimate paddy yield. The current research aims to explore the impact of chosen physical, demographic, and infrastructural elements on evaluating paddy production. This research underscores a unique method of creating a multiple linear regression model for predicting paddy yield in regions with limited data. Therefore, the three-fold objectives of the research were to estimate paddy cropped area using optical remote sensing images, to assess and map the selected factors influencing paddy crop yield, and, ultimately, to formulate a multivariate regression model for estimating paddy production. The domain of social physics seeks to understand and predict social behaviour through advanced analytical techniques. The current research combines geospatial and machine learning approaches to decode the intricate interplay between physical and socio-economic factors influencing paddy yields. This integration mirrors the methodology of contemporary social physics, which uses large-scale data and computational models to reveal hidden patterns and insights within social systems. This synergy not only enhances paddy yield prediction but also contributes to a deeper understanding of agricultural productivity and supports informed decision-making for enhancing food security and sustainable development.

2 Study Area

Located in the south of West Bengal, India, the Purba Bardhaman district is marked by its alluvial terrain. The district spans from 22° 15′ 08″ N to 23° 15′ 17″ N and 87° 13′ 17″ E to 88° 07′ 22″ E (Fig. 1) with an area of approximately 5436 sq. km. The total population is around 4,835,532 according to the 2011 Census (Pal et al., 2020). Purba Bardhaman was constituted on 7 April 2017 as the 23rd district of West Bengal (Pal et al., 2020). Three rivers encircle the district: Bhagirathi on the east, Ajay on the north, and both Damodar and Dwarakeshwar to the west. Geographically, it falls within the lower Gangetic plain, and the climate is typical of a monsoon region (Pal, 2021). The central part of the district receives a sufficient amount of rainfall ranging from 1200 to 1400 mm (Pal et al., 2020). The once unified Bardhaman district led in rice production in West Bengal, and much of that fertile land now lies within Purba Bardhaman (Mohammad et al., 2023).

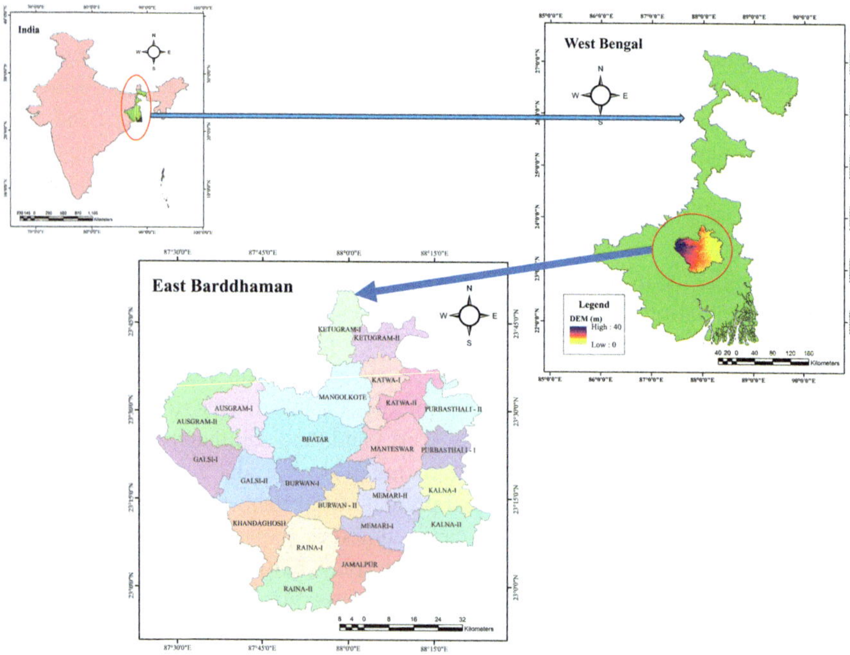

Fig. 1 Study area

3 Database and Methodology

3.1 Database and Pre-processing of Data

In the present research, a series of factor-wise databases were applied, starting from the multi-sensor satellite data to census statistical data. Sentinel 2A multispectral images from 2018, 2020, and 2022 were sourced from the ESA Copernicus Open Access Hub. Each image was consisting of 13 spectral bands with varying spatial resolution from 10 to 60 m. From the same source, Sentinel 1A SAR data were downloaded for the same temporal period. The spatial resolution of extra-wide swath data is 25 m × 40 m which is resampled to a ground resolution of 30 m. The database consists of two polarizations: vertical-vertical (VV) and vertical-horizontal (VH). Landsat 8 Operational Land Imager (OLI) and Thermal Infrared Sensor (TIRS) data for the same years were obtained from the USGS Earth Explorer portal. The spatial resolutions of optical and thermal data are 30 m and 100 m, respectively.

The precipitation data were downloaded using the NASA Earth Observing (NEO) Google Earth Engine Integrated Multi-satellitE Retrievals for GPM (Global Precipitation Measurement) (NEO IMERG) tool to monitor and analyze global precipitation patterns. It utilizes data from various satellites to offer high-resolution rainfall estimates on a worldwide level. It includes passive radar and infrared (IR) sensors in order to capture accurately the distribution and intensity of precipitation (Mo et al., 2020). Through the integration of satellite observation and ground-based rain gauge data, the dataset delivers dependable and extensive rainfall information (Sharifi et al., 2016). Atmospheric temperature data were sourced from the National Aeronautics and Space Administration (NASA) Power Data Access Viewer. The portal contains present, past, and forecasted temperature values. Users can examine temperature patterns, trends, and variability over time, visualizing the data within the tool or downloading it in different data formats (Marzouk, 2021).

Soil moisture, bulk density, and soil pH data were downloaded from NASA's SMAP and Soil Grid online platform. High-resolution and spatially explicit information on soil properties is derived by combining satellite observations, ground-based measurements, and soil modelling techniques. The spatial resolution is 250 m (Hengl et al., 2017) and was applied to evaluate soil quality, assess soil compaction, and understand soil health for diverse agricultural and environmental uses. It offers essential insights for sustainable land management, precision farming, and land use planning.

The database related to manpower involved in the agriculture sector was collected from the District Statistical Handbook of Purba Bardhaman district (Handbook Table 17.2) which is tabulated from the Census of India (2011) data. The database lists bargadars, patta holders, small and marginal farmers, and agricultural workers (Govt. of West Bengal, 2023). Besides, the number of seed stores, number of fertilizer depots, and area under different irrigation systems including canal irrigation, river lift irrigation, and irrigation through deep and shallow tube wells were collected from district headquarters and the offices of Irrigation and Waterways Department and Public Health Engineering.

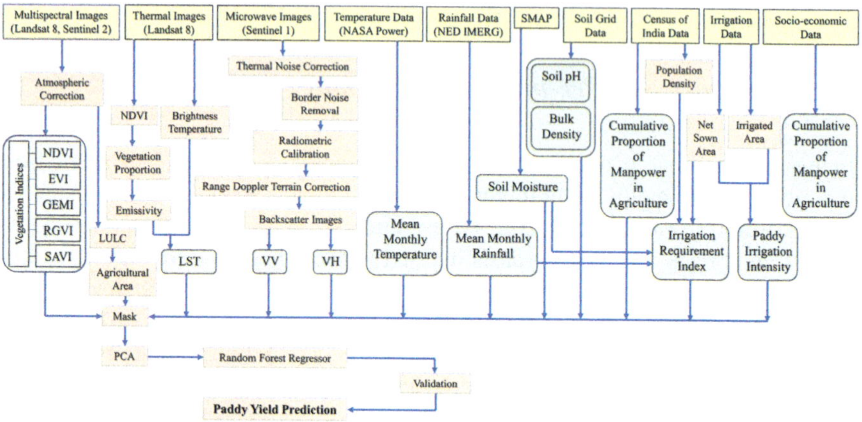

Fig. 2 Methodology flowchart

The Sentinel 2A multispectral images were atmospherically corrected using Environment for Visualizing Images (ENVI)-based Fast Line-of-sight Atmospheric Analysis of Spectral Hypercubes (FLAASH) module, a MODerate resolution atmospheric TRANsmission-4 (MODTRAN-4)-based radiative transfer model (Santra et al., 2022; Santra et al., 2019; Yadav et al., 2021). The Sentinel 1A data for all the years were first corrected from possible thermal noise and border noise. After that, the VV- and VH-polarized images were radiometrically calibrated. Then, the calibrated images were filtered using a speckle filter. Finally, the backscatter images were prepared using Range Doppler Terrain Correction in the ESA SNAP software. Soil moisture content was gauged using NASA's Soil Moisture Active Passive (SMAP) data. The L-band passive radar observations are expected to provide accurate surface soil moisture data as they possess enhanced penetration capabilities, especially in vegetated areas. This advantage sets them apart from higher-frequency observations (Zeng et al., 2022; Zhao et al., 2021). The spatial resolution of the data is 3 km. The data was sourced from the Google's Earth Engine (GEE) platform. Global researchers have used SMAP data on croplands to gauge agricultural soil moisture (Yang et al., 2017). The adopted methodology is described using a flowchart (Fig. 2). The details of these individual methods are discussed in the subsequent section.

3.2 *Methods*

Computation of Vegetation Indices

Most vegetation indices exhibit saturation at LAI >3 during crop growth estimation (Kaplan & Rozenstein, 2021). Crop biochemical properties and canopy structure primarily influence spectral reflectance in the optical region. Crop height, linked to growth and canopy development, influences spectral reflectance and indices (Kaplan et al., 2021). However, spaceborne optical sensors face limitations due to cloud cover. To address this, earlier research suggested combining data from different satellites taken at various times (Manivasagam et al., 2021; Kaplan et al., 2021; Sadeh et al., 2021). In the present research, five selected spectral indices were applied, namely, NDVI, EVI, GEMI, RGVI, and SAVI (Table 1).

Estimation of LST

In order to generate land surface temperature (LST), at first, radiance images were calculated from the raw DN values of Landsat 8 data using the following equation (Eq. 1) (Oğuz, 2016):

$$L = (\mathrm{MRF} \times \mathrm{QCal}) + \mathrm{ARF} \tag{1}$$

where

L = spectral radiance in $\mathrm{Wm^{-2}sr^{-1}\mu m^{-1}}$
MRF = multiplicative rescaling factor (3.3×10^{-4})
ARF = additive rescaling factor (0.1)
QCal = pixel DN value

The at-sensor brightness temperature (TB) was calculated from the L value computed above. The mathematical expression is mentioned below (Eq. 2) (Chander et al., 2009; Santra et al., 2022; Sekertekin & Bonafoni, 2020; Yadav et al., 2021):

$$\mathrm{TB} = \frac{K_2}{\ln\left(\frac{K_1}{L_\lambda} + 1\right)} \tag{2}$$

where

$K_1 = 774.89$ and $K_2 = 1321.08$, calibration constants for Band 10 of TIRS
TB = brightness temperature in Kelvin

The reflectance of red (R) and near-infrared (NIR) bands was calculated using Eq. 3 (Oğuz 2016):

$$\rho(\lambda) = \frac{\mathrm{RMRF} \times \mathrm{QCal} + \mathrm{RARF}}{\sin\theta} \tag{3}$$

where

RMRF = reflectance multiplicative rescaling factor
QCal = pixel DN value
θ = solar elevation angle

Table 1 Vegetation indices

Serial No.	Name	Expression	Description	References
1.	Normalized Difference Vegetation Index (NDVI)	$\mathrm{NDVI} = \frac{\mathrm{NIR} - \mathrm{R}}{\mathrm{NIR} + \mathrm{R}}$	NIR = near-infrared band R = red band	Rouse et al. (1974)
2.	Enhanced Vegetation Index (EVI)	$\mathrm{EVI} = G \times \frac{\mathrm{NIR} - \mathrm{R}}{\mathrm{NIR} + (C_1 \times \mathrm{R} - C_2 \times \mathrm{B}) + \mathrm{L}}$	B = Blue band NIR = Near Infrared band R = Red band G = Gain factor C_1 and C_2 = Aerosol resistance coefficients L = Canopy background	Liu and Huete (1995)
3.	Global Environment Monitoring Index (GEMI)	$\mathrm{GEMI} = \eta(1 - 0.25\eta)\frac{\mathrm{R} - 0.125}{1 - \mathrm{R}}$ [where $\eta = \frac{2(\mathrm{NIR}^2 - \mathrm{R}^2) + 1.5\mathrm{NIR} + 0.5\mathrm{R}}{\mathrm{NIR} + \mathrm{R} + 0.5}$]	NIR= near-infrared band R = red band	Pinty and Verstraete (1992)
4.	Rice Growth Vegetation Index (RGVI)	$\mathrm{RGVI} = 1 - \frac{(\mathrm{B} + \mathrm{R})}{(\mathrm{NIR} + \mathrm{SWIR1} + \mathrm{SWIR2})}$	B = blue band R = red band NIR = near-infrared band SWIR1 = shortwave infrared band 1 SWIR1 = shortwave infrared band 2	Nuarsa et al. (2011)
5.	Soil Adjusted Vegetation Index (SAVI)	$\mathrm{SAVI} = \frac{(\mathrm{NIR} - \mathrm{R})(1 + l)}{\mathrm{NIR} + \mathrm{R} + l}$	NIR = near-infrared band R = red band l = slope of soil line	Huete (1988)

RARF = reflectance additive rescaling factor
$P(\lambda)$ = reflectance of band λ

The fourth step of LST estimation was the computation of ground emissivity. The NDVI images computed earlier were used for this purpose. Following Eq. 4 (Carlson & Ripley, 1997), the proportion of vegetation was calculated:

$$\mathrm{PV} = \left(\frac{\mathrm{NDVI} - \mathrm{NDVI}_{\min}}{\mathrm{NDVI}_{\max} - \mathrm{NDVI}_{\min}} \right)^2 \tag{4}$$

where

PV = vegetation proportion
NDVI = image NDVI value of a specific pixel
$\mathrm{NDVI}_{\min}$ = minimum NDVI value
$\mathrm{NDVI}_{\max}$ = maximum NDVI value]

Emissivity was calculated using PV calculated above (Eq. 5) (Chakraborti et al., 2019; Santra et al., 2022; Yadav et al., 2023):

$$\varepsilon = 0.02664\,\mathrm{PV} + 0.96356 \tag{5}$$

where

ε = land surface emissivity
PV = vegetation proportion

Finally, LST values were calculated using the following equation (Eq. 6):

$$\mathrm{LST} = \frac{\mathrm{TB}}{\left[1 + \left(\frac{\lambda * \mathrm{TB}}{\rho} \right) + \ln \varepsilon \right]} - 273.15 \tag{6}$$

where

LST = land surface temperature in °C
TB = brightness temperature in Kelvin
λ = band wavelength
ε = land surface emissivity

$\rho = \frac{hc}{\sigma} = 1.438 \times 10^{-2}$ mK
c = speed of light (2.998×10^{8} ms^{-1})
σ = Boltzmann's constant (1.38×10^{-23} JK^{-1})
h = Planck's constant (6.626×10^{-34} Js)]

Categorization of Socio-economic Factors

The socio-economic variables were not directly considered. The derived indices were applied in the analysis. At first, the intensity of irrigation (II) was considered (Eq. 7) (Karunakaran & Palanisami, 1998):

$$\mathrm{II} = \frac{\mathrm{IA}}{\mathrm{NSA}} \times 100 \quad (7)$$

where

II = irrigation intensity
IA = irrigated area
NSA = net sown area

Irrigation requirement was determined by Sen (1996) by computing the Irrigation Requirement Index (Ir) at the block level, considering mainly non-climatological parameters such as percentage of net sown area, density of population in terms of NSA, present irrigation intensity, and average annual rainfall. Yet Santra (2012) adjusted the formula to include the soil's water holding capacity. However, with the advent of SMAP soil moisture data, the equation was slightly modified in the present study by replacing water holding capacity score with the soil moisture amount. The modified equation is mentioned below (Eq. 8):

$$\mathrm{IR}_{\mathrm{mod}} = \frac{\mathrm{NSA} \times \mathrm{PD}_{\mathrm{NSA}}}{R \times \mathrm{SM}} \times \frac{\mathrm{II}}{100} \quad (8)$$

where

$\mathrm{IR}_{\mathrm{mod}}$ = modified irrigation requirement index
NSA = net sown area in ha
$\mathrm{PD}_{\mathrm{NSA}}$ = population density in terms of NSA (persons per ha)
R = average annual rainfall in mm
SM = surface soil moisture derived from SMAP in mm

The workforce in the agricultural sector was assessed from the five categories of workers related to agriculture. At first, the proportions of bargadars, patta holders, small farmers, large farmers, and agricultural labourers were computed. Finally, the cumulative total proportions of the workforce, in the agriculture sector were computed to understand their relative importance with respect to the district total. Similarly, cumulative proportions were calculated taking the number of seed stores and fertilizer depots into consideration.

Statistical Analysis and Yield Prediction

Multi-variate statistical analysis proved highly efficient in the field of agriculture (Saed-Moucheshi et al., 2013). Multiple regression involves the consideration of multiple explanatory variables. However, applying multiple regression by considering all the explanatory variables as independent is not always correct. Therefore, to maximize the variance and reduce the dimensionality of the datasets, factor analysis was applied. Kendall (1950) conducted a comprehensive review of statistical methods employed by psychologists. These methods aimed to represent observed and corrected measurements through uncorrelated and hypothetical components. Kendall (1939) applied the principal component (PC) method while studying crop productivity in England. Principal component analysis (PCA) is a technique used to reduce variables in datasets with many or repetitive ones. The technique reduces data dimensionality by analysing the covariance between variables. Its primary benefit lies in effectively reducing the number of dimensions while preserving a significant amount of information (Everitt & Dunn, 2001). According to Dunetman (1989), the number of PC is limited to the number of orthogonal variables. The transformation aims to ensure that the first PC exhibits the highest variance possible, capturing the maximum data variability. Subsequent components are selected to have the greatest variance while being orthogonal (uncorrelated) to the preceding ones (Jackson, 1991).

In the present study, 17 variables were considered to predict the yield pattern at the pixel level of the image. However, the selected variables have some amount of autocorrelation. Therefore, they were not considered as independent factors to predict yield. Factor values were standardized before performing the factor analysis. After that, the factor analysis was conducted in the SPSS environment. The varimax rotation with Kaiser normalization was considered. PC method was applied to extract factors. The analysis was considered based on a correlation matrix.

After that, the multiple linear regression analysis was considered to predict yield based on the derived factor scores and known yield values. The surface of the factors was generated, and yield was predicted next using the multiple regression equation developed. Finally, the 500 random training samples were extracted to refine the model with a random forest regressor. The model validation statistics were computed subsequently taking 70% data for calibration and 30% for validation. For the rest of the years, the parameters were standardized first, and PCA was performed sequentially to reduce the data dimensionality and maximize the variance. The yield maps were predicted accordingly.

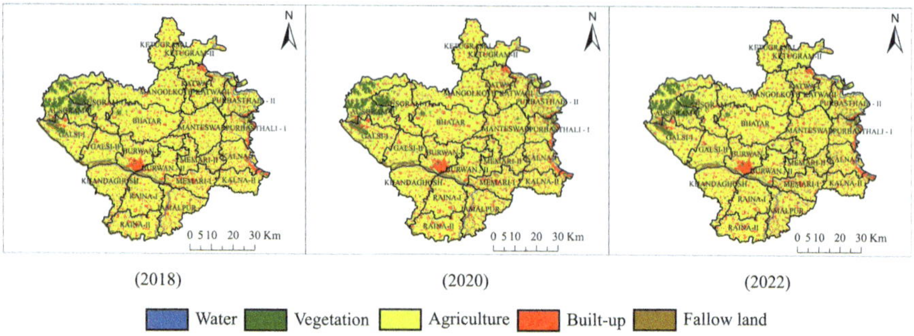

Fig. 3 LULC map of Purba Bardhaman

4 Results and Discussion

4.1 Spatio-temporal Distribution Pattern of Parameters

LULC maps provide insights into how land is used and covered on the planet. Land use refers to human activities on land, such as residential, agricultural, and commercial activities. On the other hand, land cover describes the Earth's physical features like forests, wetlands, and urban areas, showcasing the landscape's actual makeup. Comprehending LULC patterns is crucial for informed decision-making regarding land management, resource allocation, urban growth, environmental protection, and sustainable development.

The land use and land cover (LULC) maps of the district were prepared using the support vector machine (SVM) classification method. Five spectral classes were identified, namely, waterbody, vegetation, agricultural land, built-up area, and fallow land for all 3 years of analysis (Fig. 3). The overall classification accuracy came to >80% for all 3 years. Finally, the agricultural land areas were extracted, and year-wise masks were prepared. Those masks were utilized to clip the other raster images. It was observed that the cropped area increased from 2287 sq. km in 2018 to 3742 sq. km in 2022, through converting the vegetated areas and fallow land mainly into cropland. However, a substantial part of agricultural land remains unused during the colder season due to the agricultural sector's reliance on monsoon rainfall.

The crop foliage condition indicating indices remained more or less similar in all 3 years. In 2022, NDVI values varied from −0.19 to 0.81, averaging 0.45. The average NDVI value dropped from 0.65 in 2018 to 0.45 in 2022. A similar decreasing trend was observed in the case of SAVI and EVI. On the contrary, the RGVI showed an increase in mean value from 0.61 to 0.64 within the same period. GEMI maintained a nearly constant range and mean values (0.84) from 2018 to 2022. Spatial distribution patterns were almost similar for these indices over the course of time in the agricultural field. For all the vegetation indices, it was observed that the south-eastern part of the district covering Jamalpur; Memari-I, II; and Kalna-I, II showed lower index values (Fig. 4).

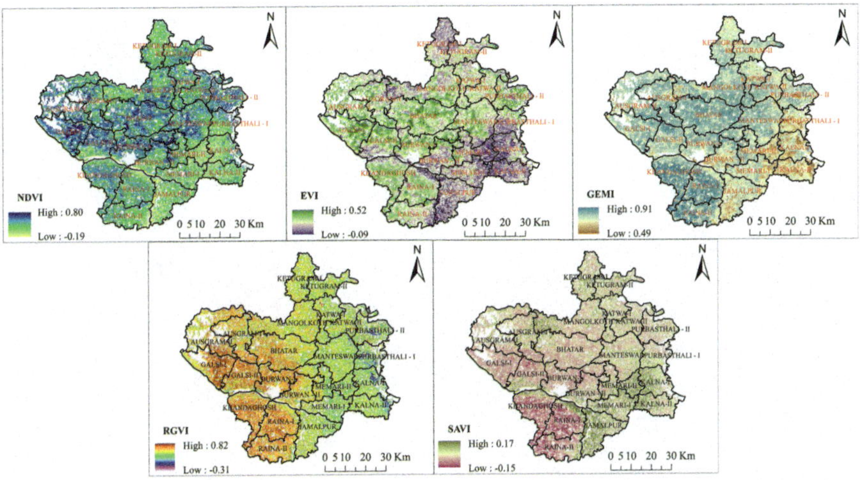

Fig. 4 Vegetation indices

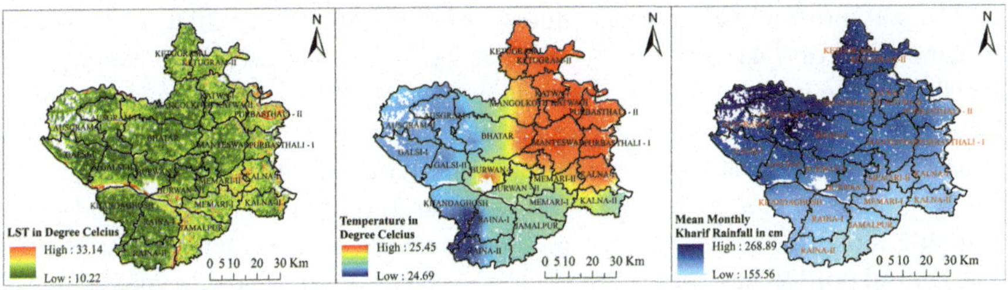

Fig. 5 Distribution pattern of temperature and rainfall

Over time, LST values increased, with the average moving from 23.48 °C in 2018 to 23.69 °C in 2022. Both the range and mean atmospheric temperature values of the district increased from 2018 to 2022. Between 2018 and 2022, the atmospheric temperature range expanded from 21.89–22.67 °C to 24.69–25.45 °C. The average value also elevated from 22.67 to 25.45 °C. The rainfall amount in the kharif season also increased from 2018 to 2022. The mean monthly kharif seasonal rainfall increased from 148.64 to 209.38 mm. All three parameters were distributed spatially following a similar pattern over the study area. The LST shows higher values again towards the SE part of the districts covering the previously mentioned blocks (Fig. 5). This clearly explains the low vegetation index values in those areas. The atmospheric temperature distribution pattern shows a clear east-west variation. Western blocks of the district have lower atmospheric temperatures compared to the eastern blocks. This could be explained by the influence of the Damodar River situated towards the west and south-west of the district. The mean monthly rainfall basically increases both towards the north and south (Fig. 5).

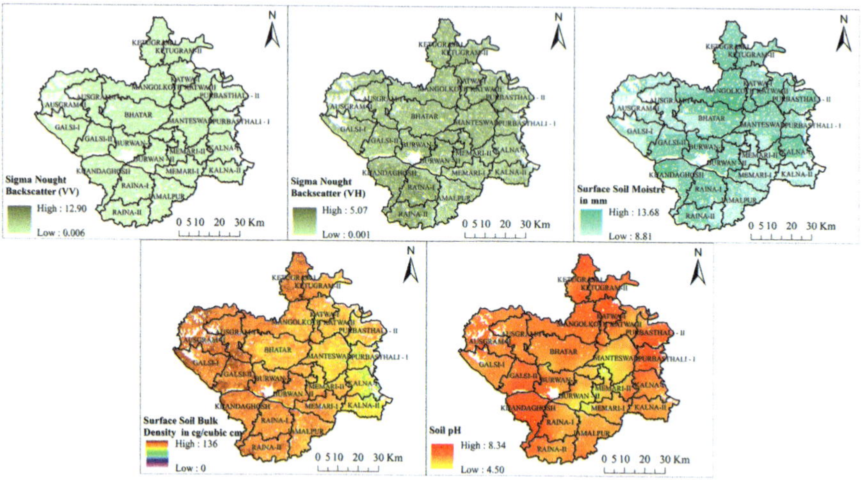

Fig. 6 Distribution pattern of soil characteristics

The soil moisture values range almost similar from 2018 to 2020 with average soil moisture of 7.41 mm to 7.64 mm, respectively. However, the average soil moisture value increased to 10.57 mm in 2022. Also, the minimum and maximum values increased to 8.81 mm and 13.68 mm, respectively. The VV and VH backscatter coefficient values showed similar spatio-temporal distribution patterns. The soil pH values range from 4.5 to 8.34 with an average of 6.35. High soil pH is common along the district's eastern, south-western, and northern edges (Fig. 6). The central portion has comparatively low soil pH values. Since the ideal range of pH values for paddy cultivation is 5.5–7.0, the soil pH of the district is suitable for growing paddy. The bulk density of the soil ranges from 0 to 136 cg/cm^3. The average bulk density was estimated at 125.9 cg/cm^3. The western district area, including Galsi-I, II; Ausgram-I, II; and Khandaghosh blocks, ranges from 130 cg/cm to 136 cg/cm^3. Conversely, the eastern region, covering Manteshwar; Memari-I, II; Purbasthali-I, II; and Kalna-I, II, primarily falls between 123 and 126 cg/cm^3 (Fig. 6).

4.2 *Yield Prediction Modelling*

Factor analysis was conducted on 17 parameters: five concerning paddy crop health, five about soil traits, three on rainfall, and four on the area's socio-economic features. As discussed earlier, before performing the actual analysis, all the block-wise datasets were standardized using *z*-score. On the other hand, zonal statistics (mean) were considered for raster-based images. The PC extraction method was considered based on the correlation coefficient matrix. The Kaiser-Meyer-Olkin (KMO) value

Table 2 Communalities

	Initial	Extraction
NDVI	1.0	.755
EVI		.900
GEMI		.871
RGVI		.959
SAVI		.879
VV		.660
VH		.794
Soil moisture		.609
Soil pH		.492
Bulk density of soil		.879
LST		.834
Temperature		.869
Rainfall		.760
Paddy irrigation intensity		.761
Irrigation requirement (IR_{Mod})		.858
Cumulative proportion of seed stores and fertilizer depots		.520
Cumulative proportion of manpower in agriculture		.662

Table 3 Total variance explained

	Extraction sums of squared loadings			Rotation sums of squared loadings		
Factors	Total	% of variance	Cumulative %	Total	% of variance	Cumulative %
1	5.850	34.412	34.412	4.258	25.047	25.047
2	3.533	20.783	55.195	3.699	21.758	46.805
3	2.276	13.389	68.584	3.266	19.211	66.016
4	1.402	8.249	76.833	1.839	10.818	76.833

was 0.529, indicating that the dataset is suitable for analysis since KMO > 0.5. KMO is employed to evaluate sampling adequacy by examining correlations and partial correlations to assess if the data are likely to converge on components. Bartlett's test of sphericity came with approximate $\chi 2 = 399.427$ with $df = 136$ (sig. value 0.000). The communality table (Table 2) represents the variance proportion. Communality represents the variance a variable has due to all components, derived from squared component loadings. The four identified factors explained 76.833% of the cumulative total variance (Table 3). The rotation converged in six iterations.

Figure 7 shows the results of factor analysis through a scree plot between components and eigenvalues. The plot is downward sloping as the eigenvalues start decreasing with the increasing number of components. The plot illustrates the amount of variation captured by each from the data. The rotated component matrix (Table 4)

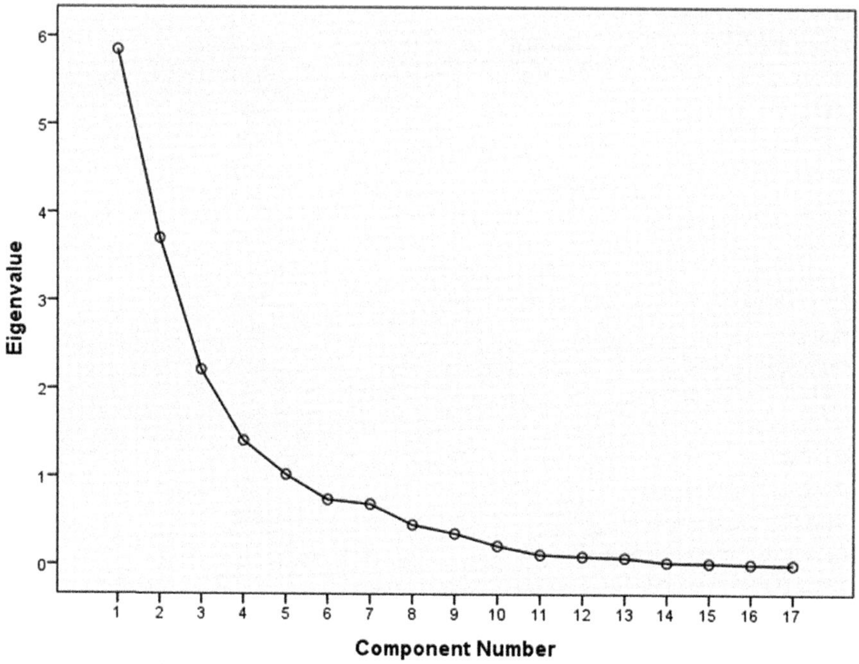

Fig. 7 Scree plot

Table 4 Rotated component matrix

	Component			
	1	2	3	4
NDVI	.831	.160		.194
EVI	.897	.297		
GEMI	.747	.551		
RGVI	.499	.819		−.193
SAVI	−.795	−.409		.280
VV	−.491		−.147	−.623
VH	.166	.273		−.831
Soil moisture	.718	−.270	−.119	
Soil pH	.387		−.512	.283
Bulk density of soil		.933		
LST	.401	.779	−.195	.167
Temperature		−.853	−.242	.282
Rainfall	.406	.222	−.447	.589
Paddy irrigation intensity	.166		.842	−.155
Irrigation requirement (IR_{Mod})			.923	
Cumulative proportion of seed stores and fertilizer depots	−.165		.682	.162
Cumulative proportion of manpower in agriculture		.171	.794	

Extraction method: principal component analysis
Rotation method: varimax with Kaiser normalization
Rotation converged in six iterations

Table 5 Component transformation matrix

Component	1	2	3	4
1	.763	.643	−.043	.038
2	−.182	.294	.886	−.311
3	.471	−.557	.459	.507
4	.403	−.435	−.055	−.803

Extraction method: principal component analysis
Rotation method: varimax with Kaiser normalization

reveals item loadings on each rotated component, highlighting which items belong to each component. The component transformation matrix (Table 5) shows correlations between components before and after rotation.

After factor analysis, the four uncorrelated components were identified for the multiple linear regression analysis. The regression was performed between the yield value and the four identified factors with multiple R^2 values that came as 0.572 with a standard error of estimate 0.65. The Durbin-Watson value came as 1.754. Table 6 shows the unstandardized and standardized coefficients. This helped to formulate the multiple linear regression equation to estimate yield.

The lack of a sufficient number of samples in multiple linear regression analysis bottlenecked the predicted yield. Therefore, a machine learning technique was applied to predict paddy yield. The random forest (RF) regressor was applied taking 500 randomly distributed training samples. A total of 350 samples were considered for calibration and 150 samples for validation. The training R^2, mean absolute error (MAE), mean squared error (MSE), root-mean-squared error (RMSE), and explained variance score came as 0.99, 0.021143, 0.000798, 0.03, and 0.99. Figure 8 highlights the pattern of yield of the testing samples. It is clear that the observed and predicted yields almost matched. Figure 9 describes the prediction error for the RF regressor. Figure 10 highlights the distribution pattern of residuals for the RF regressor model. The train and test R^2 values came as 0.998 and 0.99, respectively.

Finally, the yield maps for the 3 years were prepared (Fig. 11). The predicted average yield value remained almost the same from 5.15 tonnes/ha in 2018 to 5.18 tonnes/ha in 2022. Spatially, the yield rate decreased in the south-western blocks of the district covering Raina-I, II and Khandaghosh and the south-eastern blocks covering Memari-I and Kalna-II in the specified period.

Table 6 Standardized and unstandardized coefficients

Dependent variable: yield	Unstandardized coefficients		Standardized coefficients			95.0% confidence interval for B		Correlations			Collinearity statistics	
	B	Std. Error	Beta	*t*	Sig.	Lower bound	Upper bound	Zero-order	Partial	Part	Tolerance	VIF
(Constant)	5.087	.136		37.535	.000	4.802	5.372					
Factor1	−.297	.139	−.414	−2.143	.046	−.588	−.006	−.414	−.451	−.414	1.000	1.000
Factor 2	.148	.139	.207	1.070	.299	−.143	.439	.207	.245	.207	1.000	1.000
Factor 3	.052	.139	.073	.379	.709	−.239	.344	.073	.089	.073	1.000	1.000
Factor 4	.234	.139	.327	1.692	.108	−.057	.526	.327	.370	.327	1.000	1.000

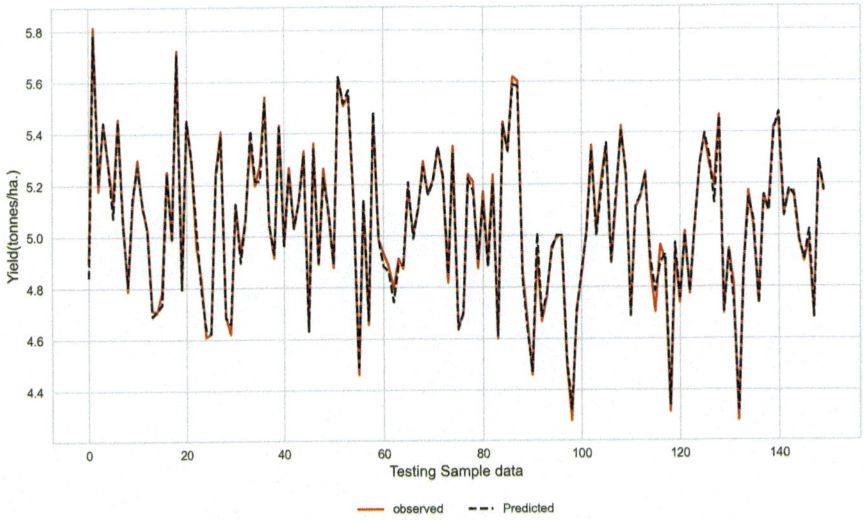

Fig. 8 Observed and predicted yield of testing samples

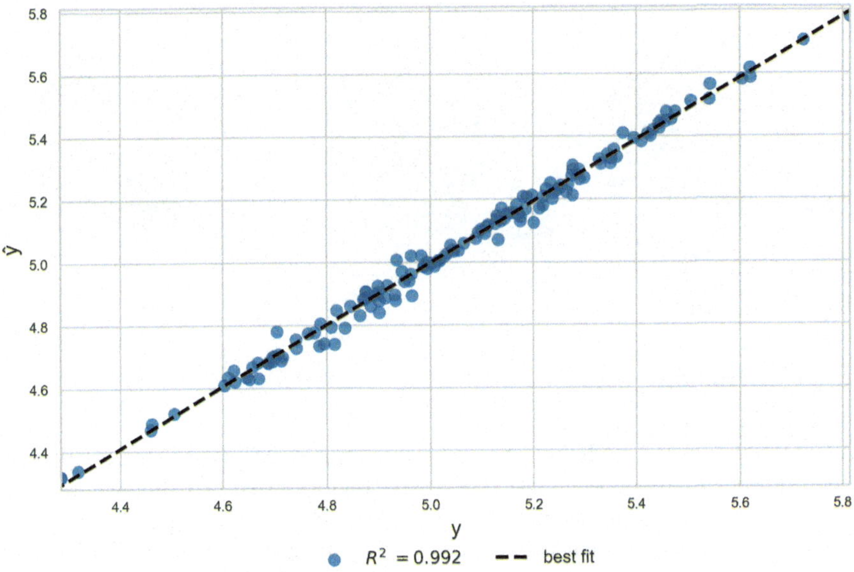

Fig. 9 Prediction error for random forest regressor

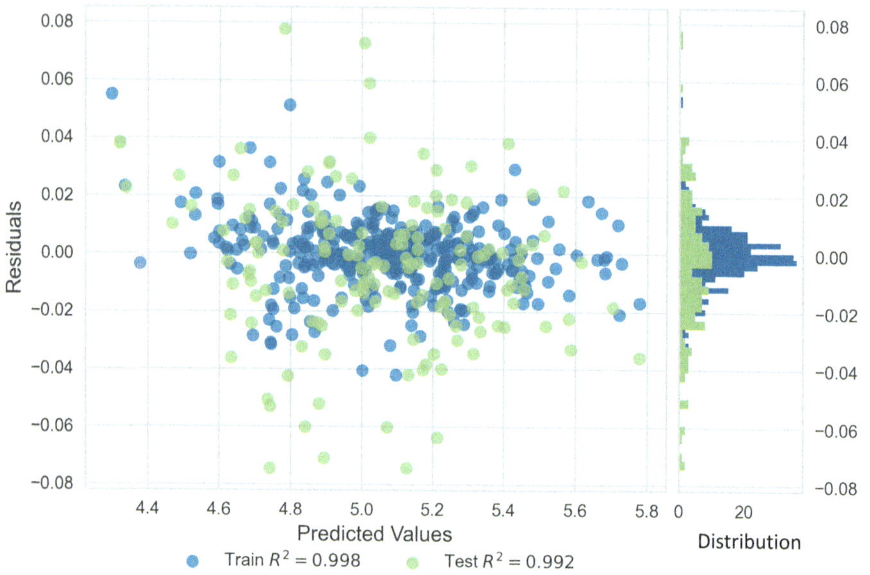

Fig. 10 Residuals for random forest regressor model

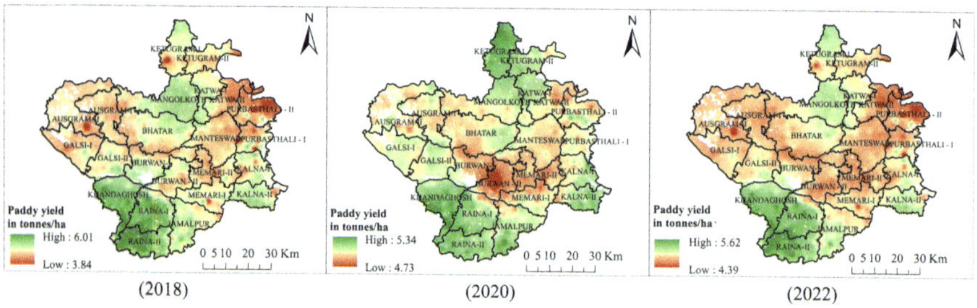

Fig. 11 Predicted yield of paddy

5 Conclusion

Yield estimation and modelling in regions with limited data pose distinct challenges and offer potential opportunities. Data-sparse regions often lack reliable crop production data, hindering accurate yield assessment and prediction. Nevertheless, progress in geospatial science and machine learning offers opportunities to address these data limitations and enhance yield estimation. The present research identified 17 factors that can contribute to the paddy yield of the district. Among those 17 factors, 13 factors were physical in nature, and the rest were socio-economic. The study highlights the importance of geospatial data for regular availability in data-sparse regions. The multi-sensor data provided a valuable source for understanding the physical characteristics of land, atmosphere, and crop health in estimating yield.

Besides, the multi-source socio-economic data provided the infrastructural and human resource support in assessing the same. The study also pinpointed the importance of machine learning techniques in paddy yield in such data-sparse regions, where regular availability of yield is not frequent. Also, the scale of data availability is not that detailed. In this regard, the random forest regressor provided satisfactory results in estimating yield.

However, the non-availability of seasonal yield data at the block and mouza level puts limitations on the applicability of the yield model in other seasons. Besides, the human resource and infrastructural data selected for the study are available only at a longer interval. This posed a certain limitation mentioned above, and the model performed successfully in estimating paddy yield in the study area. The chosen approach and model can be reused for crop yield predictions in alike geo-environmental settings.

Acknowledgement The authors are thankful to the Department of Geography, School of Basic and Applied Sciences (SoBAS), Adamas University, for providing the necessary technical and infrastructural support to complete the research.

Conflict of Interest The authors declare that there is no conflict of interest.

References

Carlson, T. N., & Ripley, D. A. (1997). On the relation between NDVI, fractional vegetation cover, and leaf area index. *Remote Sensing of Environment, 62*(3), 241–252. https://doi.org/10.1016/S0034-4257(97)00104-1

Chakraborti, S., Banerjee, A., Sannigrahi, S., Pramanik, S., Maiti, A., & Jha, S. (2019). Assessing the dynamic relationship among land use pattern and land surface temperature: A spatial regression approach. *Asian Geographer, 36*(2), 93–116. https://doi.org/10.1080/10225706.2019.1623054

Chander, G., Markham, B. L., & Helder, D. L. (2009). Summary of current radiometric calibration coefficients for Landsat MSS, TM, ETM+, and EO-1 ALI sensors. *Remote Sensing of Environment, 113*(5), 893–903. https://doi.org/10.1016/j.rse.2009.01.007

Charoen-Ung, P., & Mittrapiyanuruk, P. (2018, July 11–13). Sugarcane yield grade prediction using random forest and gradient boosting tree techniques. In *2018 15th International Joint Conference on Computer Science and Software Engineering (JCSSE)*. IEEE.

dela Torre, D. M. G., Gao, J., & Macinnis-Ng, C. (2021). Remote sensing-based estimation of rice yields using various models: A critical review. *Geo-spatial Information Science, 24*(4), 580–603. https://doi.org/10.1080/10095020.2021.1936656

Di, L., Rundquist, D. C., & Han, L. (1994). Modelling relationships between NDVI and precipitation during vegetative growth cycles. *International Journal of Remote Sensing, 15*(10), 2121–2136. https://doi.org/10.1080/01431169408954231

Dong, J., & Xiao, X. (2016). Evolution of regional to global paddy rice mapping methods: A review. *ISPRS Journal of Photogrammetry and Remote Sensing, 119*, 214–227. https://doi.org/10.1016/j.isprsjprs.2016.05.010

Dorigo, W. A., Zurita-Milla, R., de Wit, A. J. W., Brazile, J., Singh, R., & Schaepman, M. E. (2007). A review on reflective remote sensing and data assimilation techniques for enhanced agroecosystem modeling. *International Journal of Applied Earth Observation and Geoinformation, 9*(2), 165–193. https://doi.org/10.1016/j.jag.2006.05.003

Dugan, M. (2015). Rice. *Journal of Agricultural & Food Information, 16*(1), 3–10. https://doi.org/10.1080/10496505.2015.985546
Dunetman, G. H. (1989). *Principal component analysis*. Sage Publication.
Everitt, B. S., & Dunn, G. (2001). *Applied multivariate data analysis*. John Wiley & Sons, Ltd.
Fernandez-Beltran, R., Baidar, T., Kang, J., & Pla, F. (2021). Rice-yield prediction with multi-temporal Sentinel-2 data and 3D CNN: A case study in Nepal. *Remote Sensing, 13*(7), 1391.
Filippi, P., Jones, E. J., Wimalathunge, N. S., Somarathna, P. D. S. N., Pozza, L. E., Ugbaje, S. U., et al. (2019). An approach to forecast grain crop yield using multi-layered, multi-farm data sets and machine learning. *Precision Agriculture, 20*(5), 1015–1029. https://doi.org/10.1007/s11119-018-09628-4
Gandhi, N., Armstrong, L. J., Petkar, O., & Tripathy, A. K. (2016, July 13–15). Rice crop yield prediction in India using support vector machines. In *2016 13th International Joint Conference on Computer Science and Software Engineering (JCSSE)*. IEEE.
Godfray, H. C. J., Beddington, J. R., Crute, I. R., Haddad, L., Lawrence, D., Muir, J. F., et al. (2010). Food security: The challenge of feeding 9 billion people. *Science, 327*(5967), 812–818. https://doi.org/10.1126/science.1185383
Govt. of West Bengal. (2023). Purba Bardhaman. Retrieved October 5, 2023, from https://purbabardhaman.nic.in/
Hengl, T., Mendes de Jesus, J., Heuvelink, G. B. M., Ruiperez Gonzalez, M., Kilibarda, M., Blagotić, A., et al. (2017). SoilGrids250m: Global gridded soil information based on machine learning. *PLoS One, 12*(2), e0169748. https://doi.org/10.1371/journal.pone.0169748
Huete, A. R. (1988). A soil-adjusted vegetation index (SAVI). *Remote Sensing of Environment, 25*(3), 295–309. https://doi.org/10.1016/0034-4257(88)90106-X
Iizumi, T., Yokozawa, M., & Nishimori, M. (2009). Parameter estimation and uncertainty analysis of a large-scale crop model for paddy rice: Application of a Bayesian approach. *Agricultural and Forest Meteorology, 149*(2), 333–348. https://doi.org/10.1016/j.agrformet.2008.08.015
Iniyan, S., Akhil Varma, V., & Teja Naidu, C. (2023). Crop yield prediction using machine learning techniques. *Advances in Engineering Software, 175*, 103326. https://doi.org/10.1016/j.advengsoft.2022.103326
Islam, M. D., Di, L., Qamer, F. M., Shrestha, S., Guo, L., Lin, L., et al. (2023). Rapid rice yield estimation using integrated remote sensing and meteorological data and machine learning. *Remote Sensing, 15*(9), 2374.
Jackson, J. E. (1991). *A user's guide to principal components*. John Wiley & Sons, Inc.
Jin, X., Kumar, L., Li, Z., Feng, H., Xu, X., Yang, G., & Wang, J. (2018). A review of data assimilation of remote sensing and crop models. *European Journal of Agronomy, 92*, 141–152. https://doi.org/10.1016/j.eja.2017.11.002
Joshua, V., Priyadharson, S. M., & Kannadasan, R. (2021). Exploration of machine learning approaches for paddy yield prediction in eastern part of Tamilnadu. *Agronomy, 11*(10), 2068.
Kaplan, G., & Rozenstein, O. (2021). Spaceborne estimation of Leaf Area Index in cotton, tomato, and wheat using Sentinel-2. *Land, 10*(5), 505.
Kaplan, G., Fine, L., Lukyanov, V., Manivasagam, V. S., Tanny, J., & Rozenstein, O. (2021). Normalizing the local incidence angle in Sentinel-1 imagery to improve leaf area index, vegetation height, and crop coefficient estimations. *Land, 10*(7), 680.
Karunakaran, K. R., & Palanisami, K. (1998). An analysis of impact of irrigation on cropping intensity in Tamil Nadu. *Indian Economic Review, XXXIII*(2), 207–220.
Kasampalis, D. A., Alexandridis, T. K., Deva, C., Challinor, A., Moshou, D., & Zalidis, G. (2018). Contribution of remote sensing on crop models: A review. *Journal of Imaging, 4*(4), 52.
Kendall, M. G. (1939). The geographical distribution of crop productivity in England. *Journal of the Royal Statistical Society, 102*(1), 21–62.
Kendall, M. G. (1950). Part I: Factor analysis as a statistical technique. *Journal of the Royal Statistical Society. Series B (Methodological), 12*(1), 60–73.
Khanal, S., Fulton, J., Klopfenstein, A., Douridas, N., & Shearer, S. (2018). Integration of high resolution remotely sensed data and machine learning techniques for spatial prediction of soil

properties and corn yield. *Computers and Electronics in Agriculture, 153*, 213–225. https://doi.org/10.1016/j.compag.2018.07.016

Kuenzer, C., & Knauer, K. (2013). Remote sensing of rice crop areas. *International Journal of Remote Sensing, 34*(6), 2101–2139. https://doi.org/10.1080/01431161.2012.738946

Li, Y., Liao, Q., Li, X., Liao, S., Chi, G., & Peng, S. (2003). Towards an operational system for regional-scale rice yield estimation using a time-series of Radarsat ScanSAR images. *International Journal of Remote Sensing, 24*(21), 4207–4220. https://doi.org/10.1080/0143116031000095970

Li, T., Angeles, O., Marcaida, M., Manalo, E., Manalili, M. P., Radanielson, A., & Mohanty, S. (2017). From ORYZA2000 to ORYZA (v3): An improved simulation model for rice in drought and nitrogen-deficient environments. *Agricultural and Forest Meteorology, 237–238*, 246–256. https://doi.org/10.1016/j.agrformet.2017.02.025

Lingwal, S., Bhatia, K. K., & Singh, M. (2024). A novel machine learning approach for rice yield estimation. *Journal of Experimental & Theoretical Artificial Intelligence, 36*(3), 337–356. https://doi.org/10.1080/0952813X.2022.2062458

Liu, H. Q., & Huete, A. (1995). A feedback based modification of the NDVI to minimize canopy background and atmospheric noise. *IEEE Transactions on Geoscience and Remote Sensing, 33*(2), 457–465. https://doi.org/10.1109/TGRS.1995.8746027

Manivasagam, V. S., Sadeh, Y., Kaplan, G., Bonfil, D. J., & Rozenstein, O. (2021). Studying the feasibility of assimilating Sentinel-2 and PlanetScope imagery into the SAFY crop model to predict within-field wheat yield. *Remote Sensing, 13*(12), 2395.

Marzouk, O. A. (2021). Assessment of global warming in Al Buraimi, Sultanate of Oman based on statistical analysis of NASA POWER data over 39 years, and testing the reliability of NASA POWER against meteorological measurements. *Heliyon, 7*(3), e06625. https://doi.org/10.1016/j.heliyon.2021.e06625

Mia, M. S., Tanabe, R., Habibi, L. N., Hashimoto, N., Homma, K., Maki, M., et al. (2023). Multimodal deep learning for rice yield prediction using UAV-based multispectral imagery and weather data. *Remote Sensing, 15*(10), 2511.

Mo, C., Zhang, M., Ruan, Y., Qin, J., Wang, Y., Sun, G., & Xing, Z. (2020). Accuracy analysis of IMERG satellite rainfall data and its application in long-term runoff simulation. *Water, 12*(8), 2177.

Mohammad, L., Bandyopadhyay, J., Sk, R., Mondal, I., Nguyen, T. T., Lama, G. F. C., & Anh, D. T. (2023). Estimation of agricultural burned affected area using NDVI and dNBR satellite-based empirical models. *Journal of Environmental Management, 343*, 118226. https://doi.org/10.1016/j.jenvman.2023.118226

Mosleh, M. K., Hassan, Q. K., & Chowdhury, E. H. (2015). Application of remote sensors in mapping rice area and forecasting its production: A review. *Sensors, 15*(1), 769–791.

Nikitha, K., Sreenivas, G., Rani, P. L., & Madhavi, A. (2018). Performance of photo sensitive rice cv. RNR 15048 (Telangana Sona) under varied nitrogen levels and environmental conditions. *International Journal of Current Microbiology and Applied Sciences, 7*(3), 3602–3605.

Noorunnahar, M., Chowdhury, A. H., & Mila, F. A. (2023). A tree based eXtreme Gradient Boosting (XGBoost) machine learning model to forecast the annual rice production in Bangladesh. *PLoS One, 18*(3), e0283452. https://doi.org/10.1371/journal.pone.0283452

Nuarsa, I. W., Nishio, F., & Hongo, C. (2011). Spectral characteristics and mapping of rice plants using multi-temporal landsat data. *Journal of Agricultural Science, 3*(1), 54–67.

Oğuz, H. (2016). Automated land surface temperature retrieval from Landsat 8 satellite imagery: A case study of Kahramanmaraş – Turkey. In R. Efe, İ. Cürebal, A. Gad, & B. Tóth (Eds.), *Environmental sustainability and landscape management*. St. Kliment Ohridski University Press.

Pal, I. (2021). Rural settlement types in Purba Bardhaman District of West Bengal: A study at C.D. Block Level. *Shodh Sarita, 8*(29), 128–135.

Pal, S. C., Ghosh, C., & Chowdhuri, I. (2020). Assessment of groundwater potentiality using geospatial techniques in Purba Bardhaman district, West Bengal. *Applied Water Science, 10*(10), 221. https://doi.org/10.1007/s13201-020-01302-3

Palakuru, M., & Yarrakula, K. (2019). Study on paddy phenomics ecosystem and yield estimation using space-borne multisensor remote sensing data. *Journal of Agrometeorology, 21*(2), 171–175.

Panigrahy, S., Chakraborty, M., Sharma, S. A., Kundu, N., Ghose, S. C., & Pal, M. (1997). Early estimation of rice area using temporal ERS-1 synthetic aperture radar data a case study for the Howrah and Hughly districts of West Bengal, India. *International Journal of Remote Sensing, 18*(8), 1827–1833. https://doi.org/10.1080/014311697218133

Pinty, B., & Verstraete, M. M. (1992). GEMI: A non-linear index to monitor global vegetation from satellites. *Vegetation, 101*(1), 15–20. https://doi.org/10.1007/BF00031911

Rahman, A., Roytman, L., Krakauer, N. Y., Nizamuddin, M., & Goldberg, M. (2009). Use of vegetation health data for estimation of aus rice yield in Bangladesh. *Sensors, 9*(4), 2968–2975.

Rahman, A., Khan, K., Krakauer, N. Y., Roytman, L., & Kogan, F. (2012). Use of remote sensing data for estimation of Aman rice yield. *International Journal of Agriculture and Forestry, 2*, 101–107.

Ranjan, A. K., & Parida, B. R. (2019). Paddy acreage mapping and yield prediction using sentinel-based optical and SAR data in Sahibganj district, Jharkhand (India). *Spatial Information Research, 27*(4), 399–410. https://doi.org/10.1007/s41324-019-00246-4

Rouse, J. W. J., Haas, R. H., Schell, J. A., & Deering, D. W. (1974). *Monitoring vegetation systems in the Great Plains with ERTS* (Vol. 351). NASA Special Publication.

Sadeh, Y., Zhu, X., Dunkerley, D., Walker, J. P., Zhang, Y., Rozenstein, O., et al. (2021). Fusion of Sentinel-2 and PlanetScope time-series data into daily 3 m surface reflectance and wheat LAI monitoring. *International Journal of Applied Earth Observation and Geoinformation, 96*, 102260. https://doi.org/10.1016/j.jag.2020.102260

Saed-Moucheshi, A., Fasihfar, E., Hasheminasab, H., Rahmani, A., & Ahmadi, A. (2013). A review on applied multivariate statistical techniques in agriculture and plant science. *International Journal of Agronomy and Plant Production, 4*(1), 127–141.

Santra, A. (2012). *Geostatistical and quantitative approaches for resource estimation*. Firma KLM Private Limited.

Santra, A., Santra Mitra, S., Mitra, D., & Sarkar, A. (2019). Relative Radiometric Normalisation – Performance testing of selected techniques and impact analysis on vegetation and water bodies. *Geocarto International, 34*(1), 98–113. https://doi.org/10.1080/10106049.2017.1367424

Santra, A., Kumar, A., Mitra, S. S., & Mitra, D. (2022). Identification of built-up areas based on the consistently high heat-radiating surface in the Kolkata Metropolitan Area. *Journal of the Indian Society of Remote Sensing, 50*(8), 1547–1561. https://doi.org/10.1007/s12524-022-01543-6

Sekertekin, A., & Bonafoni, S. (2020). Land surface temperature retrieval from Landsat 5, 7, and 8 over rural areas: Assessment of different retrieval algorithms and emissivity models and toolbox implementation. *Remote Sensing, 12*(2), 294.

Sen, S. (1996). Need for irrigation for its spatial pattern in East and West Singhbhum District, Bihar. *Indian Journal of Landscape Systems and Ecological Studies, 19*(2), 80–91.

Shanmugan, K., & Baria, B. P. (2019). Agricultural labour productivity and its determinants in India. *The Indian Journal of Labour Economics, 62*(3), 431–449. https://doi.org/10.1007/s41027-019-00180-x

Sharifi, E., Steinacker, R., & Saghafian, B. (2016). Assessment of GPM-IMERG and other precipitation products against gauge data under different topographic and climatic conditions in Iran: Preliminary results. *Remote Sensing, 8*(2), 135.

Siagian, D. R., Shrestha, R. P., Shrestha, S., & Kuwornu, J. K. M. (2019). Factors driving rice land change 1989–2018 in the Deli Serdang Regency, Indonesia. *Agriculture, 9*(9), 186.

Sud, U. C., Ahmad, T., Gupta, V. K., Chandra, H., Sahoo, P. M., Aditya, K., et al. (2016). *Synthesis of literature and framework – Research on improving methods for estimating crop area, yield and production under mixed, repeated and continuous cropping*. ICAR – Indian Agricultural Statistics Research Institute.

Wu, S., Yang, P., Ren, J., Chen, Z., & Li, H. (2021). Regional winter wheat yield estimation based on the WOFOST model and a novel VW-4DEnSRF assimilation algorithm. *Remote Sensing of Environment, 255*, 112276. https://doi.org/10.1016/j.rse.2020.112276

Yadav, N. K., Santra, A., Samanta, A. K., Kumar, A., Mitra, S. S., & Mitra, D. (2021). Understanding the synergistic relation between land surface temperature and different biophysical parameters in Haldia industrial city of India. *Arabian Journal of Geosciences, 14*(22), 2412. https://doi.org/10.1007/s12517-021-08769-7

Yadav, N. K., Mitra, S. S., Santra, A., & Samanta, A. K. (2023). Understanding responses of atmospheric pollution and its variability to contradicting nexus of urbanization–industrial emission control in Haldia, an industrial city of West Bengal. *Journal of the Indian Society of Remote Sensing, 51*(3), 625–646. https://doi.org/10.1007/s12524-022-01649-x

Yang, Z., Crow, W., Hu, L., Di, L., & Mueller, R. (2017, July 23–28). SMAP DATA for cropland soil moisture assessment—A case study. In *2017 IEEE International Geoscience and Remote Sensing Symposium (IGARSS)*, Melbourne.

Yoshida, S. (1981). *Fundamentals of rice crop science*. The International Rice Research Institute.

Zeng, J., Shi, P., Chen, K. S., Ma, H., Bi, H., & Cui, C. (2022). Assessment and error analysis of satellite soil moisture products over the third pole. *IEEE Transactions on Geoscience and Remote Sensing, 60*, 1–18. https://doi.org/10.1109/TGRS.2021.3116078

Zhang, C., Yang, Z., Zhao, H., Sun, Z., Di, L., Bindlish, R., et al. (2022). Crop-CASMA: A web geoprocessing and map service based architecture and implementation for serving soil moisture and crop vegetation condition data over U.S. Cropland. *International Journal of Applied Earth Observation and Geoinformation, 112*, 102902. https://doi.org/10.1016/j.jag.2022.102902

Zhao, H., Zeng, Y., Wen, J., Wang, X., Wang, Z., Meng, X., & Su, Z. (2021). An air-to-soil transition model for discrete scattering-emission modelling at L-band. *Journal of Remote Sensing, 2021*. https://doi.org/10.34133/2021/3962350

Zhao, H., Di, L., & Sun, Z. (2022). WaterSmart-GIS: A web application of a data assimilation model to support irrigation research and decision making. *ISPRS International Journal of Geo-Information, 11*(5), 271.

Simran Kumari Sah, a Geoinformatics student, completed her post-graduation in Geoinformatics from Adamas University and completed graduation in Geography major from the Shri Shikshayatan College under the University of Calcutta. She is keen on doing research in the field of geoinformatics application in agricultural studies.

Prof. (Dr.) **Abhisek Santra** is presently working as a Professor in the Department of Geography, Adamas University. He did his Master's and Ph.D. in Geography from the University of Calcutta. He also completed his masters in Applied RS & GIS and Business Administration from the University of Southampton, UK, and Manipal University, respectively. As a former faculty member of Symaprasad College, Birla Institute of Technology (Mesra), and Haldia Institute of Technology, he has more than 18 years of teaching experience. He also was a Project Manager in the WBSMB (Govt. of WB). Dr. Santra is an active researcher and has published more than 70 research papers in terms of peer-reviewed indexed journal articles, conference papers, books, and book chapters. His areas of research cover the application of geoinformatics in urban and environmental studies. He has completed four R&D projects funded by UGC and SERB, India. He is a reviewer of the SERB-funded projects. Prof. Santra is also serving the research community in his field of research by reviewing research articles published by reputed SCI and Scopus-indexed journals hosted by Elsevier, Springer, Taylor and Francis, Wiley, etc.

The Advent of Tuberose Floriculture: Agricultural Growth and Its Impact on Farmers' Socio-economic Well-Being

Dipankar Biswas and Rajib Sarkar

1 Introduction

The economy plays a significant role in the development of human life. Still, the income of conventional crops (rice, paddy sugar, wheat, etc.) is gradually declining. The yield rate is also declining after repeated cultivation of the same crops. At the same time, the fertility of the soil is deteriorating. But the quality of human life is increasing day by day with higher commodity prices. In such a situation, the rural farmers have taken the initiative to cultivate those crops that provide higher production and income. In this context, tuberose cultivation has begun, and the quality of life has continued to improve.

Tuberose (*Polianthes tuberosa* L.) cultivation is now considered an extensive flower cultivation practice among farmers across many countries (Bharathi et al., 2023). This florist cultivar can be grown at any time or season of the year. It provides constant income throughout the year. It has a sober fragrance to please everyone (Sarker & Salam, 2018). Any agro-climatic condition favours this cultivar to grow, but the tropical to temperate climate is more helpful for better productivity (Usman et al., 2015). The costs of producing flowers are different compared to the other flower cultivations such as rose, gerbera, chrysanthemum, and others. The different types of tuberoses are also cultivated in India like single, semi-double, double, and variegated. The local varieties of this ornamental species are *Prajwal*, *Kolkata single*, and *Kolkata double* (Pal et al., 2019). Along with the socio-religion and aesthetic sense, it has medical value as well (Haque et al., 2012). The perfume and beverage industry also uses this ornamental species for various purposes like as an essential oil. This flower cultivation is increasing day by day in the world of

D. Biswas · R. Sarkar (✉)
Department of Geography, Adamas University, Kolkata, West Bengal, India
e-mail: rajib.sarkar@adamasuniversity.ac.in

A. Rai et al. (eds.), *Contemporary Social Physics*, Springer Geography,
https://doi.org/10.1007/978-3-031-77596-3_13

commercial trade (Swapna, 2018; Ghosh & Gupta, 2019). Moreover, it also has a massive demand for the local, national, and international markets (Haque et al., 2012: Mandal et al., 2018). The farmers and growers in the different parts of the world are producing tuberoses. Besides farming, growers also use it for landscaping, gardening, bedding, and potted plants in urban. This floriculture leads to job security for thousands of young, especially unemployed, youths and mature and older people.

The tuberose economy can change the fate of farmers' families for food, shelter, and entire livelihood (Sahu et al., 2011). The land size determines the quantities of flower production (Setu et al., 2018), influencing the growers' income (Usman & Ashfaq, 2013; Harisha, 2017). The previous literature describes that flower farmers are marginal to small according to their land size (Usman et al., 2015). However, from this small amount of land, the subsistence of every member of a family takes place, and an increment of the monthly income occurs. Other aspects of life, such as the education level, consumption of luxury goods, health expenditure, food consumption, etc., are also improved. For these reasons, flower cultivation is adopted by farmers more than any other traditional crop (Bera & Majumder, 2013; Das, 2017). Economic development occurs with the increase in agricultural production and income. In the same way, with the rise in tuberose production, economic development is taking place in areas of cultivation (Ding et al., 2022; Mitra et al., 2024; Usman & Ashfaq, 2013).

Tuberose farming has influenced farmers' socio-economic aspects in positive tune. Connecting natural or physical features with social elements, tuberose floriculture controls and enhances farmers' overall well-being, and in this way, it also signifies social physics research dealing with social phenomena in natural settings with invariable laws with big data (Tsekov, 2023; Breur, 2016).

Tuberose practice is incredibly massive for India. The Indian government has also paid particular attention to this flower production (Biswas, 2013) by providing infrastructure, training, seeds, loans, etc. Tuberose is produced in larger amounts in different provinces of India and the districts of West Bengal (Mandal et al., 2018). However, a review of the literary works reveals that the economics of tuberose farming is not connected with the socio-cultural aspects of the land. The integrated study of how tuberose production improves the socio-economic conditions of farmer's life is still lacking. Considering this drawback in this study, the objectives are framed as (1) measuring the socio-economic status of tuberose cultivators, (2) developing a model to assess the production characteristics of tuberose considering socio-economic and demographic parameters, (3) and assessing the improvement of tuberose production on farmers' well-being in the study area on the spatial platform indicating regional development (Tsekov, 2023; Breur, 2016).

2 Materials and Methods

2.1 Study Area and Sampling

Study Area

The study is carried out in Ranaghat administrative subdivision farms located right in the middle of the Nadia district in the eastern Indian state of West Bengal (Fig. 1). Nadia district produces large quantities of marigolds (Fig. 1). Geologically, the Ranaghat administrative subdivision has not much variation, nearly flat also.

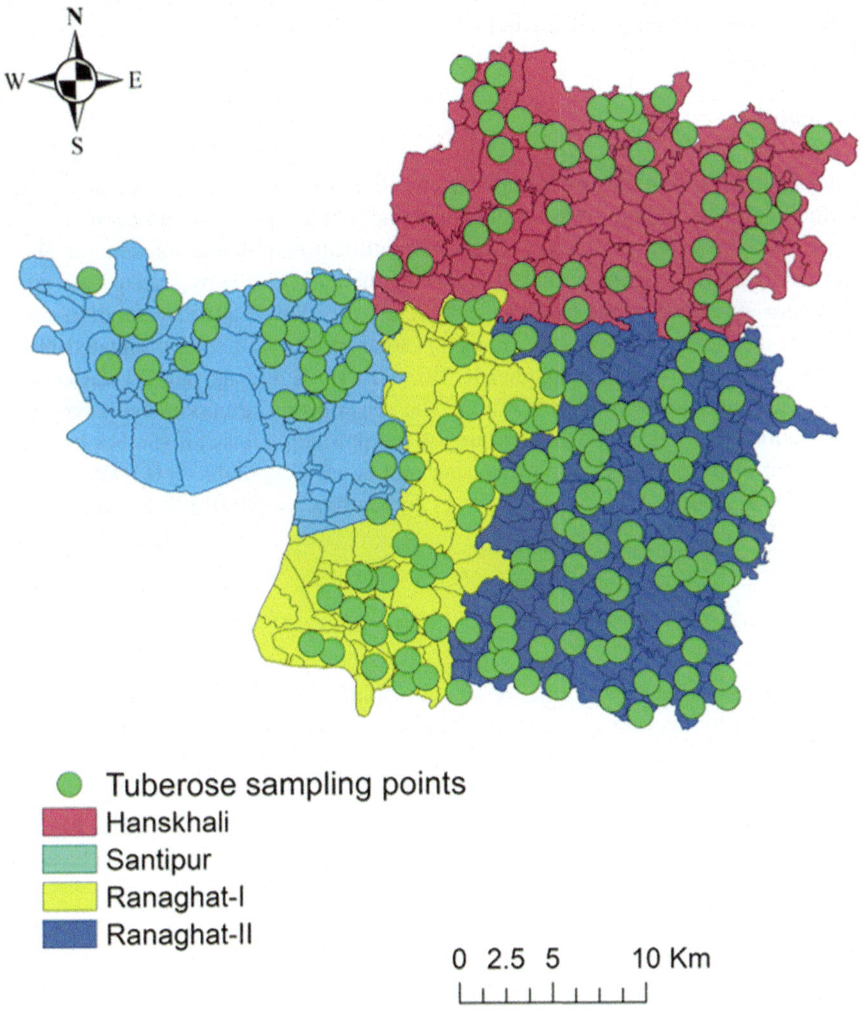

Fig. 1 Tuberose cultivation sampling points collecting from Ranaghat administrative subdivision

It consists of four C.D. blocks. They are Ranaghat-I, Ranaghat-II, Hanskhali, and Santipur. The weather and climatic conditions vary throughout the year that include hot and humid summer, monsoons with events of heavy downpours (June to September), and cold, dry winter (November to February). The total average rainfall is 1447 mm; the minimum and maximum temperatures range in between 7 and 42 °C. The soil type of the region is newer alluvium. Primarily, the economy of the region depends on the agricultural system. Flowers are predominantly grown and well irrigated by rivers (Bhagirathi, Ichhamati, and Churni), canals, and small shallow machines. The study area is considered suitable for tuberose cultivation and marigold farming (Biswas & Sarkar, 2021). This region's natural, socio-economic, and infrastructural transportation system exacerbates the total arable land into tuberose land. Tuberose species minimize the risk of land abandonment and felicitate farmers to enough employment and regular income.

Sampling

The economics of tuberose farming enriches the farmers, mainly marginal to small farmers in every village in the entire Ranaghat administrative subdivision. To estimate the socio-economic factors of tuberose cultivation, different information about farmers is collected using a random sampling method (Fig. 1) by face-to-face questionnaire survey. The survey was conducted in 2022. Ranaghat subdivision is divided into four administrative blocks. A total of 221 farmers participated in the survey—100 farmers in Ranaghat-II, 36 farmers in Ranaghat-I, 51 farmers in Hanskhali, and 34 farmers in Santipur. The selection of farmers is made based on cultivation experience. The villages are left not producing this cultivar. For the measurement of land, a local land measuring unit is used (1 'bigha' = 0.1606 ha), and for monetary purposes, a rupee is used (1 Indian rupee = $ 0.01338 US dollar).

2.2 *Statistical Analysis*

To calculate the test of normality of the sample, the methods of Kolmogorov-Smirnov (Lilliefors significance correction) and Shapiro-Wilk are used. Considering the data set being normal after processing, further analysis is carried out.

Descriptive Statistics and Box and Whisker Plot

Tuberose production improves the socio-economic conditions of farmers; to measure the statement, different methods are applied. In the first part, the descriptive analysis of the minimum, maximum, mean, and standard deviation of the measured variables to determine the variation of each social, economic, cultural, and flower production-related condition. Box and whisker plot is prepared using the quantile function.

Pearson Correlation

The correlation between tuberose-induced socio-economic-cultural (SEC) parameters was analysed with the Pearson correlation coefficient (*r*) values. It varies from +1 to −1. When *r* values are close to ±1, they indicate significant negative/positive correlations between variables (Isaac & Chikweru, 2018), and the *r* values relative to ±0 indicate no substantial correlations (Table 2).

Principal Component Analysis

The principal component analysis is carried (Isaac & Chikweru, 2018; Cohen et al., 1983; Jake et al., 2017) with all the 19 socio-economic variables of Ranaghat administrative subdivisional unit, excluding average tuberose production considering it to be the dependent variable. Exploration of the latent variable from socio-economic factors is performed to reduce the dimension of the data set. Block-wise PCA is also carried out to evaluate how these socio-economic parameters influence tuberose production.

The Stepwise Multiple Linear Regression (SMLR)

The stepwise multiple linear regression (SMLR) technique is used as the best variable selection from the socio-economic affected by tuberose yield (Fig. 2). It is a method that explains stepwise multiple linear regression, and every time it removes or excludes the weakest variables and provides the best-fit result (Weisberg, 2005). The stepwise multiple linear regression technique (Abdelhafidi et al., 2020) is the best variable selection from the socio-economic affected by floricultural yield. In this study, SMLR is performed for the entire Ranaghat subdivision, and after that, tuberose cultivation of each block has been evaluated by SMLR.

Cluster Analyses

Two versions of cluster analysis are applied to group homogenous spatial units. The first one is hierarchical cluster analysis (Words method), indicating an algorithm with similar socio-economic and flower production parameters (McIntosh et al., 2010; Hair et al., 2010). It differentiates clusters by their dissimilarities, forming a tree-like dendrogram structure, where small groups include a large group. The second one is K-means cluster analysis (Kriegel et al., 2016), where every variable is clustered with the nearest mean of cluster centres or cluster centroid making a perfect cluster. They allow to measure how advanced a farmers' group is and which feature is high among the observed or measured parameters.

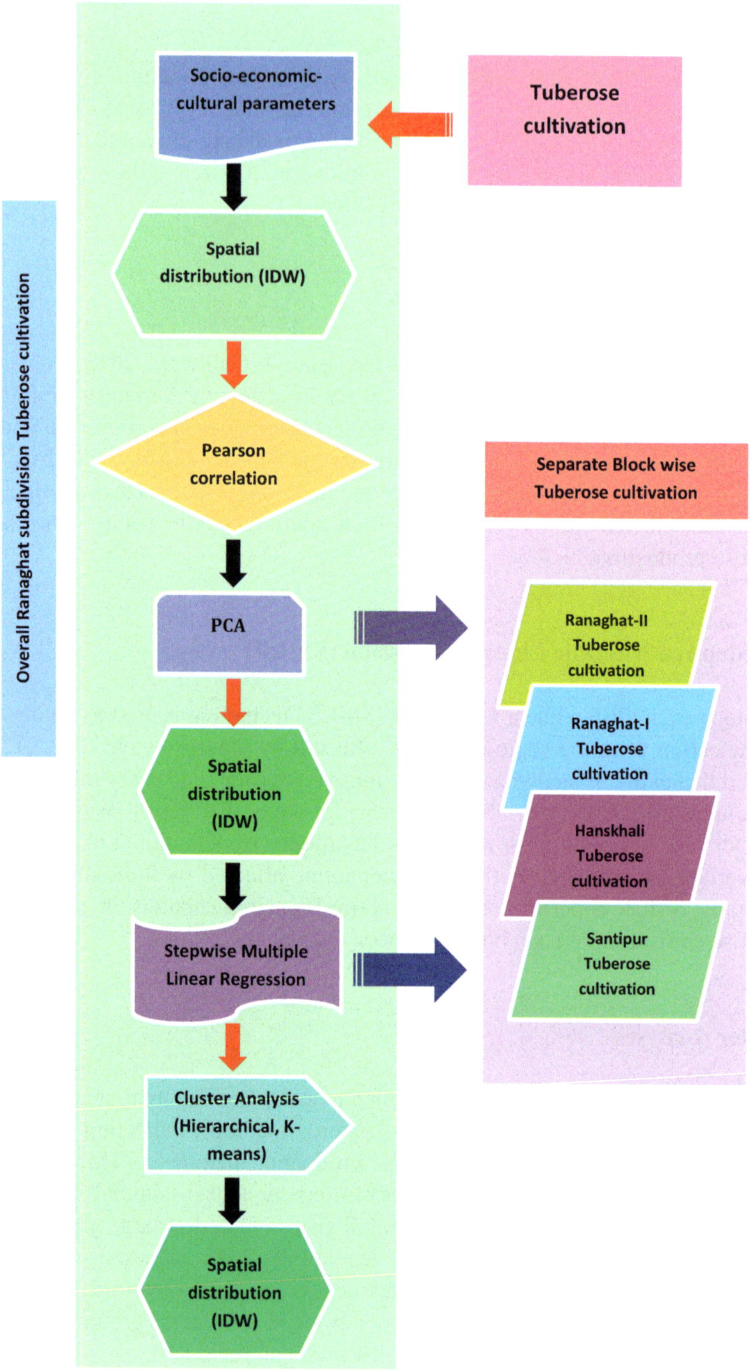

Fig. 2 Flowchart of methodology

Interpolation Method

Using tuberose location or sample points, spatial analysis is performed. IDW or inverse distance weighting is the process of deterministic interpolation. It has an application on the location of measured scatter points or values. The weighted average values of known regionalized points measure the unknown regionalized points or values. This application is used to regionalize unknown values of socio-economic parameters and tuberose average production overall area based on the known measured scatter points.

Different abbreviations of the parameters under study are used in this work. They include tuberose average production (TuberoseAvgPro, kg/month), tuberose cultivation lands (TuberoseCuLands/bigha), tuberose income (TuberoseIncome, INR in '000/month), total income (TotalIncome INR in 000/month), lands under cultivation (LandsCulti/bigha), land purchase or lease (LandPurLea/bigha), family member (FamilyMem/number), Education/years, Mobile/numbers, Internet/points, expensive goods (ExGoods/numbers), Electricity/kWh, food consumption (FoodCons/INR in '000/month), cloth expenditure (ClothEx/INR in '000/month), health expenditure (HealthEx/INR in '000/month), Furniture (number), debit and credit (Debt/Credit/INR in '000/month), working days (WorkDays/year), housing conditions (Housing), and toilet conditions (Toilet). A 6-point Likert scale is also used to underscore condition of state on certain parameters like housing condition. Precisely, they are (1) very poor, (2) poor, (3) average, (4) good, (5) very good, and (6) excellent.

3 Results

3.1 Socio-economic-Cultural Livelihood Farmers with Tuberose Production

Tuberose has been cultivated in the study area for more than four decades. Earlier, this floricultural crop was grown three to four times a year. Still, the demand for tuberose is so high that flowers are cultivated almost throughout the year, and the floral trade continues all along the year. The point of land size needs to be discussed at the initial stage of the discussion as it is of prime importance to any agricultural economic system (Sarker & Salam, 2018). It has been revealed in earlier studies that farm size has a major effect on tuberose productivity (Sarker & Salam, 2018). Most of the time, the land size of the tuberose farmers is very small to small (marginal farmers), and at other times (Table 1), the more extensive size of land also is used by producers (small farmers) (min. 0.16 to max. 1.85, ave. 0.562 bigha per capita). They produce many flowers on this stipulated amount of land and get a monthly income. This income helps to increase their total monthly income. Another aspect is that tuberose also increases the total land amount of growers to cultivate more crops.

Table 1 Descriptive statistics of tuberose farmers' socio-economic-cultural living

Variables (per capita)	*N*	Minimum	Maximum	Mean	Std. Deviation
TuberoseCuLands	221	0.16	1.85	0.562	0.288
TuberoseAvgPro	221	25.714	300	87.546	47.534
TuberoseIncome	221	1250	15,500	4403.426	2431.736
TotalIncome	221	1442.857	17,975	5294.253	2973.361
LandsCulti	221	0.525	7.5	2.104	1.228
LandPurLease	221	0.163	1.6	0.539	0.263
FamilyMem[a]	221	2	8	4.480	1.410
Education	221	5	26.667	11.708	2.608
Mobile	221	0.5	2	1.091	0.278
Internet	221	0.333	1.667	0.805	0.250
ExGoods	221	1.167	7.5	3.147	1.157
Electricity	221	8.8	28	16.792	4.112
FoodCons	221	750	2800	1371.942	394.347
ClothEx	221	141.25	833.333	376.349	106.862
HealthEx	221	108.333	525	152.755	44.789
WorkDays	221	44.375	182	88.161	31.392
Housing[a]	221	3	6	5.009	0.869
Toilet[a]	221	3	5	4.276	0.688
Furniture	221	1.375	7	3.064	1.073
Debt/Credit	221	−1.25	20	2.457	2.429

[a]They have no per capita calculation

Most of the people here are farmers. Sometimes they do other work, but most of the time, they prefer to cultivate tuberose. As a result, they bought (Table 1) some land after providing for their families' needs. Das (2017) reveals that the tuberose average family size is four, which resembles the present study.

The tuberose producers continue their family members' education facilities. Education plays a significant role in the development of tuberose production (Sarker & Salam, 2018; Das, 2017). Usman and Ashfaq (2013) have exhibited that farmers' education, age, and experience are important in this farming (Usman & Ashfaq, 2013). On the other side, income from tuberose cultivation also provides the families with the necessary socio-recreation instruments like television, fridges, motorcycles, Internet access, mobile facilities, etc. Sahu et al. provide information that floriculture improves all domestic necessities of the cultivators (Sahu et al., 2011).

Once upon a time, there was a severe food shortage in every family in the Ranaghat subdivision. However, in the present day, through floriculture, growers fulfil their food demands. Even if a farmer has other cultivation, they will cultivate tuberoses because that is the primary means of survival (Haque et al., 2012; Swapna, 2018). In most cases, family well-being depends on this farming (Gawle et al., 2012; Setu et al., 2018), such as food consumption, clothing expenditure, health expenditure, electricity, housing conditions, and toilet facilities. Since the cultivation process continues throughout the year, the farmers work all year round (Table 1).

The amount of farmers' debt is also included in this study as it explains the income level of farmers and an index of tuberose cultivation prosperity. The box and whisker plots in Fig. 3a–h graphically present all socio-economic and tuberose production data distribution by introducing five quartile summaries.

After cultivating tuberose, the quality of life of every farmer has improved. Each socio-economic and cultural factor has improved considerably to the central area of the study region and the entire south. These places are the junction or meeting place of Ranaghat-I and Ranaghat-II, Santipur, and Hanskhali. The most important villages are Nokari, Chapra, Dhantala, Bayerdangi, Nidhkur, Mollaber, Khisma, etc. where production is maximum. The amount of tuberose cultivation land is the highest (village *Nokari*) in the middle to the south in the Ranaghat administrative subdivision. In the same way, the middle and south portions also produce the average highest tuberose production and generate the highest income. The improved socio-economic factors and related villages such as cultivated lands (*Andulpota*), land under lease (*Doula*), family members (*Bahirgachi*), education (*Harishpur*), mobile access (*Matikumra*), Internet (*Daula*), electricity (*Sijda*), food consumption (*Dhantala*), etc. are in these central to south areas.

On the other hand, the northern part of Hanskhali (i.e. north of Ranaghat subdivision), part of Santipur (located in the west corner of the study area), western region of Ranaghat-I (the southwest portion of the present site), and far eastern area of Ranaghat-II have made little progress in the context of tuberose cultivation. The lowest amount is in the northern section and some villages in the southwest and eastern corner of the study area.

Family sizes of tuberose farmers are, of course, the same everywhere. The amount of loans (debit) taken by farmers is less in the middle. House and toilet conditions are much better than other socio-economic conditions covering from the middle area to outwards. The spatial distribution of all pieces of information is depicted in Figs. 3a–h, 4 and 5a–j.

3.2 Livelihood Interrelationship Through Pearson's Correlation

Correlation study is very much important in social research as it helps to identify the causal relation among variables. In this study, the correlation among socio-economic conditions of tuberose producers highlights the strength of association among variables. The detailed correlation among variables is represented in Table 2. When one variable value (socio-economic-cultural phenomenon) increases, then other variable value increases. The strong positive correlations in the correlation matrix are situated between such variables (1) TuberoseCuLands, TuberoseIncome, TotalIncome, FoodCons, and ClothEx; (2) TuberoseIncome, TotalIncome, FoodCons, and ClothEx; (3) TotalIncome, LandsCulti, FoodCons, and ClothEx; (4) LandsCulti, LandPurLease, FoodCons, and ClothEx; (5) LandPurLease, ExGoods, and Furniture; (6) FamilyMem, ExGoods, and Furniture; (7) Education, Internet,

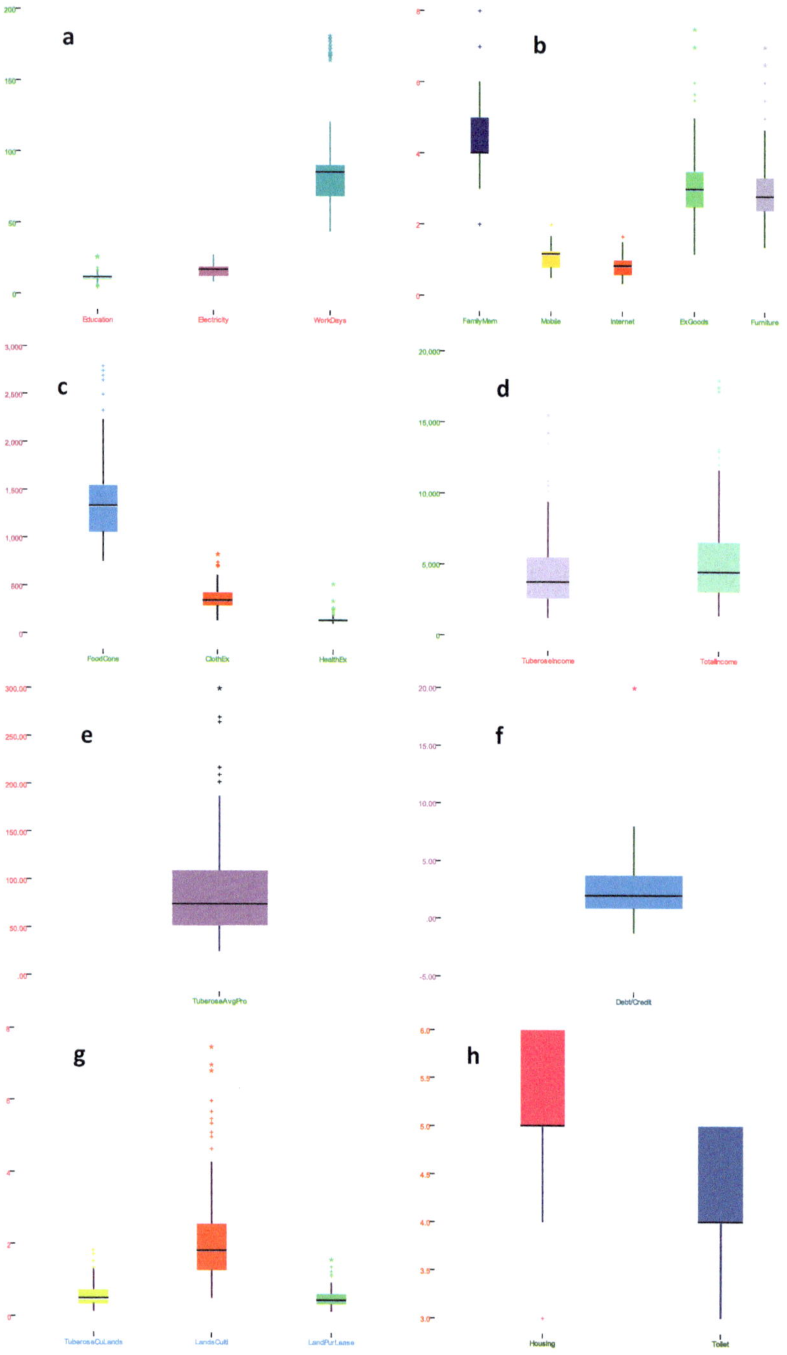

Fig. 3 (**a–h**) Socio-economic-cultural parameters of tuberose cultivation farmers

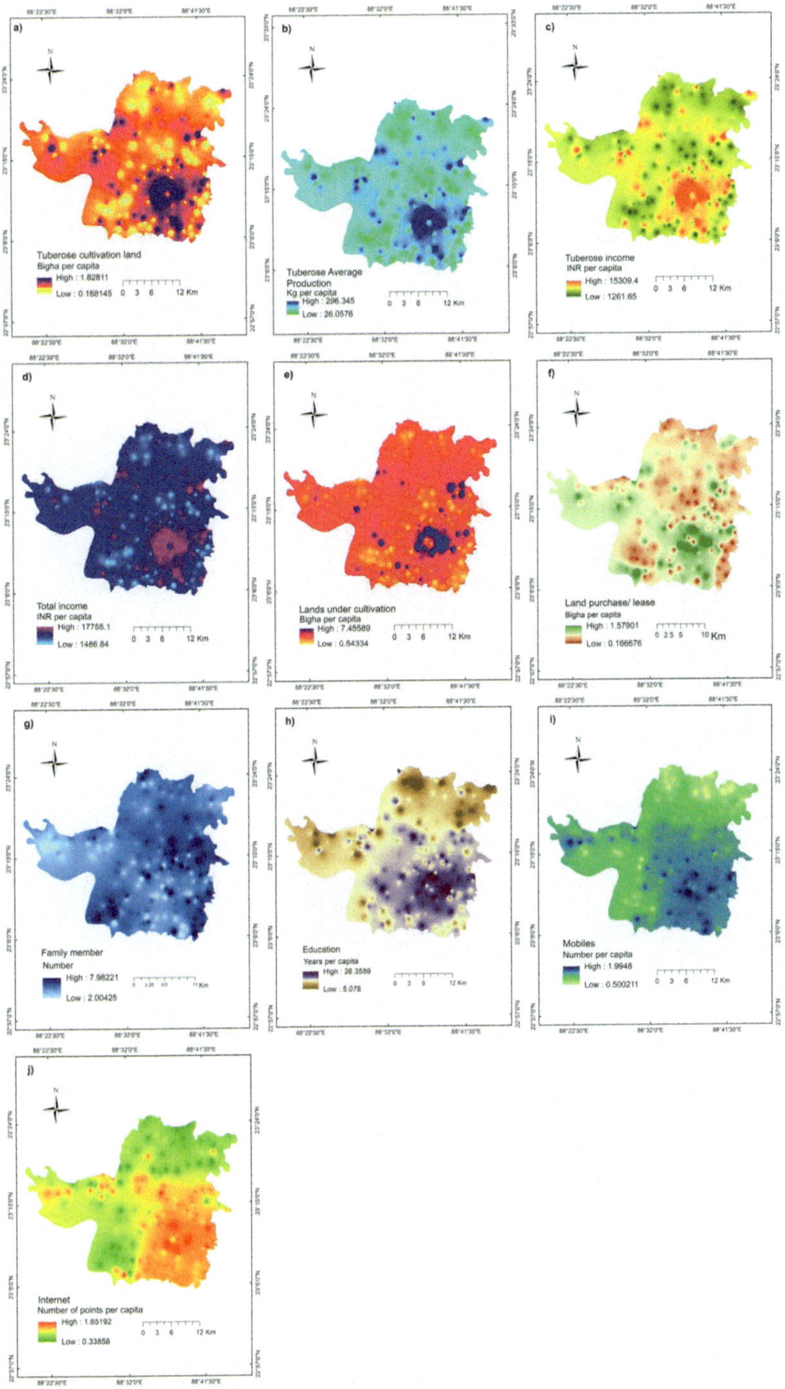

Fig. 4 (**a–j**) Map of socio-economic-cultural parameters for tuberose cultivation farmers

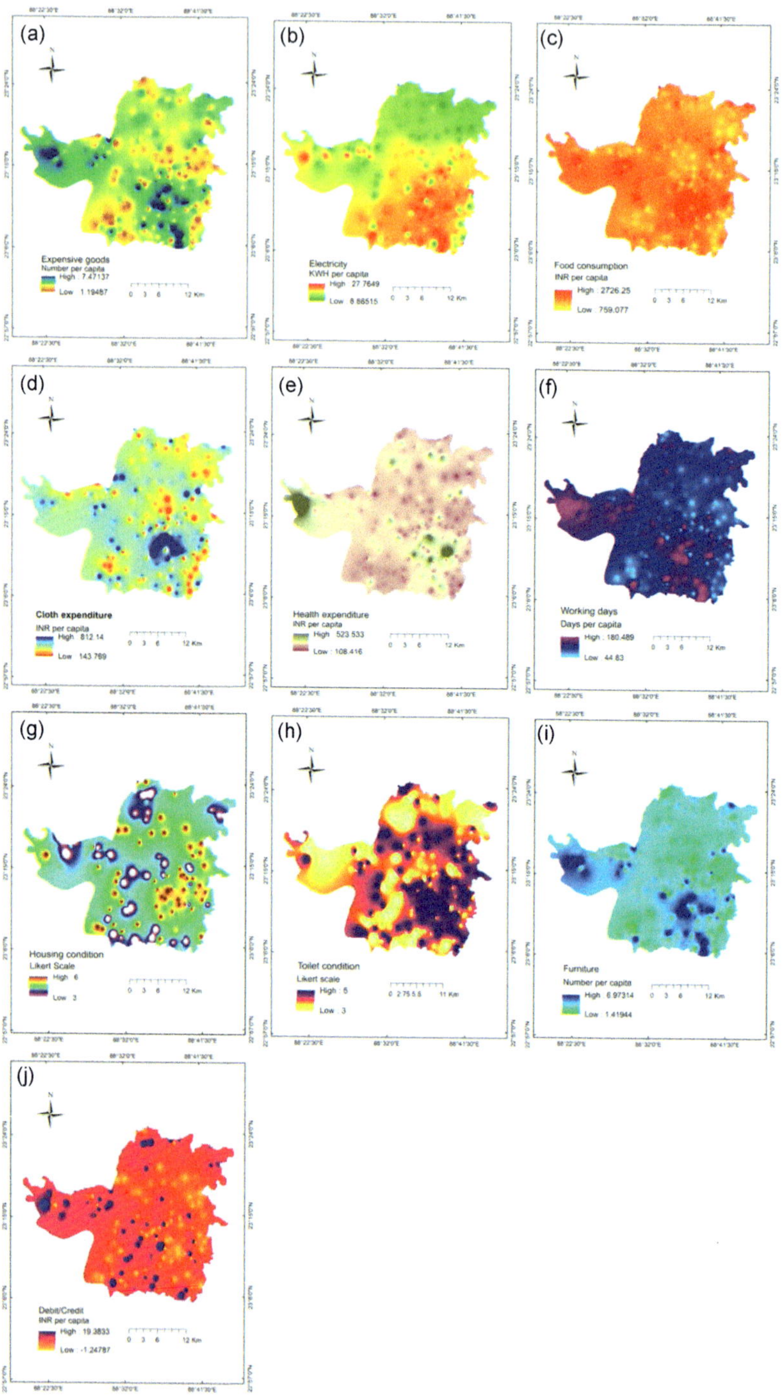

Fig. 5 (**k–t**) Map of socio-economic-cultural parameters for tuberose cultivation farmers

Table 2 Pearson correlation matrix of the socio-economic-cultural parameters of tuberose cultivation

	a	b	c	d	e	f	g	h	i	j	k	l	m	n	o	p	q	r	s
a	1																		
b	**0.989**	1																	
c	**0.973**	**0.979**	1																
d	0.732	0.739	**0.79**	1															
e	0.7	0.68	0.701	**0.76**	1														
f	−0.395	−0.35	−0.345	−0.27	−0.554	1													
g	0.381	0.388	0.396	0.49	0.494	−0.121	1												
h	0.374	0.344	0.317	0.242	0.437	−0.321	0.48	1											
i	0.25	0.226	0.226	0.224	0.351	−0.23	**0.511**	**0.798**	1										
j	0.686	0.619	0.624	0.471	**0.707**	**−0.764**	0.251	0.435	0.335	1									
k	0.442	0.417	0.408	0.377	0.571	−0.413	**0.536**	**0.711**	**0.542**	0.463	1								
l	**0.808**	**0.844**	**0.835**	**0.668**	0.677	−0.467	0.422	0.308	0.275	0.573	0.386	1							
m	**0.842**	**0.866**	**0.857**	**0.698**	0.697	−0.5	0.429	0.321	0.242	0.614	0.425	**0.895**	1						
n	0.252	0.236	0.267	0.291	0.433	−0.463	0.268	0.309	0.274	0.442	0.396	0.358	0.386	1					
o	0.464	0.409	0.398	0.251	0.581	−0.915	0.136	0.451	0.324	**0.84**	**0.5**	0.496	**0.505**	**0.511**	1				
p	0.565	0.583	0.584	0.488	0.227	0.256	0.238	−0.021	0.001	0.093	0.027	0.358	0.401	0.015	−0.23	1			
q	0.619	0.645	0.628	0.455	0.243	0.297	0.236	0.058	0.035	0.058	0.073	0.42	0.409	−0.051	−0.242	**0.802**	1		
r	0.731	0.679	0.67	0.516	**0.715**	**−0.806**	0.229	0.446	0.311	**0.93**	**0.5**	**0.666**	**0.691**	**0.484**	**0.897**	0.086	0.061	1	
s	−0.194	−0.223	−0.231	−0.263	−0.048	−0.507	−0.117	0.092	0.061	0.162	0.137	−0.057	−0.062	0.231	**0.535**	**−0.532**	**−0.579**	0.277	

(a) TuberoseCuLands, (b) TuberoseIncome, (c) TotalIncome, (d) LandsCulti, (e) LandPurLease, (f) FamilyMem, (g) Education, (h) Mobile, (i) Internet, (j) ExGoods, (k) Electricity, (l) FoodCons, (m) ClothEx, (n) HealthEx, (o) WorkDays, (p) Housing, (q) Toilet, (r) Furniture, (s) Debt/Credit and Electricity; (8) Mobile, Internet, and Electricity; (9) Internet and Electricity; (10) ExGoods, WorkDays, and Furniture; (11) Electricity, WorkDays, and Furniture; (12) FoodCons, ClothEx, and Furniture; (13) ClothEx, WorkDays, and Furniture; (14) HealthEx, WorkDays, and Furniture; and (15) WorkDays and Furniture.

The opposite side of positive relation is when one variable value (socio-economic-cultural phenomenon) decreases, then other variable value decreases. The strong negative correlations exist between such variables (1) Housing, Toilet, and Debit/Credit and (2) Toilet and Debit/Credit.

3.3 PCA

The PCA results (Table 3) based on the correlation matrix with Varimax rotation explore three principal components (Fig. 6) with eigenvalues greater than 1.00, which indicates a cumulative variance of 78.380%. The eigenvalues manifest the descending order of significance of tuberose farmers' socio-economic livelihood

Table 3 Principal components for socio-economic factors

	PC1	PC2	PC3
	Economy	Daily Necessities	Technology and Education
Eigenvalues	9.191	3.807	1.894
% of Variance	48.375	20.038	9.967
Cumulative %	48.375	68.413	78.380
Varimax Rotated Component Matrix			
TotalIncome	**0.92**	0.27	0.172
TuberoseIncome	**0.917**	0.272	0.174
TuberoseCuLands	**0.894**	0.32	0.191
Toilet	**0.802**	−0.41	
Housing	**0.779**	−0.371	
ClothEx	**0.778**	0.428	0.191
LandsCulti	**0.775**	0.168	0.247
FoodCons	**0.755**	0.407	0.193
LandPurLease	0.591	0.481	0.386
WorkDays	0.12	**0.944**	0.213
FamilyMem		**−0.93**	−0.118
Furniture	0.459	**0.82**	0.198
ExGoods	0.431	**0.757**	0.219
Debt/Credit	−0.477	0.647	
HealthEx	0.129	0.497	0.311
Internet		0.124	**0.867**
Mobile		0.242	**0.862**
Electricity	0.197	0.337	**0.743**
Education	0.333		**0.719**

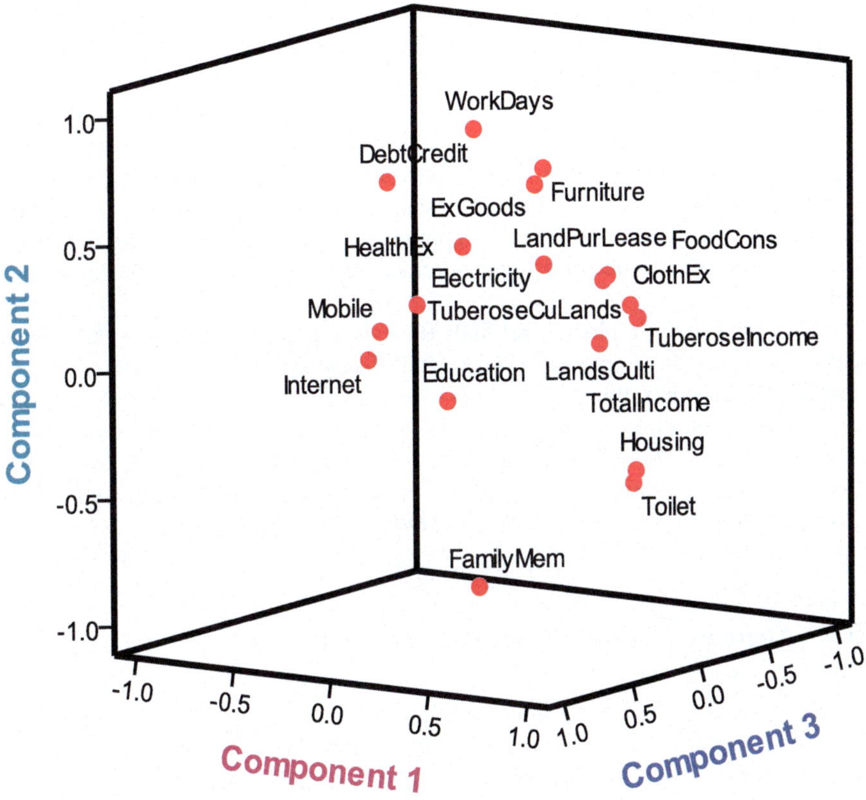

Fig. 6 Principal component plots

variables. The Kaiser-Meyer-Olkin measure of sampling adequacy is 0.883, and Bartlett's test of sphericity (approx. chi-square 6041.261, df 171, sig. 0.000) is significant. Three principal components are coming out from the dimension reduction process. The first one (PC1) correlates with many variables—TotalIncome, TuberoseIncome, TuberoseCuLands, Toilet, Housing, ClothEx, LandsCulti, and FoodCons—having high factor loading. When one of these variables is increasing, the other variable increases. On the opposite side, the low factor loading variable will decrease when the high factor loading variable increases (LandPurLease, WorkDays, FamilyMem, Furniture, ExGoods, Debit/Credit, HealthEx, Internet, Mobile, Electricity, and Education). As the one principal component contains the reduction values of each variable and the high-loading variables are economy-related, we named it 'Economy'. The second principal component (PC2) is called 'Daily Necessities' because it combines the variables' high factor loading related to daily to monthly consumption: WorkDays, FamilyMem, Furniture, and ExGoods. When each high-loading variable increases, it decreases low factor loading variables such as TotalIncome, TuberoseCuLands, and others. The third principal component (PC3) is called 'Technology and Education' as it combines the variables that are mainly related to the land amount: Internet, Mobile, Electricity, and Education. When each high-loading variable increases, it decreases low factor loading variables such as FoodCons, ClothEx, HealthEx, and others. The three components indicate stepwise multiple linear regression because we want to fit the result best.

The tuberose cultivation *Economy* (PC1) is higher in central and some eastern, western, and southern villages than in other areas (Fig. 7a). The *Daily Necessities* (PC2) is higher in central-southern, western, and northern villages (Fig. 7b). *Technology and Education* is most significant in the entire central and southern region, and it spreads outwards than extreme margins of all sides of the study area (Fig. 7c).

A separate study in block level finds Ranaghat-II has three, namely PCs: *Daily Necessities* (PC1), *Economy* (PC2), and *Technology and Education* (PC3). Ranaghat-I also has three PCs *Daily Necessities* (PC1), *Economy* (PC2), and *Technology and Education* (PC3). And at the same time, Hanskhali has PCs *Economy* (PC1), *Daily Necessities* (PC2), and *Education and Other Necessities* (PC3). However, Santipur finds four PCs *Economy* (PC1), *Technology and Others* (PC2), *Shelter* (PC3), and *Education and Others* (PC4).

3.4 Stepwise Multiple Linear Regression

Farmers' socio-economic information is estimated to prove their effectiveness on overall Ranaghat subdivision tuberose cultivation (Table 4). Stepwise multiple linear regression is performed to identify the role of individual components in the socio-economic landscape of the study region. R^2 (0.846) in model 1, R^2 (0.922) in model 2, and R^2 (0.948) in model 3 prove that independent variables can explain above 84% of the variation in model 1, 93% in model 2, and 94% in model 3. The coefficient of the *Economy* (0.920; 0.000), *Daily Necessities* (0.275; 0.000), and *Technology and Education* (0.162; 0.000) ($p < 0.05$) truthfully exhibited the flower yield. The test of null hypotheses of models 1, 2, and 3, F; 1202.200 (0.000), F; 1280.110 (0.000), and F; 1308.998 (0.000) are all significant for tuberose productivity that affects socio-economic-cultural conditions of growers. Correlations, collinearity statistics, collinearity diagnostics, and residual statistics are all in a considerable range. The growth of economic well-being at the farmers' level is the main reason for this cultivation (1–7, 27). Model 3 is the best model where a set of independent variables significantly influence the dependent variable. Every farmer in the region cultivates this flower species solely for the economy, i.e. more income and higher profit, that meets his daily needs and requirements of technology and education. This information has been proven statistically significant by SMLR. The study result has come with a novel outcome as no such study has attempted to put focus on the floricultural study and associated social well-being of cultivators (Fig. 8a). However, to drive deep into the spatial pattern of tuberose cultivation and its footprint on the socio-economic life of the communities, a block-level analysis and discussion is carried out for the study area.

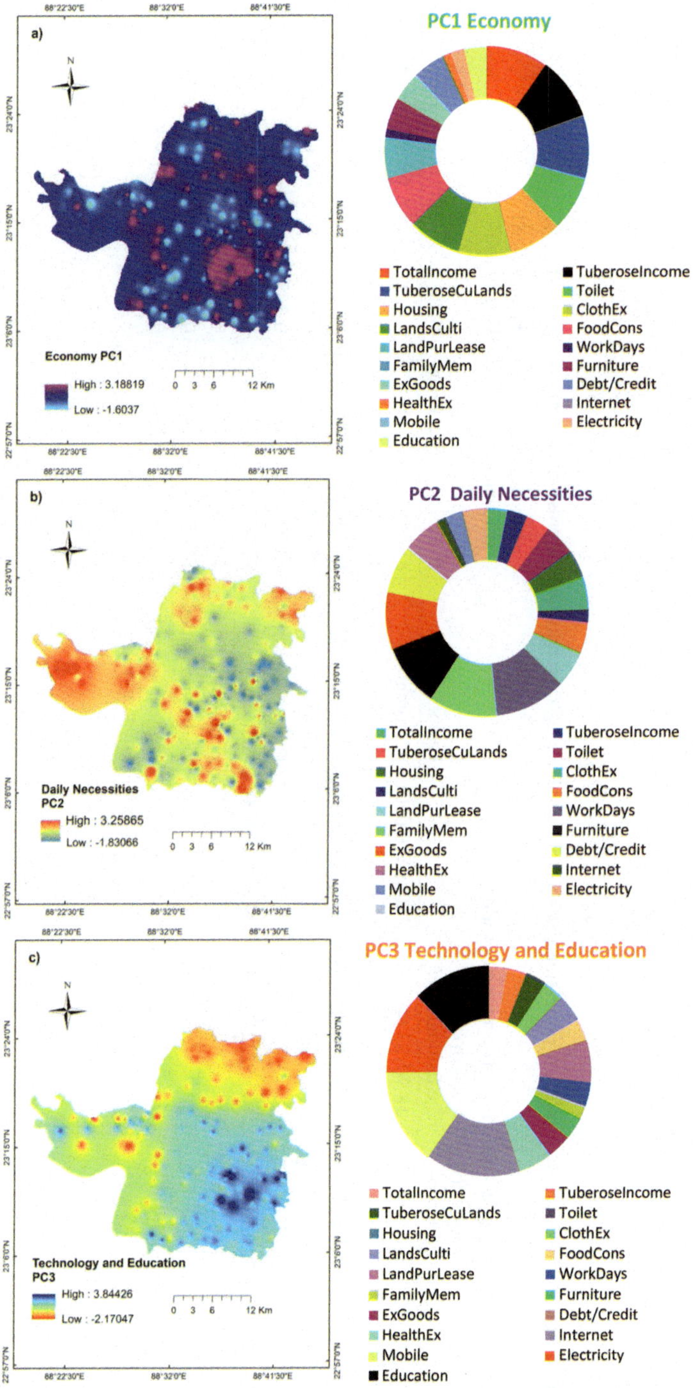

Fig. 7 (**a**–**c**) Principal component analysis spatial distribution of tuberose cultivation

Table 4 Results of the stepwise multiple regression analysis for overall Ranaghat subdivision tuberose cultivation

Model (Overall Ranaghat subdivision Tuberose cultivation)						Unstandardized Coefficients		Standardized Coefficients		
	R Square	Adjusted *R* Square	Std. Error of the Estimate	*F*	Factors	*B*	Std. Error	Beta	*t*	Sig.
1	0.846	0.845	18.70183248	1202.200 ($p = 0.000$)	(Constant)	87.546	1.258		69.591	0.000
					Economy	43.718	1.261	0.920	34.673	0.000
2	0.922	0.921	13.37606364	1280.110 ($p = 0.000$)	(Constant)	87.546	0.900		97.299	0.000
					Economy	43.718	0.902	0.920	48.478	0.000
					Daily Necessities	13.072	0.902	0.275	14.495	0.000
3	0.948	0.947	10.95222174	1308.998 ($p = 0.000$)	(Constant)	87.546	0.737		118.832	0.000
					Economy	43.718	0.738	0.920	59.207	0.000
					Daily Necessities	13.072	0.738	0.275	17.703	0.000
					Technology and Education	7.680	0.738	0.162	10.400	0.000

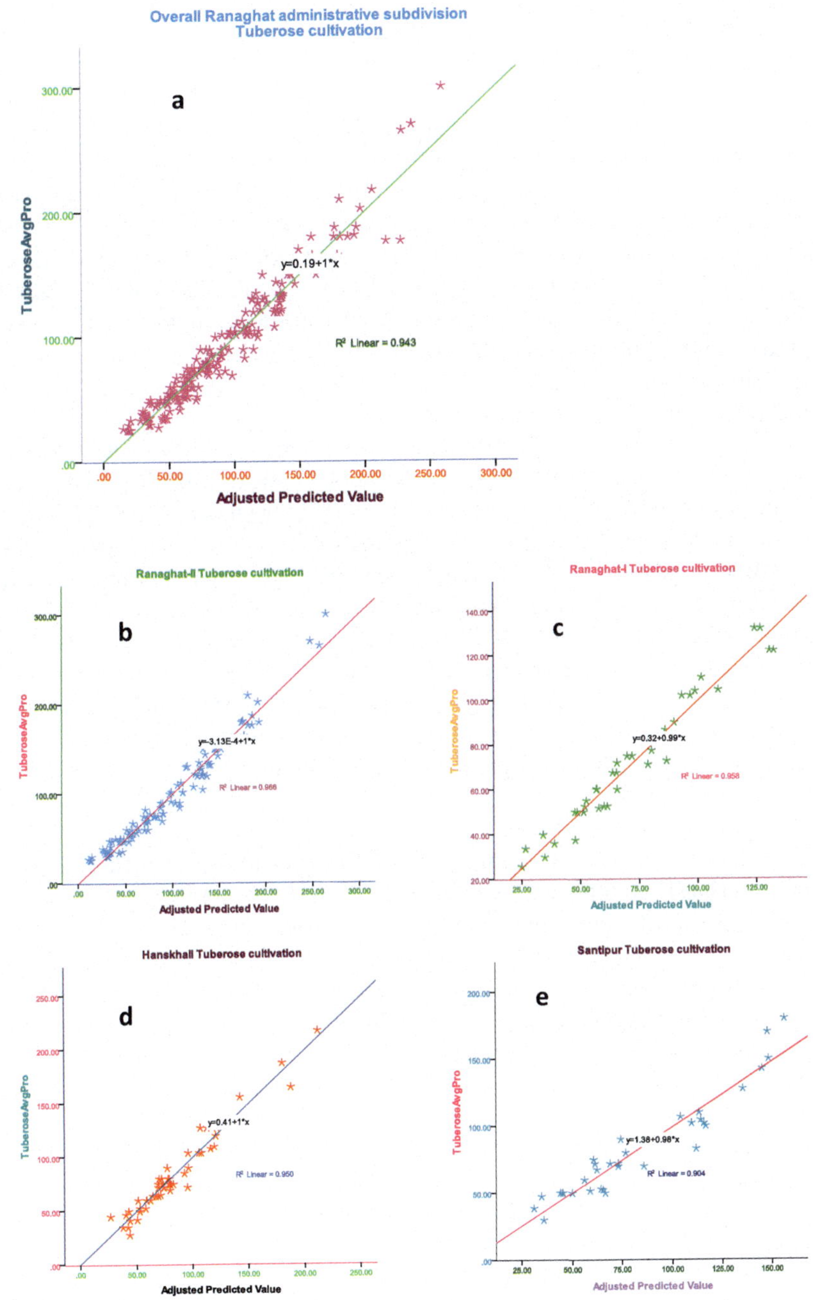

Fig. 8 Scatter plot between estimated average tuberose production using socio-economic-cultural parameters and measured average tuberose production (adjusted values) for (**a**) overall Ranaghat administrative subdivision, (**b**) for Ranaghat-II, (**c**) for Ranaghat-I, (**d**) for Hanskhali, and (**e**) for Santipur

(i) Ranaghat-II tuberose cultivation

Farmers' socio-economic information is gathered and modelled to prove their effectiveness on overall Ranaghat-II's tuberose cultivation. Table 5 shows the result of stepwise multiple linear regression. R^2 (0.507) in model 1, R^2 (0.940) in model 2, and R^2 (0.970) in model 3 show that independent variables can explain above 50% of the variation in model 1, 94% in model 2, and 97% in model 3. The coefficient of the *Economy* (0.712; 0.000), *Daily Necessities* (0.658; 0.000), and *Technology and Education* (0.174; 0.000) ($p < 0.05$) truthfully exhibited the flower production. The test of null hypotheses of models 1, 2, and 3, F; 1202.200 (0.000), F; 1280.110 (0.000), and F; 1308.998 (0.000) are all significant for tuberose productivity that affects socio-economic-cultural conditions of growers (Fig. 8b). Model 3 is the best comparing the three models under consideration.

(ii) Ranaghat-I tuberose cultivation

Farmers' socio-economic information is estimated to prove their effectiveness on overall Ranaghat-I tuberose cultivation (Table 6); SWLR is performed on the area as well. R^2 (0.943) in model 1, R^2 (0.962) in model 2, and R^2 (0.967) in model 3 prove that independent variables can explain above 94% of the variation in model 1, 96% in model 2, and 96% in model 3 as found in the earlier region of the study. The coefficients of the *Economy* (0.971; 0.000), *Daily Necessities* (−0.070; 0.037), and *Technology and Education* (0.136; 0.000) ($p < 0.05$) have explored the flower yield. The test of null hypotheses of models 1, 2, and 3, F; 1202.200 (0.000), F; 1280.110 (0.000), and F; 1308.998 (0.000) are all significant for tuberose productivity that affects socio-economic-cultural conditions of growers (Fig. 8c). In this case, as well model 3 is the best performer.

(iii) Hanskhali tuberose cultivation

Farmers' socio-economic information is measured to prove their effectiveness on Hanskhali tuberose cultivation (Table 7); R^2 (0.944) in model 1 and R^2 (0.955) in model 2 prove that independent variables can explain above 94% of the variation in model 1 and 95% in model 2. The coefficients of the *Economy* (0.972; 0.000) and *Education and Other Necessities* (0.108; 0.001) ($p < 0.05$) have truthfully explained the flower productivity. On the other way, tuberose yield improved those socio-economic issues. The tests of null hypotheses of models 1 and 2, F 844.370 (0.000) and F 532.18 (0.000), are all tuberose productivity-related socio-economic conditions of the growers (Fig. 8d). Correlations, collinearity statistics, collinearity diagnostics, and residual statistics are all in a considerable range. An excluded variable is daily necessities as it is not significant—0.026 (0.398; above p-value)—in the model. Model 2 is the best model for this region.

(iv) Santipur tuberose cultivation

Farmers' socio-economic information is estimated to prove their effectiveness on overall Santipur tuberose cultivation (Table 8). Three different models generate R^2 (0.610) in model 1, R^2 (0.886) in model 2, and R^2 (0.936) in model 3. The models show that independent variables can explain above 61% of the

Table 5 Results of the stepwise multiple regression analysis for Ranaghat-II tuberose cultivation

Model (Ranaghat-II Tuberose cultivation)						Unstandardized Coefficients		Standardized Coefficients		
	R Square	Adjusted R Square	Std. Error of the Estimate	F	Factors	B	Std. Error	Beta	t	Sig.
1	0.507	0.502	40.39673	100.749	(Constant)	96.257	4.040		23.828	0.000
				($p = 0.000$)	Economy	40.752	4.060	0.712	10.037	0.000
2	0.940	0.938	14.20112	755.620	(Constant)	96.257	1.420		67.781	0.000
				($p = 0.000$)	Economy	40.752	1.427	0.712	28.552	0.000
					Daily Necessities	37.654	1.427	0.658	26.382	0.000
3	0.970	0.969	10.06403	1035.410	(Constant)	96.257	1.006		95.645	0.000
				($p = 0.000$)	Economy	40.752	1.011	0.712	40.290	0.000
					Daily Necessities	37.654	1.011	0.658	37.227	0.000
					Technology and Education	9.969	1.011	0.174	9.856	0.000

Table 6 Results of the stepwise multiple regression analysis for Ranaghat-I tuberose cultivation

Model (Ranaghat-I Tuberose cultivation)						Unstandardized Coefficients		Standardized Coefficients		
	R Square	Adjusted *R* Square	Std. Error of the Estimate	*F*	Factors	*B*	Std. Error	Beta	*t*	Sig.
1	0.943	0.942	7.15923	565.798 ($p = 0.000$)	(Constant)	71.756	1.193		60.137	0.000
					Economy	28.785	1.210	0.971	23.787	0.000
2	0.962	0.959	5.96803	415.065 ($p = 0.000$)	(Constant)	71.756	.995		72.141	0.000
					Economy	28.785	1.009	0.971	28.534	0.000
					Technology and Education	4.026	1.009	0.136	3.991	0.000
3	0.967	0.964	5.65574	309.692 ($p = 0.000$)	(Constant)	71.756	0.943		76.124	0.000
					Economy	28.785	0.956	0.971	30.110	0.000
					Technology and Education	4.026	0.956	0.136	4.211	0.000
					Daily Necessities	−2.082	0.956	−0.070	−2.178	0.037

Table 7 Results of the stepwise multiple regression analysis for Hanskhali tuberose cultivation

Model (Hanskhali Tuberose cultivation)						Unstandardized Coefficients		Standardized Coefficients		
	R Square	Adjusted R Square	Std. Error of the Estimate	F	Factors	B	Std. Error	Beta	t	Sig.
1	0.945	0.944	9.38101	844.370 ($p = 0.000$)	(Constant)	84.681	1.314		64.465	0.000
					Economy	38.551	1.327	0.972	29.058	0.000
2	0.957	0.955	8.40702	532.18 ($p = 0.000$)	(Constant)	84.681	1.177		71.933	0.000
					Economy	38.551	1.189	0.972	32.425	0.000
					Education and Other Necessities	4.289	1.189	0.108	3.607	0.001

Excluded Variable is Daily Necessities not significant 0.026 (0.398; above p-value)

Table 8 Results of the stepwise multiple regression analysis for Santipur tuberose cultivation

Model (Santipur Tuberose cultivation)						Unstandardized Coefficients		Standardized Coefficients		
	R Square	Adjusted *R* Square	Std. Error of the Estimate	F	Factors	*B*	Std. Error	Beta	*t*	Sig.
1	0.610	0.598	23.58278	50.128 ($p = 0.000$)	(Constant)	82.944	4.044		20.508	0.000
					Economy	29.066	4.105	0.781	7.080	0.000
2	0.886	0.878	12.97258	120.206 ($p = 0.000$)	(Constant)	82.944	2.225		37.282	0.000
					Economy	29.066	2.258	0.781	12.871	0.000
					Shelter	19.524	2.258	0.525	8.646	0.000
3	0.936	0.930	9.86834	146.341 ($p = 0.000$)	(Constant)	82.944	1.692		49.009	0.000
					Economy	29.066	1.718	0.781	16.920	0.000
					Shelter	19.524	1.718	0.525	11.366	0.000
					Technology and Others	−8.340	1.718	−0.224	−4.855	0.000

Excluded Variable is Education and others not significant 0.082 (0.075; above *p*-value)

variation in model 1, 88% in model 2, and 93% in model 3. The coefficient of the -*Economy* (0.781; 0.000), *Shelter* (0.525; 0.000), and *Technology and Other Necessities* (−0.224; 0.000) ($p < 0.05$) developed the flower production. The test of null hypotheses of models 1, 2, and 3, F; 1202.200 (0.000), F; 1280.110 (0.000), and F; 1308.998 (0.000) are all significant for tuberose productivity that affects socio-economic-cultural conditions of growers (Fig. 8d). The excluded variable is *Education and Others* as it is not significant—0.082 (0.075; above *p*-value). Model 3 is the best model for this region.

From the above study, it is observed that, in Ranaghat-II, 97% of independent variables significantly define the dependent variable (tuberose cultivation) in model 3. It is much higher than other blocks.

3.5 Cluster Analysis

Clustering over PCA outcome (Fig. 9a, b) can quickly identify four patterns in farmers' social and economic environments affected by average tuberose production in the Ranaghat region. Socio-economic variables have been divided into four clusters (hierarchical clustering in the Ward method) based on PC1, PC2, and PC3 (Fig. 9c, d). Here, out of four clusters, cluster 1 (51 observations) and cluster 2 (97 observations) are forming more prominent groups, compared with cluster 3 (54 observations) and cluster 4 (19 observations) are making smaller groups. An essential aspect of the socio-economic aspect is to estimate which group is improved in terms of socio-economic conditions and tuberose yield than other groups. However, the hierarchical cluster analysis has shown the cluster distribution vividly, but group dominance is challenged in the hierarchical cluster analysis.

Apart from the cluster analysis, K-means cluster analysis is also performed to identify the group dominance among different groups (Fig. 9e, f). The clustering based on K-means algorithms shows that cluster 1 forms 107 farmers, cluster 2 forms 13 farmers, cluster 3 forms 49 farmers, and cluster 4 makes 52 farmers. Economy (F 218.122; 0.000), Economy (F 182.167; 0.000), Daily Necessities (F 109.393; 0.000), Technology and Education (F 6.702; 0.000), and TuberoseAvgPro (F 218.122; 0.000) all are significant among clusters. Cluster 1 provides high variables—TuberoseAvgPro (−0.53290), Economy (−0.37210), Daily Necessities (−0.52671), and Technology and Education (−0.29085)—in order of cluster centres. Cluster 2 has two features namely TuberoseAvgPro (2.48579), Economy (1.89419), Daily Necessities (1.51774), and Technology and Education (0.53602). Cluster 3 has variables like TuberoseAvgPro (-0.468), Economy (-0.841), Daily Necessities (1.153), and Technology and Education (0.20801). Cluster 4 has only one variable that is low: TuberoseAvgPro (0.916), Economy (1.085), Daily Necessities (−0.382), and Technology and Education (0.268). This method of clustering explores that cluster 2 is the best group of the other three groups. The main goal of tuberose farmers is to maintain the family's well-being by improving yield in tuberose farming.

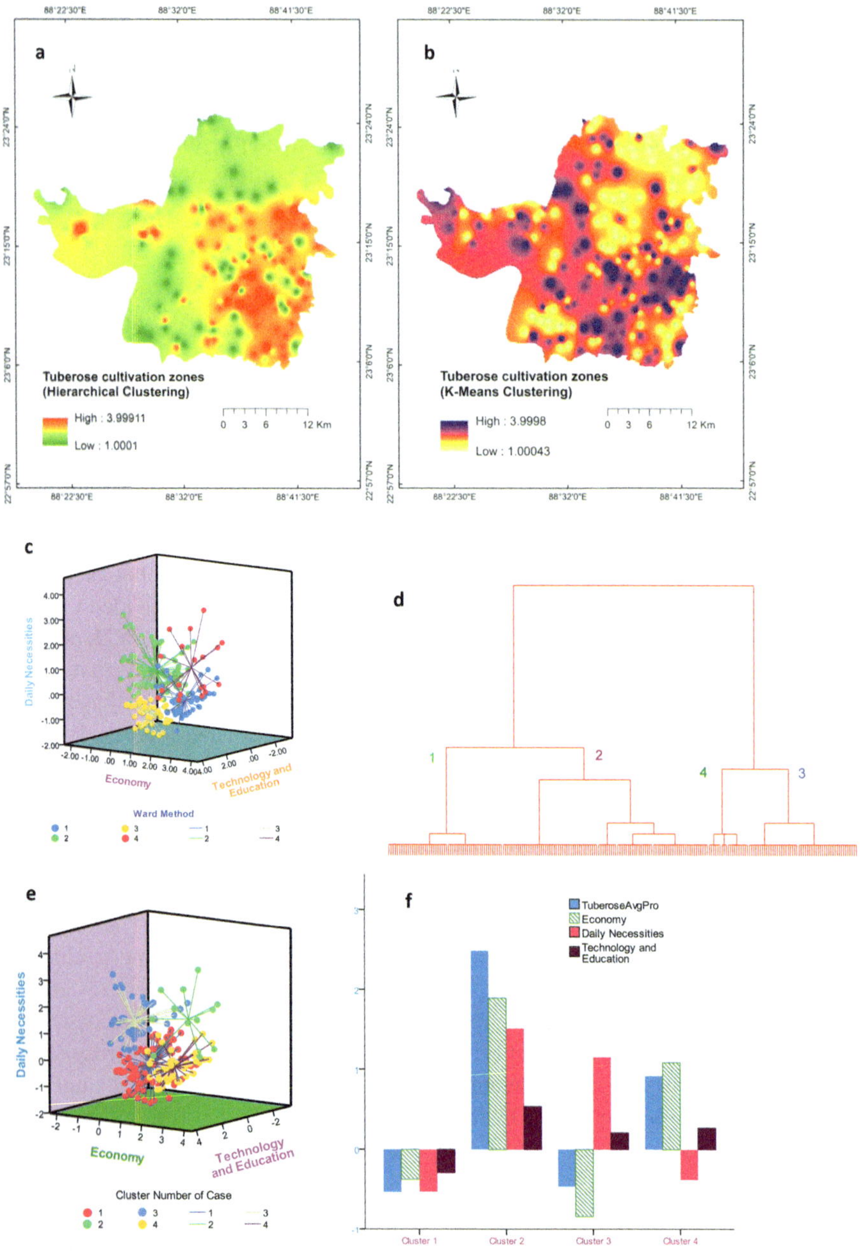

Fig. 9 Tuberose cultivation zones of (**a**), Hierarchical cluster analysis (**b**), K-means cluster analysis (**c**), 3D scatter plot of 4 individual clusters of Hierarchical cluster analysis, (**d**) Dendrogram profiles of tuberose farmers' socio-economic-cultural livings, (**e**) 3D scatter plot of 4 individual clusters of K-means cluster analysis, and (**f**) K-means cluster bar indicate socio-economic conditions and marigold production

3.6 *Tuberose Cultivation Zones*

Tuberose cultivation has spread across various administrative blocks in the entire Ranaghat administrative subdivision (Fig. 9a, b). The high floricultural zone (2–4) exists from central to southwards in the study area: (1) the whole Ranaghat-II (except a few regions of the north), (2) the northeast and south part of Ranaghat-I (except western places), (3) the southwestern part of Hanskhali, and (4) the east to the west part of Santipur (except southern areas). The causes for the improvement of tuberose cultivation in the mentioned areas (Fig. 10a, b) are developed transport systems from agricultural land to markets with low-cost vehicles, availability of fertilizer in the markets, the nearness of markets (Fig. 10a, b), home of numerous national and international flower traders, lack of alternative jobs economically, and massive flower demand and opportunity for higher income in the market every day.

Except for these areas, the tuberose cultivation of other zones (1 and very few villages) of the Ranaghat administrative subdivision is less developed because these places are low-lying areas; excess water accumulates from rivers. Due to the distance of agricultural land from the markets, farmers sell their flowers to intermediaries at low prices. As a result, growers do not have much income from tuberose. There is a shortage of other facilities (fertilizers and others) for this ornamental crop in these areas. Biswas, in 2013, discusses Nadia of West Bengal floricultural zones but does not explain flowers.

4 Discussion

Floriculture claims a prominent place in this region of West Bengal (Biswas, 2013; Biswas & Sarkar, 2021; Ghosh & Gupta, 2019). One of these flower cultivations is tuberose cultivation. It changes and enhances the level of human livelihood (Gelaye, 2022; Bharathi et al., 2023). Most senior citizens in the current area describe the lifestyle before and after tuberose cultivation have heavily changed (Biswas & Sarkar, 2022). The cultivation of flowers triggers economic development, which ensures food security and changes the agricultural establishment from traditional low-income crops to high-profit flower cultivars (Das, 2017; Harisha, 2017). However, no prominent previous studies discuss the socio-economic rise affected by tuberose cultivation. However, the results of this study reveal that tuberose cultivation is practised in most of the villages except in some lowlands. It is observed that tuberose is produced at any time of the year. Farmers have grown it every part of their small to large farmland and get higher per capita flower production. The study expressed precisely that the cultivation of the flowers has not flourished equally. Economic development has improved in those areas where the area under flowering has risen significantly. By farming tuberose, people accelerate their daily income level and expansion of agricultural land holdings on the one side; on the other side, they earn their daily household necessities such as food, textiles, shelter,

Fig. 10 (**a**, **b**) Tuberose landscapes

education, and other regular expenses. At the same, they reduce the debt load on their shoulders.

Overall, the three pillars of socio-economic conditions—Economy, Daily Necessities, and Technology and Education—are regularly improving all over the Ranaghat subdivision. Micro-level studies on the block level reveal that the farmer's well-being of Ranaghat-II, Ranaghat-I, and Hanskhali has improved by producing tuberose. This cultivation has yielded meeting of their daily necessaries, educational and technological expenses etc. However, Santipur's farmers have grown this flower to develop economy, technology and others, shelter, and education and others. They have put stress more on shelter facilities than on other blocks. Tuberose has been cultivated across various administrative blocks in the complete Ranaghat administrative subdivision. The high floricultural zones exist from central to southwards in the study area: the whole Ranaghat-II (except a few regions of the north), the north-east and south part of Ranaghat-I (except western places), the southwestern part of Hanskhali, and the east to the west part of Santipur (except southern areas). The underlying factors behind the improvement of tuberose cultivation in the mentioned areas are developed transport facilities (NH34, railways, and so on) with low-cost vehicles, availability of fertilizer in the markets, the nearness of numerous markets, places of many national and international flower traders, lack of alternative jobs economically, massive flower demand, and opportunity for higher income in the market every day. The tuberose floricultural system is socio-physical planted, dominated, and spread by farmers on the land surface for the benefit of economic enrichment. This particular viewpoint with mathematical analysis has made this study worthy to be included in the social physics domain (Tsekov, 2023; Breur, 2016).

5 Conclusion

One example of an alternative farming system to traditional agriculture is tuberose cultivation (Hasan, 1996; Maji, 2015). This farming is the primary source of income for rural farmers. Its role in the development of various aspects of farmers' lives is unique (Mou, 2012). The contribution of this paper is the impact of tuberose cultivation on socio-economic prosperity in farmers' households (Haque et al., 2012; Mandal et al., 2018). This model is valid by effectively using backward elimination of uncorrelated socio-economic parameters and thus reduces the dispersion of observations to be solved. It has been found that the model developed better relationships than did by any other literature. This paper offers a new statistical application of tuberose production formulation, and it is hoped that it will stimulate further research on numerical solutions for missing socio-economic-cultural problems. This cultivation can be the primary tool of poverty alleviation in the world in the coming days, considering the contribution of this farming to the overall development of a farming family. Flower farming serves the supply of work continuously every day, which accelerates the cycle of economic growth. It fills the life of farmers with food, clothing, shelter, pleasure, and recreation while preserving soil and sustainability using crop rotation (Lichtfouse et al., 2009). The farming of tuberose can be cited as an example of social learning, a significant component of contemporary social physics by modifying farmers' role with the study of natural to social features.

Acknowledgement The authors thank the Department of Geography of Adamas University, India, for valuable suggestions. The authors also thank Bithika Biswas Dafadar for her excellent assistance during the whole work process.

Declarations Funding

None

Conflicts of Interest/Competing Interests

We declare that we have no conflict of interest.

Ethical Approval

The authors completed all fieldwork and all procedures of this research. This article does not contain any studies with human participants or animals like biological applications performed by any authors.

Ethical Standards

The research did not breach any ethical standards.

Data Availability

The data that support the findings of this study are available upon reasonable request from the corresponding author, Biswas, D. The data are not publicly available due to them containing information that could compromise research participant privacy/consent and confidentiality of financial information.

References

Abdelhafidi, N., Bachari, N. E. I., & Abdelhafidi, Z. (2020). Estimation of solar radiation using stepwise multiple linear regression with principal component analysis in Algeria. *Meteorology and Atmospheric Physics*. https://doi.org/10.1007/s00703-020-00739-0

Bera, J., & Majumder, S. (2013). Economics of tuberose cultivation and its marketing – A case study in Purba Medinipur District of West Bengal, Vidyasagar University. *Journal of Economics, XVII*, 161–168. ISSN – 0975-8003.

Bharathi, T. U., Kumar, R., Nair, S. A., Umamaheswari, R., Sonavane, P., Kalaivanan, D., & Rao, V. K. (2023). Evaluation of tuberose genotype IIHR 17-23SP-08 (IC0642158) for flower yield, quality and response to biotic stress. *Journal of Horticultural Sciences, 18*(1), 60–66. https://doi.org/10.24154/jhs.v18i1.2148

Biswas, N. K. (2013). Floriculture concentration zone of Nadia District, West Bengal. *I.O.S.R. Journal of Humanities and Social Science (IOSR-JHSS), 7*(1), 17–23.

Biswas, D., & Sarkar, R. (2021). Impact of soil chemical properties on marigold production in Ranaghat-II, Nadia, India. *Spatial Information Research, 29*, 207–221. https://doi.org/10.1007/s41324-020-00347-5

Biswas, D., & Sarkar, R. (2022). Rise of marigold floriculture, a new stirring door walk through economic, social, and entertainment factors in Eastern India: A combined approach of multi-group structural equation modeling and cluster analyses. *Quality and Quantity*. https://doi.org/10.1007/s11135-022-01347-3

Breur, T. (2016). Statistical power analysis and the contemporary "crisis" in social sciences. *Journal of Marketing Analytics, 4*(2–3), 61–65. https://doi.org/10.1057/s41270-016-0001-3. ISSN 2050-3318. Palgrave Macmillan, London, England.

Cohen, P., Cohen, P., West, S. G., & Aiken, L. S. (1983). *Applied multiple regression/correlation analysis for the behavioral sciences* (2nd ed.). Psychology Press. https://doi.org/10.4324/9781410606266

Das, A. (2017). An anthropological study on the socio-economic life of the tuberose cultivators. *Proshikhyan, Bangladesh Society for Training and Development (BSTD), Bangladesh, 25*(02), 1–18.

Ding, X., Lu, Q., Li, L., Sarkar, A., & Li, H. (2022). Evaluating the impact of institutional performance and government trust on farmers' subjective well-being: A case of urban–rural welfare gap perception and family economic status in Shaanxi, Sichuan and Anhui, China. *International Journal of Environmental Research and Public Health, 20*(1), 710. https://doi.org/10.3390/ijerph20010710

Gawle, S. K., Chaturvedi, M. K., & Yadaw, K. N. (2012). Analysis of socio-economic profile of marigold grower in adoption of improved marigold production technology. *Journal of Plant Development Sciences, 4*(2), 225–230.

Gelaye, Y. (2022). The status and natural impact of floriculture production in Ethiopia: A systematic review. *Environmental Science and Pollution Research, 30*(4), 9066. https://doi.org/10.1007/s11356-022-24279-9

Ghosh, P. S., & Gupta, K. (2019). Impact of socio-economic factor on floriculture in Kolaghat C. D. Block, Purba Medinipur District. *International Journal of Innovative Studies in Sociology and Humanities (IJISSH), 4*(3), 61–69.

Hair, J., Black, W., Babin, B., & Anderson, R. (2010). *Multivariate data analysis* (7th ed.). Prentice-Hall, Inc..

Haque, M. A., Monayem Miah, M. A., Hossain, S., & Sharifuzzaman, S. M. (2012). Agro-economic analysis of tuberose cultivation in selected areas of Bangladesh. *Bangladesh Journal of Agricultural Research, 37*(3), 457–464.

Harisha, B. N. (2017). An economic analysis of floriculture in India. *International Journal of Academic Research and Development, 2*(6), 225–231.

Hasan, M. A. (1996). *Study on production and marketing of tuberose in some selected areas of Bangladesh* [M.S. Thesis]. Department of Cooperation and Marketing, Bangladesh Agricultural University, Mymenshing.

Isaac, O. E., & Chikweru, A. E. (2018). Test for significance of Pearson's correlation coefficient (r). *International Journal of Innovative Mathematics, Statistics & Energy Policies, 6*(1), 11–23.

Jake, L., Martin, Z., & Naomi, A. (2017). Points of significance: Principal component analysis. *Nature Methods, 14*. https://doi.org/10.1038/nmeth.4346

Kriegel, H. P., Schubert, E., & Zimek, A. (2016). The (black) art of runtime evaluation: Are we comparing algorithms or implementations? *Knowledge and Information Systems, 52*(2), 341–378. https://doi.org/10.1007/s10115-016-1004-2. ISSN 0219-1377. S2CID 40772241.

Lichtfouse, E., Navarrete, M., Debaeke, P., Souchere, V., Alberola, C., & Menassieu, J. (2009). Agronomy for sustainable agriculture. A review. *Agronomy for Sustainable Development, 29*, 1–6. https://doi.org/10.1051/agro:2008054

Maji, S. (2015). Changing scenario of present floriculture system & associate problems in Bagnan Blocks I and II, Howrah District. *I.O.S.R. Journal of Humanities and Social Science, 20*(11), 07–12. https://doi.org/10.9790/0837-201130712

Mandal, M., Maitra, S., & Mahata, D. (2018). Production technology of tuberose (*Polianthes tuberosa L.*) cultivation. *Journal of Pharmacognosy and Phytochemistry, 7*(6), 2360–2364.

McIntosh, A. M., Sharpe, M., & Lawrie, S. M. (2010). *Companion to psychiatric studies* (8th ed., pp. 157–198). Research methods, statistics and evidence based practice. Elsevier.

Mitra, S., Datta, A., Rashid, A. D., & Khatun, N. (2024). Weather index-based agricultural insurance for flower farmers: Willingness to pay, sales, and profitability perspectives. *Open Agriculture, 9*(1). https://doi.org/10.1515/opag-2022-0315

Mou, N. H. (2012). Profitability of flower production and marketing system of Bangladesh. *Bangladesh Journal of Agricultural Research, 37*(1), 77–95. https://doi.org/10.3329/bjar.v37i1.11179

Pal, A., Adhikary, R., Bera, M., Garnayak, R., & Roy, R. (2019). Growing of tuberose with different irrigation treatment under alluvial soil of West Bengal. *Journal of Pharmacognosy and Phytochemistry, 8*(3), 4534–4538.

Sahu, A., Chatterjee, S., & Paul, A. K. (2011). Growth of floriculture and its role in promoting socio-economic status of floriculturists of Panskura and Kolaghat Blocks, Purba Medinipur. *Indian Journal of Geography & Environment, 12*, 01–08.

Sarker, J. R., & Salam, M. A. (2018). Economics and variety adoption of tuberose: A farm level survey in Bangladesh. *African Journal of Agricultural Research, 13*(45), 2579–2586. https://doi.org/10.5897/AJAR2018.13497

Setu, M. W. A., Islam, I. B., Ali, M. S., Sujan, M. H. K., & Islam, H. M. S. (2018). Impact of flower cultivation on farmers livelihood in some selected area of Bangladesh. *Review of Agricultural and Applied Economics, 21*(2), 10.15414/raae.2018.21.02.03-11.

Swapna, B. (2018). Cultivation and marketing of tuberose (lily) flowers by the rural poor: An approach of swot analysis. *International Journal of Agriculture, Environment and Biotechnology, 11*(1), 137–140. https://doi.org/10.30954/0974-1712.2018.00178.18

Tsekov, R. (2023). *Social thermodynamics 2.0.* arXiv:2307.05984.

Usman, M., & Ashfaq, M. (2013). Economics analysis of tuberose production in Punjab, Pakistan. *Sarhad Journal of Agriculture, 29*(2), *147*(1–4), 169–195. https://doi.org/10.1016/0022-1694(93)90080-S. ISSN 0022-1694.

Usman, M., Ali, A., & Hassan, M. (2015). An efficiency analysis of tuberose cut-flower: A case study. *Journal of Animal and Plant Sciences, 25*(6), 1699.

Weisberg, S. (2005). *Applied linear regression* (Vol. 528). Wiley.

Dr. Dipankar Biswas, a former research scholar in the Department of Geography at Adamas University, is recognized for his academic dedication and contributions to geographical research. With a Ph.D. in Geography from Adamas University, Dr. Biswas specialized in environmental geography, focusing on the impacts of agricultural led socio-environmental change. During his research tenure, Dr. Biswas worked on interdisciplinary projects that addressed pressing environmental challenges, particularly in the areas of floricultural practice rural development and natural resource conservation. His work provided valuable insights into the socioeconomic implications of geographical phenomena, bridging the gap between theoretical frameworks and practical solutions. He has published research papers in prominent academic journals, furthering the discourse on environmental sustainability and development policies. In addition to his research, Dr. Biswas was actively involved in academic seminars and conferences, where he presented his findings to the broader academic community. His scholarly achievements and analytical approach have earned him recognition within the geography research community. As an academic, Dr. Biswas is passionate about contributing to meaningful solutions for environmental issues and continues to engage with ongoing research, aiming to create a positive impact on both local and global scales.

Dr. Rajib Sarkar is an esteemed academic in the Department of Geography at Adamas University. With a Ph.D. in Geography, Dr. Sarkar's research interests span a wide range of subjects including geomorphology, environmental studies, urban geography, and climate change. His work frequently explores the intersection of geographical phenomena and societal impacts, significantly contributing to sustainable development and disaster management research. Dr. Sarkar has authored numerous publications in reputed national and international journals, showcasing his in-depth knowledge and innovative approaches to contemporary geographical issues. He is actively engaged in field research and has led several interdisciplinary projects, which have garnered attention for their practical applications in policy formulation and regional development planning. At Adamas University, Dr. Sarkar is a dedicated educator deeply committed to fostering academic excellence among his students. His teaching methods emphasize critical thinking and real-world applications, encouraging students to develop a comprehensive understanding of geographical concepts. He is also an active participant in academic conferences and workshops, contributing to the growth of geography as a discipline. Dr. Sarkar's academic leadership and research continue to inspire and shape future generations of geographers.

Part IV
Dynamics of Social Behavior in Spatial Contexts

Spatial Heterogeneity and Pandemic Dynamics: Geospatial Insights from Kolkata During Post-Lockdown COVID-19 Spread and Management

Sanjana De and Anu Rai

1 Introduction

The COVID-19 pandemic, with its far-reaching impacts, underscored the necessity of comprehending the intricate relationships between social dynamics and public health. This relationship is particularly pertinent in urban settings, where dense populations and diverse socio-economic conditions create a complex landscape for disease spread and management. Integrating social physics with geospatial analysis may offer a robust framework for understanding these complexities and devising effective public health interventions during and after a crisis.

Social physics, a concept that applies mathematical tools and theories from physics to study human social behavior, provides a unique lens through which public health issues can be analyzed. This approach considers how social interactions, networks, and behaviors influence the spread of diseases (Jusup et al., 2022). In the context of COVID-19, social physics can be instrumental in analyzing how patterns of movement, social distancing practices, and other behavioral factors contributed to the virus's transmission. Studies have shown that variations in these behaviors—often influenced by economic, cultural, and social factors—play a significant role in determining the rate and extent of disease spread (Pentland, 2014; Kawachi & Ransome, 2024). For example, adherence to social distancing measures has been observed to be uneven across different socio-economic groups, with marginalized communities often facing greater challenges in complying due to their living conditions and economic constraints (Kraemer et al., 2020). Understanding these dynamics through the lens of social physics allows for the development of targeted public

S. De · A. Rai (✉)
Department of Geography, School of Basic and Applied Sciences, Adamas University, Kolkata, West Bengal, India
e-mail: anu.rai@adamasuniversity.ac.in

A. Rai et al. (eds.), *Contemporary Social Physics*, Springer Geography,
https://doi.org/10.1007/978-3-031-77596-3_14

health interventions that address specific behaviors and vulnerabilities within different communities.

Social dynamics, particularly in urban environments, are a critical factor in the spread of infectious diseases. The COVID-19 pandemic provided a stark illustration of how interconnected human behaviors and social structures contribute to disease transmission (Rai et al., 2024). Factors such as population density, mobility patterns, and the nature of social interactions within neighborhoods were found to significantly influence the speed and breadth of the virus's spread (Liu et al., 2021). In densely populated cities of Asia like Kolkata, where social interactions are frequent and often unavoidable, the virus spread more rapidly compared to less dense regions. Furthermore, the pandemic highlighted the impact of social inequalities on public health outcomes. Disparities in access to healthcare, economic resources, and living conditions exacerbated the pandemic's effects on vulnerable populations, leading to higher infection and mortality rates in marginalized communities (Patel et al., 2020). The spatial distribution of these social factors within a city can significantly affect the outcomes of public health interventions, making it crucial to consider social dynamics in the planning and implementation of these strategies.

Geospatial analysis provides a powerful tool for understanding the spatial dimensions of disease spread and the effectiveness of public health interventions. By mapping the distribution of COVID-19 cases and overlaying this data with socio-economic and demographic information, researchers and policymakers can gain insights into the factors driving the spread of the virus and identify areas that require targeted interventions (Rendana et al., 2021). Neighborhood-level geospatial analysis is particularly valuable in this regard, as it allows for a detailed understanding of how different areas within a city are affected by the pandemic and respond to public health measures. In Kolkata, this approach was essential for identifying high-risk areas, assessing the effectiveness of lockdowns and other containment measures, and ensuring that resources were allocated efficiently.

The use of geospatial data in public health management during the COVID-19 pandemic in Kolkata provided several critical insights. For instance, mapping the spatial distribution of cases revealed significant variations in infection rates across different neighborhoods, which were often correlated with socio-economic factors such as population density and income levels (Jalali & Etemadfard, 2024). This information was crucial for directing resources to areas with the greatest need, such as neighborhoods with high infection rates and limited access to healthcare facilities (Liu et al., 2021). Additionally, geospatial analysis allowed for the evaluation of the effectiveness of public health interventions by tracking changes in infection rates over time and across different areas. For example, studies have shown that areas with stricter enforcement of lockdown measures and higher compliance with social distancing guidelines experienced slower rates of infection (Hsiang et al., 2020). These findings underscore the importance of incorporating geospatial data into public health planning and decision-making, particularly in complex urban environments.

By considering the social dynamics that influence disease spread and using geospatial tools to map and analyze these factors, policymakers can develop more effective strategies for controlling the spread of diseases and mitigating their impact on vulnerable populations. The experience of Kolkata during the post-lockdown phase of the COVID-19 pandemic illustrates the value of this approach, highlighting the need for targeted, data-driven interventions that address the specific challenges faced by different communities within a city.

2 Aim and Objective of the Study

2.1 *Research Aim*

This study will explore how human interactions and societal structures impact the spread of infectious diseases, drawing insights from social physics. The aim is to propose a comprehensive action plan for managing pandemic scenarios, with a focus on the social and behavioral aspects of pandemic response, and adopting a case study approach that analyzes the spatio-distribution of post-lockdown containment zones within the Kolkata Municipal Corporation area. The following objectives set to achieve the aim of the study:

2.2 *Objectives*

- To conduct a spatio-temporal analysis of COVID-19 within the Kolkata Municipal Corporation area, identify key hotspots and post-lockdown containment zones, and understand their relation to social and behavioral factors
- To estimate the direction and magnitude of COVID-19 spread in the study area beyond the 2020 lockdown, with a focus on how social interactions and community structures influenced the virus's propagation
- To map the perceptions, needs, and concerns of the local population through geospatial perception mapping, integrating these insights into an action plan that addresses the social and behavioral dynamics of pandemic management

3 Study Area

Kolkata, often referred to as the "City of Joy," is the capital of West Bengal and a major urban center in Eastern India. Located on the eastern bank of the Hooghly River, approximately 75 kilometers west of the Bangladesh border, Kolkata covers an area of 206.08 square kilometers (Fig. 1). The city serves as the primary

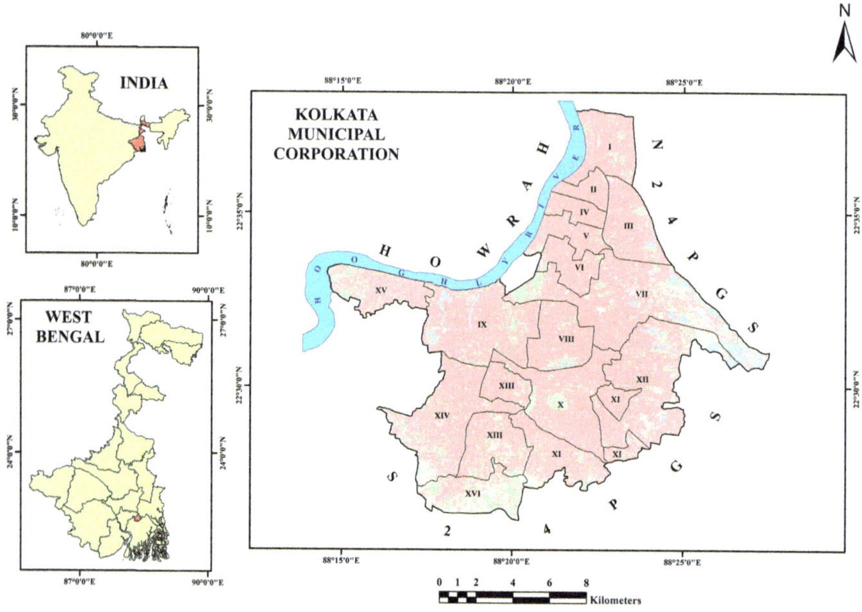

Fig. 1 Study area

business, commercial, and financial hub of Eastern India and is a critical port for communication with Northeast India. The Kolkata Municipal Corporation (KMC) administers the core of the city, covering 187.33 square kilometers. The city is divided into 144 wards, grouped into 16 boroughs, with the "added area" (boroughs 11–16) being incorporated into KMC in 1985. This expansion included South Suburban Municipality, Garden Reach Municipality, and Jadavpur, further complicating the city's administrative and social structure. Kolkata's geographical coordinates range from 22°28′00″ to 22°37′30″ N latitude and 88°17′30″ to 88°25′00″ E longitude. According to the 2011 Census, the city's population was approximately 4.5 million, with a high population density of 24,252 per square kilometer, which presents significant challenges in managing public health crises.

The city's dense population, coupled with its socio-economic diversity, makes it an ideal case for analyzing the spread and management of infectious diseases such as COVID-19. The varied social fabric, which includes both densely populated slums and affluent neighborhoods, offers a representative setting for studying how social interactions and human behavior influence the spread of pandemics. Moreover, Kolkata's role as a historical trading hub and its strategic importance as a gateway to Northeast India (Rai, 2019) further highlight the relevance of this study in understanding the broader implications of social dynamics on public health.

Kolkata's administrative framework, with its division into wards and boroughs, provides a structured basis for conducting detailed spatio-temporal analyses. This structure allows for targeted public health interventions, which are crucial in managing pandemics in urban environments. Previous studies have emphasized the importance of understanding local social dynamics and using geospatial data to identify and address high-risk areas during pandemics (Saeed et al., 2024).

By focusing on Kolkata, this study aims to contribute to the broader discourse on pandemic management in densely populated urban areas. The insights gained from this case study can inform future public health strategies, particularly in understanding the interplay between social dynamics and disease spread in similar urban settings.

4 Methodology

4.1 Spatial Dynamics of COVID-19 in Kolkata

To effectively map the spatial distribution of public health crises in Kolkata during the COVID-19 pandemic, this study utilized a combination of geospatial techniques, owing to the lack of readily available borough-wise COVID-19 data within the Kolkata Municipal Corporation (KMC). Instead, month-wise data on containment zones, spanning from May 2020 to January 2021, was sourced from the West Bengal Health Portal. This data, initially in tabular form, was geocoded into point data using zip codes with the help of BatchGeo, an online mapping tool.

The methodology employed the following steps:

Data Conversion and Geocoding

The containment zone data was transformed into spatial point data through geocoding, using zip codes to pinpoint the exact locations within the KMC area. This conversion allowed for a detailed spatial analysis of the containment zones, which is crucial for understanding the spread and concentration of COVID-19 cases across the city (ESRI, 2020).

Point Density Analysis

Using the student version of ArcGIS, a point density analysis was conducted to visualize the density of containment zones across Kolkata. This analysis highlighted areas with a higher concentration of COVID-19 cases, providing critical insights into the regions most affected by the pandemic. Point density maps are a standard tool in spatial epidemiology, offering a clear representation of disease clusters (Bivand et al., 2013).

Cumulative Change Detection

To assess temporal changes in the spatial distribution of containment zones, a cumulative change detection analysis was performed. The Raster Calculator tool in ArcGIS was used to subtract the number of containment zones from 1 month with those from the subsequent month. By summing the results across the study period, significant shifts in the spatial patterns of COVID-19 containment were identified. This approach is effective in monitoring the progression of infectious diseases over time, allowing for the identification of emerging hotspots (Tobler, 1970; Wong & Lee, 2005).

Hotspot Identification

The results of the cumulative change detection were further analyzed to identify spatial hotspots or areas that experienced significant changes in containment zone density. Hotspot analysis is crucial in epidemiological studies as it helps in targeting public health interventions to areas most in need (Getis & Ord, 1992; Anselin, 1995).

Spatial Pattern Analysis

To determine whether the distribution of containment zones was random or influenced by underlying geographical factors, the average nearest neighbor analysis was conducted. This analysis measures the distance between each point (containment zone) and its nearest neighbor, providing a statistical measure of spatial clustering or dispersion. Such analyses are fundamental in identifying patterns in disease distribution that may be influenced by environmental or social factors (Clark & Evans, 1954; Rogerson, 2001).

Spatial Autocorrelation and Moran's I Analysis

To further explore spatial patterns, Moran's I statistic and spatial autocorrelation analysis were performed using the Joins Count method. Moran's I measures the degree of spatial autocorrelation, indicating whether nearby locations are similarly affected by the pandemic. This analysis is pivotal in understanding the spatial dependencies of disease spread, particularly in complex urban environments like Kolkata (Moran, 1950; Cliff & Ord, 1981).

These methodological steps provided a robust framework for analyzing the spatial dynamics of COVID-19 in Kolkata. By combining geospatial analysis with statistical tools, this study identified key hotspots and trends in the pandemic's spread.

4.2 Mapping Community Perception on Crisis Response

To understand the community perception on crisis response during the post-lockdown phase of the COVID-19 pandemic, a cross-sectional survey was conducted over nearly 1 month, from April 7 to May 5, 2021. Given the health risks associated with physical contact during this period, the survey was administered online to ensure the safety of both respondents and researchers. This approach is consistent with best practices in public health research during pandemics, where online surveys offer a feasible and efficient alternative to in-person data collection, especially when timely data is required (Ritter & Sue, 2017; Saleh & Bista, 2017).

Survey Criteria

The survey targeted Indian nationals who were adults, able to understand English, and willing to participate after reading the information provided on the first page of the questionnaire. This criterion ensured that participants could comprehend the survey content and provided informed consent, a critical ethical consideration in survey research (Dillman et al., 2014).

Validation of Survey Tool

A self-administered questionnaire was carefully designed and validated to ensure reliability and accuracy. The survey tool underwent face validity checks and reliability assessments, with a pilot survey conducted to test its effectiveness. Face validation, also known as content validity, was achieved by consulting experts in public health, who reviewed the questionnaire through the Delphi method—a structured communication technique that gathers consensus through iterative rounds of surveys (Hsu & Sandford, 2007). This rigorous validation process is essential for ensuring that the survey tool accurately captures the intended data and is reliable across different populations (Polit & Beck, 2006).

Survey Tool

The survey comprised eight sections covering a wide range of variables relevant to the pandemic. These sections included demographic information, COVID-19-related experiences, treatment satisfaction, infrastructure capacity, cost of treatment, and transportation issues. Questions were predominantly designed using a 5-point Likert scale to measure the respondents' satisfaction and perceptions, a common approach in health research that facilitates the quantification of attitudes and experiences (Joshi et al., 2015). The comprehensive nature of the questionnaire allowed for the collection of nuanced data on the public's experience and satisfaction with the healthcare system during the pandemic.

Data Collection and Sampling Method

Data collection was carried out using an online survey, distributed via various digital platforms, including WhatsApp, Telegram, Facebook, and Gmail. A snowball sampling method was employed, where initial respondents were asked to forward the survey link to others in their network. This non-probability sampling technique is often used in online surveys to reach a broad and diverse population when the target population is difficult to access (Atkinson & Flint, 2001; Baltar & Brunet, 2012). The first page of the survey included detailed information about the study's background, objectives, and expected outcomes, ensuring transparency and informed consent.

Data Analysis

Upon completion of data collection, the responses were analyzed using statistical software. This analysis involved the application of various statistical methods to interpret the data, providing insights into the public's experiences and the effectiveness of the healthcare response during the pandemic. Statistical analysis in survey research allows for the identification of patterns, correlations, and potential areas for policy intervention, making it a critical component of public health research (Pallant, 2020).

4.3 Evaluating the Effectiveness of Local Interventions Through Geographically Weighted Regression (GWR)

The effectiveness of local interventions during the post-lockdown phase of COVID-19 management in Kolkata was evaluated using geographically weighted regression (GWR). GWR is a spatial analysis technique that allows for the examination of how relationships between dependent and independent variables vary across different geographical locations within the study area. Unlike traditional regression models that assume a uniform relationship across all locations, GWR accounts for spatial heterogeneity by allowing the coefficients of the explanatory variables to vary across space (Fotheringham et al., 2002; Páez et al., 2011). This approach allows for a more nuanced understanding of how the strength and direction of relationships vary across different geographical locations, making it particularly valuable for studies where spatial context is critical.

In this study, GWR was applied to understand the spatial factors contributing to the spread of COVID-19 and the effectiveness of containment measures within the Kolkata Municipal Corporation (KMC) area. The data used for this analysis included the containment zone information from May 2020, the borough-wise population density, the total number of Below Poverty Line (BPL) households, the number of slums, the cases of acute respiratory infection (ARI), the number of community latrines, and the number of dug wells and hand wells. These variables were chosen based on their potential influence on the spread of the virus and the effectiveness of public health interventions (Hu et al. 2024; You et al., 2020). The application of GWR in this context enabled the identification of spatial variation in the influence of these factors on the spread of COVID-19. The resulting coefficient distribution provided insights into where and how much variation was present across different boroughs within the KMC area. This variation is critical for understanding the localized nature of the pandemic and the need for tailored interventions.

Statistically significant global variables that exhibited little regional variation indicated factors that could inform region-wide policies. For instance, if population density was found to be a consistent predictor of COVID-19 spread across all boroughs, it would suggest that interventions targeting population density could be effective at a city-wide level. On the other hand, variables that showed strong regional variation suggested the need for localized policy interventions. For example, the number of slums might be a significant predictor of COVID-19 spread in some boroughs but not in others, indicating that slum-focused interventions should be tailored to specific areas (Cellmer et al., 2020). In this study, the GWR analysis revealed a strong regional variation, suggesting that each borough within the KMC area is influenced by different factors. This finding underscores the importance of developing local prediction models that can address the unique challenges faced by each borough. Such models are crucial for creating effective neighborhood action plans that can prevent the spread of disease and manage future public health crises more effectively (Gollini et al., 2015; Jasim et al., 2022). This case study of Kolkata demonstrates the utility of GWR in capturing the spatial nuances of disease spread and management, ultimately contributing to a more comprehensive understanding of how social and environmental factors interact to shape public health outcomes in urban settings.

5 Results and Discussion

5.1 Spatio-temporal Pattern of Containment Zone Distribution

According to the Union Ministry for Health and Family Welfare, containment zones are designated geographic areas where a significant number of COVID-19 cases are detected (What is Containment Zone?, 2021). The primary objective of establishing containment zones is to map the local transmission of the disease and prevent its further spread. Spatio-temporal analysis, which involves collecting data across both space and time, is a crucial tool in this context. It allows for the mapping of changes in the number and distribution of containment zones over time, providing insights into which areas are most severely affected and the patterns of contagion.

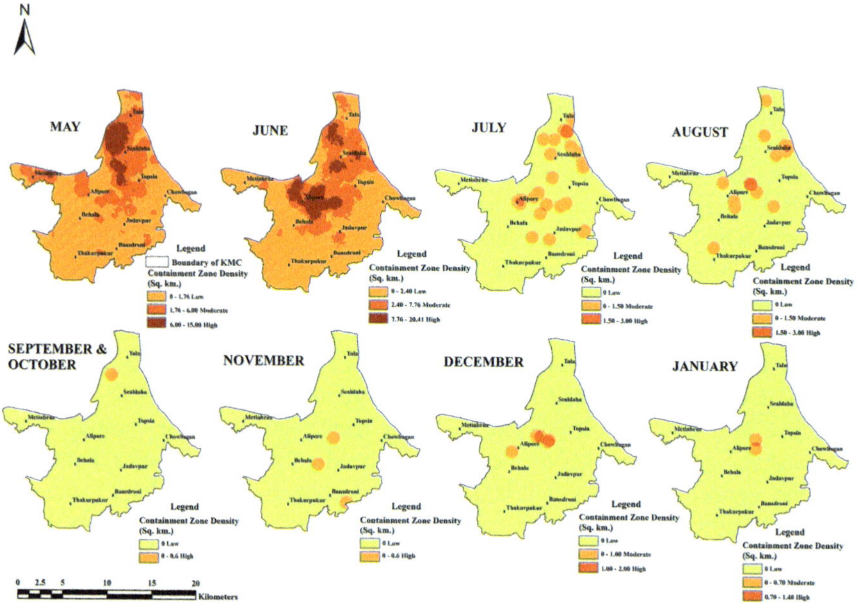

Fig. 2 Density mapping of COVID-19 containment zone (May 2020–Jan 2021) using point density method

In a study by Bag et al. (2020), hotspot analysis revealed that the core clustering of highly affected COVID-19 districts in India was predominantly in the western regions, encompassing 60 districts. Initially, the spread of COVID-19 clusters was observed along the southern, western, and north-central regions of the country, affecting 139 districts. This finding underscores the importance of spatial analysis in understanding the spread of the pandemic.

Figure 2 in this study illustrates the temporal evolution of containment zones within the Kolkata Municipal Corporation (KMC) area from May 2020 to January 2021. In May 2020, the northern part of KMC, particularly around Tala and Sealdah, exhibited a higher concentration of containment zones compared to the southern regions near Alipore, Behala, and Jadavpur. However, by June 2020, this pattern shifted, with an increase in containment zones in the central part of KMC and a decrease in the northern areas. From July 2020 onward, the situation evolved continuously, with a gradual reduction in the number of containment zones. By September and October 2020, there was only one containment zone near Tala, followed by a slight increase to five zones in December 2020. By January 2021, the number of containment zones had again decreased to just two.

The spatial distribution of containment zones during this period was categorized into three density levels: high density, moderate density, and low density. In May 2020, high-density zones were concentrated in the northern part of KMC, particularly to the west of Tala and south of Sealdah, with a smaller high-density patch in the eastern part near Metiabruz. Surrounding these high-density zones were

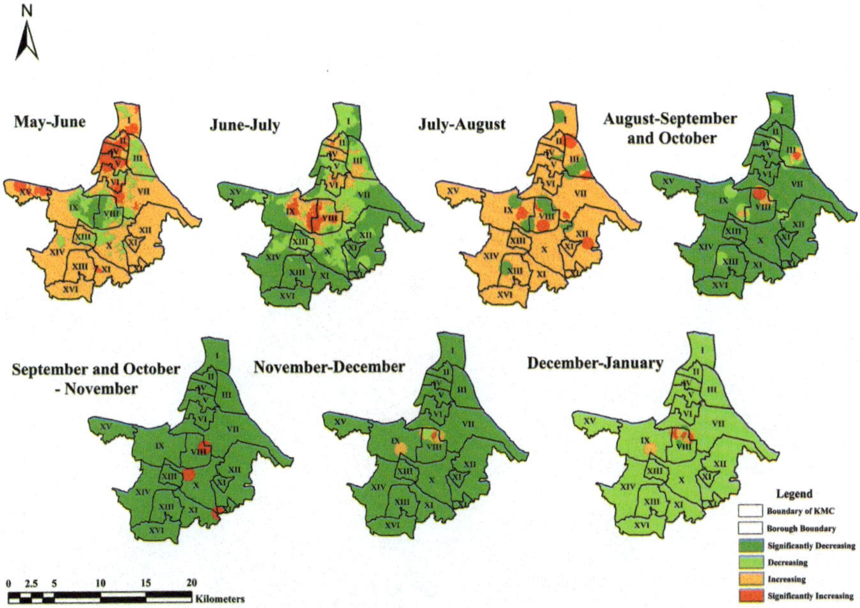

Fig. 3 Change detection in containment density (May 2020–Jan 2021)

moderate-density zones, while the southern part of KMC predominantly exhibited low-density zones. In June 2020, the high-density zone shifted to the central part of KMC near Alipore, with low-density zones dominating the eastern, western, and southern regions. In July 2020, small patches of moderate-density zones appeared in the central and northern parts, with high-density zones near Alipore and Tala. This pattern persisted into August 2020, albeit with fewer moderate-density patches. By September to November 2020, moderate-density zones had disappeared, and most of the area fell under low-density zones. In December 2020 and January 2021, high- and moderate-density zones were concentrated in the central part of KMC, with the rest of the area exhibiting low density.

The change detection analysis, mapped in Fig. 3, reveals regions with specific characteristics:

- *Significantly Increasing*: From May to July 2020, the northern and central parts of KMC exhibited a significant increase in the number of containment zones. From November 2020 to January 2021, this significant increase was confined to the central part of KMC.
- *Increasing*: From May to June and July to August 2020, most areas fell under the increasing category, with other months showing small patches in this category around significantly increasing areas.
- *Decreasing*: The decreasing category was mainly observed from December 2020 to January 2021, with small patches in other months.

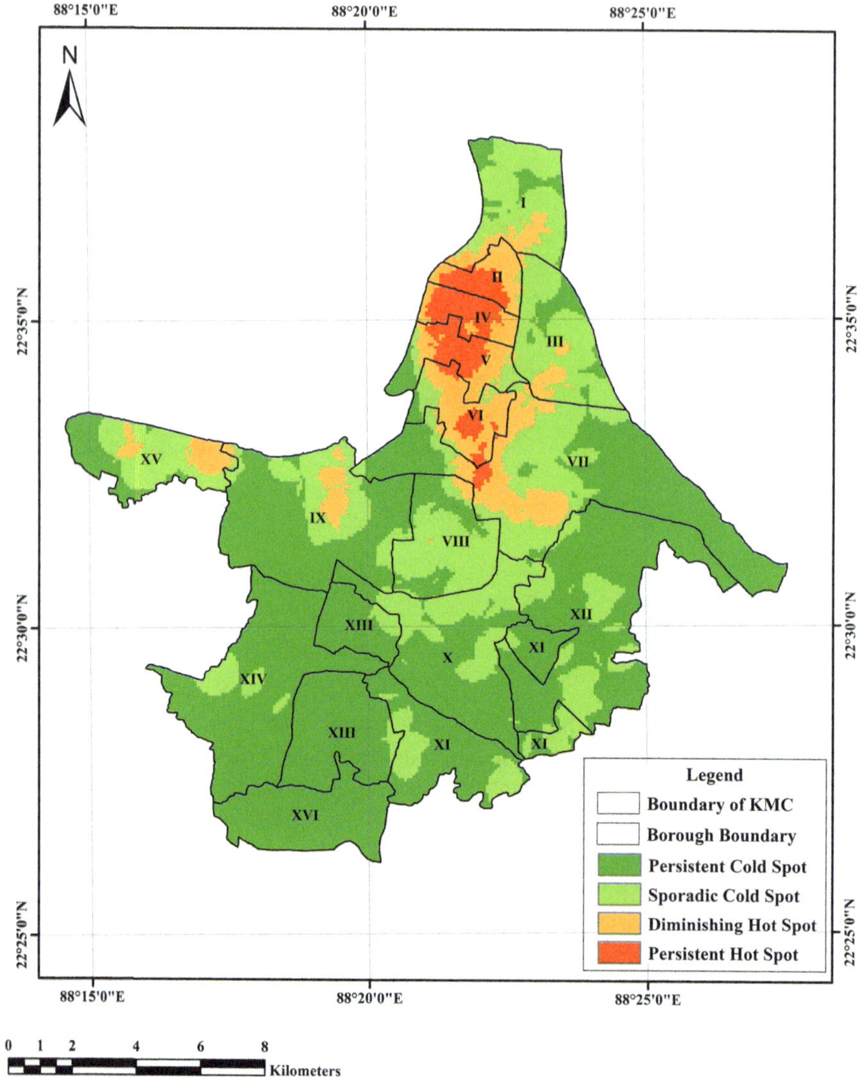

Fig. 4 Pandemic hotspot in Kolkata

- *Significantly Decreasing*: From August to November 2020, there was a significant decrease in the number of containment zones.

Figure 4 identifies the persistent hotspots and cold spots of containment zones. The northern part of KMC emerged as a persistent hotspot, while most of the area was classified as a persistent cold spot, indicating the absence of containment zones. Small scattered patches of sporadic cold spots were also present throughout the study area.

Table 1 Summary of nearest neighbor ratio (NNR) of COVID-19 distribution in Kolkata during post-lockdown

Month	NNR	Z-score	P-value
May	0.653023	12.094841	0.0000[b]
June	0.537018	−20.139122	0.0000[b]
July	1.000602	0.004604	0.9963[a]
August	0.767995	−2.348590	0.0188[b]
September	2.092698	−4.674292	0.0000[b]
October	3.409233	7.98376	0.0000[b]
November	2.092698	4.674292	0.0000[b]
December	3.409233	7.983076	0.0000[b]
January	63.42724	168.8962	0.0000[b]

[a]*P*-value >0.005 at a 1% significant level, representing the dispersed/clustered pattern of COVID-19 distribution and it could be the result of random chance

[b]*P*-value <0.005 demonstrates that our result of NNA is significant at a 1% level and point distribution is clustered/dispersed at a 99% significant level

Table 1 presents the results of the nearest neighbor analysis (NNA), which indicates whether the distribution of containment zones is significantly clustered or dispersed. In May 2020, the *p*-value was less than 0.05, rejecting the null hypothesis and indicating a clustered distribution pattern. This clustered pattern persisted in June and July. In contrast, November, December, and January showed a dispersed point distribution pattern.

The borough-wise spatio-temporal analysis revealed significant variations in the number of containment zones across different boroughs. In May 2020, Borough VIII had the highest number of containment zones (51), followed by Boroughs IV, V, and VI with 38, 35, and 29 containment zones, respectively. Borough XVI, on the other hand, had no containment zones. By June 2020, Borough IX had the highest number of containment zones (99), followed by Boroughs VIII, VII, and III with 90, 61, and 46 containment zones, respectively. Boroughs XIV, XV, and XI had the lowest numbers, with 13, 10, and 7 containment zones, respectively. In July 2020, Borough III had the highest number of containment zones (9), with several boroughs, including II, XI, and XIII, reporting zero containment zones. By August 2020, Borough VIII had the highest number of containment zones (5), followed by Borough III with four. In this month, the number of boroughs with zero containment zones increased. September and October 2020 saw only one containment zone in Borough IV. In November 2020, Boroughs X and VIII had two and one containment zones, respectively. In December 2020, Boroughs VIII and IX had four and one containment zones, respectively, and by January 2021, only Borough VIII had two containment zones.

The spatial association among containment zones was also tested, with the results presented in Table 2. The analysis revealed a statistically significant clustered pattern for May 2020, particularly in the northern part of KMC, which is characterized by high population density. As Mishra et al. (2020) observed,

Table 2 Summary of global Moran's I statistics of COVID-19 distribution in Kolkata during post-lockdown

	Moran's index		
Month	Moran's *I*	*Z*-score	*P*-value
May	0.520466	3.8706	0.0001[b]
June	0.128151	1.32140	0.003[b]
July	−0.067512	−0.006477	0.99483[a]
August	−0.13591	−0.47806	0.63260[a]
September	−0.185262	−1.673817	0.09416[a]
October	−0.185262	−1.673817	0.09416[a]
November	0.096272	1.43771	0.15051[a]
December	0.002587	0.79407	0.42715[a]
January	−0.078585	−0.168204	0.86642[a]

[a]*P*-value >0.05 indicates that the pattern does not appear to be significantly different than random
[b]*P*-value <0.05 denotes that there is a less than 5% likelihood that pattern could be the result of random chance

"residents living in areas with high population density, such as big or metropolitan cities, have a higher probability of coming into close contact with others, and consequently, any contagious disease is expected to spread rapidly in dense areas." This observation is further supported by Bhadra et al. (2021), who noted the increased risk of disease transmission in densely populated urban areas.

In conclusion, the spatio-temporal analysis of containment zones in KMC highlights significant spatial and temporal variations in the spread and containment of COVID-19. The findings indicate that the central and northern parts of KMC were the most significant hotspots for COVID-19, necessitating further analysis to address the specific challenges in these regions. A critical area for future research is evaluating whether health infrastructure and facilities in these regions are adequate and accessible to manage the pandemic effectively.

5.2 *Accessibility and Availability to Healthcare Services: Mapping Neighborhood Experience*

Perception mapping is an essential tool for understanding how individuals perceive and respond to crises like the COVID-19 pandemic. This method provides valuable insights into public sentiment, knowledge, and behavior, which are crucial for effective policy implementation and pandemic management. By analyzing people's perceptions, decision-makers can better address public needs, enhance communication strategies, and improve service delivery during such unprecedented times. Risk perception significantly influences public behavior during a pandemic. Individuals with direct experience of COVID-19, or those who have received information from trusted sources such as family and friends, tend to perceive higher levels of risk than those without such experiences (Dryhurst et al., 2020). This indicates that personal experience and reliable information sources are critical in shaping how people view and respond to health crises. Moreover, differences in risk perception across countries are often linked to the varying severity of the pandemic in each region (Nair et al., 2022).

In this study, the demographic profile of respondents reveals a higher participation rate among males, with a majority identifying as Hindu and belonging to the General category. The sample predominantly consists of students, reflecting the educational background of the respondents. To better understand their experiences during the pandemic, respondents were categorized into two groups: direct victims, those infected with COVID-19, and proxy victims, those not infected but with indirect experiences related to the pandemic (Table 3). Direct victims were further classified based on their hospitalization status: hospitalized and non-hospitalized. Among the hospitalized respondents, many sought treatments at multiple hospitals, with Ruby General Hospital being the most frequently visited (Table 4). This variability in healthcare access and utilization among COVID-19 patients suggests the need for a more streamlined approach to healthcare provision during crises.

The study found that most respondents received information about available COVID-19 services from family, friends, or relatives, with moderate satisfaction levels reported by both direct and proxy victims (Fig. 5). Despite some respondents utilizing the West Bengal Government's helpline services, satisfaction with these services was generally low. Additionally, the availability of doctors was found to be higher than that of beds or ICU facilities in hospitals and nursing homes, highlighting gaps in critical healthcare infrastructure (Saeed et al., 2024). The cost burden for supplementary services was reported to be excessively high, adding financial strain on patients. While satisfaction with insurance services for COVID-19 treatment was

Table 3 Demographic profile of the respondents ($N = 204$)

Descriptive statistics of the respondents			
Total: 204		*N*	%
Gender	Male	108	52.9
	Female	96	47.1
Religion	Hindu	192	94.1
	Muslim	4	2.0
	Jain	4	2.0
	Others	4	2.0
Caste	General	176	86.3
	SC	8	3.9
	OBC	20	9.8
Occupation	Business	16	7.8
	Govt. or Pvt. employee	72	35.3
	Student	112	54.9
	Housewife	4	2.0
Family member affected with COVID-19	Direct victim	44	21.6
	Proxy victim	160	78.4

Table 4 COVID-19 profile of the respondents ($N = 204$)

COVID-19 profile (direct victim)			
Total: 44		*N*	%
Number of family members who have been detected with the COVID-19-positive cases	1	16	36.4
	1–2	8	18.2
	3–4	16	36.4
	>4	4	9.1
Avail hospital services	Hospitalized	28	63.6
	Not hospitalized	16	36.4
Treated in one hospitals or multiple hospitals	Single	16	57.1
	Multiple	12	42.9
Hospital visit for treatment	BP Poddar Hospital	4	14.3
	Angel Nursing Home	4	14.3
	Calcutta Medical College Hospital	4	14.3
	Columbia Asia Hospital	4	14.3
	Bangur Multisuperspeciality Hospital	4	14.3
	Ruby General Hospital	8	28.6
	Peerless Hospital	4	14.3

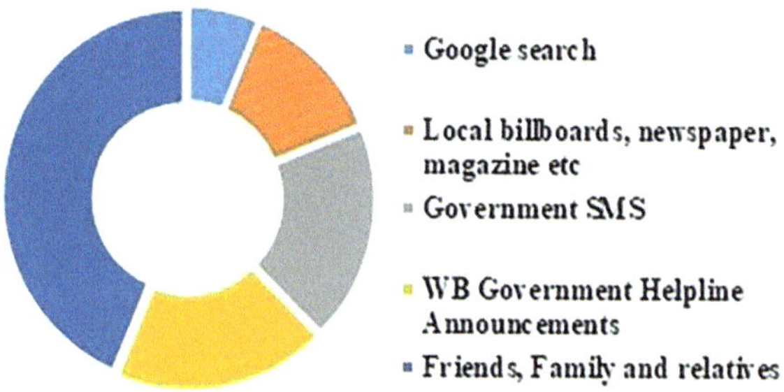

Fig. 5 Sources of information about the availability of facilities. (Source: Field Survey, 2021)

moderate, the overall availability of insurance was inadequate, further complicating the financial challenges faced by patients (Table 4). This underscores the need for more accessible and affordable healthcare services during pandemics, particularly for vulnerable populations. Most respondents used personal vehicles to access healthcare services, with only a few relying on public transport, which remained largely available. The distance to healthcare facilities was generally manageable, and transportation costs were considered bearable. However, the availability of COVID-19-specific ambulances was severely lacking, posing significant challenges for those in need of urgent care (Table 5). This points to the necessity of enhancing emergency transport services during health crises to ensure timely access to care. For those treated at home, the availability of doctor services and necessary medications was rated as average, indicating a need for better support for non-hospitalized

Table 5 Perception mapping on information sources, availability of services in hospital/nursing home, and cost burden ($N = 204$)

Parameters	Direct victim, 44 (hospitalized, 28, and non-hospitalized, 16)						Proxy victim, 160					
		N	Mean	Std. dev	Min	Max		*N*	Mean	Std. dev	Min	Max
Information sources	Satisfaction level with the availability of COVID-19 information	44	3.27	0.87	2	5	Rate of experience about availability of information on corona help desk	160	3.10	1.02	1	5
	Satisfaction level on the helpline services offered by the Government of West Bengal	44	2.91	1.01	1	5						
Availability of services in hospital/ nursing home	Bed	28	3.29	0.22	2	5	Rate of experience about availability of facilities in the hospitals/nursing homes	160	3.2	1.09	1	5
	Doctor		3.86	0.19	2	5			3.3	1.05	1	5
	ICU/ventilation		3.29	0.25	2	5			3.1	0.90	1	5
Cost burden and insurance benefit	Cost of hospital services (bed and medicine)	28	3.14	1.38	1	5	Rate your experience about the cost burden of COVID-19 treatment	160	3.08	1.11	1	5
	Cost of additional services (PPE kit, mask, sanitizer, gloves, etc.)	28	2.71	1.70	1	5	NA					
	Availability of insurance service	28	2.00	0.77	1	3	NA					
	Satisfaction level of provision of insurance services	28	3.86	1.01	2	5	NA					

patients during the pandemic (Table 6). This finding aligns with other studies highlighting the challenges faced by patients managing COVID-19 symptoms at home, often due to limited access to healthcare resources and professional guidance (Rao et al., 2021).

The study also revealed that both direct and proxy victims exhibited low levels of awareness and interest in local vaccination services (Tables 5 and 6). This lack of engagement with vaccination efforts underscores a critical area for public health intervention, as increasing vaccination coverage is essential for controlling the spread of COVID-19. Effective communication and community outreach are necessary to improve public awareness and participation in vaccination programs (Kawuki et al., 2023).

The perception mapping exercise reveals significant dissatisfaction with various aspects of healthcare services during the COVID-19 pandemic, including inadequate health infrastructure, unreliable information, and the unavailability of essential services (Table 6). These deficiencies have exacerbated the challenges faced by the population during the crisis. Addressing these gaps through improved infrastructure, better information dissemination, and enhanced service availability is crucial for effective pandemic management.

5.3 *Spatial Variability of Factors Influencing COVID-19 Spread Within the Kolkata Municipal Corporation (KMC) Area*

This study employs geographically weighted regression (GWR) to assess the spatial variability of factors influencing COVID-19 spread within the Kolkata Municipal Corporation (KMC) area. The northern and central parts of KMC exhibit a higher concentration of COVID-19 cases compared to the southern regions, suggesting that different regions face unique challenges that necessitate localized intervention strategies. By utilizing GWR, the study identifies these regional differences and offers a granular understanding of how various socio-economic and environmental factors contribute to the spread of COVID-19.

The GWR model provides localized regression coefficients, allowing us to observe how the relationship between the dependent variable (COVID-19 containment zones) and independent variables (such as population density, number of slums, BPL households, ARI cases, community latrines, dug wells, and hand wells) varies across the KMC area (Fig. 6). The model's goodness-of-fit is assessed using local R^2 values, which indicate the proportion of variance explained by the model at each location. In this study, the local R^2 values vary significantly across different boroughs, ranging from 0.0 to 1.0. High local R^2 values, particularly in Boroughs I, XII, and XV, suggest that the GWR model effectively captures the local dynamics of COVID-19 spread in these areas (Fig. 7). This indicates that the selected independent variables are strong predictors of containment zone locations in these boroughs.

Table 6 Perception mapping on transportation facilities ($N = 204$)

Parameters	Direct victim (hospitalized, 28)						Proxy victim, 160					
		N	Mean	Std. dev	Min	Max		*N*	Mean	Std. dev	Min	Max
Mode of transportation	Private	20	1.29	0.46	1	2	Rate your experience about the availability of transport to reach the hospitals and the COVID-19 ambulance services	160	3.13	1.10	1	5
	Public	8										
Travel time (surface transport—Car speed)	Less than 15 min	4	2.14	0.65	1	3	NA					
	15–30 min	16										
	More than 30 min	8										
Distance from house to hospital/nursing home	–	28	3.43	1.20	2	5	NA					
Availability of transport facilities (public)	–	8	3.5	1.60	2	5	NA					
Cost of transportation	–	28	3.71	1.05	2	5	NA					
Availability of COVID-19 ambulance	–	28	2.57	1.20	1	5	NA					

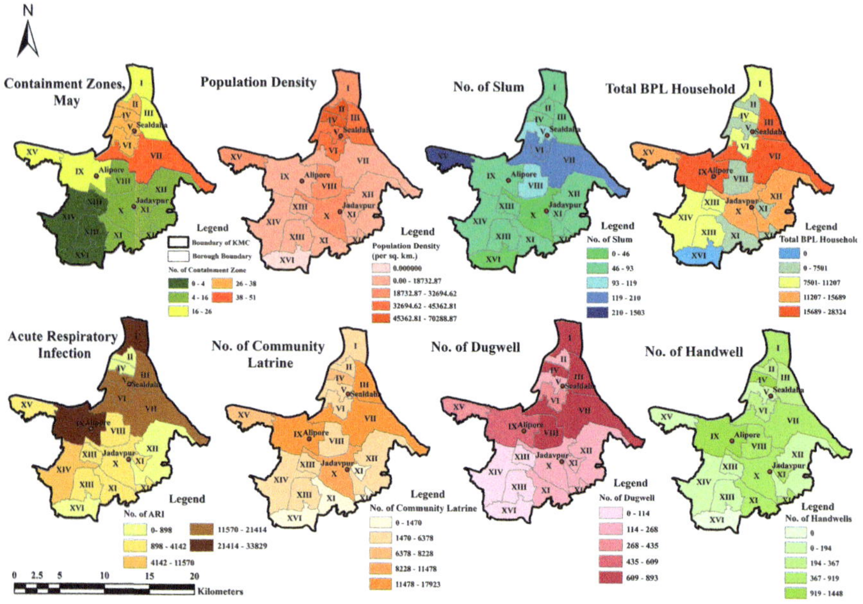

Fig. 6 Item selected for geographically weighted regression

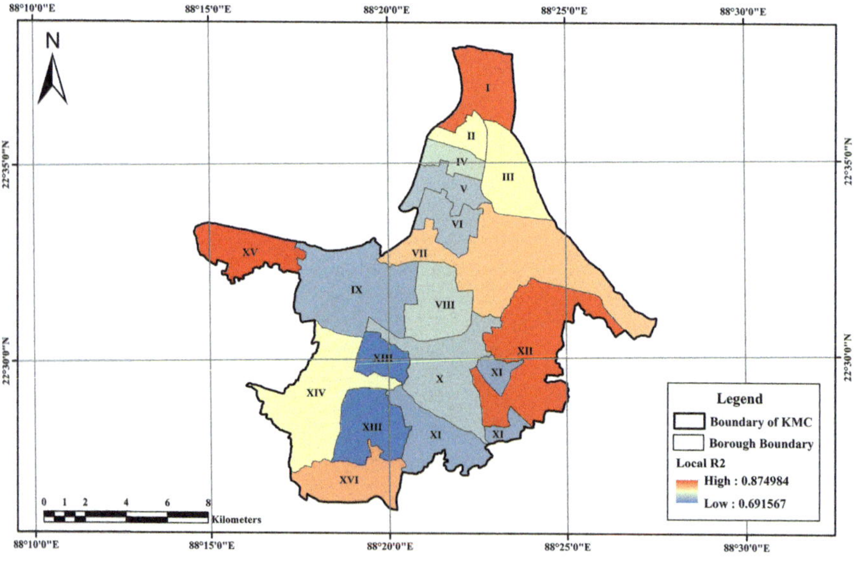

Fig. 7 Geographical distribution of R^2 values

For Boroughs IV, VI, VIII, X, XI, and XVI, the standard residuals vary between −0.5 and 0.5. This range indicates that the GWR model predictions closely match the observed values in these areas. The low residuals suggest that the model's explanatory variables, such as population density and the presence of slums or community latrines, are effective predictors for the spread of COVID-19 in these boroughs. The close alignment between predicted and observed values implies that the factors influencing the spread are relatively homogeneous across these areas and that local interventions based on these predictors are likely to be effective. Conversely, in Borough VII, the model exhibits high standard residuals, indicating that the GWR model's predictors are less effective in this area, possibly due to the borough's large size and heterogeneous population, which introduces complexity and variability in the data. In Borough VII, the standard residuals are more than 2.5, which signifies a substantial discrepancy between the model's predictions and the actual observed values. This high residual indicates that the variables used in the model are not sufficient to explain the variability in COVID-19 cases within this borough (Fig. 8). Several factors could contribute to this discrepancy, such as unaccounted-for local conditions, social behaviors, or other epidemiological factors not included in the model. The heterogeneity within this borough, possibly due to its larger size and diverse population, suggests that additional or alternative variables may be necessary to capture the dynamics of COVID-19 spread accurately. For instance, variables like mobility patterns, specific local health interventions, or unique socio-economic conditions might need to be included to improve model accuracy.

The GWR analysis provides coefficient estimates for each independent variable across different locations, allowing for a detailed understanding of how these factors influence the spread of COVID-19 (Fig. 9). For instance, population density and the number of slums are positively correlated with the number of containment zones in certain boroughs, highlighting the role of crowded living conditions in exacerbating the spread of the virus. In contrast, other boroughs may show a weaker or even negative relationship between these variables and COVID-19 spread, indicating that other factors, such as the availability of healthcare services or local economic conditions, may play a more significant role in these areas. The coefficient standard error, which measures the reliability of each coefficient estimate, also varies across the study area. Boroughs with low coefficient standard errors indicate a strong and reliable relationship between the independent variables and the dependent variable. In contrast, boroughs with high coefficient standard errors suggest that the relationship is less stable and more influenced by unobserved factors, necessitating caution when interpreting these results.

Examining the coefficient distribution highlighted the need for localized intervention strategies:

- Statistically significant global variables with minimal regional variation should guide region-wide policies.
- Global variables with strong regional variation should inform localized policies.
- Variables with no global significance but with regional relevance should be used to tailor specific interventions.

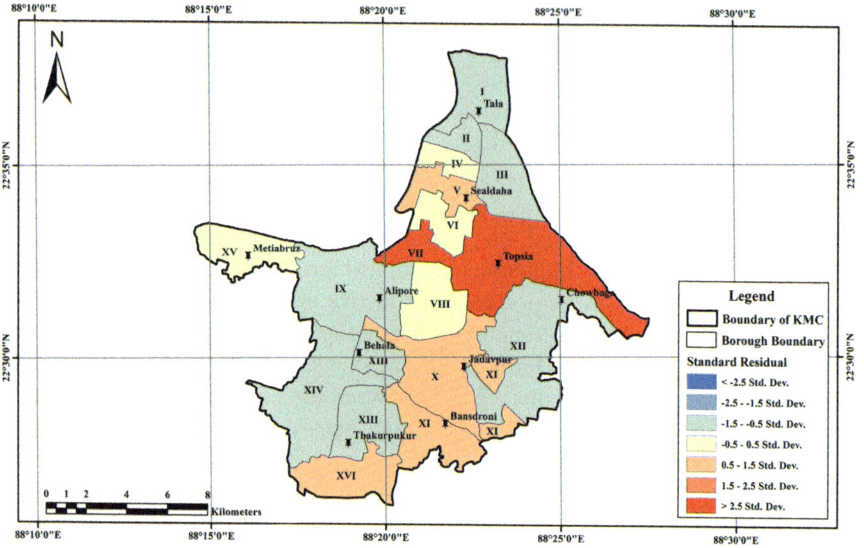

Fig. 8 Geographical distribution of standard residual

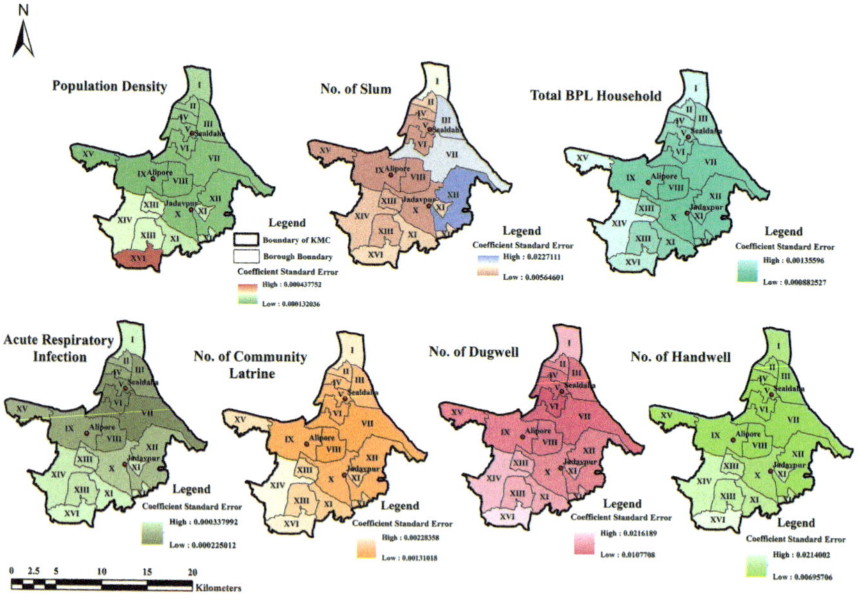

Fig. 9 Coefficient standard error map of independent items

In this study, significant regional variations were observed, underscoring the importance of localized prediction models for effective disease control. For example, boroughs with high COVID-19 hotspots, population density, and slums require targeted awareness campaigns and strict enforcement of social distancing measures. In areas with a high concentration of BPL households, government support is essential to meet basic needs and improve health conditions. Boroughs with a high incidence of acute respiratory diseases and COVID-19 cases should have increased oxygen facilities. Community latrines, particularly in slum areas, need frequent sanitation to prevent the virus's spread. Additionally, ensuring access to water and enforcing social distancing are crucial in these vulnerable areas. Detailed borough-wide recommendations are outlined in Table 7.

The findings from this study provide valuable insights for local authorities in the KMC area to adopt localized action plans to enhance preparedness and response to the COVID-19 pandemic and similar future challenges. These recommendations can also be extended to other megacities in India facing similar pandemic-related difficulties.

Recent studies have demonstrated the effectiveness of GWR in understanding the spatial heterogeneity of COVID-19 transmission and guiding targeted interventions ; Hu et al., 2024). The application of GWR in similar studies has shown its potential to uncover nuanced spatial patterns that traditional global models might overlook, thus contributing significantly to the field of social physics in pandemic management.

5.4 *Implications for Future Public Health Crises*

The analysis of containment zones in the Kolkata Municipal Corporation (KMC) during the COVID-19 pandemic reveals critical insights into the spatial dynamics of disease transmission, contributing to the broader field of social physics. Social physics, which explores the quantitative laws governing social phenomena, provides a framework to understand how human behavior and interactions, influenced by spatial factors, impact the spread of infectious diseases.

The spatio-temporal analysis conducted in this study demonstrates that containment zones are not merely administrative tools but reflect underlying social and physical structures that drive the transmission of COVID-19. The shift in containment zones from the northern to the central parts of KMC over time suggests that disease spread is closely linked to population density, mobility patterns, and socio-economic characteristics of different regions. This is consistent with findings from other studies that highlight the role of urban density in accelerating the spread of COVID-19 (Zhou et al., 2020; Liu et al., 2020).

In the context of social physics, this study underscores the importance of spatial heterogeneity in understanding pandemic dynamics. The persistent hotspot in the northern part of KMC, coupled with the high population density and lower socio-economic status of this area, suggests that certain urban regions are more vulnerable

Table 7 Borough-wise COVID-19 response matrix: forecasts, perceptions, key issues, and policy recommendations

Borough	Forecast	Perception	Key issues	Future policies
I	High	Moderately satisfied	Moderate population density, high slum areas, high BPL households, high acute respiratory infections	Strengthen awareness campaigns on infection prevention in slum areas Enforce strict social distancing protocols during any contagious disease outbreak. Improve sanitation in public areas. Expand healthcare services with oxygen support
II	Moderate	Moderately satisfied	Moderate population density, moderate slums, high BPL households, moderate acute respiratory infections	Focus on public hygiene campaigns. Enforce strict social distancing protocols during any contagious disease outbreak. Regularly sanitize community latrines. Establish emergency oxygen support facilities in health centers
III	High	Moderately satisfied	Low population density, high slum areas, moderate BPL households, low acute respiratory infections	Increase health education efforts. Enforce strict social distancing protocols during any contagious disease outbreak. Provide financial and healthcare support to slum communities Ensure emergency oxygen availability
IV	High	Satisfied	Low population density, low slum areas, high BPL households, high acute respiratory infections	Continuously educate the public about infection control. Maintain strict social distancing protocols during outbreak of any contagious disease. Prioritize healthcare and economic support for BPL households
V	High	Moderately satisfied	Low population density, low slums, moderate acute respiratory infection risk	Strengthen law enforcement for pandemic-related regulations. Increase public sanitation efforts. Provide oxygen support in healthcare facilities

VI	High	Satisfied	Low population density, moderate BPL households, low acute respiratory infections	Support vulnerable groups with economic and healthcare aid. Increase focus on public space sanitation. Ensure emergency oxygen availability
VII	Very high	Moderately satisfied	Low population density, high slum areas, low BPL households, low acute respiratory infections	Strengthen policing for health protocols. Provide financial aid and healthcare services for slum populations. Ensure proper sanitation in public spaces. Improve respiratory healthcare services
VIII	Very high	Dissatisfied	Low population density, low BPL households, low respiratory infection risk	Launch trust-building public health campaigns to address dissatisfaction. Increase enforcement of public health and distancing regulations during outbreak. Prioritize sanitation efforts in public spaces and slum areas
IX	Very high	Dissatisfied	Low population density, moderate acute respiratory infections	Increase awareness and health education efforts. Strengthen police enforcement of pandemic regulations. Sanitize public spaces and water sources. Ensure healthcare access with oxygen support for respiratory cases
X	High	Satisfied	Low population density, moderate BPL households, moderate respiratory infection risk	Regularly sanitize public latrines. Provide financial aid to vulnerable groups. Enforce strict social distancing measures during outbreak
XI	Low	Dissatisfied	Moderate population density, moderate acute respiratory infection risk	Enforce strict social distancing protocols during any contagious disease outbreak. Increase public health awareness campaigns Prioritize sanitation in public spaces

(continued)

Table 7 (continued)

Borough	Forecast	Perception	Key issues	Future policies
XII	Moderate	Moderately satisfied	Low population density, high slum population, moderate acute respiratory infections	Strengthen police enforcement of public health guidelines. Provide economic support for low-income households. Focus on improving sanitation in slum areas
XIII	Low	Moderately satisfied	Moderate population density, moderate BPL households, high respiratory infection risk	Ensure strict adherence to social distancing norms. Focus on sanitation in public and private spaces. Launch public health education campaigns
XIV	Low	Dissatisfied	Moderate population density, high BPL households, high respiratory infection risk	Conduct trust-building health awareness campaigns. Strengthen enforcement of public health guidelines. Focus on economic and healthcare support for vulnerable groups
XV	Moderate	No data	Low population density, high BPL households, moderate respiratory infection risk	Increase public health education and hygiene awareness efforts. Enforce strict social distancing measures. Provide healthcare aid to vulnerable populations
XVI	No data	No data	High population density, high BPL households, high acute respiratory infection risk	Prioritize government aid for economic and healthcare support. Focus on sanitation in high-risk areas. Ensure healthcare facilities are equipped with oxygen support for respiratory emergencies

to rapid disease transmission. This aligns with the principles of social physics, which posit that human behavior in densely populated areas can lead to complex contagion effects, where the probability of transmission increases not just with the number of contacts but also with the intensity and frequency of interactions (Moussaïd et al., 2020).

Furthermore, the temporal changes in containment zones reflect the dynamic nature of pandemic spread and the effectiveness of public health interventions. The significant decrease in containment zones from August to November 2020 indicates that mitigation measures, such as lockdowns and social distancing, were effective in reducing transmission in certain areas. However, the resurgence of containment zones in December 2020 suggests a potential relaxation of these measures or an increase in mobility due to social and economic pressures. This phenomenon is well-documented in studies that examine the impact of public compliance and government interventions on pandemic outcomes (Olney et al., 2021).

The spatial association analysis, which revealed clustered patterns of containment zones in high-density areas, further highlights the relevance of social physics in pandemic management. The clustering indicates that COVID-19 transmission is not uniformly distributed but is concentrated in specific hotspots, driven by social interactions and the built environment. Understanding these patterns is crucial for designing targeted interventions that can effectively curb the spread of the virus, especially in urban settings where population density and social connectivity are high (Cartenì et al., 2021).

Recent studies have emphasized the need to integrate spatial analysis with epidemiological models to better predict and manage pandemics (Brockmann & Helbing, 2013). The findings from this study contribute to this growing body of knowledge by providing empirical evidence on the spatial dynamics of COVID-19 in an urban context. By mapping containment zones and analyzing their spatio-temporal evolution, this study offers valuable insights into the social and physical determinants of disease spread, which are critical for developing effective public health strategies. The insights gained from this study can inform future research on the interplay between human behavior, spatial organization, and pandemic dynamics, ultimately contributing to a more comprehensive understanding of how to manage and mitigate the impact of global health crises.

Despite widespread awareness, knowledge alone doesn't always lead to positive attitudes or behaviors. Even well-informed individuals might be reluctant to follow preventive measures, highlighting the need for continuous education. Factors like age, education, and healthcare profession significantly influence knowledge and attitudes toward COVID-19 (Hsiang et al., 2020). This demographic insight is essential for understanding varied responses to the pandemic. From a social physics perspective, analyzing these patterns helps optimize healthcare resource distribution. This study found that most respondents relied on family and friends for information, with moderate satisfaction reported. This reliance aligns with social physics theories that suggest human behavior in crises is influenced by peer networks (Pentland, 2014). Gaps in healthcare service delivery, such as limited beds and high costs, highlight the need for improved resource allocation during pandemics (Saeed et al., 2024). The perception mapping exercise reveals significant dissatisfaction with healthcare services during the COVID-19 pandemic, pointing to gaps in health infrastructure, information dissemination, and service availability. By understanding how human behavior and social networks influence the spread of disease and the public's response to health crises, social physics can play a crucial role in pandemic management, helping to mitigate the impact of future outbreaks.

To analyze the spatial variability of factors influencing the spread of COVID-19, geographically weighted regression (GWR) was employed. GWR allows for an in-depth understanding of how relationships between the dependent variable (COVID-19 containment zones) and independent variables—such as population density, number of slums, total Below Poverty Line (BPL) households, number of acute respiratory infection (ARI) cases, and availability of community latrines, dug wells, and hand wells—vary across different locations. The analysis found that 97% of the variability in containment zones could be explained by the selected independent variables.

The GWR results indicated that local R^2 values ranged from 0.0 to 1.0, reflecting how well the local regression models fit the observed data. High local R^2 values in Boroughs I, XII, and XV demonstrated strong model performance in these areas, while Borough VII exhibited a high standard residual value, indicating poor model performance due to the population's heterogeneity. The coefficient standard error values further suggested that in areas with lower errors, the independent variables were more effective in predicting the dependent variable, while areas with higher errors indicated less predictive power.

Recent studies have demonstrated the effectiveness of GWR in understanding the spatial heterogeneity of COVID-19 transmission and guiding targeted interventions. For instance, highlighted the utility of GWR in analyzing spatially varying relationships, emphasizing its role in uncovering local patterns that global models might overlook. Similarly, Hu et al. (2024) applied GWR to explore the socio-economic factors influencing COVID-19 transmission, demonstrating how spatial analysis can inform region-specific public health strategies.

The application of GWR in similar studies has shown its potential to uncover nuanced spatial patterns that traditional global models might overlook. This approach contributes significantly to the field of social physics in pandemic management, offering a framework for understanding and addressing the complex interactions between social, economic, and environmental factors that drive disease transmission.

In this study, significant regional variations were observed, underscoring the importance of localized prediction models for effective disease control. For example:

- Boroughs with high COVID-19 hotspots, population density, and slums require targeted awareness campaigns and strict enforcement of social distancing measures.
- Areas with a high concentration of BPL households need government support to meet basic needs and improve health conditions.

- Boroughs with a high incidence of acute respiratory diseases and COVID-19 cases should increase oxygen facilities.
- Community latrines, particularly in slum areas, should be frequently sanitized to prevent virus spread.
- Ensuring access to water and enforcing social distancing are crucial in these vulnerable areas.

These findings provide valuable insights for local authorities in the KMC area to adopt localized action plans to enhance preparedness and response to the COVID-19 pandemic and similar future challenges. The recommendations from this study can also be extended to other megacities in India facing similar pandemic-related difficulties.

6 Conclusion

Incorporating spatial pattern analysis alongside socio-demographic and risk perception data enables a comprehensive understanding of how diverse factors contribute to the pandemic's spread. Recent literature supports this approach, with studies such as those by Hu et al. (2024) emphasizing the necessity of considering spatial variability in disease modeling and intervention planning. Further underline the value of GWR in uncovering nuanced spatial patterns that broader models might miss, reinforcing the relevance of this approach in pandemic management.

The practical applications of this research are significant. By identifying areas with high disease spread and examining socio-demographic factors, local authorities can tailor public health strategies to address specific needs and challenges. For instance, high-density urban areas and economically disadvantaged neighborhoods might require targeted awareness campaigns and increased healthcare resources. The perception study of the population's risk awareness and response further informs the development of effective communication strategies and interventions, ensuring they resonate with the public's experiences and concerns.

Future research should expand on these findings by integrating additional variables such as mobility patterns, access to healthcare, and longitudinal changes in risk perception. Multi-scale analyses could provide deeper insights into how disease dynamics evolve over time and space. Additionally, exploring the interplay between social behaviors and spatial patterns can enhance our understanding of pandemic responses and improve public health strategies. By continuing to bridge the gap between spatial data, social behavior, and disease dynamics, future studies can contribute significantly to the field of social physics and the broader public health domain.

AI Declaration This chapter has undergone grammatical proofreading and language refinement using AI-based tools. However, the core research work, including data collection, data analysis, and result representation, has been conducted solely by the authors. The intellectual content, conclusions drawn, and interpretations presented within this chapter are entirely the product of the authors' own efforts and research. The AI was not involved in the generation, validation, or interpretation of the research data or findings.

References

Anselin, L. (1995). Local indicators of spatial association—LISA. *Geographical Analysis, 27*(2), 93–115.

Atkinson, R., & Flint, J. (2001). Accessing hidden and hard-to-reach populations: Snowball research strategies. *Social Research Update, 33*, 1–4.

Bag R, Ghosh M, Biswas B, Chatterjee M. (2020). Understanding the spatio-temporal pattern of COVID-19 outbreak in India using GIS and India's response in managing the pandemic. *Regional Science Policy and Practice, 12(6), 1063–1103.* https://doi.org/10.1111/rsp3.12359

Baltar, F., & Brunet, I. (2012). Social research 2.0: Virtual snowball sampling method using Facebook. *Internet Research, 22*(1), 57–74. https://doi.org/10.1108/10662241211199960

Bhadra, A., Mukherjee, A., & Sarkar, K. (2021). Impact of population density on Covid-19 infected and mortality rate in India. *Modeling Earth Systems and Environment, 7(1), 623–629.* https://doi.org/10.1007/s40808-020-00984-7

Bivand, R. S., Pebesma, E., & Gomez-Rubio, V. (2013). *Applied spatial data analysis with R*. Springer.

Brockmann, D., & Helbing, D. (2013). The hidden geometry of complex, network-driven contagion phenomena. Science, 342(6164), 1337-1342. https://doi.org/10.1126/science.1245200. Erratum in: Science. 2014 Feb 14;343(6172):730

Cartenì, A., Di Francesco, L., & Martino, M. (2021). The role of transport accessibility within the spread of the coronavirus pandemic in Italy. *Safety Science, 133*, 104999. https://doi.org/10.1016/j.ssci.2020.104999

Cellmer, R., Cichulska, A., & Bełej, M. (2020). Spatial analysis of housing prices and market activity with the geographically weighted regression. *ISPRS International Journal of Geo-Information, 9(6), 380.* https://doi.org/10.3390/ijgi9060380

Clark, P. J., & Evans, F. C. (1954). Distance to nearest neighbor as a measure of spatial relationships in populations. *Ecology, 35*(4), 445–453.

Cliff, A. D., & Ord, J. K. (1981). *Spatial processes: Models and applications*. Pion.

Dillman, D. A., Smyth, J. D., & Christian, L. M. (2014). *Internet, phone, mail, and mixed-mode surveys: The tailored design method*. John Wiley & Sons.

Dryhurst, S., Schneider, C. R., Kerr, J., Freeman, A. L. J., Recchia, G., van der Bles, A. M., Spiegelhalter, D., & van der Linden, S. (2020). Risk perceptions of COVID-19 around the world. *Journal of Risk Research, 23*(7–8), 994–1006. https://doi.org/10.1080/13669877.2020.1758193

ESRI. (2020). *ArcGIS Pro: Advanced analysis for 2D, 3D, and 4D GIS*. Environmental Systems Research Institute.

Fotheringham, A. S., Charlton, M. E., & Brunsdon, C. (2002). *Geographically weighted regression: The analysis of spatially varying relationships*. John Wiley & Sons.

Getis, A., & Ord, J. K. (1992). The analysis of spatial association by use of distance statistics. *Geographical Analysis, 24*(3), 189–206.

Gollini, I., Lu, B., Charlton, M., Brunsdon, C., & Harris, P. (2015). GWmodel: An R package for exploring spatial heterogeneity using geographically weighted models. *Journal of Statistical Software, 63*(17), 1–50.

Hsiang, S., Allen, D., Annan-Phan, S., Bell, K., Bolliger, I., Chong, T., et al. (2020). The effect of large-scale anti-contagion policies on the COVID-19 pandemic. *Nature, 584*(7820), 262–267.

Hsu, C. C., & Sandford, B. A. (2007). The Delphi technique: Making sense of consensus. *Practical Assessment, Research, and Evaluation, 12*(1), 1–8.

Hu, N., Zhang, Z., Duffield, N., et al. (2024). Geographical and temporal weighted regression: Examining spatial variations of COVID-19 mortality pattern using mobility and multi-source data. *Computational Urban Science, 4*, 6. https://doi.org/10.1007/s43762-024-00117-1

Jalali, R., & Etemadfard, H. (2024). Spatio-temporal analysis of COVID-19 lockdown effect to survive in the US counties using ANN. *Scientific Reports, 14, 19608.* https://doi.org/10.1038/s41598-024-70415-5

Jasim, I. A., Fileeh, M. K., Ebrahhem, M. A., et al. (2022). Geographically weighted regression model for physical, social, and economic factors affecting the COVID-19 pandemic spreading. *Environmental Science and Pollution Research, 29, 51507–51520.* https://doi.org/10.1007/s11356-022-18564-w

Joshi, A., Kale, S., Chandel, S., & Pal, D. K. (2015). Likert scale: Explored and explained. *British Journal of Applied Science & Technology, 7(4), 396.*

Jusup, M., Holme, P., Kanazawa, K., Takayasu, M., Romić, I., Wang, Z., Geček, S., Lipić, T., Podobnik, B., Wang, L., Luo, W., Klanjšček, T., Fan, J., Boccaletti, S., & Perc, M. (2022). Social physics. *Physics Reports, 948*, 1–148. https://doi.org/10.1016/j.physrep.2021.10.005

Kawachi, I., & Ransome, Y. (2024). Social capital, social cohesion, and COVID-19. In D. T. Duncan, I. Kawachi, & S. S. Morse (Eds.), The social epidemiology of the COVID-19 pandemic (New York, 2024; online edn, Oxford Academic, 23 May 2024). https://doi.org/10.1093/oso/9780197625217.003.0015

Kawuki, J., Chen, S., Fang, Y., Liang, X., Chan, P. S., & Wang, Z. (2023). COVID-19 vaccine acceptance, attitude and perception among slum and underserved communities: A systematic review and meta-analysis. Vaccines (Basel), 11(5), 886. https://doi.org/10.3390/vaccines11050886

Kraemer, M. U., Yang, C. H., Gutierrez, B., Wu, C. H., Klein, B., Pigott, D. M., et al. (2020). The effect of human mobility and control measures on the COVID-19 epidemic in China. *Science, 368*(6490), 493–497.

Liu, Y., Gu, Z., Xia, S., Shi, B., Zhou, X. N., Shi, Y., & Liu, J. (2020). What are the underlying transmission patterns of COVID-19 outbreak? An age-specific social contact characterization. *EClinicalMedicine, 22*, 100354. https://doi.org/10.1016/j.eclinm.2020.100354

Liu, W., Wang, D., Hua, S., et al. (2021). Spatiotemporal analysis of COVID-19 outbreaks in Wuhan, China. *Scientific Reports, 11, 13648.* https://doi.org/10.1038/s41598-021-93020-2

Mishra, S. V., Gayen, A., & Haque, S. M. (2020). COVID-19 and urban vulnerability in India. *Habitat International, 103, 102230.* https://doi.org/10.1016/j.habitatint.2020.102230

Moran, P. A. P. (1950). Notes on continuous stochastic phenomena. *Biometrika, 37(1/2), 17–23.*

Moussaïd, M., Eltahir, R., & Helbing, D. (2020). Social influence and the collective dynamics of opinion formation. *PLoS One, 8*(11), e78433. https://doi.org/10.1371/journal.pone.0078433

Nair, N., Selvaraj, P., & Nambudiri, R. (2022). Culture and COVID-19: Impact of cross-cultural dimensions on behavioral responses. *Encyclopedia, 2*(3), 1210–1224. https://doi.org/10.3390/encyclopedia2030081

Olney, A. M., Smith, J., Sen, S., Thomas, F., & Unwin, H. J. T. (2021). Estimating the effect of social distancing interventions on COVID-19 in the United States. *American Journal of Epidemiology, 190(8), 1504–1509.* https://doi.org/10.1093/aje/kwaa293

Páez, A., Farber, S., & Wheeler, D. (2011). A simulation-based study of geographically weighted regression as a method for investigating spatially varying relationships. *Environment and Planning A: Economy and Space, 43(12), 2992–3010.* https://doi.org/10.1068/a44111

Pallant, J. (2020). *SPSS survival manual: A step by step guide to data analysis using IBM SPSS.* Routledge.

Patel, J. A., Nielsen, F. B. H., Badiani, A. A., Assi, S., Unadkat, V. A., Patel, B., Ravindrane, R., & Wardle, H. (2020). Poverty, inequality and COVID-19: The forgotten vulnerable. *Public Health, 183*, 110–111. https://doi.org/10.1016/j.puhe.2020.05.006

Pentland, A. (2014). *Social physics: How social networks can make us smarter. Penguin Books.*

Polit, D. F., & Beck, C. T. (2006). The content validity index: Are you sure you know what's being reported? Critique and recommendations. *Research in Nursing & Health, 29*(5), 489–497.

Rai, A. (2019). Kolkata: A medical tourism destination in eastern India. In *Medical tourism in Kolkata, eastern India. Global perspectives on health geography*. Springer. https://doi.org/10.1007/978-3-319-73272-5_2

Rai, A., Routh, D., & Bhunia, G. S. (2024). Measuring social vulnerability indicators to CoVID pandemic: A GIS based analysis in North 24 Parganas, West Bengal, India. *IOP Conference Series: Earth and Environmental Science, 1382*, 012013. https://doi.org/10.1088/1755-1315/1382/1/012013

Rao, K D., Kaur, J., Peters, M A., Kumar, N., & Nanda, P. (2021). Pandemic response in pluralistic health systems: A cross-sectional study of COVID-19 knowledge and practices among informal and formal primary care providers in Bihar, India. *BMJ Open, 11*(4), e047334. https://doi.org/10.1136/bmjopen-2020-047334

Rendana, M., Idris, W. M. R., & Rahim, S. A. (2021). Spatial distribution of COVID-19 cases, epidemic spread rate, spatial pattern, and its correlation with meteorological factors during the first to the second waves. *Journal of Infection and Public Health, 14(10), 1340–1348.* https://doi.org/10.1016/j.jiph.2021.07.010

Ritter, L. A., & Sue, V. M. (2017). *Using online surveys in evaluation. John Wiley & Sons.*

Rogerson, P. A. (2001). *Statistical methods for geography*. Sage Publications.

Saeed, S., Niazi, F., Kaur K. N., Rana, S., Singhal, M., & Singh, H. (2024). Assessing patient satisfaction among COVID-19 survivors in Northeast India: A cross-sectional study. *Frontiers in Medicine, 11*, 1308288. https://doi.org/10.3389/fmed.2024.1308288

Saleh, A., & Bista, K. (2017). Examining factors impacting online survey response rates in educational research: Perceptions of graduate students. *Journal of MultiDisciplinary Evaluation, 13(29), 63–74.*

Sarkar, A., Chouhan, P., & Singh, P. (2021). A spatial-temporal analysis of COVID-19 and its impact on urban systems in India: A case study of Delhi-NCR. *Journal of Urban Management, 10*(4), 263–279.

Tobler, W. R. (1970). A computer movie simulating urban growth in the Detroit region. *Economic Geography, 46*, 234–240.

What is containment zone? (2021). *Union Ministry for Health and Family Welfare.* [Online]. Available at: https://www.mohfw.gov.in. Accessed Aug 2021.

Wong, D. W., & Lee, J. (2005). *Statistical analysis of geographic information with ArcView GIS and ArcGIS*. Wiley.

You, H., Wu, X., & Guo, X. (2020). Distribution of COVID-19 morbidity rate in association with social and economic factors in Wuhan, China: Implications for urban development. *International Journal of Environmental Research and Public Health, 17*(10), 3417. https://doi.org/10.3390/ijerph17103417

Zhou, Y., Jiang, H., Wang, Q., Yang, M., Chen, Y., Jiang, Q., … & Tong, Z. (2020). Effects of human mobility restrictions on the spread of COVID-19 in Shenzhen, China: A modelling study using mobile phone data. *The Lancet Digital Health, 2*(8), e417–e424. https://doi.org/10.1016/S2589-7500(20)30165-5

Ms. Sanjana De is an alumnus of Department of Geography in School of Basic and Applied Sciences (SoBAS) of Adamas University. Being an urban and regional planning specialist, Ms. De is concerned about the development of any place. As there are many dimensions of development and health is one of them, it is important to know the health status of a place, so how to manage the hustle condition during an epidemic and how to confirm the benefit of all levels of people in the society is an important aspect of her work.

Dr. Anu Rai is an Assistant Professor of Geography in School of Basic and Applied Sciences (SoBAS) of Adamas University. Being specialised in Regional Planning and Development, Dr. Rai is much interested in exploring the anatomy of spatial events from the perspective of neighbourhood analyst. Mapping infrastructure accessibility and availability for vulnerable communities; modelling spatial interactions and resultant space reconstruction and developing more viable, transparent and real option strategies are her core domain of investigation.

With keen interests in the issues of healthcare accessibility, availability and patient-migration in search of better opportunity for healthcare services, Dr. Rai has authored book entitled "Medical Tourism in Kolkata, Eastern India" published by Springer Nature Switzerland AG, Nov 2019. This book examines the global influence and scope of medical tourism with an emphasis on the city of Kolkata in Eastern India as an emerging destination at the regional scale. Through a geographical research perspective, the book discusses the importance of the phenomenon of medical tourism including recent trends, policies, and scale studies to develop sustainable strategies for medical tourism at particular micro destinations.

Impact of Evolving Contemporary Social Behaviour on Geoheritage Site from a Geolocation Perspective: A Tale of Queen City of Himalayas—Darjeeling

Joydip Datta, Prathama Paul, and Sayan Dey

1 Introduction

Over the past few decades, the preservation of geosites for scientific purposes along with educational purposes has received a lot of attention in the field of geoconservation. Simultaneously, there has also been an increasing acknowledgement of the societal and aesthetic significance of geoheritage, particularly in connection with the growth of geotourism (Brilha, 2016) as well as UNESCO Global Geoparks, which are mandated to provide geoconservation and a host of other educational, financial, cultural and societal advantages (Mc Keever & Zouros, 2005; Brilha, 2016). During the 1990s, as ecotourism and sustainable growth were becoming more and more popular, geotourism was born out of the need to raise consciousness about geoheritage and its standards outside of the geoscience sector and to generate encouragement for geoconservation. Tourism centred on the visual enjoyment of the natural terrain and unique geologic "wonders" represents not a contemporary phenomenon; rather, it is a forerunner to modern geotourism. It has a history that spans over 2000 years in Western Europe as well as quite a while in Asia in different forms (Mccabe et al., 2015).

Geoheritage and geoconservation are the two concepts that take action of preservation and protection of natural characteristics on the planet. Rock monuments, geology, fossil parks and land formations are all covered. Geoheritage also comprises archaeological sites and cultural sites of unsegregated importance in response to a specific area. Geotourism's major goal is to preserve natural assets and cultural values through education, entertainment and interpretation (Gordon, 2018;

J. Datta (✉)
Manager (Pre Sales) — ESRI India, Kolkata, West Bengal, India

P. Paul · S. Dey
Department of Geography, University of Calcutta, Kolkata, India

A. Rai et al. (eds.), *Contemporary Social Physics*, Springer Geography,
https://doi.org/10.1007/978-3-031-77596-3_15

Panneerselvam et al., 2019). Dowling (2014) said emphatically that geotourism is a sustainable form of tourism that mainly concentrates on the experience of the geological characteristics of the Earth in a way that promotes environment, culture and conservation and is beneficial locally. Farid (2015) and Panizza (2009) said all natural qualities on the exterior of the Earth, such as a gorge, a hilltop, or a sea cliff, have cultural worth because they are part of a certain cultural region. Natural heritage sites can sometimes become a hallowed location for humans, and they are recognized as an important component of geo-cultural knowledge (Binoy, 2011). The landscape is frequently used to communicate architectural design, writings, poetry and artwork (Lindberg & Johnson, 1997). By the reason for territorial resource geoconservation, all of these factors are incorporated into the idea of "geoheritage". Geotourism aids in the identification of geoheritage resources, which are then attempted to be used for long-term economic development with the active engagement of the local population (Brown et al., 2016; Halder & Sard, 2021). In the recent past, many academics have conducted scientific analyses addressing prospective geoheritage areas in various sections of the state. Suryawanshi and Ranyewale (2018) compared the geomorphosite potential of beach sites along the Sindhudurg coast (Maharashtra). The relevance of Assam's Majuli Island, the core of Neo-Vaishnavite ethics in India, has been explained by Roy et al. (2020) since it sustains a large number of individuals who are frugally dependent on it. It is recognized as India's heritage of geo-culture. Bhosale et al. (2021) assessed the empiric, conservation and educational ethics for diverse Mesozoic and Cenozoic fossil flora and fauna as a prospective geopark in Western India's Kachchh basin. In India, there are various mountain railways, which are continuing, i.e. (a) Kalka-Shimla Railway, (b) Kangra Valley Railway, (c) Darjeeling Himalayan Railway, and (d) Nilgiri Mountain Railway (Jahan, 2019). Darjeeling is one of the most spectacular hill resorts in the world. Blazing rhododendrons, effervescent milky magnolias and yards of undulating mountain sides covered with green tea trees, the silver fir forests—all under a blanket of a cerulean sky speckled with flecks of the clouds compelling to consider the district as "Queen of Hill Stations" (Dasgupta, 1999). The mountaintop of Kanchenjunga shining in the first dawn light supports it truly (Roy & Hannam, 2013). The foremost and the most magnificent example of mountain commuter railway is the Darjeeling Himalayan Railway (DHR). In 1879, to reduce the haulage charges in the hill section and due to the commercial viability of tea export, the railway came into existence. Primarily, it was known as Darjeeling Steam Tramway Co. and subsequently renamed as Darjeeling Himalayan Railway Co. (September 15, 1881). DHR was built in 1886, to permit the line to cross the tightly quirk way through mountains, A 2 foot gauge was constructed of the Darjeeling line (600 mm). The official starting date of the DHR, beginning at Siliguri to Kurseong, is August 23, 1880. Coming from Kurseong to Sonada, the date was August 1, 1881; Sonada to Jorebunglow April 5, 1881; Jorebunglow to Darjeeling July 4, 1881; and Darjeeling to Darjeeling Bazar June 16, 1886. On December 2, 1999, UNESCO announced DHR as the World Heritage Site (Coulls et al., 1999). Later, a couple of additional railway cords were added. The site became well-known as the mountain railways of India (Lancashire, 2020). Most importantly, DHR has no tunnels, and

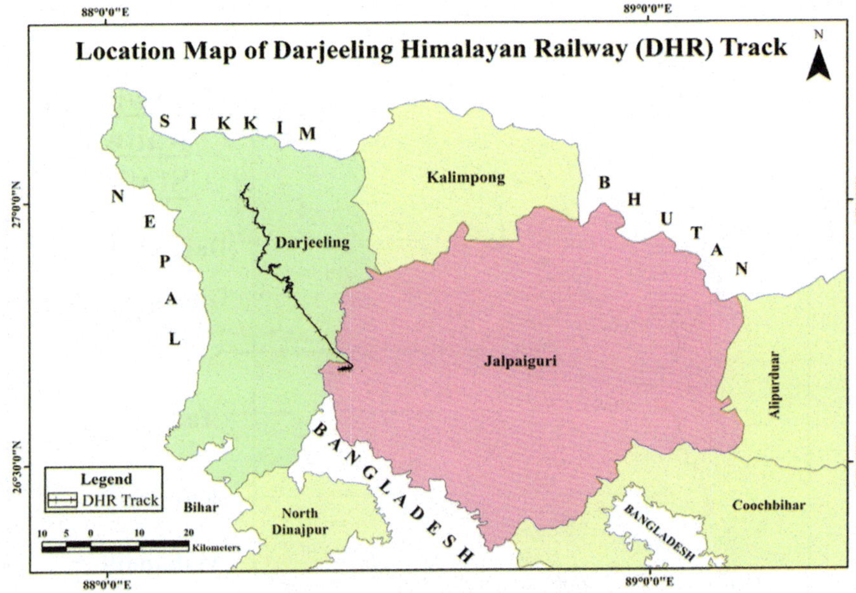

Fig. 1 Location map of Darjeeling

the total number of stations is 13 and the entire length of DHR is 88 km. The operational speciality of DHR is the gauge, which is about 2–00 feet (610 mm), with a total distance of 87.48 km. There are 14 B Class Steam Locos and 4 NDM6 Class Diesel Locos. There is no fixed signal from Sukna to Darjeeling. Hand signals and written order control trains. Trains can follow each other quickly, and two trains are crossed in mid-section and intermediate siding (Thapa, 2017a) (Figs. 1, 2 and 3).

Darjeeling is distinctive from natural eco-perception. The topography differs from hundred feet to more than 8500 feet. Various meteorological zones with unique natures and imperiled animals such as red panda and memory orchids and medicinal herbs are within easy reach. In this area, the rock structure is relatively new. It has a direct influence on landslides. Heavy monsoonal precipitation is a widespread reason for colonial disasters (Dikshit et al., 2018) (Fig. 4).

Depending on height, slope angle, negative cover and lithology, soils of these regions are also different. The natural erosion system in the mountain becomes more complex with human interference (Sharma, 2015). The mountains act as the storehouse of resources for the people residing in the mountains and the plains. So the environmental exploitation is extensive. More specifically is applied to the extraction of wood; forest goods, mining and cultivation are taken into consideration (Giri, 2019; Cajee, 2018). As the total population increases (Chhetri & Tamang, 2013), forests are depleted to extend agricultural lands, roadways, new settlements, etc. Earth has started behaving to the accumulated mankind guilt due to the unbalanced growth of the people during the last few decades in Darjeeling. Landslides in the rainy season are a basic factor for the residents. The rugged

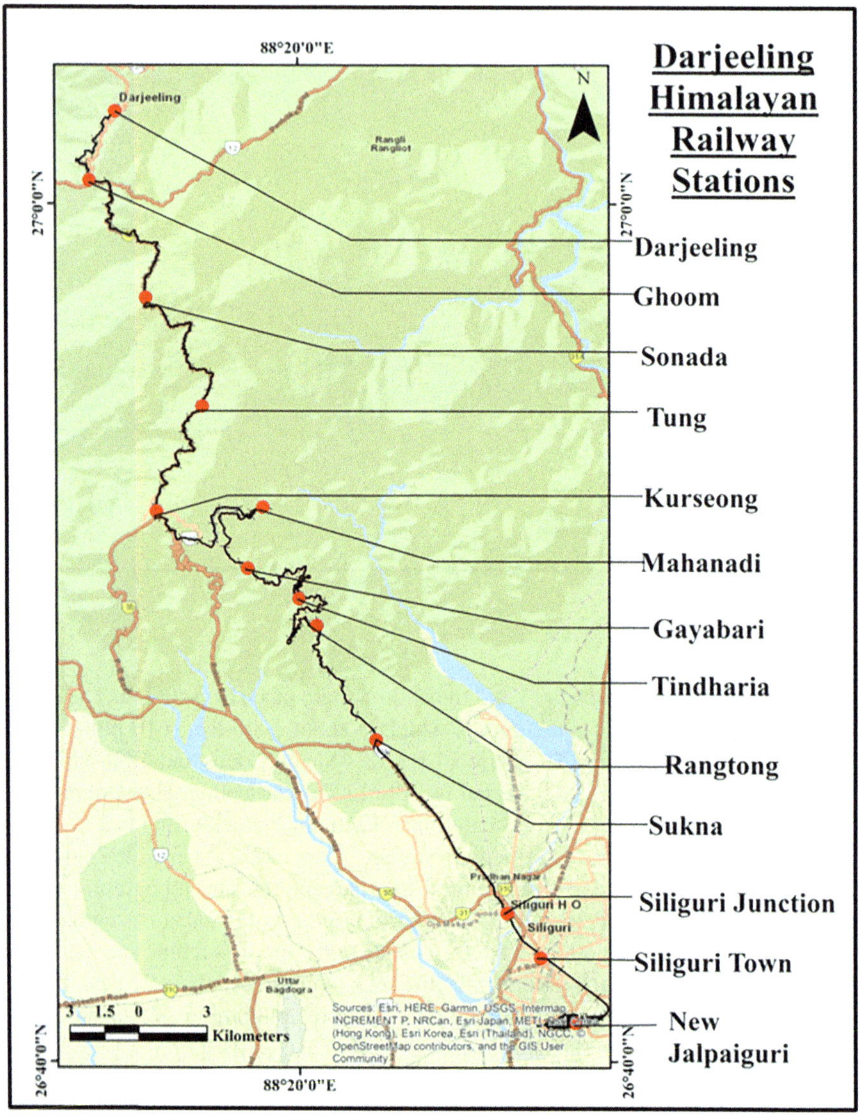

Fig. 2 Location of DHR stations

Kalimpong Hill is decadent by entirely dropping gullies and water bodies and contributing to Tista and Jaldhaka River system. In 2 years (1879–1881) at the cost of INR 2.8 million by Gillanders Arbuthnot & Company.

This paper examines how the connections between geoheritage and the arts can be created to improve the experience of visitors and grow geoconservation. The impact of evolving contemporary social behaviour on geoheritage sites will be analysed through the geolocational perspective.

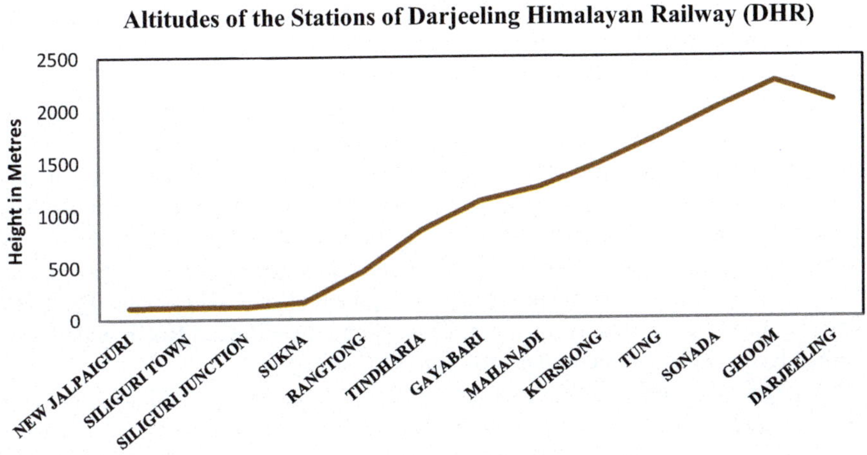

Fig. 3 Altitude of DHR stations

Fig. 4 Logo of DHR

2 History of Construction and Development of the DHR

In 1878, at the Himalayan foot, Siliguri was connected by a wide-gauge railway to Calcutta (now Kolkata). Tonga is located halfway between Darjeeling and Siliguri. Services were provided by single cart road (which is still in use today). Franklin Prestage is an Eastern Bengal Agent. The Railway Company made an application to the Government. The construction of a steam tramway between Siliguri and Kolkata has been proposed. The Lieutenant Governor of Bengal, Sir Ashley Eden, convened a committee to assess the situation. Following a favourable report, the idea was accepted in 1879.

The construction was delegated to Gillanders Arbuthnot & Company. The railroad was completed to Tindharia by March 1880. The first viceroy to visit Darjeeling, Lord Lytton, was transported to Tindharia by train. On August 23, 1880, the section between Siliguri and Kurseong was completed. On July 4, 1881, the Siliguri-Darjeeling railway was opened. The Darjeeling Himalayan Railway Company was quickly renamed. The train was first aligned with the Hill Cart Road. However, this became clear that the route's steepness in some parts was too much for the locomotives to handle. To make the gradient easier, four reverses (zigzags) and four loops were built between Gayabari and Sukna in 1882. In 1886, the line has been extended to Darjeeling Bazar a fourth mile. The Darjeeling railway station was modernizing in 1891, and by 1896, Kurseong was given a new house and storehouse facility. An earthquake struck the DHR in 1897, followed by a strong storm in 1899.

In the middle of 1909–1910, the Darjeeling Himalayan Railway beard 174,000 commuters and 47,000 bunch of cargo per year. The first trucks were put into operation, replacing the more primitive four-wheeled carriages. By 1914, Kishanganj had been completed with DHR expansion lines and, then in 1915, up to Gielkhola. The railway operations at Tindharia were located from in back of the loco shed to a new and bigger area. The loop of Batasia was designed by 1919 to reduce problems on the conquest from Darjeeling through a development of smoother grades. The DHR began competing on the bus route that began along the Hill Cart Road, which consumed lesser time than the railway to reach Darjeeling.

A large earthquake in Bihar struck Northeast India in 1934. Many structures were severely collapsed, as was the railway, which recovered quickly and enacted a major contribution in the provision of repair substances. During the Second World War, the Darjeeling Himalayan Railway was instrumental in delivering defence personnel and providing various cantonments in Darjeeling and Ghum.

After independence, it was bought by the Government of India and included in Government Railways concern (Wenner, 2021; Middleton, 2021). Assam Railways organization took over the management of DHR. Assam Railway, comprising DHR, in 1952, was appointed to the North Eastern Railway. At Siliguri, the railway was readjusted in 1962 and expanded to New Jalpaiguri (NJP) by roughly 4 miles (6 km) for the new broad-gauge line. In the same year, it opened for freight, and in 1964, it opened for people. Siliguri Junction's loco shed and carriage depot was moved to NJP. During the Gorkhaland Movement's hostile era in 1998–1999 (Rappaport, 2021), DHR stayed shuttered for 18 months. UNESCO designated DHR as an ancestry in 1999, making it just the world's second track to receive the honour just after Austria's Semmering Railway in 1998.

3 Contemporary Social Behaviour on Geoheritage Sites

The COVID-19 pandemic's detrimental effects have made sustainable growth an unfamiliar obstacle for modern human life in all spheres. The extensive international travel bans caused a sharp drop in both arrivals and profits for tourism; plus the current dire circumstances have also had a negative impact on heritage travel (UNWTO, 2011). A new normal paradigm should be sought by the tourism sector, which is a vital component of society, to address the catastrophe that a global epidemic has brought upon mankind, communities and society at large (Bae & Chang, 2020). With complete or partial closures due to the ongoing pandemic issue, legacy tourism is going through a very trying period in history. This has had a severe negative impact on economic growth and the cultural fabric of towns and nations.

4 The Planned Behaviour Theory

The concept of planned behaviour (TPB) is frequently employed in the field of social psychology to forecast human decisions or behaviour because of its strong predictive ability (Sommer, 2011). It is a more comprehensive theoretical framework for Ajzen's (1991) theory of reasoned action (TRA). TPB has additionally been successfully used in studies on eco-friendly decision-making in the tourist and hospitality sectors (Hyun, 2017). Within the settings of drone delivery services for food, festival tourism, the cruise industry and other related contexts, numerous researchers have investigated extended theoretical frameworks for combining norm activation model along with TPB to forecast the growth of pro-social behaviour (Chi, X., & Han, H., 2020; Chi, X., & Han, H., 2021; Chi et al., 2021). In these models, two crucial components of the TPB method theory variables are consistently emphasized when paired with norm activation model (NAM). However, since the years following the pandemic, there has been minimal research on merging both the TPB and the NAM in historical tourism.

5 Data Source and Methodology

This research has followed qualitative, quantitative and ethnographic research techniques. It is rooted on both primary and secondary data sources. It has been formed on empirical consideration available from books, journals, research reports and web sources. A compilation of available literature observed the situation in the study area through distinct observations. Primary surveys were done in the years of 2014 (on December), 2017 (on December) and 2019 (on December) by a field visit to understand the condition of DHR from the ground reality. The observation is made through one-to-one interaction with the residents, businessmen related to tourism and tourists to understand their perception about DHR and its impact on the local economy and tourism (Sharma & Sharma, 2017). Discussions with some of the administrators, NGOs, travel agencies, etc. who are assiduously working on tourism sector made it more significant (Vijayalakshmi & Milton, 2020).

Data analysis is represented through elementary percentage method. Collected information are computed, analysed and represented with the help of MS Word and MS Excel.

6 Result and Discussions

The breath-taking journey of the DHR toy trains travels on the narrow path along fascinating scenic mountain roads. There are several small stations along its way. At the elevation of 7404 ft., Ghoom, the apex of the course, is the uppermost train station in Asia and the second highest in the Earth. In 1896, when Mark Twain visited Darjeeling, he said after his toy train ride, "It is the most enjoyable day I have spent on earth". These toy trains depart from New Jalpaiguri (NJP) railway station and travel approximately 88 km to Darjeeling. It clutched more than 10 h to get from Darjeeling to the steam engine toy train, whereas the loco-hauled diesel train lasted approximately 7 h and 15 min (Thapa, 2017b).

The Darjeeling Himalayan Railway, designated as a UNESCO's heritage site by 1999, began operations from the 1880s. It marked the start of a contemporary era in the surrounding communities' economic and social lives (Viu et al., 2008). The trains, which connected villages, carried thousands of people and motivational stories and songs that are quiet fresh in everyone's minds. Besides being a unique industrial artefact, mountain railways have such a special place in India's national heritage (Shankar, 2015).

6.1 Present Status of the Line and the Coaches

The NG-DJ segment currently runs for 87.48 km from New Jalpaiguri. This piece of track features 860 curves, 3 loops, 6 reverses, 557 bridges/culverts and 177'C′ Class U/Manned L/X-ings, with a maximum curve angle of 120 degrees and a total length of 61 km. This segment currently has three 13 stations departing from New Jalpaiguri. In addition, between stations, there are 13 sidings and 5 unattended halts (which are no longer in operation).

The following three examples demonstrate DHR's inventiveness:

- Over 150 crossings along the Hill Cart Road.
- Running on loops, the train does a complete circle to finish 20 feet higher.
- Reversing on a "Z" arrangement—forward, reverse and forward at a greater height.

DHR climbs 6812 ft. from New Jalpaiguri, starts at Sukna, allows continuation to Ghum at 7407 ft. and then goes down the last 5 miles towards Darjeeling. After freedom of India, DHR was absorbed into Assam Railways, and only its extension track to Kishanganj was turn into metre gauge. The institution of the Assam-Bengal connection line was shut. The other extended track to Kalimpong station was destroyed by flooding. In 1952, the DHR reopened and united in 1952 with the North Eastern Railway, and in 1958, it became the Northeast Frontier.

6.2 Types of Train Services in DHR

There are six types of train services available in DHR in the Darjeeling district. The Joy Ride Special Train passes beautiful scenes as it mounts fearsome 1 in 20 slopes and loops to increase height. These are divided into Joy Ride Special Trains (Diesel Loco) and Joy Ride Special Trains (Steam Loco). A circuit from Siliguri and

Rongtong station on Saturday and Sunday Jungle Safari Special Train runs. The train goes through Mahananda Wildlife Sanctuary and attracts a large number of tourists visiting Siliguri and the Dooars. The special train, Red Panda, leaves Darjeeling for Kurseong every Tuesday, Thursday and Sunday, with two first-class trains behind the B class locos every week. The passenger trains run between New Jalpaiguri-Darjeeling and Kurseong-Darjeeling, and the charter trains are mainly for tourists coming in bulk, travel agencies, corporate houses, film shootings, etc. Charter trains have a steam loco and two first-class coaches. It is provided throughout the year on a single-trip and round-trip basis with a facility to opt for any station as an originating station/destination station within DHR as per the requirement. Besides that, special coach booking facilities are also available. Passengers can also book individual coaches as well. The Joy Ride of DHR departs from Darjeeling and returns to the same station via Ghum. Mainly two types of engines are there: diesel and steam. Nine trains are run through the day. This facility is only available in first class. The journey duration approximately comprises 2 h and some few minutes.

The train goes from Darjeeling to Ghum and back to Darjeeling on a round-trip basis. A number of renowned sights and attractions are crossed by Joy Ride Special Trains as they climb up fearsome 1 in 20 gradients and ride high above them. Joy Ride Trains are, namely, divided into two categories—Joy Ride Special Trains (Diesel Loco) and Joy Ride Special Trains (Steam Loco). The train stops in 10 min at the Batasia Loop to admire the view from the passenger and then in Ghum to the DHR Railway Museum for 30 min. Tickets to the museum are included. Charter trains for tourists, travel agencies, companies and movie shootings are offered. The compositions of a steam loco and two first-class trains are used by charter trains. All year round, they are offered on a single trip and a round trip with the facility to choose any station within DHR as their originating/destination station. Passengers can book individual coaches, viz. Heritage Coach—124 built in 1917, Heritage Coach—14, Heritage Coach—149 (dinning car), first class coach, second class coach and open and covered wagons.

7 Impact of DHR on Tourism

Tourism creates jobs in a variety of economic sectors. Tourism contributes to the development of long-term income in a community (AUCTUS_ADVISORS, 2019; Rosická et al., 2009). When locals participate in tourism, new sources of income are generated for the community as a whole as well as for individuals. DHR is an interconnected roadside tram that is far larger than other trains. Darjeeling spans across an area of about 1215 square miles, while the hill town of Darjeeling is only 4 square miles. The population of the Darjeeling district was 1,846,823 (census 2011). The current population has grown to about 1.4 lakhs. Local tourism and travel industries are the majority of locals. Many of them have turned residential homes into guest houses, houses or hotels, while several have also joined the DHR as the

Fig. 5 (1, 2 and 3) Locomotive, carriage and wagon workshop (NFR Tindharia) and its layout. (4) Loco shed. (5) Parts of the toy trains. (6) Interior of the old coaches

station manager. The attraction of Darjeeling town has been shown in Fig. 5. It is clear from the diagram that according to local people and the outside tourists, the main attraction is the toy train of DHR. The other interests of the tourists are the scenic beauty, the pleasant climate and the monastery.

DHR bounds the social structures of the whole area. It affects the entire north Bengal tourism as well as West Bengal tourism. DHR works for the welfare of the local people of hills as they depend on the Himalayan Railway a lot. That is way the heritage, the Darjeeling Himalayan Railway, should be maintained, polished and restructured again for the betterment of its glory as well as the people of that state.

7.1 Economic Importance of the Darjeeling Himalayan Railway in Local Level

Tourism is a major economic activity for Darjeeling, which generates local people incomes and jobs (Doganer, 2013). The area receives about 50,000 foreign tourists a year and about 3.5 lakhs of domestic tourists. Thirty per cent of the total tourism sector in the region is generated, with an annual value of around Rs 350 crore. This expenditure on tourism intersects with the economy and creates other economic activities. The multiplier effect is based on the concept of interdependence between various economic sectors. Changes in host economy levels of output, income, jobs, public revenues and exchange flows are greater than the original value of the change. In this sector, employment levels have significant effects on the labour-intensive nature of tourism industries. The most obvious positive impacts of tourism and the local economy are revenue and employment generation (Alexandrakis et al., 2019; A Study on Young Tourist Preferences towards Heritage Attraction: Case Study of Agra and Delhi, 2019). Almost 90% of the people said that DHR has dramatically influenced the tourism industry. DHR has a big part in flourishing the tourism industry of Darjeeling. The opinion of local peoples, hotel owners and tourists are taken for evaluating the contribution of DHR in economic benefit for local people where almost 60–70% of people said that DHR is providing economic benefits to local people and 25–35% of people gave negative comments.

8 Economic Importance of the Darjeeling Himalayan Railway in Regional Level

The Darjeeling Himalayan Railway plays a significant role at the regional level as it is unique in itself. The promotion of toy trains through the advertisements by West Bengal tourism has accrued a lot of importance in the regional economy (Viu et al., 2008). Tourists prefer to carry these toy trains and increase their importance both culturally and economically. Therefore, there is an interlinked relationship between DHR and regional level economy (Farid, 2015; Álvarez-García et al., 2019).

9 Economic Importance of the Darjeeling Himalayan Railway in National Level

Being a UNESCO World Heritage Site, the Darjeeling Himalayan Railway provides tremendous importance towards the national level. The fund provided by UNESCO helped the national railways to develop this DHR sector (Biswas, 2020). Also, the UNESCO tag of DHR attracts many tourists, contributing a lot of GDP to the Indian economy. Not only has this but the climatic condition of Darjeeling also provided an excellent base for shooting films. A good number of Indian movies were already shooting in Darjeeling, where DHR got a different significance. Ultimately, these films earned a lot from the national and international markets and contributed to the national economy.

10 Train Type-Wise Earnings of DHR

The train type of wise earning has been shown in Fig. 6. It can be seen that the share of making is maximum in the joy ride and the shares of others are very much lower. Among all the types, passenger train stays in the first position followed by charter train, and the contribution of others is very low. So it is clear that people prefer to train only for tourism purposes.

11 Cumulative Passenger Flow and Seasonal Earnings of DHR

In Fig. 7, cumulative passenger flow of 6 years and seasonal earning flow Fig. 8 have been shown that the passenger flow and earnings are high is mainly summer season and the lowest in the rainy season. In a hot and humid country like India,

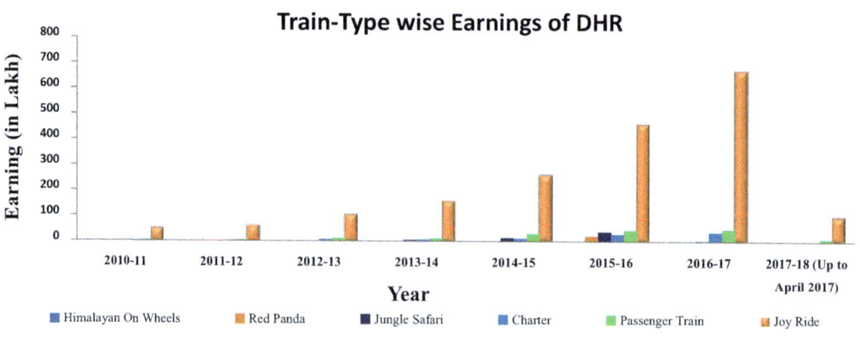

Fig. 6 Train type-wise earnings of DHR

people especially choose to spend their summer holidays in the pleasant weather of Darjeeling in summer and avoid rainy months due to the chances of landslides and blockage of a road in the hilly area (Baker, 2014). Figure 6 shows a temporal fluctuation in the income of DHR. A base year (2011–2012) and a target year (2016–2017) have been selected for the analysis. The month-wise fluctuation of 2011–2012 was more or less the same. But in the present scenario (2016–2017), the month-wise flux is very high. The revenue is increased than the previous days because the infrastructural quality and the publicity are higher than the last days. Local people engaged with the hotel industry, and the tourism industry of Darjeeling condemns that the earnings are gradually increasing day by day because of the increasing flow of tourists, and DHR plays the most.

12 Impacts of Landslides in DHR Service

Landslides have been a problem for the DHR throughout its history, and slope stabilization is a major task for the Railway's continued operation. Another major issue that causes frequent landslides along the roads is slope instability caused by heavy automobile movements along the main thoroughfares, especially during the wet season. The common reason for continual landslides in the Himalayan area is the evolution or rising of young mountains. This comprises unstable geology, tectonic disruptions and aligned setting of Himalayan forward deeps.

Gneiss, schists and phyllites have dominated the soils of the Darjeeling hill areas. Soil erosion has increased especially in recent time as a consequence of heavy deforestation. An analysis of land-use patterns and their changing attributes over the past 150 years reveals a significant transformation in forest cover. This change is largely attributable to the rapid and unsustainable expansion of cultivated land, as

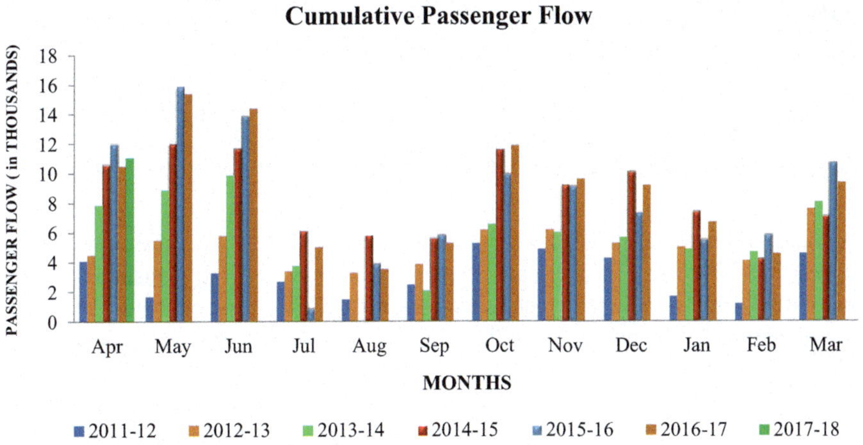

Fig. 7 Cumulative passenger flow

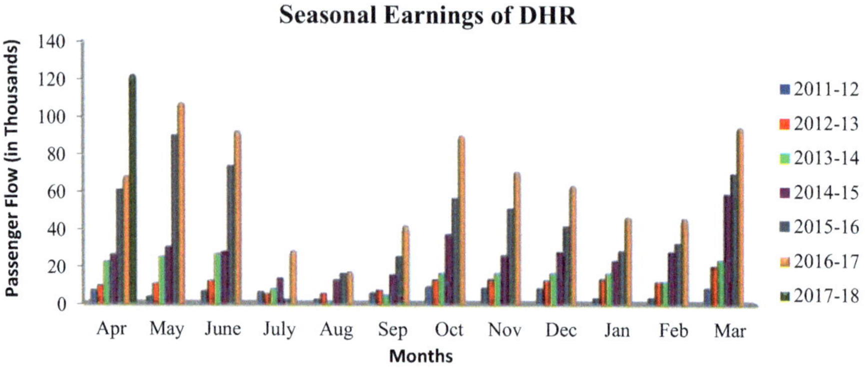

Fig. 8 Seasonal earnings of DHR

well as the extensive development and construction of road networks, which have collectively placed forest ecosystems in a precarious state. Over the last two decade, an unparalleled increase in populace in this hilly area is noticed, particularly in the conurbation area. The detonation was quickly gone after by a growth in automotive movement. Continual horizontal shivering along the roadways gradually undermines the already unsteady slopes and geology. Last but not the least, the stipulation of water is also increased. Deforestation and surface exposure of the hilly terrain are the causes of groundwater depletion, and consequently the streams remain dry for several months in a year.

Field surveys in each year could not be done from 2014 to 2020 due to landslides. In June to July 2015, landslides occurred in various places of Darjeeling. The Siliguri-Darjeeling connecting bridge was washed out at Nimbujhora. NH 10 was also disconnected. This landslide hindered the track of toy trains as well. Later on, Paglajhora faced a massive landslide on August 2018, which disrupted the DHR's tracks again. Due to this situation primary surveys were conducted in 2014 and 2017, with a subsequent field visit undertaken in 2020, adhering to the established consistency in temporal intervals between observations.

13 Relationship Between Earning and Expenditure of DHR

A consumption schedule is a term used to describe the relation between earning and expenditure. To take out information on customer trends within a business, a company should appraise the connection between consumption and economizing. There are three types of income, i.e. regular, irregular and benefit-in-kind. Expenditure is also having three classifications like fixed, irregular and discretionary. Here, in the case of DHR finance, a positive relation between cost and revenue has been shown in Fig. 9. So it is clear that whenever the cost of capital has been invested in the infrastructure, the benefit of income is also reciprocated.

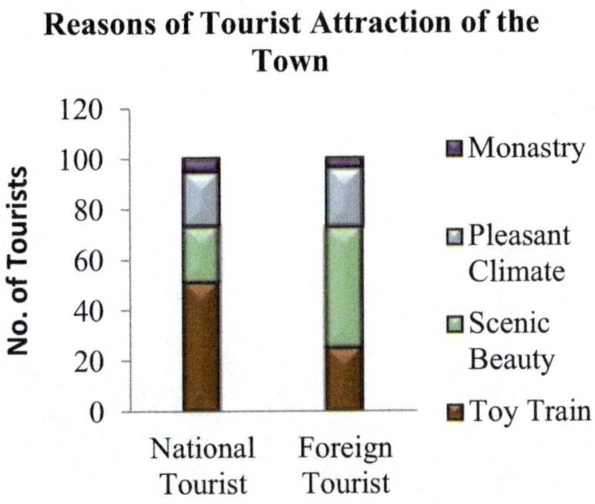

Fig. 9 Reasons of tourist attraction of the town

14 People's Perception About DHR

Local people, persons related to the Darjeeling tourism industry and the tourists have their consents on the Darjeeling Himalayan Railway and toy trains. Residents and businessmen of Darjeeling agreed that toy trains and DHR have major contributions to the GDP of the district. The primary surveys have been done in 2014, 2017 and 2020. These selected survey years have significant influence in the study. In 2014, the joyride was unavailable for a long period of time due to a major landslide in Tindharia. Again in 2017, the journey started without any physical hardship. And in 2020, the wheels of DHR again got stopped because of the lockdown for the spread of COVID-19 pandemic. Inconsistent answers were expected from the respondents due to the fragile availability of the DHR services in the three surveyed years, but surprisingly, the answers were more or less gone on the favour of the Darjeeling Himalayan Railway. The interdependence between Darjeeling tourism and the Darjeeling Himalayan Railway (DHR) is well-recognized, with both national and international tourists acknowledging their intrinsic connection (Raj, 1947). There were various questions regarding the state and Darjeeling Himalayan Railway that help to assess the importance of DHR and regarding heritage as well (Ahamed, 2017). The legacy of DHR is unexplainable. About 56% of tourists who come to Darjeeling with an interval of every 2–3 years said that they love to visit Darjeeling only for the toy trains (Fig. 10).

They also think that the initiative of the Government isn't enough to energize the DHR. According to the repliers, DHR faces economic problems as well as its identity crisis (Ramalakshmi, 2016). Though it is a heritage of UNESCO, lack of maintenance is the main hindrance of the wheels. People are now concurring to raise the fare of DHR for its existence.

About 90% of tourists in 2014, 81% in 2017 and 92% in 2020 admitted that the Darjeeling Himalayan Railway has a huge importance on the area-based economy of the state, whereas 97% of local people said the same thing in 2014 and 2020, and

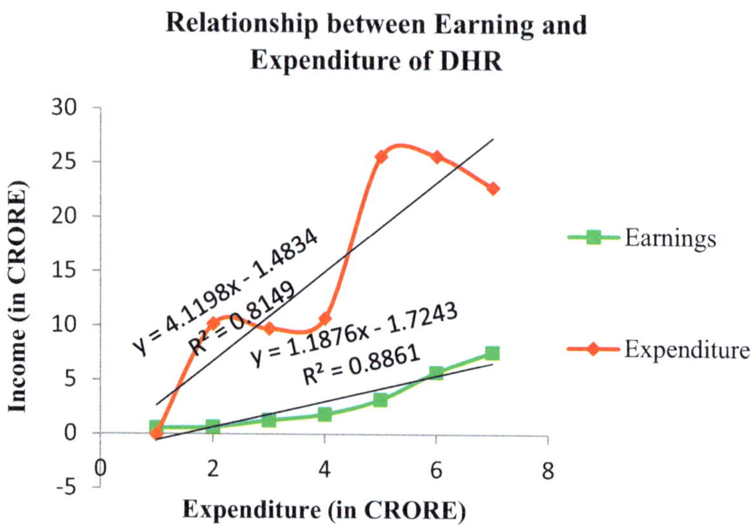

Fig. 10 Relationship between earning and expenditure of DHR

only 2–3% of businessmen denied on this matter. More than half of the respondents stated that DHR is mainly famous for its joy rides; nearly about one fourth of the respondents in every surveyed year went for its heritage and the remaining respondents said for its daily transportation values. Though people utter about DHR, the railway faces an existence crisis due to lack of maintenance and low earnings. At the time of 2014, only 40% agreed to raise the price of the tickets, which increases 75% in 2020 as common people understood that the financial support is one of the important push factors to continue the journey (Figs. 11, 12, and 13).

15 Conclusion

During the agitation of the Gorkhaland issue, DHR service was fully interrupted. Crossing many hurdles, the condition of the DHR is now improving. Recently, A.C. coaches have also been major challenges. Due to mass deforestation and excessive cultivation of roots, soil degradation has risen considerably in the present day (Siddhi et al., 2020). Even in the last two decades, the unprecedented growth of the population is noticed. Technology-related challenges are another (Kumar, 1996; Technology, 2002). In the present-day context, there is a lack of intellect and skills found. Day by day, the competition with other transportations is increasing (Raghu, 2016). There are plenty of private taxis, buses and jeeps available outside the New Jalpaiguri railway station and Bagdogra Airport at a reasonable fare. The speed of the trains is another issue.

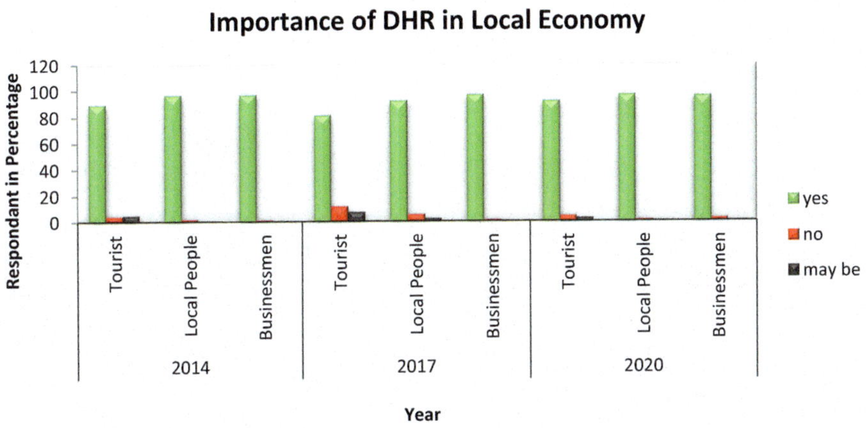

Fig. 11 Importance of DHR in local economy

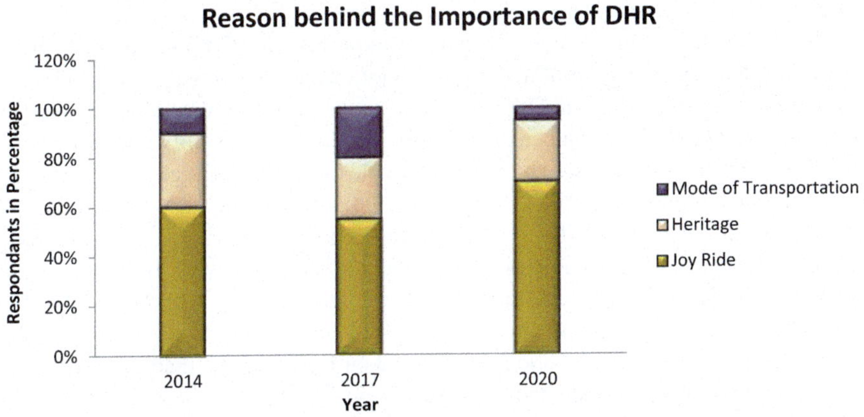

Fig. 12 Reason behind the importance of DHR

The journey from New Jalpaiguri takes approximately 7 h. The operating cost, concerning earnings, is very high. The annual price of running is as high as Rs. 23 crores, whereas the income is about Rs. 7.65 crores only. Thus, it faces the problems of profit maximization. Less local participation is also noticed. The developmental activities promote this railway. Government of India & UNESCO aims to improve the development of the railway sector, leading to local development. Though the NF Railway took some measures in recent years to enlighten the Darjeeling Himalayan Railway like DHR Freight wagon used in Darjeeling Tea (Wenner, 2020) and Tourist Festival, restoration of old heritage car and inspection car, restoration of the tracks in the landslide prone region, interior renovation of NG Coaches, preservation of Old DHR photographs etc. through digitization and archiving, centralized the control office at Kurseong, clearance of garbage from the tracks regularly and etc. still

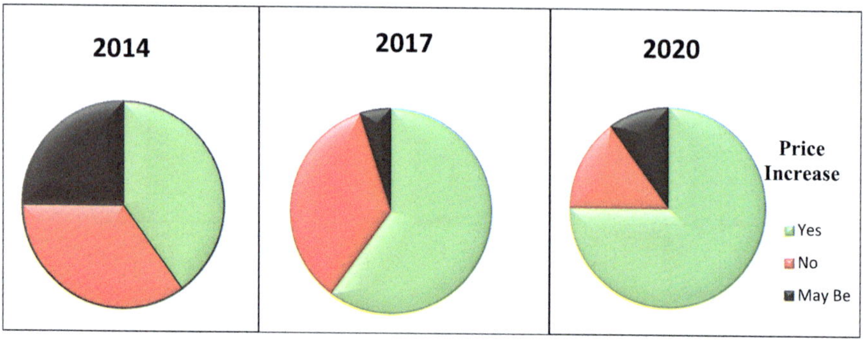

Fig 13 Price increase

it faces natural problems like landslides and other obstacles such as lack of fund, lack of proper planning and maintenance, lack of promotion, frequent political disturbance, insufficient infrastructure (Saha & Chakraborty, 2017; Khawas, 2002; Nilsson, 2018; Shankar, 2015).

The Darjeeling Himalayan Railway is an important piece of our country's heritage. This toy train has a lot of history and emotion associated with it. Darjeeling is synonymous with a toy train. Darjeeling tour is incomplete unless you ride the toy train. It is the pride of the Queen of the Hills, but due to its fragility and inadequate service, UNESCO has decided to remove the title of World Heritage Site. So we can say that our toy train, our pride, is in danger. Other mountain railways in India run more smoothly than the DHR. If the toy train service works properly, it will have a significant economic impact on both the local and national levels. If the passenger service becomes more frequent, then school, college and office workers will be able to use it as before. Its ongoing development will benefit Darjeeling's tourism industry. As a result, immediate action must be taken to locate the matters in order to preserve our symbol of pride and heritage (Story, 2017).

References

A Study on Young Tourist Preferences towards Heritage Attraction: Case Study of Agra and Delhi. (IJREAM) ISSN : 2454–9150 Vol-05, Issue-01, April 2019. https://ijream.org/papers/IJREAMV05I0149062.pdf

Ahamed, M. (2017). Cultural heritage tourism – An analysis with special reference to West Bengal, India. *IJRTBT, 1*(October), 55–62.

Ajzen, I. (1991). The theory of planned behavior. *Organizational Behavior and Human Decision Processes, 50*, 179–211. https://doi.org/10.1016/0749-5978(91)90020-T

Alexandrakis, G., Manasakis, C., & Kampanis, N. A. (2019). Economic and societal impacts on cultural heritage sites, resulting from natural effects and climate change. *Heritage, 2*(1), 279–305. https://doi.org/10.3390/heritage2010019

Álvarez-García, J., Maldonado-Erazo, C. P., de la Río-Rama, M., Del, C., & Castellano-Álvarez, F. J. (2019). Cultural heritage and tourism basis for regional development: Mapping of scientific coverage. *Sustainability (Switzerland), 11*(21), 1–21. https://doi.org/10.3390/su11216034

AUCTUS_ADVISORS. (2019). *Sustainable heritage tourism in India knowledge report. March.*

Bae, S. Y., & Chang, P.-J. (2020). The effect of coronavirus disease-19 (COVID-19) risk perception on behavioral intention towards 'untact' tourism in South Korea during the first wave of the pandemic (March 2020). *Current Issues in Tourism, 24*, 1017–1035. https://doi.org/10.1080/13683500.2020.1798895

Baker, J. C. T. (2014). Mobility, tropicality, and landscape: The Darjeeling Himalayan Railway, 1881–1939. *Journal of Historical Geography, 44*, 133–144. https://doi.org/10.1016/j.jhg.2013.11.003

Bhosale, S., Chaskar, K., Pandey, D. K., Lakhote, A., Thakkar, A., Chauhan, G., Bhandari, S., & Thakkar, M. G. (2021). Jurassic geodiversity and geomorphosite of Kanthkot area, Wagad, Kachchh, Western India. *International Journal of Geoheritage and Parks, 9*(1), 51–68. https://doi.org/10.1016/j.ijgeop.2020.12.008

Binoy, T. A. (2011). Archaeological and heritage tourism interpretation: A study. *South Asian Journal of Tourism and Heritage, 4*(1), 100–105.

Biswas, S. (2020). Process and prospects of Darjeeling hill development. *Golden Research Thoughts*, May.

Brilha, J. (2016). Inventory and Quantitative Assessment of Geosites and Geodiversity Sites: a Review. *Geoheritage, 8*, 119–134. https://doi.org/10.1007/s12371-014-0139-3

Brown, T., Ganguly-Scrase, R., & Scrase, T. J. (2016). Urbanization, rural mobility, and new class relations in Darjeeling, India. *Critical Asian Studies, 48(2), 235–256.* https://doi.org/10.1080/14672715.2016.1164392

Cajee, L. (2018). Physical aspects of the Darjeeling Himalaya: Understanding from a geographical perspective. *IOSR Journal of Humanities and Social Science, 23*(3), 66–79. https://doi.org/10.9790/0837-2303016679

Census of India (2011). National Population Register & Socio Economic and Caste Census

Chhetri, B., & Tamang, L. (2013). Population growth and associated problems: A case study of Darjeeling town. *International Journal of Humanities and Social Science Invention, 2*(5), 63–67.

Chi, X., & Han, H. (2020). Exploring slow city attributes in Mainland China: Tourist perceptions and behavioral intentions toward Chinese Cittaslow. *Journal of Travel & Tourism Marketing, 37*, 361–379. https://doi.org/10.1080/10548408.2020.1758286

Chi, X., & Han, H. (2021). Emerging rural tourism in China's current tourism industry and tourist behaviors: The case of Anji County. *Journal of Travel & Tourism Marketing, 38*, 58–74. https://doi.org/10.1080/10548408.2020.1862026

Chi, X., Han, H., & Kim, S. (2021). Protecting yourself and others: Festival tourists' pro-social intentions for wearing a mask, maintaining social distancing, and practicing sanitary/hygiene actions. *Journal of Sustainable Tourism*, 1–22. https://doi.org/10.1080/09669582.2021.1966017

Coulls, A., Divall, C., Lee, R., & ICOMOS. (1999). Railways as world heritage sites. *Report, 33*, 1. http://www.international.icomos.org/studies/railways.pdf

Dasgupta, A. (1999). Ethnic problems and movements for autonomy in Darjeeling. *Social Scientist, 27*(11/12), 47. https://doi.org/10.2307/3518047

Dikshit, A., Sarkar, R., & Satyam, N. (2018). Probabilistic approach toward Darjeeling Himalayas landslides: A case study. *Cogent Engineering, 5(1), 1–11.* https://doi.org/10.1080/23311916.2018.1537539

Doganer, S. (2013). Cultural heritage tourism research: A sustainable community-based design project for the San Antonio Mission Historic District. *WIT Transactions on Ecology and the Environment, 173*, 219–230. https://doi.org/10.2495/SDP130181

Dowling, R. K. (2014). Global geotourism – An emerging form of sustainable tourism. *Czech Journal of Tourism, 2*(2). https://doi.org/10.2478/cjot-2013-0004

Farid, S. M. (2015). Tourism management in world heritage sites and its impact on economic development in Mali and Ethiopia. *Procedia - Social and Behavioral Sciences, 211*, 595–604. https://doi.org/10.1016/j.sbspro.2015.11.078

Farid, S. (2015). *Tourism in world heritage sites and its impact on economic development: Some African countries case studies*. 721–737.

Giri, N. (2019). A study of forest resources for sustainable rural development in Darjeeling Hills of West Bengal. *Think India Journal, 35*, 1051–1068.

Gordon, J. E. (2018). Geoheritage, geotourism, and the cultural landscape: Enhancing the visitor experience and promoting geoconservation. *Geosciences (Switzerland), 8*(4), MDPI AG. https://doi.org/10.3390/geosciences8040136

Halder, S., & Sarda, R. (2021). Promoting intangible cultural heritage (ICH) tourism: Strategy for socioeconomic development of snake charmers (India) through geoeducation, geotourism, and geoconservation. *International Journal of Geoheritage and Parks, 9*(2), 212–232. https://doi.org/10.1016/j.ijgeop.2021.02.008

Han H., & Hyun, S. (2017). Impact of hotel-restaurant image and quality of physical-environment, service, and food on satisfaction and intention. *International Journal of Hospitality Management, 63*, 82–92. https://doi.org/10.1016/j.ijhm.2017.03.006

Jahan, R. (2019). Heritage tourism in South Tamil Nadu – India. *Journal of Tourism & Hospitality, 8(01), 1–4*. https://doi.org/10.35248/2167-0269.19.8.399

Kaur, G., Makki, M. F., Avasia, R. K., Bhusari, B., Duraiswami, R. A., Pandit, M. K., Fareeduddin, B., & Kad, S. (2019). The late cretaceous-Paleogene Deccan traps: A potential global heritage stone province from India. *Geoheritage, 11*(3), 973–989. https://doi.org/10.1007/s12371-018-00342-1

Khawas, V. (2002). *Environment and rural development in Darjeeling Himalaya: Issues and concern*. November.

Kumar, R. B. (1996). *Indian heritage tourism: Challenges of identification and presentation*. 120–135. http://www.publishingindia.com/GetBrochure.aspx?query=UERGQnJvY2h1cmVzfC80NDQucGRmfC80NDQucGRm

Lancashire, C. (2020). *A critical analysis of dissonant heritage and dark tourism in India: The case of the 'Wall of Truth' memorial*. January.

Lindberg, K., & Johnson, R. L. (1997). Modeling resident attitudes toward tourism. *Annals of Tourism Research, 24*(2), 402–424. https://doi.org/10.1016/s0160-7383(97)80009-6

Mc Keever, P., & Zouros, N. (2005). Geoparks: Celebrating Earth heritage, sustaining local communities. *Episodes, 28*, 274–278. https://doi.org/10.18814/epiiugs/2005/v28i4/006

Mccabe, S., Li, C., & Chen, Z. (2015). Time for a Radical Reappraisal of Tourist Decision Making? Toward a New Conceptual Model. *Journal of Travel Research, 55*, https://doi.org/10.1177/0047287515592973

Middleton, T. (2021). Becoming-after: The lives and politics of quinine's remains. *Cultural Anthropology, 36(2), 282–311*. https://doi.org/10.14506/ca36.2.05

Nilsson, P. Å. (2018). Impact of cultural heritage on tourists: The heritagization process. *Athens Journal of Tourism, 5(1), 35–54*. https://doi.org/10.30958/ajt.5.1.3

Panizza, M. (2009). The geomorphodiversity of the Dolomites (Italy): A key of geoheritage assessment. *Geoheritage, 1*(1), 33–42. https://doi.org/10.1007/s12371-009-0003-z

Panneerselvam, R., Management, H., & Nagar, P. (2019). Heritage tourism in Tamilnadu: Potential and challenges. *Journal Name, 6*(1), 45–49.

Raghu, A. (2016). To study the various factors affecting heritage tourism in India. *Journal Name, 1*, 841–847.

Raj, B. (1947). In search of a 'new home': Anglo-Indians in the Darjeeling hills, 1900-1947. *Journal Name, 12*, 1–8.

Ramalakshmi, V. (2016). A study on the problems and prospects of e-commerce in India with special reference to the retail industry. *ACADEMIA: An International Multidisciplinary Research Journal, 6*(10), 53. https://doi.org/10.5958/2249-7137.2016.00073.2

Rappaport, J. (2021). The problem of caste. In *The disappearing mestizo* (pp. 205–226). https://doi.org/10.2307/j.ctv11smwgb.12

Rosická, Z., Bednářová, D., & Beneš, L. (2009). The development and economic impact of cultural tourism and sustainable heritage management. *Transactions on Transport Sciences, 2*(1), 27–30. https://doi.org/10.5507/tots.2009.005

Roy, S., & Hannam, K. (2013). Embodying the mobilities of the Darjeeling Himalayan Railway. *Mobilities, 8*(4), 580–594. https://doi.org/10.1080/17450101.2012.745695

Roy, N., Pandey, B. W., & Rani, U. (2020). Protecting the vanishing geo-cultural heritage of India: Case study of Majuli Island in Assam. *International Journal of Geoheritage and Parks, 8*(1), 18–30. https://doi.org/10.1016/j.ijgeop.2020.03.003

Saha, B., & Chakraborty, G. (2017). Understanding the demand for self-rule in the Darjeeling hills. *The Wire*. https://thewire.in/politics/darjeeling-hills-self-rule-gorkhaland

Shankar, S. (2015). Impact of heritage tourism in India – A case study. *International Journal of Innovative Research in Information Security, 6*(2), 2014–2016.

Sharma, J. (2015). Producing Himalayan Darjeeling: Mobile people and mountain encounters. *Himalaya, 35*(2), 87–101.

Sharma, A., & Sharma, S. (2017). Heritage tourism in India: A stakeholder's perspective. *Tourism and Travelling, 1*(1), 20–33. https://doi.org/10.21511/tt.1(1).2017.03

Siddhi, E., Vakharia, R., & Bhagat, S. S. (2020). Proposal for one-day tourist circuit in Orchha: Heritage tourism. *Journal Name*, 6804–6808.

Sommer, L. (2011). The Theory Of Planned Behaviour And The Impact Of Past Behaviour. *The International Business & Economics Research Journal*. https://doi.org/10.91-110.10.19030/iber.v10i1.930

Story, C. (2017). How politics subsumed the popular sentiment for Gorkhaland. *The Caravan*, 7–10.

Suryawanshi, R. S., & Ranyewale, S. K. (2018). Tourism potential of geomorphosites: A comparative assessment of selected beach sites in Malvan Tahsil, Sindhudurg Coast of Maharashtra (India). *Transactions of the Institute of Indian Geographers, 40(2), 285.*

Technology, I. (2002). Tourism and heritage. *Journal Name, 8*(1), 666–676.

Thapa, S. (2017a). Darjeeling Himalayan Railways – Steam locomotives and their problems. *International Journal of Advanced Research and Publications, 88*(2), 88–91. http://www.ijarp.org/published-research-papers/aug2017/Darjeeling-Himalayan-Railways-Steam-Locomotives-And-Their-Problems.pdf

Thapa, S. (2017b). Darjeeling Himalayan Railways – Steam locomotives and their problems. *International Journal of Advanced Research and Publications, 88*(2) http://www.ijarp.org/published-research-papers/aug2017/Darjeeling-Himalayan-Railways-Steam-Locomotives-And-Their-Problems.pdf

Vijayalakshmi, K., & Milton, T. (2020). Tourism in Chennai – A special reference to heritage tourism. *International Journal of Management, 11*(12), 1170–1174. https://doi.org/10.34218/ijm.11.12.2020.108

Viu, J. M., Fernández, J. R., & Caralt, J. S. (2008). The impact of heritage tourism on an urban economy: The case of Granada and the Alhambra. *Tourism Economics, 14*(2), 361–376. https://doi.org/10.5367/000000008784460481

Wenner, M. (2020). Towards an alternative Indian tea economy: Examples of producer cooperatives from Darjeeling. *Economic and Political Weekly, 55*(45), 53–60.

Wenner, M. (2021). Trajectories of hybrid governance: Legitimacy, order and leadership in India. *Development and Change, 52*(2), 265–288. https://doi.org/10.1111/dech.12624

World Tourism Organization (2011). 'Resolution 593(XIX) of 13 October 2011', Resolutions Adopted by the General Assembly at Its Nineteenth Session, online available at: https://www.unwto.org/archive/global/event/generalassembly-nineteenth-session [16–12–2021]

Joydip Datta is a Geospatial Practitioner as well as a Geographer. He has completed his graduation in Geography from Vidyasagar College under the University of Calcutta and Masters in Geography from Delhi School of Economics under Delhi University. To achieve specialization in the field of geospatial domain, he completed M. Tech in GIS from NIIT University. His interests are in the field of urban land and infrastructure management, disaster risk reduction, environmental impact assessment, agricultural risk assessment and socio-political analysis using geospatial capabilities. He has over 6 years of experience in the geospatial analytics sector. He started his geospatial career with Sony Engineers' Co-operative Society Ltd. followed by Farm Guide as Agricultural

Risk Analyst. After that, he got an opportunity to be associated with India Smart City Fellowship, under the Ministry of Housing and Urban Affairs. From June 2021, he joined ESRI India as an Assitant Manager and in April 2024 he has been promoted to Manager and taking care of several states of the eastern and northeastern India.

Prathama Paul is a 5 years of experienced Senior Research Analyst from Kolkata, India, with expertise in geography and geospatial sciences. She holds a Master's degree in Geography from the University of Calcutta, specializing in regional planning. Besides that, she carries certificates of virtual diplomas in Smart City technologies and GIS from the University of Toronto and EPFL. Her research focuses on urban planning, urban market analysis, human-environment relationships, socio-economic parameters and sustainable management, with published work on smart transportation, integrated approach to assess the urban growth pattern and loss in urban green space in Kolkata and human rights in Jammu and Kashmir. Skilled in modern technologies, she has potential experience in geospatial analysis, fieldwork, analysis and writing.

Sayan Dey is currently working as PGT in the subject of Geography at DAV Public School, Pandaveswar. He has completed his Master of Science in Geography from Vidyasagar College, University of Calcutta, and has a specialization in regional planning. He has also completed Bachelor's in Education from West Bengal University of Teachers' Training Education Planning and Administration. He is railway enthusiastic and also has interest to solve urban problems.

Unraveling Crime in Kolkata: A Socio-economic Perspective

Neha Arong, Kasturi Mukherjee, Paramita Seth, and Supratim Karmakar

1 Introduction

Urban crime is a significant concern in cities worldwide, with complex social, economic, and environmental factors influencing its occurrence. Urban crime refers to unlawful activities that occur within cities, ranging from petty theft and vandalism to violent crimes such as assault and murder. The factors contributing to urban crime are multifaceted, including socio-economic disparities, unemployment, population density, inadequate urban planning, and lack of access to education and social services (Schleimer et al., 2022). Crime in urban areas often concentrates in certain neighborhoods, creating "hotspots" where criminal activity is more prevalent (Petrič, 2024; Rosenfeld, 2024). As urbanization intensifies, understanding the location and factors contributing to crime becomes increasingly crucial for mitigating crime. One emerging approach to study and address urban crime is through the lens of social physics, a multidisciplinary field that applies principles from physics and mathematics to analyze human behavior and social phenomena (Hälterlein, 2021).

The foundational principles of classical criminological theories have established a framework for elucidating the rational decision-making processes that individuals undertake when involved in criminal activities. According to Shaw and McKay's social disorganization theory, neighborhoods characterized by social instability—such as poverty and ethnic heterogeneity—are linked to high crime rates (Shaw & McKay, 1942). In slum areas, weakened social controls may contribute to higher crime rates as a result of the disorganized social environment (Errol et al., 2021). Merton's strain theory suggests that individuals turn to criminal behavior when they

N. Arong
Department of Geoinformatics, Symbiosis Institute of Geoinformatics, Pune, Maharashtra, India

K. Mukherjee (✉) · P. Seth · S. Karmakar
Department of Geography, Adamas University, Kolkata, West Bengal, India

A. Rai et al. (eds.), *Contemporary Social Physics*, Springer Geography,
https://doi.org/10.1007/978-3-031-77596-3_16

face frustration in attaining culturally approved objectives through lawful methods (Merton, 1938). Cohen and Felson's routine activity theory suggests that crime occurs when motivated offenders encounter suitable targets in the absence of capable guardians. In areas with literacy challenges or high unemployment, disruptions in routines may elevate the likelihood of criminal activities (Cohen & Felson, 2010).

Building upon classical theories, contemporary criminological perspectives offer a more intricate understanding of the intersections between crime and specific socio-economic determinants (Weisburd et al., 2014). Agnew's general strain theory develops from Merton's strain theory, incorporating various negative experiences, such as poverty and unemployment, as sources of strain (Agnew & White, 1992). Individuals facing these strains may turn to criminal behavior as a coping mechanism. Cultural deviance theory posits that certain cultural values in disadvantaged communities may promote criminal behavior. In areas with low literacy rates and high poverty levels, unique subcultures may emerge, legitimizing criminal activities as a means of survival or social status.

Recent literatures also indicate that poverty exerts a significant influence on crime rates through multiple mechanisms, including constrained opportunities, heightened social disorganization, economic strain, and increased prevalence of substance abuse (Santiago et al., 2011). The absence of access to essential resources such as education, healthcare, and viable employment options propels individuals and communities toward illicit pursuits, encompassing activities such as theft and drug trafficking. Economic strain amplifies feelings of desperation and frustration, elevating the propensity for engaging in criminal behaviors. Concurrently, substance abuse becomes both a consequence and a catalyst for criminal activities, particularly contributing to a surge in drug-related crimes. This cyclical relationship between poverty and crime establishes a self-reinforcing loop, where elevated crime rates in impoverished areas act as a deterrent to investment and hinder broader economic development (Kivivuori et al., 2015). Studies (Teltscher, 2002) have emphasized the ongoing correlation between low literacy levels and involvement in criminal activities. The findings underscore the persistent need for educational interventions to address literacy gaps, aligning with global efforts that highlight the role of education in crime prevention. Unemployment remains a crucial socio-economic factor influencing crime, and recent research reinforces the connection between unemployment rates and street crimes in Kolkata. This recognizes the effect of financial instability on crime, emphasizing the importance of tailored employment strategies in urban areas (Paul, 2011).

Labeling theory argues that individuals who are marked as deviant are at an increased likelihood of participating in criminal activities (Gehring & Batista, 2022). In slum environments, negative labels and stereotypes may contribute to a self-fulfilling prophecy, reinforcing criminal behavior as a response to societal expectations. Spatial concentration theory suggests that certain forms of crime may be spatially concentrated in specific areas characterized by poverty, unemployment, and inadequate living conditions. Slum environments, with their clustering of these factors, may create criminogenic spaces where criminal opportunities are more prevalent (Farrell, 2015).

The intricate interplay of poverty and the prevalence of slums creates conditions conducive to heightened illiteracy, fostering an environment that correlates with increased criminal activities (Crespo et al., 2023). The ability to read and write is essential for the development of critical thinking and problem-solving skills, but a lack of literacy can impede effective communication, making it more difficult for people to articulate their needs and ask for assistance. When illiterate parents struggle to give quality education, the cycle of poverty and criminality continues, increasing the likelihood that the following generation will also experience difficulties. By giving people the necessary skills, knowledge, and opportunities, education and literacy programs can help break this cycle and lower crime rates. This will make it simpler for people to find employment, make informed decisions, and actively participate in their communities, which will ultimately reduce the likelihood that they will commit crimes.

The existing body of literature demonstrates the multifaceted nature of crime, emphasizing the significant influence of socio-economic and demographic factors such as poverty, unemployment, illiteracy, and the spatial concentration of disadvantaged populations. The concentration of large numbers of people within confined spaces in cities increases the likelihood of deviant behavior, leading to a higher incidence of criminal activities compared to rural areas (Ward et al., 2018). Furthermore, the anonymity among urban residents exacerbates the crime rate, as individuals in densely populated areas are more prone to engage in criminal behavior. Social disorganization, driven by factors such as illiteracy, poverty, and the presence of slum populations, emerges as a critical component in creating crime-prone neighborhoods in cities (Wickes, 2017; Bellair, 2017). The spatial concentration of disadvantaged populations contributes to the formation of environments where economic crimes, such as theft and dacoity, are prevalent due to the sense of relative deprivation and power imbalance (Kubrin and Mioduszewski, 2019). These conditions are particularly pronounced in slum areas, where illiteracy and unemployment further amplify the propensity for criminal activities (Ha et al., 2020; Rigon, 2022; Abdel-Rahman, 2024).

The classical, contemporary theories and recent literatures provide a robust framework for understanding crime. However, most of these methods often address the symptoms of crime rather than its root causes. To develop effective strategies, it is essential to understand the underlying social dynamics that drive criminal behavior (Burt, 2023; Hafez, 2024). Social physics models can analyze large datasets, such as mobile phone records, social media activity, and urban infrastructure data, to predict where and when crimes are likely to occur. By identifying patterns in human movement and social interactions, these models can help law enforcement agencies allocate resources more effectively and prevent crime before it happens. For example, predictive policing uses algorithms to forecast crime hotspots by analyzing historical crime data, socio-economic indicators, and patterns of human activity (Reichert et al., 2024). Here remains a gap in the application of advanced geospatial techniques to analyze the spatiotemporal dynamics of crime in the context of urbanizing cities. This study aims to address this gap by leveraging geoinformatics and geostatistics to explore the intricate relationship between demographic and socio-economic parameters and crime occurrences in Kolkata. In very few literatures where the relationship of socio-economic factors and crime in Kolkata has been discussed, it has been found that the prevalence of slums in Kolkata introduces challenges (Mukherjee and Mondal, 2022; Mukherjee et al., 2023) that underscore the correlation between slum density and crimes (Ray, 2016). This also aligns with global perspectives, emphasizing the importance of targeted interventions in urban planning and social development (Loftin et al., 1975). Aligned with the theme of social physics, this study seeks to contribute data-driven insights into the social dynamics that shape criminal behavior, ultimately guiding the development of targeted interventions for safer urban environments.

2 Study Area

The Kolkata Police Station area has been chosen as an area of interest for this current study. Kolkata Police is the first Commissionerate in the country having 243 km^2 of area coverage comprising all 144 wards of the Kolkata Municipal Corporation area. The headquarter of the Kolkata police is situated at Lalbazar. It has an illustrious legacy of policing from all aspects, starting from preservation of the orderliness of the city to dealing traffic, crime detection and prevention, and also organizing various citizen-friendly activities, which in turn increase the awareness of crime reporting among the Samaritans of the city.

The recent data from the National Crime Records Bureau (NCRB, 2022) highlighted it as one of the safest metropolitan areas in India, boasting the lowest crime rates among 19 cities for the second consecutive year. This overall safety can be attributed to improved law enforcement, increased police visibility, and the implementation of community policing initiatives. Additionally, technology such as CCTV surveillance and mobile policing apps has played a crucial role in crime prevention and quicker response times (Karmakar & Saha, 2016). However, this general safety belies certain disturbing trends, such as high-profile violent crimes and persistent issues like crimes against women, which have caused public outrage and highlighted the disconnect between statistics and perceived safety (Karmakar, 2016; Mukherjee et al., 2024).

The Kolkata police oversee 81 police stations, including 6 dedicated to women. The jurisdiction is divided into nine subdivisions: Central, East, East Suburban, North and North Suburban, Port, South, Southwest, Southeast, and South Suburban divisions. Preventing crime is a significant challenge for the Kolkata police due to the area's diverse population. The nature of crime is complex and constantly changing, encompassing everything from marginalized, rundown areas like the Port to bustling commercial zones with large transient populations such as Burrabazar, Hare Street, and Bowbazar (https://kolkatapolice.gov.in/units/).

Currently, the Kolkata police aim to deter crime by effectively combining human resources with advanced technological tools. Unthinkable technological developments over current years have impacted all walks of life. Along with conservative crimes like murder and theft, the uprising of eve-teasing and drugging are some new additions of the past few years in Kolkata. Kolkata police have also gone through such changes to cope with these issues with razor-sharp professionalism.

3 Data and Method

3.1 Data

In this study, the authors have categorized crime into two major types, crime against property and crime against the body. Crime against property refers to offenses that involve the destruction and damage of property like theft, snatching, and pickpocketing. Crime against body refers to offenses that involve violence or harm against a person, such as murder, crime against women, and drugging. Crime data have been obtained from Lalbazar Headquarter, crime record section for the year 2020 for 81 police stations including the women's police station. As the women police station and the main branch share the same location, for analysis purposes, the authors have merged the data of the women police station with the main branch. The present count of police station points is 71 (Fig. 1). Temporal data for 30 years from the year 1990 to 2020 have also been obtained from Lalbazar Headquarters, crime record section. However, before 2015, as the police station-wise data are not available for all types of crimes considered for this study, this data can only be used to analyze the temporal variation of different types of crime in KMC. The police station-wise socio-economic data (population density, illiterate population, population below poverty line (%), unemployed population and slum population) have also been obtained from Lalbazar Headquarter for the year 2011.

3.2 Method

Trend Analysis of Criminal Activities

To analyze crime trends from 1990 to 2020, the Mann-Kendall test was employed to identify patterns in criminal activities. The Mann-Kendall test (MKT) is a widely used non-parametric method for detecting nonlinear trends. This test produces test statistics referred to as Kendall's τ statistic (Kendall, 1975). It utilizes rank-based analysis, making it robust for skewed data and effective at minimizing the impact of outliers (Partal & Kahya, 2006). In MKT, the null hypothesis H0 assumes that the deseasonalized data ($x_1, x_2, ..., x_n$) are a sample of n independent, identically distributed random variables. Conversely, the alternative hypothesis H1 posits that for any $k, j \leq n$ where $k \neq j$, the distributions of x_k and x_j differ. According to Hirsch and Slack (1984), Eqs. (2) and (3) are used to calculate the test statistic S, while Eq. (1) is employed to determine variances.

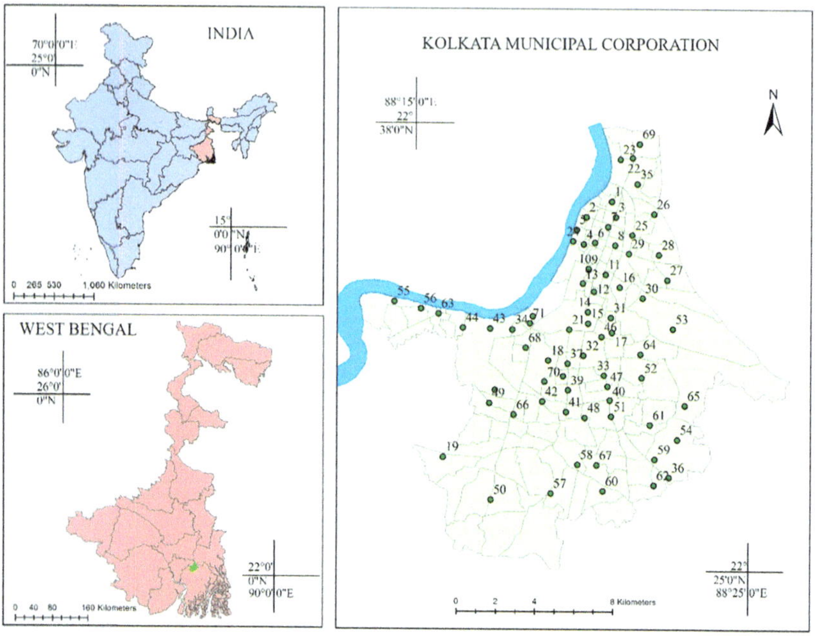

1. Shyampukur, 2. Jorabagan, 3. Burtolla, 4. Burrabazar, 5. Posta, 6. Jorasanko, 7. Girish Park, 8. Amherst Street, 9. Hare Street, 10. Bowbazar, 11. Muchipara, 12. Taltala, 13. New Market, 14. Park Street, 15. Shakespeare Sarani, 16. Entally, 17. Topsia, 18. Alipore, 19. Sarsuna, 20. Hastings, 21. Maidan, 22. Cossipore, 23. Chitpur, 24. North Port, 25. Manicktala, 26. Ultadanga, 27. Beliaghata, 28. Phoolbagan, 29. Narkeldanga, 30. Tangra, 31. Beniapukur, 32. Ballygunge, 33. Gariahat, 34. South Port, 35. Tala, 36. Panchasayar, 37. Bhowanipur, 38. Kalighat, 39. Tollygunge, 40. Lake, 41. Charumarket, 42. New Alipore, 43. West Port, 44. Gardedn Reach, 45. Taratala, 46. Karaya, 47. Rabindra Sarovar, 48. Golf Green, 49. Parnasree, 50. Thakurpukur, 51. Jadavpur, 52. Kasba, 53. Pragati Maidan, 54. Purba Jadavpur, 55. Nadial, 66. Rajabagan, 57. Haridevpur, 58. Regent Psrk, 59. Survey Park, 60. Bansdroni Ps, 61. Garfa, 62. Patuli, 63. Metiabruz, 64. Tiljala, 65. Anandapur, 66. Behala, 67. Netaji Nagar, 68. Ekbalpur, 69. Sinthi, 70. Chetla, 71. Watgunge

Fig. 1 Location map of the study area with police stations

$$S = \sum_{k=1}^{n-1} \sum_{j=k+1}^{n} \operatorname{sgn}\left(x_j - x_k\right) \tag{1}$$

$$\operatorname{sgn}\left(x_j - x_k\right) = \begin{cases} +1 \ \text{if}\left(x_j - x_k\right) > 0 \\ 0 \ \text{if}\left(x_j - x_k\right) = 0 \\ -1 \ \text{if}\left(x_j - x_k\right) < 0 \end{cases} \tag{2}$$

$$\operatorname{Var}(S) = \left[n(n-1)(2n+5) - \sum_t t(t-1)(2t+5) \right] / 18 \tag{3}$$

Here, t represents the extent of any given tie, and the summation over all t indicates the total sum of all ties. For a sample size (n) greater than 10, the standard normal variate (z) is calculated using Eq. (4) (Douglas et al., 2000):

$$z = \begin{pmatrix} \frac{S-1}{\sqrt{\operatorname{Var}(S)}} & \text{if } S > 0 \\ 0 & \text{if } S = 0 \\ \frac{S+1}{\sqrt{\operatorname{Var}(S)}} & \text{if } S < 0 \end{pmatrix} \tag{4}$$

Sen's estimator:

A simple non-parametric technique is used to estimate the trend in a time series that is computed as true slope (change per unit time) (Sen, 1968). Equation (5) is used to estimate the slope and N pairs of data as follows:

$$Q_i = \frac{x_j - x_k}{j-k} \quad \text{for } i = 1,\ldots,N \tag{5}$$

where x_j and x_k are data values at times j and k, respectively (with $j > k$) and Q_i represents Sen's estimator of the slope, calculated as the median of these N values. For Sen's estimator, Qmed is determined as $Q(N + 1)/2$ when N is odd and as $[QN/2 + Q(N + 2)/2]/2$ when N is even. The significance of Qmed is tested using a two-sided test at the $100(1 - \alpha)\%$ confidence interval.

Partial Least Squares Regression

Partial least squares regression (PLSR) is employed in this study as a robust statistical technique to model the complex relationships between socio-economic factors and various crime types. PLSR is particularly well-suited for scenarios where the number of predictor variables exceeds the number of observations or when multicollinearity is present among the predictors. In the context of this research, the socio-economic parameters used to analyze their correlation with crime types are likely to be interrelated, making PLSR an ideal choice (Guo et al., 2021; Wang et al., 2022; Gélinas et al., 2024).

PLSR reduces the predictor variables to a smaller set of uncorrelated components, or latent variables, that capture the maximum variance in both the predictors and the response variable. This characteristic of PLSR makes it robust in situations where predictor variables are highly correlated, effectively negating the problem of multicollinearity (Latan et al., 2017; Avkiran, 2018; Okwan & Kovacs, 2023). By doing so, PLSR not only simplifies the model but also enhances its predictive power, particularly in social sciences where the dependent variable is often related to complex attitudes and behaviors.

In the context of analyzing the impact of socio-economic factors on crime, PLSR is used to identify specific factors that have the greatest impact on crime levels. The variable importance in the projection (VIP) scores are a key feature of PLSR, as they quickly identify which explanatory variables contribute the most to the models, thereby highlighting the impact of socio-economic factors on crime. This analysis enables a detailed understanding of the influence of individual factors on specific types of crime, such as murder, theft, or crimes against women. The factor with the highest VIP score is indicative of a greater impact on a particular crime type, guiding targeted policy improvements in those areas.

To implement this, the PLS regression technique was carried out using XLSTAT, a comprehensive tool for modeling and forecasting data in Excel. The analysis focused on the correlation between socio-economic factors and six different crime types—pickpocketing, snatching, theft, drugging, murder, and crime against women (CAW)—for the year 2020. The use of PLSR in this study not only allows for a nuanced understanding of the socio-economic determinants of crime but also aids in the development of data-driven strategies for crime prevention and urban policy planning (Mathlin et al., 2021; Bhagat & Jha, 2024). Figure 2 presents the method in its complete form.

4 Result

4.1 Temporal Variation of Crime

The occurrences of crimes can change over time due to a complex interaction of economic, social, cultural, and law enforcement factors. The nature and frequency of crimes have changed in Kolkata significantly as a result of changes in surveillance. The Mann-Kendall test has been used to examine each type of crime separately in order to determine whether there is a noteworthy trend or pattern in the temporal variation of those crimes.

For every type of crime like snatching, pickpocketing, crime against women, and murder, the results also indicate a significant positive trend within the time span of 30 years (1990–2020). Due to the unavailability of data, drugging has been analyzed for 20 years' span, from 2000 to 2020. The results indicate that for all crime types examined, the null hypothesis is rejected, as the p-value is less than 0.0001. This indicates that the temporal variance of these crimes throughout the examined period exhibits a notable trend or pattern. For example, the Mann-Kendall statistic for theft is 371, and Kendall's tau is 0.853, indicating a strong positive correlation between time and the number of thefts. The variance (S) is 3141.667, which suggests that the data is moderately dispersed. The p-value is <0.0001, demonstrating that the trend is statistically significant, and the null hypothesis is rejected (Table 1).

Crime against women (CAW) has the highest Kendall's tau coefficient, which means a crime against women has a strong correlation with time. With the changing

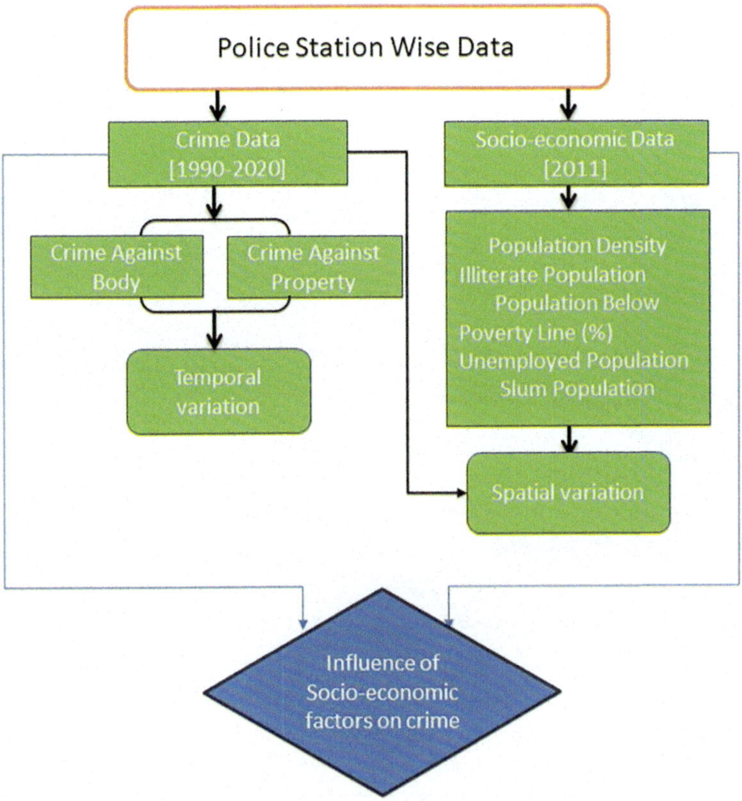

Fig. 2 Methodology

Table 1 Mann-Kendall statistics and model interpretation

Crime	Mann-Kendall statistics	Kendall's tau	Variance (S)	*P*-value	Model interpretation
Theft	371	0.853	3141.667	<0.0001	Reject H0
Snatching	401	0.922	3141.667	<0.0001	Reject H0
Pickpocket	383	0.880	3141.667	<0.0001	Reject H0
Drugging	113	0.542	1093	0.001	Reject H0
Crime against women	412	0.948	3140.667	<0.0001	Reject H0
Murder	327	0.755	3137.667	<0.0001	Reject H0

time scenario, the crime against women is changing in the number of occurrences. The probable explanation of this kind of correlation will be the introduction of new legal supports for the victimized women and increasing awareness among the women to report the crimes that occur to them on a regular basis. The police surveillance in the streets has become stronger, and the deployment of more women police

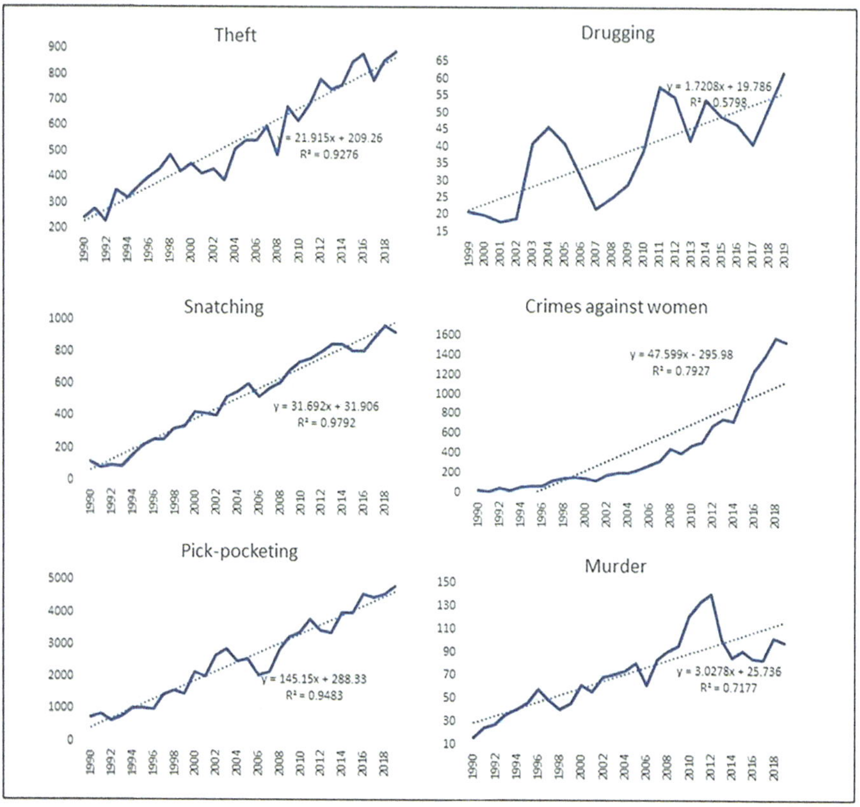

Fig. 3 Trend analysis of different types of crime

in the streets of Kolkata made the reporting of crimes easier for the women victims. Snatching is on the second rank, in terms of the correlation with changing time and occurrence. Snatching is totally a street-based crime that needs to be restricted with very efficient and proper surveillance (Fig. 3).

On a deeper level, the commitment of more snatching with the passage of time unveils the grim picture of socio-economic deprivation of a certain section of people who are involved in these kinds of petty economic crimes. Snatching sometimes can be a form of organized crime also. Gangs operate within the city and commit this sort of criminal offense. From the year 2015, there is a sharp increase in the numbers of snatching, which is evident that the city police are somehow lagging behind to manage the offenses of snatching. The incidences of snatching are followed by the incidences of pickpocketing. It is also showing a strong correlation with the change of time, but not as much as snatching. Snatching and pickpocketing are both very similar in commitment. The same sort of socio-economic causes leads to economic crimes like snatching and pickpocketing. Both are street crimes, and the target victims are almost the same in nature, in terms of their class for both crimes. The higher Kendall's tau coefficient for both economic crimes put the efficiency of the police to maintain surveillance in public spaces under the radar. The activity spaces of the criminals must be monitored on a regular basis, and proper tight guardianship measures must be taken.

Looking at personal issue-based crimes like murder, it is showing the lowest Kendall's tau coefficient and the lowest variance value. It signifies that murder is not that strongly correlated with time in comparison to the other forms of crime. However, the occurrences of murders are also showing a significant increasing trend. As it is different in nature from the other selected forms of crime, the factors leading to murder are much more diversified. As it is a personal issue-oriented crime, the individual personality and experiences of the offender matter a lot in case of commitment of murder. Looking at the overall crime scenario of Kolkata where all the forms of crime are showing a significantly increasing trend, there are lots of scopes for the betterment of the police system.

4.2 Spatial Variation of Socio-economic Determinants and Crime

For detailed spatial analysis, the different crimes reported in stations and the socio-economic parameters were visualized using proportional symbols. This approach involved representing the magnitude of each factor with appropriately sized symbols, providing a visual representation of the varying trends across the study area (Figs. 4 and 5).

Population Density
The dataset illustrates the distribution of population density across different police stations in Kolkata, shedding light on the concentration of residents in each jurisdiction (Fig. 4). Noteworthy variations emerge, with New Alipore PS, Chetla PS, and Tollygunge PS exhibiting notably high population densities exceeding 10,000 individuals per square kilometer. This signals significant urban crowding and potential resource strain. Conversely, police stations like Parnasree PS and Haridevpur PS report lower population densities, reflecting relatively less crowded living conditions.

Illiteracy
A comprehensive examination of illiteracy across Kolkata's police stations reveals distinct educational disparities in various areas of the city (Fig. 4). West Port PS and South Port PS stand out with the highest illiteracy rates, while Survey Park PS and Kasba PS report lower figures. This spatial distribution unveils clusters of high illiteracy, particularly in the southern and western regions.

Poverty
The dataset highlights the population below the poverty line across Kolkata's police stations, emphasizing diverse economic conditions within the city (Fig. 4).

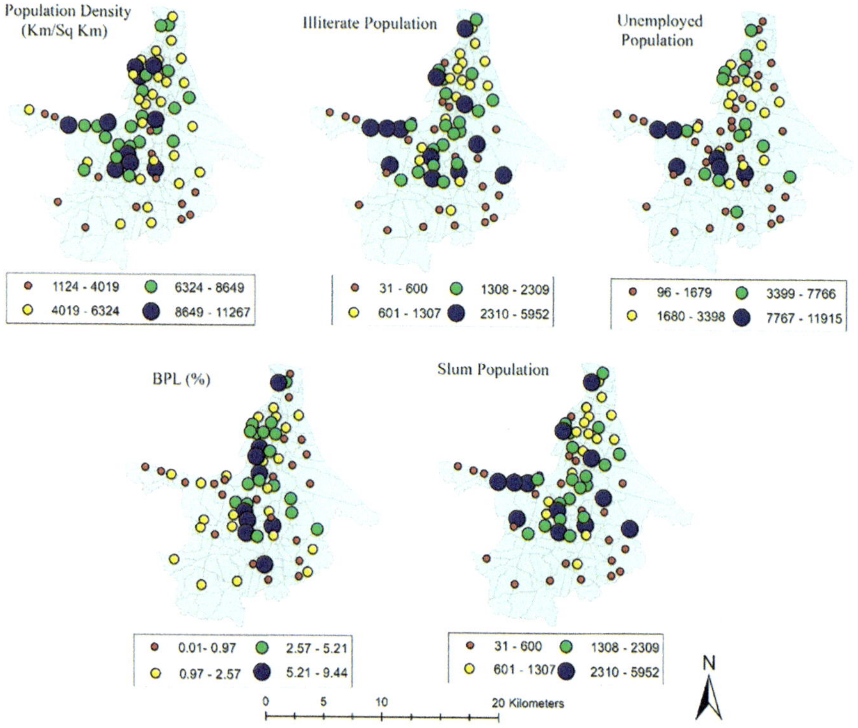

Fig. 4 Spatial variations of socio-economic determinants

Bhowanipur PS, Kalighat PS, and Tollygunge PS exhibit higher percentages, indicating a prevalence of poverty in these regions. Conversely, areas like Regent Park PS and Survey Park PS demonstrate exceptionally low percentages.

Unemployment

Insights into unemployment across Kolkata's police stations provide a nuanced perspective on economic challenges in different neighborhoods (Fig. 4). West Port PS, Tollygunge PS, and Garden Reach PS show elevated unemployment figures, indicating substantial joblessness. In contrast, Maidan PS and Hastings PS report relatively lower unemployment rates.

Slum Population

A comprehensive overview of slum populations across various police stations in Kolkata highlights the urban challenges associated with informal settlements (Fig. 4). Striking observations include substantial slum populations in West Port PS, South Port PS, and Garden Reach PS, emphasizing the prevalence of informal housing. Conversely, police stations like Survey Park PS and Golf Green PS report comparatively lower slum populations. These disparities underscore the spatial heterogeneity of living conditions in the city, calling for targeted urban development and slum improvement programs.

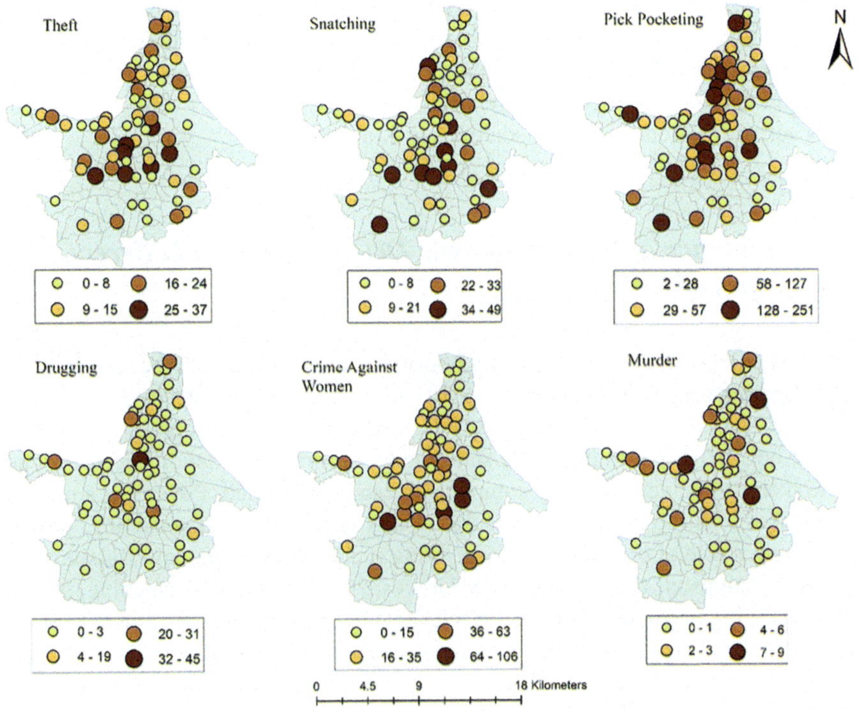

Fig. 5 Spatial variations of crime

Crime

In 2020, the Kasba police station recorded the highest number of theft cases in Kolkata, reporting 37 incidents, with Charu Market and Lake police stations following closely with 34 and 31 cases, respectively. Notably, areas like Thakurpukur, Purba Jadavpur, and Behala police stations reported a relatively high incidence of snatching, recording 49, 47, and 46 cases, respectively. New Market police station topped the list for pickpocketing cases, reporting 258 incidents. These findings suggest that commercial areas in Kolkata are more susceptible to pickpocketing crimes (Fig. 5).

While drugging cases were comparatively lower in number, the potential harm they pose to society underscores the need for attention. Lake police station recorded the highest number of drugging cases (29), followed by Park Street police station with 22 cases.

Addressing crimes against women (CAW) is crucial, focusing both on safety within homes and on the streets and workplaces. Tollygunge reported the highest number of CAW cases (83), followed by Lake police station (72) and Kalighat police station with 61 cases, emphasizing the need for targeted interventions to ensure women's safety in public spaces.

Murder, being the most severe crime, demands special attention. Watgunge police station had the highest number of reported murder cases, followed by Kolkata Leather Complex, North Port, Metiabruz, and Tiljala police stations. These findings highlight the imperative for comprehensive strategies to address varying crime patterns across different police jurisdictions in Kolkata.

4.3 *Influence of Socio-economic Determinants on Different Types of Crime*

The partial least squares (PLS) regression VIP score table unveils the nuanced relationship between socio-economic factors and various crime categories in Kolkata (Table 2). A VIP score exceeding 0.8 is deemed indicative of a substantial contribution (Abdi, 2010). From the above stated socio-economic model of selected crimes in Kolkata, some very relevant components can be identified. Poverty is a relevant factor that leads to pickpocketing. As a petty crime like pickpocketing, poverty plays a vital role of provocation of satisfying sudden need of money (Bursik, 1986; Byrne & Sampson, 1986; Sampson & Groves, 2017). In case of snatching, poverty and unemployment came as the most relevant components as controlling factors. In case of theft, which is also an economic crime, poverty, unemployment, illiteracy, and slum population percentage turn out to be the most relevant controlling factors (Felson & Cohen, 1980; Browning & Jackson, 2013). From the partial least squares analysis, it can be figured out that poverty and unemployment are the two most relevant components for the economic crimes. These two factors have the most numbers of VIP scores that are higher than 1. Illiteracy is also having an important role in shaping the crime scenario of Kolkata. Drugging and murder are both personal issue based and a form of personality disorder. For drugging, poverty plays a crucial role according to the analysis. Poverty leads to illegal selling of drugs and also abuse of substance, which are meant for addiction. Poverty-stricken areas are also the zone of urban subculture where this sort of substance abuse is considered to be a part and parcel of lifestyle (Kane, 2002). The urban subculture acts as a catalyst to the crimes like drugging. Drugging needs money, so that the offenders of drugging go for pickpocketing, snatching, and, in extreme level, theft.

Table 2 VIP scores to show the impact of socio-economic parameters on different types of crime

Socio-economic factors	Pickpocketing	Snatching	Murder	Theft	Drugging	Crime against women (CAW)
Population density	0.598	**0.847**	0.651	0.588	**0.925**	0.550
Illiteracy	0.386	0.168	**1.237**	**1.083**	0.709	**0.812**
Poverty	**1.990**	**1.787**	0.385	**1.132**	**1.673**	**1.522**
Unemployment	0.620	**1.016**	**1.169**	**1.013**	0.581	**1.031**
Slum population	0.386	0.168	**1.237**	**1.083**	0.710	**0.812**

After being successful in committing these crimes, they go for substance abuse with the money. There is a circular occurrence of crime bonded by cause-and-effect relationship. Crime against women (CAW) is also a very critical type of crime to analyze, as it is showing illiteracy, poverty, unemployment, and slum population residence as relevant factors (Hardie & Wikström, 2021). The slum areas are the zone of dilapidation, which can be marked as gray neighborhood of the urban centers (Badom and Ndeeue, 2023). These findings underscore the intricate interplay between socio-economic indicators and crime dynamics, providing a valuable framework for targeted interventions and policy formulations to address the root causes of crime in specific areas of the city.

5 Discussion

The tendency of criminality is very often said to be an outcome of the social and economic deprivation. Social and economic isolation of certain section of population from the main stream makes feeling of deprivation more pronounced that propels some people to commit offences to meet the social and economic desired ends. Illiteracy, poverty, and unemployment are some of the key factors that lead to the regression of certain section of population and give rise to slum areas in urban centers. These slum areas or areas with high rate of poverty, unemployment, and illiteracy create ideal zones of deprivation (Fissel & Wilcox, 2021). These give rise to social disorganization, which in turn gives birth to offensive behavior (Goodson and Bouffard, 2020).

The Mann-Kendall test reveals a significant rise in various crimes over the past 30 years, emphasizing the necessity for adaptive law enforcement strategies. The intricate spatial patterns uncovered in the socio-economic indicators and crime statistics across various police stations in Kolkata reveal compelling correlations between population characteristics and criminal activities. The population density, a key metric reflecting the concentration of individuals in a given area, exhibits substantial variations across police stations. High-density regions, exemplified by Beniapukur PS, Tollygunge PS, and New Alipore PS, with over 10,000 individuals per square kilometer, coexist with areas of lower density, such as Parnasree PS and Haridevpur PS. This divergence signifies diverse living conditions and infrastructural challenges across different sectors of the city.

When delving into illiteracy rates, a discernible trend emerges, linking higher illiteracy levels to police stations with elevated population densities. This suggests a potential correlation between population density and educational disparities, indicative of the intricate dynamics shaping socio-economic factors. Moreover, regions grappling with higher poverty rates, including Tollygunge PS and New Alipore PS, align with those experiencing heightened illiteracy rates, implying a complex interplay between economic conditions and educational opportunities.

Unemployment rates mirror these patterns, showcasing that police stations characterized by greater population density tend to grapple with higher unemployment rates. Notably, Tollygunge PS, New Alipore PS, and Garden Reach PS emerge as hotspots with both elevated population density and increased unemployment rates, pointing to potential economic challenges and job scarcity in these localities.

The correlation extends to slum populations, demonstrating a concentrated presence in areas characterized by higher population densities, as evident in Beniapukur PS, Tollygunge PS, and Garden Reach PS. This underscores the urgent need for targeted urban development strategies and social interventions in these regions to address the socio-economic disparities and associated crime challenges. The interwoven nature of these factors necessitates comprehensive approaches for effective policy formulation and community development initiatives in Kolkata.

The PLS regression VIP score table illuminates key insights into the intricate relationship between socio-economic factors and crime dynamics in Kolkata. It highlights major findings such as the substantial influence of higher population density and increased poverty levels on pickpocketing cases. Similarly, snatching incidents are also strongly correlated with both population density and poverty. The analysis underscores the pivotal roles of illiteracy and unemployment in murder cases, with broader implications for theft and drugging. Additionally, illiteracy and poverty emerge as influential factors impacting crime against women.

In recent dates, various studies have explored the socio-economic determinants of crime, highlighting diverse factors that contribute to criminal activities. Blakeney (2023) identified key socio-economic factors such as unemployment, income inequality, poverty, hereditary factors, and the influence of peer groups as primary drivers of crime. Canepa (2023) focused on the socio-economic determinants of environmental crimes, particularly wildfire crimes in Italy, using a panel data approach at the regional level. The study highlighted the role of material deprivation and economic downturns in driving wildfire crimes, with a particular emphasis on the southern regions where unemployment and income inequality are more pronounced. While the study's use of advanced econometric models provides robust insights into environmental crimes, it primarily focuses on regional disparities without offering a detailed analysis of urban-specific dynamics. Maldonado-Guzmán et al. (2024) explored the relationship between overtourism and crime rates in Malaga, Spain, specifically analyzing the impact of Airbnb accommodation on crime density. This study employed spatial models to control for traditional crime-related factors and found significant correlations between Airbnb concentration, nightlife, and crime density. Although this research offers a novel perspective on the intersection of tourism and crime, it does not address broader socio-economic factors such as poverty or unemployment that may also influence crime rates in urban areas.

Despite the valuable contributions of these studies, they have certain limitations, particularly in their application to urban crime dynamics. This work builds on these foundations by examining the relative impacts of specific socio-economic variables—population density, illiterate population, population below poverty line (%), unemployed population, and slum population—on crime in Kolkata. This approach not only fills the gap left by previous studies in understanding urban-specific factors but also provides a more granular analysis of the socio-economic determinants of crime in a rapidly urbanizing Indian city. By focusing on these specific variables, this research offers a comprehensive understanding of how different socio-economic factors contribute to crime, along with the spatiotemporal variations. The major findings provide a complete understanding, offering valuable perspectives for targeted interventions and policy formulations to address the root causes of crime in specific areas of the city.

6 Conclusion

By employing geospatial analysis and PLSR, this research not only uncovers the spatial distribution of crime but also provides a scientific basis for understanding the socio-economic dynamics that drive it. In doing so, this chapter contributes to the broader academic discourse on crime prevention and urban planning, offering valuable perspectives that can inform policy formulation and targeted interventions in urban settings like Kolkata. These findings underscore the need for holistic interventions that address socio-economic root causes, necessitating a multidimensional approach for effective policy formulation and community development initiatives in the city. Moreover, this study exemplifies the potential of integrating social science theories with advanced analytical techniques, thus pushing the boundaries of traditional crime analysis. By situating crime within the broader context of socio-economic disparities, this research contributes to the ongoing efforts in academia to address complex social issues with rigor and precision. The insights gained from this study not only enhance our understanding of urban crime but also provide a robust framework for future research in social physics and related fields. Future research could expand on these findings by incorporating real-time data and predictive modeling to further refine strategies for crime prevention and urban development.

AI Declaration This paper includes the use of artificial intelligence (AI) tools, specifically OpenAI's language model, ChatGPT, to assist in the drafting, editing, and refinement of the manuscript. AI was utilized to generate text based on provided input, suggest relevant literature, and improve the clarity and coherence of arguments while ensuring adherence to academic writing standards. However, the content, ideas, and conclusions presented in this paper are the sole responsibility of the authors. The AI tools were used strictly as supplementary aids, with all final decisions regarding the paper's content made by the authors. The authors affirm that the research, analysis, and interpretations presented herein are original and the product of their own intellectual efforts.

References

Abdel-Rahman, S. (2024). High unemployment and crime rates threaten slum dwellers in Egypt. *Journal of Poverty*, 1–15.

Abdi, H. (2010). Partial least squares regression and projection on latent structure regression (PLS Regression). *Wiley Interdisciplinary Reviews: Computational Statistics, 2*(1), 97–106.

Agnew, R., & White, H. R. (1992). An empirical test of general strain theory. *Criminology, 30*(4), 475–500. https://doi.org/10.1111/j.1745-9125.1992.tb01113.x

Avkiran, N. K. (2018). Rise of the partial least squares structural equation modeling: An application in banking. In *Partial least squares structural equation modeling: Recent advances in banking and finance* (pp. 1–29). Springer.

Badom, P. M., & Ndeeue, K. (2023). Impact of urban slum violence on work in slum areas of Port Harcourt, Nigeria. *European Review of Applied Sociology, 16*(27), 71–86.

Bellair, P. (2017). Social disorganization theory. In *Oxford research encyclopedia of criminology and criminal justice*. Oxford University Press.

Bhagat, G., & Jha, K. N. (2024). Exploring the impact of the integrity climate on integrity violations in public procurement: A partial least square structural equation modeling approach. *Journal of Legal Affairs and Dispute Resolution in Engineering and Construction, 16*(1), 04523042.

Blakeney, M. (2023). Impacts of counterfeiting: Commercial, fiscal and socio-economic. In *Counterfeit goods and organised crime* (pp. 30–41). https://doi.org/10.4337/9781035318735.00009

Browning, C. R., & Jackson, A. L. (2013). The social ecology of public space: Active streets and violent crime in urban neighborhoods. *Criminology, 51*(4), 1009–1043.

Bursik, R. J. (1986). Delinquency rates as sources of ecological change. In *The social ecology of crime* (pp. 63–74). Springer.

Burt, C. H. (2023). Irreducibly social: Why biocriminology's ontoepistemology is incompatible with the social reality of crime. *Theoretical Criminology, 27*(1), 85–104.

Byrne, J. M., & Sampson, R. J. (1986). Key issues in the social ecology of crime. In *The social ecology of crime* (pp. 1–22). Springer.

Canepa, A. (2023). Socio-economic risk factors and wildfire crime in Italy: A quantile panel approach. *Empirical Economics, 66*(1), 431–465. https://doi.org/10.1007/s00181-023-02462-2

Cohen, L. E., & Felson, M. (2010). Social change and crime rate trends: A routine activity approach (1979). In *Classics in environmental criminology* (pp. 203–232). Routledge.

Crespo, A., Brito, J., Ajala, S., Amaro, I. R., & Castillo, Z. (2023, April). Multivariate statistical techniques to analyze crime and its relationship with unemployment and poverty: A case study. In *Computer science on-line conference* (pp. 180–192). Springer International Publishing.

Crime in India Report. (2022). *In National Crime Records Bureau*. National Crime Records Bureau.

Douglas, E. M., Vogel, R. M., & Kroll, C. N. (2000). Trends in floods and low flows in the United States: Impact of spatial correlation. *Journal of Hydrology, 240*(1–2), 90–105. https://doi.org/10.1016/s0022-1694(00)00336-x

Errol, Z., Madsen, J. B., & Moslehi, S. (2021). Social disorganization theory and crime in the advanced countries: Two centuries of evidence. *Journal of Economic Behavior & Organization, 191*, 519–537.

Farrell, G. (2015). Crime concentration theory. *Crime Prevention and Community Safety, 17*(4), 233–248. https://doi.org/10.1057/cpcs.2015.17

Felson, M., & Cohen, L. E. (1980). Human ecology and crime: A routine activity approach. *Human Ecology, 8*, 389–406.

Fissel, E. R., & Wilcox, P. (2021). Social ecology. In *The encyclopedia of research methods in criminology and criminal justice* (Vol. 1, pp. 217–222). Wiley.

Gehring, K. S., & Batista, M. R. (2022). *Labeling theory*. Oxford University Press.

Gélinas, A., Profili, J., Fotouhiardakani, F., Caceres Ferreira, W. M., Laurent, M., Ravichandran, S., & Laroche, G. (2024). Toward a better interpretation of the partial least squares regression models for fluoropolymers treated by dielectric barrier discharges at atmospheric pressure. *Plasma Processes and Polymers, 21*(2), 2300098.

Goodson, A., & Bouffard, L. A. (2020). The rural/urban divide: Examining different types of assault through a social disorganization lens. *Journal of Interpersonal Violence, 35*(17–18), 3530–3553.

Guo, Y., Xiang, H., Li, Z., Ma, F., & Du, C. (2021). Prediction of rice yield in East China based on climate and agronomic traits data using artificial neural networks and partial least squares regression. *Agronomy, 11*(2), 282.

Ha, O. K., Andresen, M. A., & Davies, G. (2020). The temporal (in)stability of the unemployment and crime relationship. *International Journal of Offender Therapy and Comparative Criminology, 64*(8), 840–859.

Hafez, A. I. (2024). Anticipating and addressing unconventional future crimes: A mechanistic analysis in the context of technological advancements. *Police Thought, 33*(1).

Hälterlein, J. (2021). Epistemologies of predictive policing: Mathematical social science, social physics and machine learning. *Big Data & Society, 8*(1), 20–43.

Hardie, B., & Wikström, P. O. (2021). Space–time budget methodology: Facilitating social ecology of crime. In *The encyclopedia of research methods in criminology and criminal justice* (Vol. 1, pp. 126–143). Wiley.

Hirsch, R. M., & Slack, J. R. (1984). A nonparametric trend test for seasonal data with serial dependence. *Water Resources Research, 20*(6), 727–732.

https://kolkatapolice.gov.in/units/

Kane, R. J. (2002). The social ecology of police misconduct. *Criminology, 40*(4), 867–896.

Karmakar, S. (2016). Emergency of smart crimes: Hidden reality of smart cities. *Indian Journal of Regional Science, 48*(1), 128–136.

Karmakar, S., & Saha, A. (2016). Socio-spatial explanation of crime: A study on Kolkata metropolitan area. *Transactions, 38*(2), 191.

Kendall, M. (1975). *Multivariate analysis*. Charles Griffin & Company Ltd..

Kivivuori, J., Savolainen, J., & Aaltonen, M. (2015). The revenge motive in delinquency. *Acta Sociologica, 59*(1), 69–84. https://doi.org/10.1177/0001699315607969

Kubrin, C. E., & Mioduszewski, M. D. (2019). Social disorganization theory: Past, present and future. In *Handbook on crime and deviance* (pp. 197–211). Springer.

Latan, H., Noonan, R., & Matthews, L. (2017). Partial least squares path modeling. In *Partial least squares path modeling: basic concepts, methodological issues and applications*. Springer.

Loftin, C., Clinard, M. B., & Abbott, D. J. (1975). Crime in developing countries: A comparative perspective. *Social Forces, 54*(1), 306. https://doi.org/10.2307/2576122

Maldonado-Guzmán, D. J., Chamizo-Nieto, F. J., & Reyes-Corredera, S. (2024). Home sharing or crime sharing? Evidences of the relationship between Airbnb, crime and structural factors in Malaga, Spain. *GIScience & Remote Sensing, 61*(1), 2384330. https://doi.org/10.1080/15481603.2024.2384330

Mathlin, G., Freestone, M., Taylor, C., & Shaw, J. (2021). Offenders with personality disorder who fail to progress: A case-control study using partial least squares structural equation modeling path analysis. *JMIRx Med, 2*(4), e27907.

Merton, R. K. (1938). Social structure and anomie. *American Sociological Review, 3*(5), 672. https://doi.org/10.2307/2084686

Mukherjee, K., & Mondal, D. (2022). Spatial equity in urban facility versus residents' satisfaction: Challenge for the unplanned cities—A case study of Barasat City, India. *GeoJournal, 87*(S4), 931–949. https://doi.org/10.1007/s10708-022-10674-9

Mukherjee, K., Roy, K., & Pal, B. (2023). Gender parity in urbanizing city: A case study of Barasat City, India. *Global Social Welfare, 10*(4), 313–333. https://doi.org/10.1007/s40609-023-00296-5

Mukherjee, K., Saha, S., Karmakar, S., & Dash, P. (2024). Uncovering spatial patterns of crime: A case study of Kolkata. *Crime Prevention and Community Safety, 26*(1), 47–90.

Okwan, F., & Kovacs, P. (2023). Examining the causal effect of social development on maternal mortality in sub-Saharan Africa using partial least squares (PLS) structural equation modeling (SEM). *Pure and Applied Mathematics Journal, 10*(1), 23–33.

Partal, T., & Kahya, E. (2006). Trend analysis in Turkish precipitation data. *Hydrological Processes, 20*(9), 2011–2026. https://doi.org/10.1002/hyp.5993

Paul, T. (2011). Space, gender, and fear of crime: Some explorations from Kolkata. *Gender, Technology and Development, 15*(3), 411–435. https://doi.org/10.1177/097185241101500305

Petrič, K. (2024). An empirical study of interdisciplinary crime research. *European Journal of Economics, Law and Social Sciences, 8*(1), 1–21.

Ray, B. (2016). Quality of life in selected slums of Kolkata: A step forward in the era of pseudo-urbanisation. *Local Environment, 22*(3), 365–387. https://doi.org/10.1080/13549839.2016.1205571
Reichert, K. G., Reichert, N. C. M., & Reichert, R. A. (2024). Neuroscience and social sciences intersections in the legal-criminal field: Ethical, legal, political-social and health implications. In *Substance use, public policy and harm reduction: Social and health studies and interventions* (p. 11). Ethics International Press.
Rigon, A. (2022). Diversity, justice and slum upgrading: An intersectional approach to urban development. *Habitat International, 130*, 102691.
Rosenfeld, R. (2024). *Crime dynamics: Why crime rates change over time*. Cambridge University Press.
Sampson, R. J., & Groves, W. B. (2017). 8 Community structure and crime: Testing social-disorganization theory 1. In *Social, ecological and environmental theories of crime* (pp. 93–122). Routledge.
Santiago, C. D., Wadsworth, M. E., & Stump, J. (2011). Socioeconomic status, neighborhood disadvantage, and poverty-related stress: Prospective effects on psychological syndromes among diverse low-income families. *Journal of Economic Psychology, 32*(2), 218–230. https://doi.org/10.1016/j.joep.2009.10.008
Schleimer, J. P., Pear, V. A., McCort, C. D., Shev, A. B., De Biasi, A., Tomsich, E., et al. (2022). Unemployment and crime in US cities during the coronavirus pandemic. *Journal of Urban Health, 99*(1), 82–91.
Sen, P. K. (1968). Estimates of the regression coefficient based on Kendall's tau. *Journal of the American Statistical Association, 63*(324), 1379–1389.
Shaw, C. R., & McKay, H. D. (1942). *Juvenile delinquency and urban areas*. University of Chicago Press.
Teltscher, K. (2002). India/Calcutta: City of palaces and dreadful night. In *The Cambridge companion to travel writing* (pp. 191–206). https://doi.org/10.1017/ccol052178140x.012
Wang, H., Chu, X., Chen, P., Li, J., Liu, D., & Xu, Y. (2022). Partial least squares regression residual extreme learning machine (PLSRR-ELM) calibration algorithm applied in fast determination of gasoline octane number with near-infrared spectroscopy. *Fuel, 309*, 122224.
Ward, K. C., Kirchner, E. E., & Thompson, A. J. (2018). *Social disorganization and rural/urban crime rates: A county level comparison of contributing factors*. https://doi.org/10.18061/1811/86153
Weisburd, D., Groff, E. R., & Yang, S. M. (2014). The importance of both opportunity and social disorganization theory in a future research agenda to advance criminological theory and crime prevention at places. *Journal of Research in Crime and Delinquency, 51*(4), 499–508.
Wickes, R. (2017). Social disorganization theory: Its history and relevance to crime prevention. *Preventing Crime and Violence*, 57–66.

Neha Arong holds a bachelor's degree in Geography from Vivekananda College and a master's degree in Geoinformatics from Symbiosis Institute of Geoinformatics. She has experience as a Junior Analyst at Deduce Technologies and completed a 2-month summer internship at Adamas University, where she analyzed various physical and socio-economic factors influencing crime. Additionally, during her master's, she interned with Tech Mahindra, contributing to a project focused on Renewable Energy Sources Location Mapping.

Dr. Kasturi Mukherjee is an Associate Professor and Head, Department of Geography, School of Basic and Applied Sciences, Adamas University, and Visiting Researcher in Stockholm Resilience Centre, Stockholm University. She has received her Ph.D. in Urban Climate, from the University of Calcutta. Dr. Mukherjee has completed a post-graduation degree in Geography with a specialization in Environmental Geography and Post-Graduation Diploma in Geoinformatics. She is a gold medalist for securing first rank in the university in post-graduation, and she has received the best paper award two times in 2010 and 2018. Her research interest is focused on urban climate, sustainable development, and peri-urban development. This year, she has received the prestigious Core Research Grant from the Science and Engineering Research Board (SERB), DST as CoPI.

Paramita Seth holds a bachelor's degree in Geography from Sarat Centenary College and a master's degree in Geoinformatics from Adamas University. She has experience as a Junior GIS Analyst at Stesalit Systems Ltd. During her master's program, she completed a 3-month internship at Stesalit Systems Ltd., where she contributed to the Bihar Master Plan project, focusing on mapping, land use analysis, cadastral mapping, and pre-and post-GT surveys. As part of her master's dissertation, Paramita Seth worked on the Assessment of Spatial Pattern of Urban Street Crime: A Case Study of Kolkata Metropolitan Area.

Dr. Supratim Karmakar is currently an Assistant Professor in the Department of Geography at Adamas University. He earned his Ph.D. in Geography from Visva Bharati in 2017. Before his current role, Dr. Karmakar served as Post-Graduation Coordinator at Bag Bazar Women's College and as a Guest Lecturer at Vidyasagar College. He has published 17 research articles in peer-reviewed journals and contributed 11 book chapters. In 2023, he co-edited a book on Health Geography. His research interests include geography of crime, gender studies, urban ecology, and folk culture studies.

Analysis of Cyber Risk in Finance: The State-of-the-Art and Future Research Trends

Bikram Ghosh and Riya Sil

1 Introduction

Cybersecurity, also known as information technology security, constitutes the body of technical processes and practices (Akhtar, 2021). Its primary purpose is to safeguard devices, networks, data, and programs from potential threats such as unauthorized access, damage, and cyberattacks (Armin et al., 2015). The digital landscape of 2022–2023 is riddled with alarming statistics that underscore the critical need for enhanced cybersecurity measures.

Daily, over 30,000 websites fall victim to hacking attempts globally. Furthermore, an alarming 64% of companies worldwide have experienced one or more cyber threats, showcasing the pervasiveness of the issue (Bouveret, 2018). Even within supposedly secure networks, an astonishing 90 records go missing every second, exposing persistent vulnerabilities (Cashell et al., 2004). It is becoming increasingly evident that any business relying on computers and the Internet is at risk of cyber threats (Munk, 2022).

By the end of 2020, as the world grappled with the peak of the COVID-19 pandemic, cybersecurity analysts sounded a warning about a looming global cyber pandemic in the financial sector. This alarm was raised due to the sudden surge in remote work and electronic exchange (Curran et al., 2007). The financial sector witnessed a staggering 238% increase in cyberattacks worldwide from February 2020 to the end of April. Alarmingly, 80% of financial organizations reported a surge in cyberattacks, signifying the gravity of the situation (Kayabaş & Tuna, 2023).

B. Ghosh
Department of Cyber Science & Technology, Brainware University, Kolkata, India

R. Sil (✉)
Department of Computer Science & Engineering, Brainware University, Kolkata, India
e-mail: drs.cse@brainwareuniversity.ac.in

A. Rai et al. (eds.), *Contemporary Social Physics*, Springer Geography,
https://doi.org/10.1007/978-3-031-77596-3_17

Recent months have seen a spate of significant global digital attacks (Kalinich, 2017). Notably, the Orion system of the information technology firm SolarWinds was compromised at the end of 2020, affecting almost 18,000 customers (Joshi, 2020). Additionally, in 2021, Microsoft disclosed that hackers were exploiting "zero-day" vulnerabilities in its Exchange Server mail and calendar software to infiltrate company network systems. These attacks exploited various vulnerabilities in Microsoft Exchange Server for purposes like ransomware deployment and espionage (Ros & Schaanning, 2020).

The COVID-19 pandemic accelerated digitization, forcing an increasing number of individuals to work from home. With this shift, innovation and technology became even more integral in personal and professional lives (Hafkamp & Steenvoorden, 2009). Research indicates that those working from home have a higher susceptibility to cybercrimes, with scams and frauds accounting for 47% of these incidents. The absence of a safe digital workspace in the organizations they work for is a key contributing factor. Consequently, the remote work environment and electronic exchange systems are in a precarious state due to the surging vulnerabilities of this digital era (Tumkevič, 2017).

Analytically, the number of digital crimes per organization surged by 31% in 2021 compared to the previous year. It took organizations an average of 280 days to detect and respond to cyber espionage. In 2021, a successful cyber espionage attack could cost a firm around $3.6 million. Thus, cyberattacks and data breaches have become a critical concern, impacting approximately 55% of organizations worldwide. Fraud activities constitute 24% of these crimes, while ransomware attacks account for 20% (Monteith et al., 2021). Ransomware attacks, social engineering, and insider malware activity are the top 3 types of cyberattacks. Cybersecurity experts are most concerned about the potential infrastructural breakdown resulting from these cyberattacks (Jayasuriya, 2015).

Amid the ever-evolving landscape of the digital era, against the backdrop of the tumultuous years 2022–2023, the financial industry teeters on the brink of an increasingly perilous journey. Recent cyberattacks on financial institutions have sent alarm bells ringing across the sector, highlighting the urgent need for a profound transformation in cybersecurity measures. As we navigate through the intricate realm of cyber risk within the financial sector, we encounter a multifaceted array of threats, spanning from data breaches to fraudulent activities and disruptive breaches in business operations (Lusthaus & Varese, 2017).

Our analysis leaves no stone unturned, shedding light on emerging trends by dissecting a rich tapestry of diverse datasets. Despite financial institutions having led the charge in cybersecurity, they now face treacherous waters, beset by mounting challenges. The surge in both the frequency and sophistication of cyber intrusions has sounded a clarion call, necessitating substantial investments dedicated to bolstering cybersecurity and enhancing risk management strategies (Sonett & Colburn, 1968).

This chapter exposes the contemporary landscape of cyber risk within the financial domain, thrusting into sharp relief the pressing urgency for proactive measures.

As the financial world grapples with an environment riddled with digital perils, we emphasize the strategic initiatives undertaken by financial institutions, regulatory bodies, and government agencies. Their collective goal is to fortify resilience, preventing the recurrence of catastrophic financial crises akin to the tumultuous events experienced just a decade ago.

Moreover, 2023 marks a pivotal juncture where the financial sector contends not only with cyber threats but also with the lasting effects of the COVID-19 pandemic. The global crisis accelerated the digitization of financial processes and pushed organizations to adopt remote work models (Sood & Enbody, 2014). However, this rapid transformation introduced new vulnerabilities, rendering the financial sector more susceptible to cyberattacks. It's a clarion call for unwavering commitment to fortify the digital ramparts of the financial world, as we navigate through this intricate web of challenges. This is a moment that necessitates not only comprehensive action but also collaborative resilience, urging us to embark on a paradigm shift towards a fortified cyber defence infrastructure to safeguard the very core of the global financial fabric (Banga et al., 2019).

Section 2 provides a detailed discussion on the types of cyberattacks. Section 3 provides a clear view on the financial market ecosystem. Section 4 discusses on the analysis of malware distribution channels and implications for user protection. Section 5 discusses on the preventions and countermeasures in the financial market ecosystem post-pandemic (COVID-19) through 2023. Section 6 concludes the chapter and discusses the future scope of work.

2 Types of Cyberattacks

In this section, we will discuss the various cybersecurity threats that pose substantial risks, offering practical suggestions to help entities reduce overall risk in the years 2022–2023.

(i) Ransomware Attacks: In 2022 and 2023, ransomware attacks continue to plague organizations worldwide. These attacks have evolved to become more frequent and sophisticated, with cybercriminals employing new tactics to extort victims. Ransomware remains a significant threat, with each successful incident potentially costing organizations substantial sums. The attacks often involve demands for cryptocurrency payments to unlock compromised systems (Das, 2023).

(ii) Remote Work Vulnerabilities: Remote work arrangements persist in 2022 and 2023, making organizations vulnerable to cyber threats. The expanded use of home networks for professional purposes continues to increase the attack surface for corporations (Batur Şahin, 2022). Cybercriminals are keen to exploit security weaknesses associated with remote work, including unsecured home networks and personal devices used for business tasks.

(iii) Advanced Persistent Threats (APTs): Advanced persistent threats are cyberattacks characterized by a prolonged and targeted approach. In 2022–2023, APTs have become more sophisticated and persistent. These threats often aim to infiltrate organizations and exfiltrate sensitive data over extended periods, making them challenging to detect and mitigate (Mahmoud et al., 2023).

(iv) Zero-Day Exploits: Cyberattackers are increasingly leveraging zero-day vulnerabilities in software and systems. These vulnerabilities are previously unknown to the vendor and, therefore, unpatched (Halabi & Zulkernine, 2023). In 2022–2023, attackers are quick to exploit these weaknesses, posing a severe risk to organizations.

(v) Social Engineering and Phishing: Social engineering attacks, including phishing, remain a prevalent threat. In 2022–2023, cybercriminals are capitalizing on social engineering tactics to deceive individuals and gain unauthorized access to sensitive information. These attacks are often disguised as legitimate communications or enticing offers (Tomičić, 2023).

(vi) Supply Chain Attacks: Supply chain attacks continue to grow in complexity and scope. In recent years, cybercriminals have targeted third-party vendors and software providers, compromising the systems of multiple organizations downstream. These attacks present a significant challenge for organizations that rely on external services and software for their operations (Nygård et al., 2022).

(vii) Fileless Malware: Fileless malware attacks have evolved to bypass traditional detection methods (Mohanta & Saldanha, 2020). In 2022–2023, these threats have become more sophisticated, as cybercriminals utilize trusted platforms and existing technologies to infiltrate systems. Fileless malware often aims to compromise critical systems and exfiltrate data without leaving a trace.

(viii) Insider Threats: Insider threats, whether intentional or accidental, continue to be a concern in 2022–2023. Organizations must remain vigilant about protecting against internal actors who may compromise data or systems, either maliciously or unintentionally.

(ix) IoT Vulnerabilities: The proliferation of Internet of Things (IoT) devices introduces new cybersecurity challenges. In 2022–2023, cybercriminals are exploring vulnerabilities in IoT devices to gain unauthorized access to networks and data, posing a risk to both individuals and organizations.

(x) Artificial Intelligence in Cyberattacks: Cyberattackers are increasingly incorporating artificial intelligence (AI) and machine learning into their tactics. AI-driven attacks can adapt and evolve rapidly, making them challenging to detect and defend against (Reveraert & Sauer, 2020).

In summary, the landscape of cyber threats is continually evolving, with cybercriminals adapting to new technologies and vulnerabilities. Organizations must stay vigilant and implement robust cybersecurity measures to protect against these evolving threats in 2022–2023.

3 Financial Market Ecosystem

The increased integration of sophisticated digital systems and the extensive reliance on innovation and technology, coupled with the growing utilization and interconnectivity of third-party products and services, have heightened the susceptibility of financial services to cyber threats. At the close of 2021, financial analysts identified cyberattacks as the most significant menace to the global financial market. Presently, hackers exhibit a high level of innovation and possess considerable resources, allowing them to meticulously identify potential avenues for pilfering funds from their targets (Calandro & Hoffmire, 2022). Ransomware attacks, often accompanied by demands for ransom in the form of cryptocurrencies such as Bitcoin, have become increasingly prevalent. Cybercriminals are progressively exploiting vulnerabilities within the supply chain and through third-party vendors with the ultimate aim of compromising systems or initiating data breaches, culminating in substantial ransom demands. The landscape of cyberattacks is evolving, becoming more sophisticated and agile, thereby amplifying their potential impact. Threats to supply chain networks have become a cause for concern for global IT service providers. Hackers are directing their focus towards these service providers to gain access to other infrastructures utilizing their services or software. A striking example occurred in February 2021, when a significant data breach unfolded within Air India's passenger service system. This breach resulted in the loss of data belonging to 4.5 million passengers, affecting Air India's flyers globally. The service provider responsible for this system was SITA, a company headquartered in Geneva, Switzerland (Abdelli et al., 2013). Air India reported that data recorded between August 26, 2011, and February 20, 2021, had been compromised. The compromised data encompassed personal information such as names, dates of birth, contact details, passport information, ticket information, credit card details, and more. The extended 2-year period of pandemic-induced lockdowns has significantly strained business organizations and employees, impacting both human and financial systems. This period has witnessed a massive shift towards remote work across various sectors, encompassing businesses, educational institutions, and government offices. Despite the advancements made in security protocols, communications, and crisis management capacities within business and information technology organizations, the surge in cyber threats and security crises during this time has presented unprecedented challenges (McCusker, 2017). Cybercriminals are capitalizing on the immense pressure experienced by these organizations, propagating malware through COVID-related news and exploiting essential data repositories. This exploitation is often followed by coercive tactics, pressuring these entities to pay ransomware to ensure business continuity throughout the ongoing pandemic crisis. Particularly vulnerable are critical infrastructure providers such as those in healthcare, energy, and financial services (Głogowski, 2022).

The twenty-first century has seen the realization of a global village where digital technology interconnects with the world's economies, cultures, and populations. India, with over 560 million Internet users at the onset of 2020, stands as the second-largest Internet user base worldwide. While this heightened connectivity through the Internet promises vast advancements on a global scale, it also exposes our digital societies to new vulnerabilities (Eisenbach et al., 2022). For instance, intellectual property theft alone accounts for roughly 25% of the costs associated with digital crime and poses a threat to national security, including military technologies and innovations (Lusthaus & Varese, 2017). Over the past 15 years, digital crimes within the country have inflicted significant financial harm on Indian organizations (Peihani, 2022).

(i) The Prevalence of Cyber Risk: The expansion of remote work and digital finance solutions has presented a double-edged sword to the financial industry. While these innovations have enhanced accessibility and efficiency, they have also significantly increased the sector's vulnerability to cyber threats. Cyberattacks and data breaches have surged in both frequency and sophistication, affecting not only financial institutions but also the broader ecosystem, including fintech startups, payment processors, and regulatory bodies (Crovini & Marchini, 2023).
(ii) Effects of COVID-19: The pandemic acted as a catalyst for these changes. Financial organizations rapidly adopted remote work solutions, online transactions, and digital communication tools to ensure business continuity. However, this hasty transition often resulted in the oversight of cybersecurity measures, leaving sensitive financial data exposed. The data suggests that the financial sector has experienced a substantial increase in cyberattacks since the pandemic's onset, with notable breaches and incidents targeting high-profile institutions.
(iii) Financial Market Vulnerabilities: The financial market ecosystem encompasses a vast network of stakeholders, including traditional banks, investment firms, fintech companies, and government regulatory bodies. Each entity faces distinct cyber threats, ranging from data breaches and financial fraud to ransomware attacks. Moreover, the interconnected nature of this ecosystem means that an attack on one part can have far-reaching consequences, affecting market stability, investor confidence, and economic resilience.
(iv) Data and Statistics: This analysis is supported by compelling data and statistics. For example, research reveals that the number of digital crimes per financial association increased by 31% in 2021 compared to the previous year. It now takes an average of 280 days for organizations to detect and respond to cyber espionage, with each successful espionage attack costing approximately \$3.6 million. The impact is pervasive, with cyberattacks or data breaches affecting around 55% of organizations worldwide in 2021. Fraud activities account for 24% of these crimes, while ransomware attacks contribute 20%. These statistics underscore the urgent need for robust cybersecurity strategies in the financial market ecosystem (Pandey, 2021).
(v) Additional Data: According to a 2022 report by a leading cybersecurity firm, the financial sector has been the primary target of cyberattacks, accounting for 23% of all reported attacks. A study conducted by a cybersecurity research

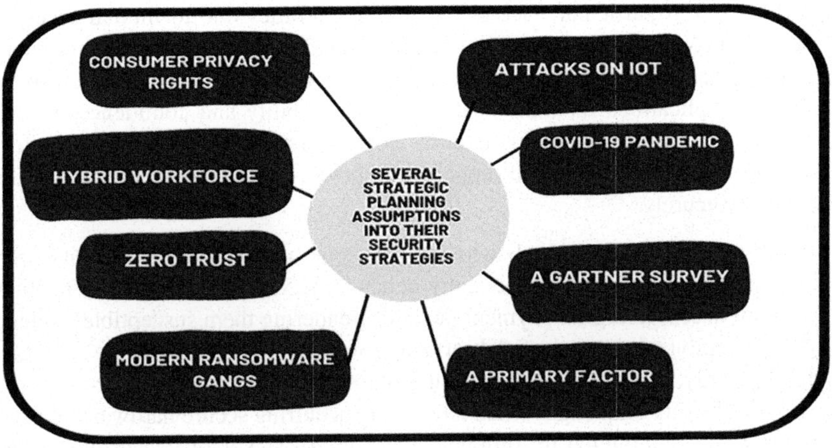

Fig. 1 Top predictions for 2022 and beyond

group found that during the COVID-19 pandemic, there was a 600% increase in phishing attacks targeting financial organizations. These attacks aimed to steal sensitive customer data and gain unauthorized access to financial systems. The Monetary Authority of Singapore (MAS) reported a 150% increase in cyber threats to financial institutions in the post-pandemic period, underlining the growing risk in the financial market ecosystem. Regulatory bodies such as the US Securities and Exchange Commission (SEC) have seen a surge in cyber incident reporting by financial organizations. In 2022, there was a 67% increase in reported incidents compared to the previous year (Calandro & Hoffmire, 2022).

In Fig. 1, authors have provided a clear idea on the most significant strategic planning assumptions into their security strategies for the year 2022 and beyond.

4 Analysis of Malware Distribution Channels and Implications for User Protection

The distribution of malware has evolved into a sophisticated and growing threat posing significant risks to users, especially those in low-income or new-to-tech categories. This analysis delves into the various channels and techniques through which malware is distributed, shedding light on the implications for user protection. The presence of a digital security divide creates a susceptibility to malware and fraud that disproportionately affects low-income individuals and those who are new to technology (Tsuchiya, 2016). This divide extends beyond access to technology, encompassing disparities in digital literacy and experience, leaving newer users at a higher risk of falling victim to online scams and malware attacks. Low-income communities, in particular, face elevated vulnerabilities due to limited access to robust security measures, amplifying the likelihood of becoming targets for cybercriminals. This digital security gap not only poses immediate threats but also has enduring implications for individuals' financial stability and confidence in using technology. Bridging this divide requires not only providing access to technology but also fostering education and support to empower all users to navigate the digital landscape securely.

(i) To begin with, individuals who are both low-income and relatively inexperienced with technology frequently depend on external assistance to facilitate their access to digital payment methods, rendering them susceptible to potential exploitation by unofficial service providers.
(ii) Secondly, it's worth noting that low-income individuals sometimes face affordability challenges when it comes to acquiring secure hardware and software, as highlighted in Anthony's work in 2023. Notably, security concerns tend to be more pronounced in the context of lower-cost Android phones, as observed by Morrison in 2020. This is primarily attributed to the fact that numerous budget-friendly phones are produced by lesser-known manufacturers who may not adhere to standardized vetting processes, as underscored in Morrison's research. Additionally, low-income users are often compelled to use older devices that are no longer eligible for routine software updates, consequently raising the risk of malware infiltration and exposing them to heightened security threats.
(iii) Further, scammers may no longer depend on users to disclose detailed information, as they now employ malware to pilfer data from users' devices. Typically, most malware necessitates only a brief interaction with the user to secure access to a device. Despite the user's installation of a malicious trojan app, the fraudster requires initial authorization from the user to grant permissions, enabling the malware to infiltrate the device. However, this permission-granting stage often marks the final interaction the banking trojan has with the user. Subsequently, upon acquiring these initial permissions and privileges, the malware can autonomously obtain additional permissions without necessitating further authorization from the user.
(iv) Moreover, it's worth noting that malware often employs tactics to hide its icon from the device screen, as reported by McAfee in the later 2020s. This clandestine behaviour enables the illicit extraction of information from the user's device without their knowledge of the malware's presence. Additionally, banking trojans frequently disguise themselves as seemingly unrelated applications, bearing no visible connection to payment or banking services. This clever disguise can make it challenging for users to pinpoint these deceptive applications as the source of any financial losses they may incur. Even when users exercise caution and refrain from sharing their credentials and PINs with potential impostors, they remain vulnerable to the pernicious impact of malware attacks.

(v) Additionally, certain malware has the potential to exploit vulnerabilities within applications. While the majority of banking trojans generally rely on user deception rather than exploiting inherent operating system vulnerabilities, some trojans may seek to leverage security weaknesses in third-party applications installed on the device. For instance, the Android-based banking trojan known as Ginp, as highlighted in IBM Security Trusteer's findings, specifically targets vulnerabilities within select banking apps. It employs a technique that superimposes counterfeit login screens over legitimate ones, deceiving unsuspecting users into believing they are interacting with genuine applications, ultimately resulting in financial losses.

(vi) Probably, the one-click fraud instances reported by the individuals in our primary study were likely a result of malware infiltration. Studies conducted by "Dvara Research" elsewhere indicate that the permissions requested by apps to access various types of data are often presented in extensive terms and agreements, creating a concerning scenario. Moreover, users often tend to accept these terms and conditions almost reflexively, without considering them as a noteworthy event. Consequently, users might have innocuously clicked on a link and agreed to terms and conditions without actively divulging any sensitive financial information, yet found themselves in a situation where they incurred financial losses. As elaborated earlier, a majority of malware is disseminated through social engineering techniques like phishing and malvertisements, methods that users may not immediately recognize as suspicious or dubious.

(vii) One-click frauds, which do not rely on social engineering, become a more realistic scenario when cybercriminals identify weaknesses in an operating system's security features. In such instances, malware can obtain the necessary permissions without requiring user interaction. As an illustrative case, let's consider the "Towelroot Exploit" from 2016, where a vulnerability within the Android operating system allowed malware to take control of a device without the need for specific permissions or user involvement, as documented by Threatpost in 2016. It's essential to acknowledge that these kinds of vulnerabilities are relatively rare, and they are typically swiftly addressed by device manufacturers and software developers to bolster system security, a practice that remains especially critical in the context of the year 2023.

4.1 Channels for Distributing Malware

This section discusses and provides a clear view on the various channels for distribution of malwares:

(i) Phishing Links: Phishing links are a prevalent method for malware distribution, with approximately 25% of fraud cases linked to this approach. Fraudsters send deceptive links via SMS, instant messaging apps, emails, and social media, disguising them as messages from trusted entities like banks or regulators. Also, clicking on these links results in the installation of malware on users' devices.

(ii) Malvertisements (Malvertising): These are online advertisements embedded with malicious code, exploiting vulnerabilities in users' browsers or operating systems. It can deliver various types of malware, including adware, spyware, ransomware, or trojans. Users are often tricked into clicking on these ads, believing them to be legitimate.

(iii) Obtaining Applications from Unreliable Sources: "Trojan malware" frequently masquerades as legitimate apps and is distributed through third-party app stores. Users unknowingly download apps that are infected with malware, compromising the security of their devices.

(iv) Juice Jacking: This method involves malevolent actors exploiting shared charging stations to inject malware into users' mobile devices while they are being charged, posing a significant security threat. This unauthorized access to users' devices during the charging process can result in the installation of harmful software or the extraction of sensitive information without the users' knowledge or consent. It's a growing concern in public spaces where individuals often resort to using shared charging stations, inadvertently exposing their devices to potential security risks.

(v) Insecure or Fake Wi-Fi Networks: Fraudsters engage in the creation of deceptive Wi-Fi networks that mimic legitimate ones, luring unsuspecting users into connecting to these rogue networks. Once users connect to these deceptive networks, malicious actors exploit this connection to deploy and distribute malware, thereby compromising the security of the connected devices. This technique is a prevalent cybersecurity threat, as individuals are often enticed by seemingly genuine Wi-Fi networks, unaware that they are inadvertently granting attackers access to their devices and data. Such deceptive networks are often found in public spaces, adding to the concern of cybersecurity in today's interconnected world.

(vi) Exploitation by Technology Assistants: Individuals new to technology often require assistance with digital payments, making them vulnerable to malware distributed by unofficial helpers. Inexperienced users seeking guidance may unknowingly expose themselves to malicious software, putting their personal and financial information at risk. Awareness and education on cybersecurity are vital for this user group.

4.2 Implications for User Protection

In this section, authors have provided a clear idea about the implications for user protection:

(i) Digital Security Divide: Low-income and new-to-tech users are disproportionately vulnerable to malware due to their reliance on assistance for digital payments. Access to secure hardware and software can be financially prohibitive for these individuals. Low-cost Android devices, which are commonly used by low-income users, often lack rigorous vetting processes, exacerbating security concerns. Older devices, commonly used by low-income users, no longer receive regular software updates, exposing them to increased security risks.

(ii) Malware Accessibility and Lowered Cost: Changes in the cybercrime landscape have made malware more accessible and affordable, increasing its supply. The average price of information-stealing malware has dropped to as low as $5, expanding the pool of potential attackers. The emergence of Malware-as-a-Service (MaaS) offerings enables individuals with minimal cybersecurity expertise to engage in malicious activities. Malware authors have evolved from merely selling their products to providing mentoring services and detailed playbooks on malware usage.

(iii) Malware Evasion and Data Theft: Malware can steal data from users' devices without requiring extensive information disclosure. Malware operates covertly, hiding its presence from users and often disguising itself as unrelated applications. Certain malware exploits vulnerabilities in third-party apps, deceiving users into engaging with seemingly legitimate applications.

4.3 User Protection Measures

In this section, authors have discussed on the various protection measures that should be taken by users to avoid the spread of malware threats.

(i) Education and Awareness: Users should be educated about malware threats and encouraged to exercise caution when clicking on suspicious links or downloading apps from unverified sources.

(ii) Device Security: Users should keep their devices updated, install reputable antivirus software, and connect to secure and verified Wi-Fi networks.

(iii) Behavioural Caution: Users should refrain from sharing sensitive information, even if they believe they've only clicked on links without sharing data.

(iv) Vulnerability Patching: Certain malware exploits vulnerabilities in operating systems to gain permissions without user interaction.

Device manufacturers and software developers actively patch these vulnerabilities to enhance user security.

The distribution of malware continues to be an ever-evolving threat landscape, with cybercriminals employing diverse tactics to exploit user vulnerabilities. Users, particularly those in low-income and new-to-tech categories, must remain vigilant in protecting themselves against these threats. Awareness, education, and secure practices are essential to mitigate the risks associated with malware distribution. As the cybercrime ecosystem evolves, proactive measures and collaboration among users, device manufacturers, and service providers are paramount to safeguarding digital ecosystems.

5 Preventions and Countermeasures in the Financial Market Ecosystem Post-Pandemic (COVID-19) Through 2023

Addressing the heightened cyber risks in the financial market ecosystem post-pandemic requires a proactive and comprehensive approach. Here are key prevention and countermeasure strategies to fortify the financial sector against cyber threats:

(i) Enhanced Employee Training: Invest in continuous and robust cybersecurity education and training programmes for employees at all levels within financial institutions. Ensure that they understand the evolving threat landscape and the critical role they play in cybersecurity (Głogowski, 2022).

(ii) Strengthened Network Security: Implement advanced network security measures, including real-time threat detection, intrusion prevention systems, and network segmentation. Regularly update and patch network devices and systems to close vulnerabilities.

(iii) Rigorous Application Security: Incorporate secure coding practices into the software development lifecycle to identify and mitigate vulnerabilities. Utilize application security tools, such as penetration testing and code review, to proactively identify and address application-specific threats.

(iv) Robust Information Security Framework: Develop and enforce a comprehensive information security framework that includes data encryption, access controls, data classification, and regular security audits. Safeguard sensitive information through data loss prevention tools and encryption mechanisms.

(v) Cloud Security Best Practices: Implement cloud security best practices to protect virtualized assets, data, and services hosted in the cloud. Establish stringent access controls, encryption, and continuous monitoring of cloud infrastructure.

(vi) Data Loss Prevention Protocols: Develop and implement data loss prevention protocols that minimize the risk of data breaches. This includes regular data backups and secure data recovery plans.

(vii) Collaboration and Threat Sharing: Foster industry-wide collaboration for sharing threat intelligence among financial institutions, regulatory bodies, and government agencies. Create a unified framework for sharing real-time threat data and coordinated responses to cyber incidents.

(viii) Harmonized Regulatory Framework: Promote the development of standardized cybersecurity regulations and guidelines applicable to all financial entities. Encourage close collaboration between regulatory bodies and the financial sector to ensure compliance and effective cyber resilience.

(ix) Frequent Cyber Resilience Testing: Regularly assess and enhance cyber resilience through penetration testing, simulated cyberattack exercises, and tabletop incident response drills. And also identify and address weaknesses and vulnerabilities in real time.

(x) Global Cyber Coordination: Facilitate cross-border coordination to address cyber threats that transcend national boundaries—promote harmonized international responses to global cybersecurity challenges.
(xi) Embrace Technological Innovation: Stay at the forefront of technological advancements and assess their potential impact on the financial sector—develop controls and regulations to mitigate risks while allowing innovative technologies to thrive.
(xii) Incident Response and Recovery Planning: Develop and continually update incident response and recovery plans to ensure swift and effective responses to cyber incidents—include communication strategies, legal compliance, and data breach notifications in these plans.

As the financial market ecosystem navigates the complexities of 2023 and beyond, these comprehensive prevention and countermeasure strategies are essential to fortify its cybersecurity posture. By taking a proactive approach, enhancing employee awareness, fostering collaboration, and staying updated on emerging threats, the financial sector can better protect sensitive financial data and mitigate the risks of cyberattacks.

6 Conclusion

The ever-growing landscape of cyber risks in the financial sector presents an escalating threat to global economic stability, marked by over 200 cyber incidents focused on financial institutions since 2019. The surge in cyber casualties is underscored by the evolving nature of cyber threats, reflecting a landscape increasingly fraught with sophisticated and destructive attacks. State-supported cyberattacks on financial institutions have become more frequent and sophisticated, posing a severe challenge to the integrity and security of the financial ecosystem. The interconnectedness of financial systems, driven by technological advancements and intricate supply chains, has exposed vulnerabilities that cybercriminals are exploiting to infiltrate networks and compromise critical data. The ransomware demands resulting from these breaches have only exacerbated the severity of the situation, underscoring the critical need for robust cybersecurity measures to safeguard the financial world. In this digitally advancing era, the evolution of cybercriminal tactics outpaces conventional security protocols, demanding proactive and innovative defence strategies. Protecting against cyber risks is not only the responsibility of financial institutions but also a collective effort that necessitates widespread awareness and active participation from individuals in safeguarding their devices and data. Enhancing personal cybersecurity habits such as using secure networks, employing robust passwords, and enabling two-factor authentication are pivotal in fortifying one's defences against cyber threats.

In essence, proactive measures at both the institutional and individual levels are indispensable to mitigate the expanding scope and sophistication of cyber risks in the financial sector. A resilient and secure future hinges on a collaborative commitment to fortify cybersecurity frameworks, fostering vigilance, awareness, and adaptive measures against the dynamic landscape of cyber threats. Only through collective efforts and continuous adaptation can we fortify the financial ecosystem against the escalating cyber risks that threaten the foundation of global financial stability.

References

Abdelli, A., Dahani, Z., Dalli, K., Benabbou, L., & Berrado, A. (2013). A classification framework of issuers in the Moroccan financial market. In *2013 8th international conference on intelligent systems: Theories and applications (SITA)*. https://doi.org/10.1109/sita.2013.6560810

Akhtar, N. (2021). Latest trends in the cybersecurity after the solar wind hacking attack. *Foundation University Journal of Engineering and Applied Sciences, ISSN 2706–7351, 1*(2), 14–24. https://doi.org/10.33897/fujeas.v1i2.347

Armin, J., Thompson, B., Ariu, D., Giacinto, G., Roli, F., & Kijewski, P. (2015). 2020 cybercrime economic costs: no measure no solution. In *2015 10th international conference on availability, reliability and security*. https://doi.org/10.1109/ares.2015.56

Banga, A., Gupta, D., & Bathla, R. (2019). Towards a taxonomy of cyber attacks on SCADA system. In *2019 international conference on intelligent computing and control systems (ICCS)*. https://doi.org/10.1109/iccs45141.2019.9065534

Batur Şahin, C. (2022). Learning optimized patterns of software vulnerabilities with the clock-work memory mechanism. *European Journal of Science and Technology*. https://doi.org/10.31590/ejosat.1159875

Bouveret, A. (2018). Cyber risk for the financial sector: A framework for quantitative assessment. *SSRN Electronic Journal*. https://doi.org/10.2139/ssrn.3221231

Calandro, J., & Hoffmire, J. (2022). Cyber risk. In *The business of personal finance* (pp. 41–43). https://doi.org/10.4324/9781003215417-7

Cashell, B., Jackson, W. D., Jickling, M., & Webel, B. (2004). The economic impact of cyber-attacks. In *Congressional research service documents, CRS RL32331, Washington, DC, 2.*

Crovini, C., & Marchini, P. L. (2023). Managing cyber risk in the financial sector: Insights from a case study. *Financial Reporting, 1*, 97–125. https://doi.org/10.3280/fr2023-001004

Curran, K., Concannon, K., & McKeever, S. (2007). Cyber terrorism attacks. In *Cyber warfare and cyber terrorism* (pp. 1–6). https://doi.org/10.4018/978-1-59140-991-5.ch001

Das, R. (2023). A technical review of ransomware attacks. In *Ransomware* (pp. 41–73). https://doi.org/10.1201/9781003431633-3

Eisenbach, T. M., Kovner, A., & Lee, M. J. (2022). Cyber risk and the U.S. financial system: A pre-mortem analysis. *Journal of Financial Economics, 145*(3), 802–826. https://doi.org/10.1016/j.jfineco.2021.10.007

Głogowski, A. (2022). Systemic and cyber risk: The two monsters of financial system. In *Digital finance and the future of the global financial system* (pp. 199–212). https://doi.org/10.4324/9781003264101-16

Hafkamp, W., & Steenvoorden, R. (2009). Experience from the financial sector with consumer data and ICT security. In *Securing electricity supply in the cyber age* (pp. 159–169). https://doi.org/10.1007/978-90-481-3594-3_9

Halabi, T., & Zulkernine, M. (2023). The ultimate battle against zero-day exploits: Toward fully autonomous cyber-physical defense. In *2023 IEEE international conference on software services engineering (SSE)*. https://doi.org/10.1109/sse60056.2023.00041

Jayasuriya, D. (2015). Compliance issues in the financial sector. In *Research handbook on international financial crime*. https://doi.org/10.4337/9781783475797.00034

Joshi, V. C. (2020). Cyber-risk management. In *Digital finance, bits and bytes* (pp. 131–150). https://doi.org/10.1007/978-981-15-3431-7_9

Kalinich, K. (2017). Treating cyber risks—Using insurance and finance. In *The cyber risk handbook* (pp. 143–158). https://doi.org/10.1002/9781119309741.ch10

Kayabaş, H., & Tuna, G. (2023). Cyber wars and cyber threats against mobile devices. In *Advances in digital crime, forensics, and cyber terrorism* (pp. 85–107). https://doi.org/10.4018/978-1-6684-6741-1.ch005

Lusthaus, J., & Varese, F. (2017). Offline and local: The hidden face of cybercrime. *Policing: A Journal of Policy and Practice, 15*(1), 4–14. https://doi.org/10.1093/police/pax042

Mahmoud, M., Mannan, M., & Youssef, A. (2023). APTHunter: Detecting advanced persistent threats in early stages. *Digital Threats: Research and Practice, 4*(1), 1–31. https://doi.org/10.1145/3559768

McCusker, R. (2017). Transnational organised cyber crime: Distinguishing threat from reality. In *Transnational financial crime* (pp. 415–432). https://doi.org/10.4324/9781315084572-23

Mohanta, A., & Saldanha, A. (2020). Fileless, macros, and other malware trends. In *Malware analysis and detection engineering* (pp. 721–767). https://doi.org/10.1007/978-1-4842-6193-4_20

Monteith, S., Bauer, M., Alda, M., Geddes, J., Whybrow, P. C., & Glenn, T. (2021). Increasing cybercrime since the pandemic: Concerns for psychiatry. *Current Psychiatry Reports, 23*(4). https://doi.org/10.1007/s11920-021-01228-w

Munk, T. (2022). Cyber attacks, means, and methods. In *The rise of politically motivated cyber attacks* (pp. 47–77). https://doi.org/10.4324/9781003126676-3

Nygård, A., Sharma, A., & Katsikas, S. (2022). Reverse engineering for thwarting digital supply chain attacks in critical infrastructures: Ethical considerations. In *Proceedings of the 19th international conference on security and cryptography*. https://doi.org/10.5220/0011384300003283

Pandey, P. (2021). Cyber risk, risk transfer, and financial innovation. In *Encyclopedia of cryptography, security and privacy* (pp. 1–4). https://doi.org/10.1007/978-3-642-27739-9_1617-1

Peihani, M. (2022). Regulation of cyber risk in the banking system: A Canadian case study. *Journal of Financial Regulation, 8*(2), 139–161. https://doi.org/10.1093/jfr/fjac006

Reveraert, M., & Sauer, T. (2020). Redefining insider threats: A distinction between insider hazards and insider threats. *Security Journal, 34*(4), 755–775. https://doi.org/10.1057/s41284-020-00259-x

Ros, G., & Schaanning, E. (2020). The making of a cyber crash: A conceptual model for systemic risk in the financial sector. *SSRN Electronic Journal*. https://doi.org/10.2139/ssrn.3664808

Sonett, C. P., & Colburn, D. S. (1968). The principle of solar wind induced planetary dynamos. *Physics of the Earth and Planetary Interiors, 1*(5), 326–346. https://doi.org/10.1016/0031-9201(68)90027-7

Sood, A. K., & Enbody, R. (2014). Why targeted cyber attacks are easy to conduct? In *Targeted cyber attacks* (pp. 113–122). https://doi.org/10.1016/b978-0-12-800604-7.00007-3

Tomičić, I. (2023). Social engineering aspects of email phishing: An overview and taxonomy. In *2023 46th MIPRO ICT and electronics convention (MIPRO)*. https://doi.org/10.23919/mipro57284.2023.10159691

Tsuchiya, M. (2016). Cyber security of financial sectors in Japan, South Korea and China. In *Managing cyber risk in the financial sector* (pp. 96–111). https://doi.org/10.4324/9781315675930-5

Tumkevič, A. (2017). Cybersecurity in Central Eastern Europe: From identifying risks to countering threats. *Baltic Journal of Political Science, 5*(5), 73. https://doi.org/10.15388/bjps.2016.5.10337

Bikram Ghosh is working as an Assistant Professor in the Department of Cyber Security and Technology, Brainware University, Kolkata, India. He holds a B.Tech and M.tech degree in Computer Science and Engineering from Adamas University, Kolkata. He has a total professional experience of about 2 years, including academics and industry. His research focuses on cybersecurity, malware analysis, cyber forensics, and cloud computing. He has served as reviewer for various international conferences.

Dr. Riya Sil is currently serving as Associate Professor in the Department of Computer Science and Engineering, Brainware University, Kolkata, India. Dr. Riya Sil holds a B.Tech and M.tech degree in Computer Science and Engineering from Birla Institute of Technology and a PhD degree with a specialization in Legal Analytics. She has a total professional experience of about 9 years, including 7 years in academics and 2 years as Software Developer at Cognizant Technology Solutions (CTS). During her academic tenure, she has served as Assistant Professor in Computer Science and Engineering department in various institutions like Techno India University, Adamas University, and Kristu Jayanti College. She is currently engaged as an Associate in the Department of Computer Science and Engineering, Brainware University, Kolkata, India. Her research focuses on machine learning, natural language processing (NLP), deep learning, artificial intelligence (AI), and cybersecurity. She serves as a reviewer for various reputable journals and international conferences. She has about 40 publications to her credit, which include peer-reviewed SCI and Scopus-indexed journals, conferences, and book chapters.

Google Scholar: https://scholar.google.com/citations?user=s2LV5GEAAAAJ&hl=en&oi=ao

Part V
Fostering Sustainability Through Social Physics

Aligning Peri-Urban Areas Toward Sustainable Development Goals: A Case of Pratapsasan—A Peri-Urban Area of Bhubaneswar City

Sudha Panda and Soumyendu Shankar Ray

1 Introduction

Social physics, which has evolved into a multidisciplinary area of research interests focusing on human sociality embedded in social network and societal structures, can help urban planners and policymakers understand the dynamics of urban and peri-urban systems. It is capable of revealing the structure and behavioral patterns of social systems at different scales from individual to societal level as well as capturing the long-term evolution of the society (Pentland, 2015).

Peri-urban areas are not suburbs but instead represent a "third space" ("*tiers espace*") and an interconnected *continuum* (Camagni & Brunetta, 1994) between urban and rural areas and beyond administrative boundaries (Iaquinta & Drescher, 2000; Rauws & de Roo, 2011; Qvistrom, 2013; Hoggart, 2016; Wandl & Magoni, 2017). To be defined as peri-urban, it should be "an area between urban settlement areas and their rural hinterland. Larger peri-urban areas can include towns and villages within an urban agglomeration. Such areas are often fast-changing, with complex patterns of land use and landscape, fragmented between local or regional boundaries." More precisely, this area is also defined as "discontinuous built development, containing settlements of less than 20,000 inhabitants, with an average density of at least 40 persons per km^2 (averaged over 1 km^2 cell)" (Cattivelli, 2021). This structure reflects a low population density and a preference among new inhabitants (pensioners, families, and migrants) for a housing model based on a housing-family property (Dekkers, 2010).

Since these areas are at the fringes of cities and adjacent to the rural areas, they are in constant state of flux in terms of land use change and economic activities.

S. Panda (✉) · S. S. Ray
School of Architecture and Planning, KIIT University, Bhubaneswar, Odisha, India
e-mail: sudha.pandafar@kiit.ac.in; soumyendu.ray@kiit.ac.in

A. Rai et al. (eds.), *Contemporary Social Physics*, Springer Geography,
https://doi.org/10.1007/978-3-031-77596-3_18

In India, peri-urban areas may be classified as villages or census towns depending on the population. However, the governance of these peri-urban areas is by rural local government (panchayats) and notified area councils though some of the census towns display urban characteristics. Around 37% of new census towns declared in 2011 are concentrated around major metropolises (Pradhan, 2017). Urban and rural development and its governance is a state subject in India; hence, state government classifies these peri-urban areas according to their own set of criteria. However, if a settlement is classified as urban, it has to abide by building by laws and other development controls prescribed by the urban local body, which the residents are not in favor of. At the same time, the panchayats are unequipped to deal with the rapidly changing land use and activities brought about by urbanization. The sudden increase in migrants to peri-urban areas causes a spurt in demand for land and housing, which in turn influences the land use changes that take place without any planning or regulation (Fig. 1).

As there are no prescriptive planning principles for these transition areas, the objective of this research is to frame the guidelines to formulate the planning principles of peri-urban areas so that it is in alignment with the UN Sustainable Development Goals taking the case study of Pratapsasan, a peri-urban area of Bhubaneswar, the capital of Odisha. Bhubaneswar is a tier 2 city of India and is

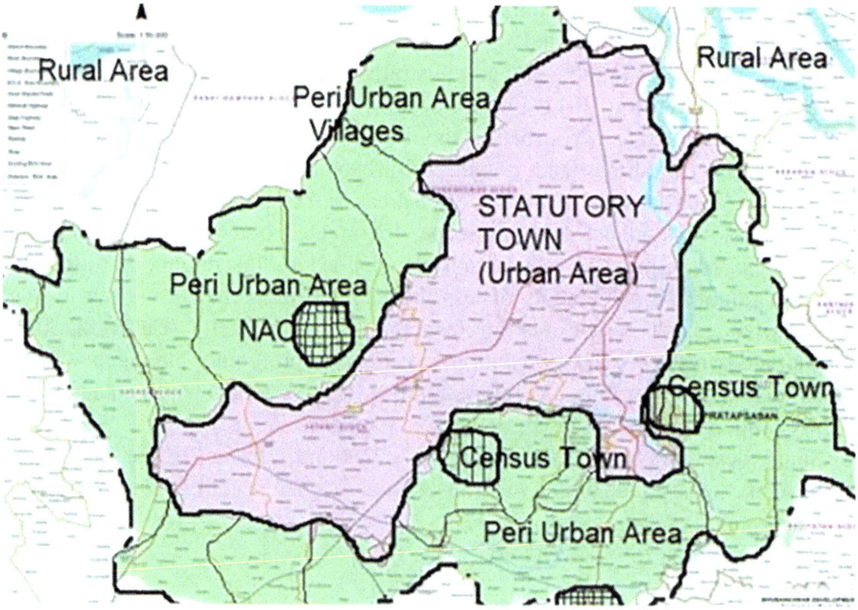

Fig. 1 A visualization of Urban and Peri-Urban Areas. (Source: Authors 2021)

undergoing sudden massive urbanization and rapid economic growth in a very short span of time.

The UN Sustainable Development Goal 11 aims to promote sustainable cities and communities by supporting positive economic, social, and environmental links between urban, peri-urban, and rural areas. In the context of a rapidly urbanizing world, to decrease the developmental disparity between urban and rural areas, 195 countries have committed to strengthening urban-rural linkages under New Urban Agenda (NUA). The rationale for strengthening these linkages is spatial flows of people, products, services, culture, and knowledge, which is central to linking public spaces and health and incorporating food and nutrition sensitivity into spatial planning. Peri-urban areas are vulnerable to environmental hazards, which increases the health vulnerability of urban areas indirectly by disruption of food, water, and ecosystems, transport pollution, and congestion. It also adds to social health vulnerability, by disruption of livelihoods and public services, growing social inequality, and exclusion. A case study in Melbourne shows the vision of the city as a food bowl by effective urban and rural linkages. Perishable fruits and vegetables are grown closer to the consumption areas, which is the city itself, thus bypassing the need of cold storage. The city itself becomes the food bowl by communities growing their own food. Value-added products are made by food processing from perishable items (Carey et al., 2015).

2 Research Questions

The objective of the study is to find solutions to some serious questions that confront the development in peri-urban areas:

- What are the founding principles for planning our peri-urban areas as a part of the future cities? How do we link these future cities with the UN Sustainable Development Goals?
- Can the aspiration, wants, and concerns of communities residing in these areas govern the planning of the peri-urban regions?
- How do we balance the market forces and governing authorities to create sustainable planning?
- How to make food production be an integral part of the city?
- How do we preserve the crafts and traditions of the peri-urban areas? Can it be integrated in an existing tourist circuit?
- How do we integrate geomorphology, ecosystem, riverbeds, and watersheds while planning the road network in peri-urban areas?

3 Significance of the Study

A look at some of the common issues faced by peri-urban areas shows:

3.1 Indiscriminate Land Use Conversion and Reduced Food Production

Land designated for agriculture in peri-urban areas is being converted to built-up areas, which adversely affects the local supply of food and also degrades the ecosystem. In contrast, if food is brought in from the rural hinterland, the cost of food increases due to the additional transportation and warehousing cost component. Hence, food security of urban areas is being compromised by the rapid conversion of agricultural lands into built areas in peri-urban areas.

3.2 Unregulated Development

As described earlier, most of the peri-urban villages and census towns are governed by rural local government (panchayats) and notified area councils who are not vigilant or equipped to check building control violations and land monitoring systems leading to unplanned development of structures that are beyond the permissible densities. Real estate developers and influential builders persuade the landowners to carry out illegal subdivisions and sell the agricultural land for building activity at attractive prices (Aijaz, 2019).

Uncontrolled growth at the periphery of cities is occurring, as seen in the fast depleting open spaces, increasing population densities and haphazard development of built structures (residential, commercial, industrial) that fail to meet building safety norms. The ambiguity in governance boundaries creates further complications in administrative extent of peri-urban areas. Such deficiencies in planning and governance leave these areas open to the rampant construction of illegal structures by individuals and groups. There are cases of peri-urban areas being colonized by private builders and public sector agencies that cater only to the interests of influential groups. There are displacement of local population, loss of neighborhood structure, and loss of biodiversity too (Aijaz, 2019).

3.3 Poor Access to Basic Services

Most peri-urban areas generally remain uncovered by piped and metered water supply. There is no collaboration between the urban and rural development authorities due to jurisdictional and land ownership issues. Water required in peri-urban areas for livestock and agriculture results in depletion of surface water sources like rivers and lakes. Lack of proper storm water drainage causes waterlogging on the roads during the rainy season. Such conditions affect daily activities such as walking and cycling to work. Besides causing harm to the road surface, the waterlogged areas also become breeding grounds for mosquitoes and impact public health.

Solid waste collection in most peri-urban areas is negligible. To add to the woes, urban governments see vacant lands in peri-urban areas as ideal places for dumping solid waste. The untreated waste dumped on agricultural land, water bodies, and roads by the peri-urban population or city agencies contaminates soil and groundwater, which impacts public health. In some peri-urban areas, there is poor accessibility and inadequate transport network (Aijaz and Observer Research Foundation, 2019).

4 Materials and Methods

This paper focuses on formulating sustainable planning principles of peri-urban areas taking the case of Pratapsasan, a fringe area of Bhubaneswar city, which is the capital of the eastern state of Odisha, in India. Bhubaneswar has been selected as a study area because it is a fast urbanizing town and its peri-urban areas (shown as green in Fig. 2) are facing the problems of indiscriminate land use conversion, unregulated development, and poor access to basic services.

Only 17% of the state is urbanized, and the twin city of Bhubaneswar and Cuttack forms a major part of this urbanization. Bhubaneswar, which was originally planned by Koenisberger for 40,000 people in 1960 on 20 sq. km, today accommodates more than 9 lakhs people within a development area of 233 sq. km. As seen in Fig. 2, Pratapsasan (hatched) is a large peri-urban area that is at the periphery of the

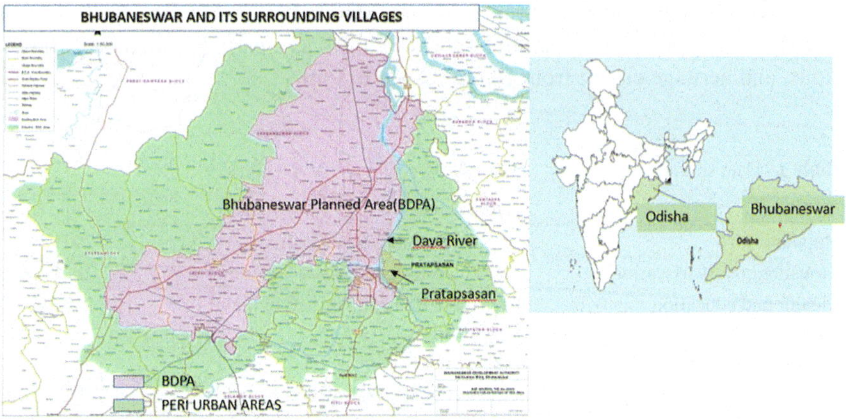

Fig. 2 Bhubaneswar and its surrounding Peri-Urban areas. (Source:www.bda.gov.in)

Bhubaneswar Development Authority boundary (marked in pink) and stands on the eastern bank of the Daya River.

4.1 Data Sources and Methods

Most of the data sources were from field investigation as seen in Table 1. Supporting data was taken from secondary sources like Census (2011).

The research steps followed are shown graphically in Fig. 3. With a vision to have a sustainable development of the craft village, the social, economic, and environmental dimensions are underpinned on the pivots of tourism, community participation, and promoting the heritage and culture. Each of the three dimensions is further elaborated in the hierarchical model to include social well-being like quality of life, socioeconomic development, economic upliftment by developing handicraft, and agriculture and ecological balance by taking care of land use changes and infrastructure provisions.

4.2 Regional Setting

Pratapsasan enjoys a very strategic position in the religious tourist trail and Odisha's handicraft circuit. It is located off NH 316, connecting Bhubaneswar to Puri. It also falls on the route of State Highway 13, connecting Bhubaneswar to Konark. Lying at the banks of the River Kuakhai, which splits into its distributaries Daya and Bhagirathi, it has a massive potential of getting leveraged as a tourist node as you can see in Fig. 4.

Since it also lies on the handicraft circuit of Pipili (famous for Chandua/appliqué work), Raghurajpur (artisan village famous for pattachitra paintings), Nimapada (famous for Nimapada sweets), and most importantly Puri (for holiday and religious tourism), it has enormous potential to be developed as a part of the handicraft, religious, and heritage circuit from Bhubaneswar to Puri.

Table 1 Data sources

Areas	Data sources	Time frame
Population	Projected from Census (2011)	2021
Housing quality assessment	Field investigation	2021–2022
Health and education infrastructure	Census (2011) and verified by field investigation	2021–2022
Built use distribution	Field investigation	2021–2022
Access to basic amenities	Census (2011) and verified by field investigation	2021–2022
Livelihood	Field investigation and questionnaire	2021–2022

OBJECTIVE

1. Frame the planning principles for Peri-Urban areas by linking with UN Sustainable Development Goals
2. Preserve craft and tradition by leveraging on tourism
3. Create a Community Centric Planning
4. Develop market linkages in governance
5. Make food production an integral part of the future city

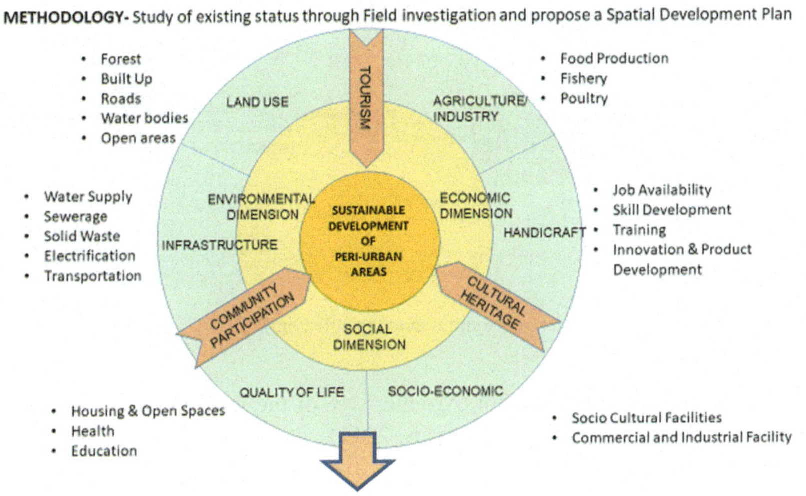

To attain SDG 11 with development in economic, environmental and social dimension through the underpinnings of Tourism, Cultural Heritage and Community Participation

Fig. 3 Research steps

4.3 Demographic

Pratapsasan is well known for brass and bell metal handicrafts since 1400 AD though details of its history traced from 1887 AD. Now it is a census town of 6.65 sq. km with a total population of 12,830 (6634 males and 6196 females). The sex ratio in Pratapsasan city is approximately 934, which is slightly lower than the average sex ratio of 979 for the state of Odisha. The literacy rate in Pratapsasan city stands at 77.79%, with 82.74% of males and 72.48% of females being literate. In terms of social demographics, 15.03% of the population belongs to the Scheduled Caste (SC) category, while the Scheduled Tribe (ST) population comprises 1.16% of the total population in Pratapsasan city. In Pratapsasan, out of the total population, 4405 were engaged in work activities. Around 28.47% of workers describe their work as main work, while 8.94% were involved in marginal activity. Of 4405 workers engaged in main work, 139 were cultivators (owner or co-owner), while 319 were agricultural laborers. Eleven percent of the workers are engaged as

Fig. 4 Proximity to tourist places. (Source: maps.google.com)

cultivators and agricultural labor. Around 89% of the workers are involved in the household industry (bell metal) (Census, 2011).

4.4 Sustainability Assessment of Pratapsasan

Social Sustainability

Housing and Open Spaces for Community Interaction

At present, there are 2576 households. Projecting the housing requirement (assuming an arithmetical increase in population) till 2051, there is a requirement of 4788 houses implying the need for 2375 more houses by 2051 (calculation based on Planning Commission Estimation of Rural Housing Shortage for 12th Five-Year Plan). Taking a proposed density of 60 plots/hectare (as per Rural Area Development Plan Formulation and Implementation (RADPFI) guidelines), there is a requirement of 40 hectares of land for housing (Table 2).

Housing quality assessment (Fig. 5) by age, height, building condition, and building type showed that 80% of the houses were pucca houses with good construction quality made within the last decade and generally two-floored structures.

Housing conditions are good but lack infrastructure services. Houses are abutting the streets with no setbacks. There is no street lighting, and storm water

Table 2 Projections of housing shortage

Year	Households	Acceptable housing (growth at 2% every year)	Congestion (6.5% of no. of households)	Obsolete (4.3% of acceptable housing: permanent + semi-permanent)	Total housing shortage	Housing requirement
2011	2576	2413	167	103	433	2846
2021	3221	2941	209	126	615	3556
2031	3640	3558	236	152	470	4028
2041	4113	4306	267	185	259	4565
2051	4648	4934	302	212	228	5162

drainage, solid waste collection, and sewerage network provision are not there in residential areas.

Ample open spaces (as per RADPFI norms) are present but not developed/ equipped as playgrounds and parks. There should be play equipment, seating places, and a designed landscape in open spaces, so they do not become dumping grounds for garbage. Areas around ponds can be developed for gatherings on religious occasions and recreational spaces. It will prevent the ponds from being converted into built spaces due to pressure of rising land values and lead to drainage issues as they are natural water sinks of the area. At present, all congregation activities happen on the roads, which disrupts the traffic. There is potential for riverfront development along banks.

Health

As can be seen from Table 3, health infrastructure is good and sufficient till 2051.

Education

As can be seen from Table 4, education infrastructure is also sufficient for population growth till 2051.

Environmental Sustainability

Land Use

The current land use (see Fig. 6, which has been obtained by overlaying the Tehsil map with current uses, contours, and gram panchayat boundary) shows a built-up distribution, which has doubled up since the start of the millennium (as seen in the satellite images).

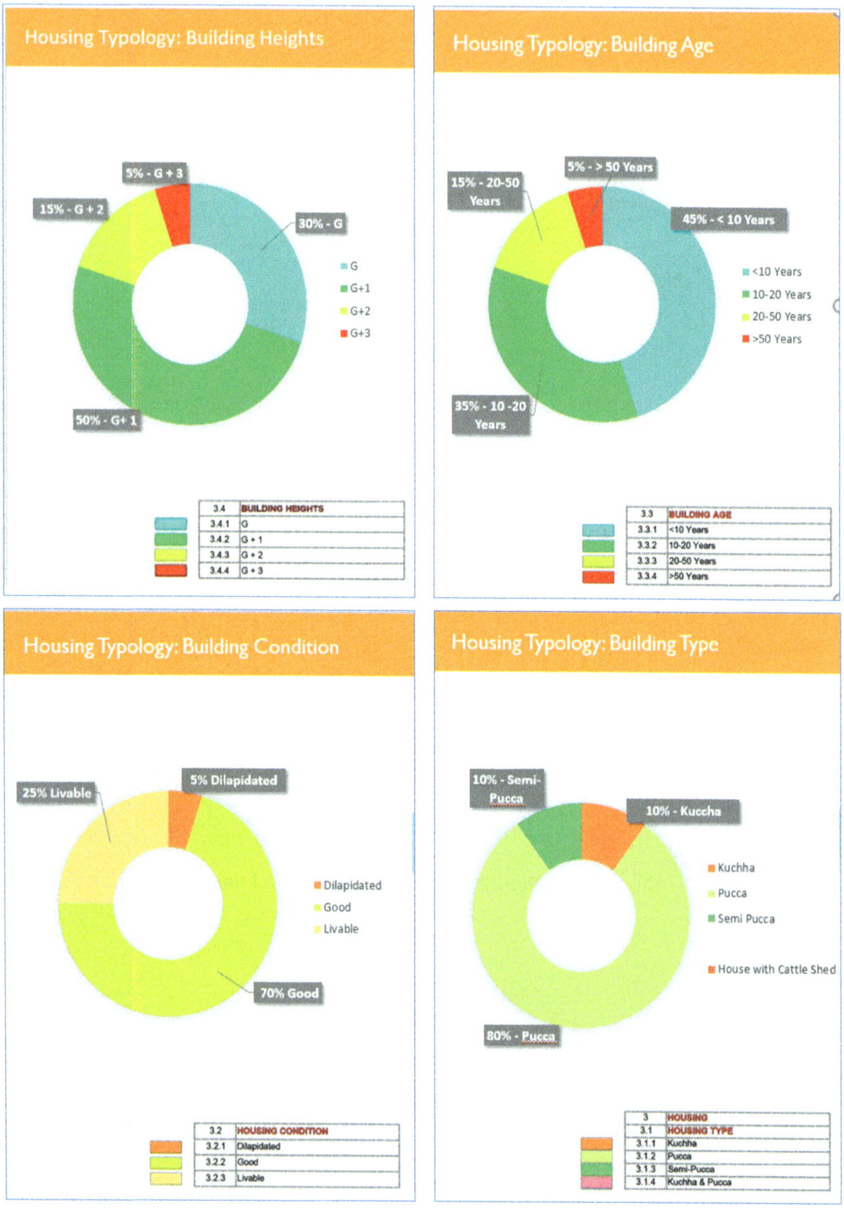

Fig. 5 Housing quality assessment. (Source: Authors, 2021)

A look at the map shows that the residential settlement (indicated by yellow) is densely placed along the older areas of Rathijema in the north (which has long linear plots) and in the west. The blue areas represent the commercial development, which is along the Banamalipur Road and the road that meets the Bainchua Road.

Table 3 Health infrastructure assessment

Infrastructure	Status (2011)	Requirements till 2051(as per RADPFI guidelines)
Primary health sub-center	1 health center with 2 beds with 1 doctor and 2 paramedics	Sufficient
Primary health center	1 health center with 6 beds with 1 doctor and 4 paramedics	
Maternity and child welfare	1 health center with 8 beds with 1 doctor and 1 paramedic	
Hospital (community health center)	1 with 6 beds, 5 doctors, and 25 paramedical staff	

Source: Census (2011)

Table 4 Education infrastructure assessment

Infrastructure	Status (2011)	Requirements till 2051(as per RADPFI guidelines)
Anganwadi	5	1 more by 2051
Primary school	7 (govt.) + 2 (private)	Sufficient
Middle school	3 (govt.) + 1 (private)	Sufficient
Secondary school	2 (govt.) + 1 (private)	1 more by 2051
Sr. secondary school (+2)	1 (govt.) + 1 (private)	Sufficient
Colleges	1 govt. degree college (art and science only) +1 govt. and 1 private degree college (art, science, and commerce)	Sufficient
Special school for disabled	1 govt.	–

Source: Census (2011)

There are several ponds as can be seen by light blue. The green shows mixed-use development (residential plus commercial). Gray areas are utility areas. State Highway 13 has been shown in dark gray.

The areas marked in red are public and semi-public areas, which house the educational and health institutions, private and government offices (like police stations and panchayat office), and other sociocultural amenities. The schools and anganwadis are distributed across the residential lots.

Figure 7 shows the distribution of various built uses in the existing land use plan.

Access to Basic Amenities

Water Supply

The groundwater availability at Pratapsasan is very good (less than 30 m depth in most places). At present, there is 5.13 lakh liters of water storage in overhead tanks. In addition, there are two tanks with 6 lakh liters under construction, which is sufficient till 2021.

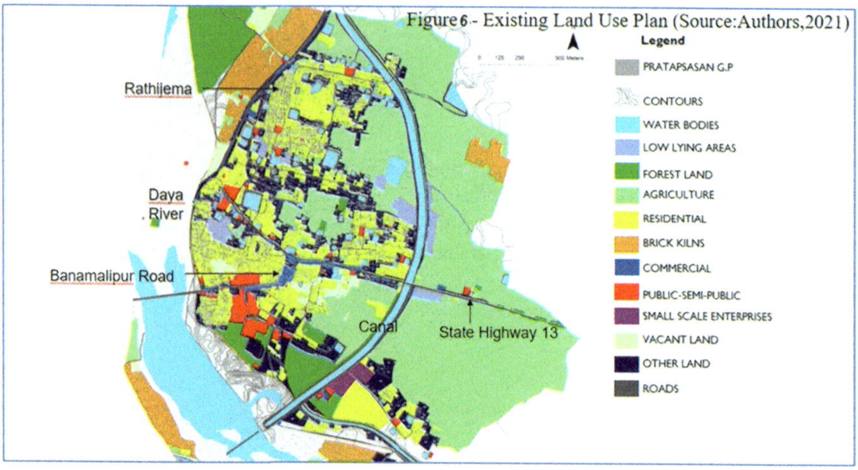

Fig. 6 Existing Land Use Plan. (Source: Authors, 2021)

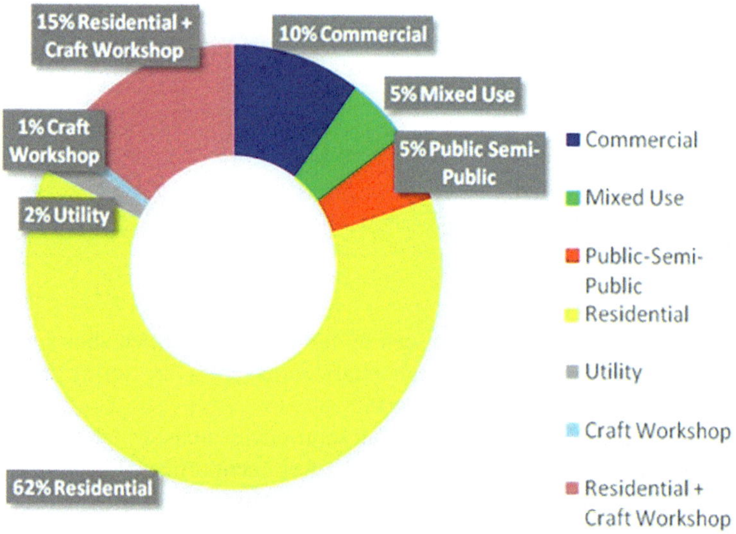

Fig. 7 Built Use distribution

Storm Water Drainage and Sewerage System

The sewerage and liquid waste disposal is mainly through a system of soak pits and septic tanks. Several houses have availed the system of toilet construction through Swachh Bharat Mission. However, there is no proper disposal of wastewater, which is released into the street from the houses. There is only one community toilet at the village haat (market) on Banamalipur Road, which is also inadequate. There are no storm water drains along the roads.

Solid Waste Management

At present, there is not enough manpower for door-to-door solid waste collection. Hence, all the solid waste is dumped in open areas, which are overflowing with garbage. This is very unsanitary and unhygienic. One such landfill site is under construction at Nuapatna, and there is a proposal for another at Rathijema. Since 1 hectare of composting site and landfill can handle 83.33 tons, 1 acre of land is sufficient to handle the 5.2 tons/day (till 2051).

Transportation

While planning the road network in Pratapsasan, an effort has not been made to integrate geomorphology, ecosystem, ponds, and watersheds. There are several ponds, largest being the Chandan Pokhari and Raja Pokhari. Roads and buildings are coming in the path of the natural drainage, which accentuates the problem of urban flooding.

The main arterial road experiences heavy traffic throughout the day. Ring roads (Bandha roads) are narrow and non-motorable. There is no provision of shoulders in the CC road sections; as a result, the edge drop was found to be very high at places raising safety concern for the vehicles plying on the roads during crossing and overtaking maneuvers, and also the edges are prone to damage in such conditions. Speed breakers are not provided on the busy roads. Encroachment was found along the major arterial roads making the roads inaccessible for the emergency services and also leaving no space for construction of drains. Newly developed residential areas and the fringe areas were not provided with concrete roads and mostly accessible by semi-kaccha or kaccha roads. Way finding was difficult because of absence of signage and legible directions to destinations.

Economic Sustainability

Livelihood

Due to its proximity to Bhubaneswar and its infrastructure, Pratapsasan enjoys the advantages of accessibility to facilities and services available in the capital city. The residents are involved in a mix of agricultural activities, small-scale industries (bell metal artisans), informal activities, and services. The aspiration, wants, and concerns of communities residing in this area should be the governing factors to decide the spatial planning.

Brass and Bell Metal

Pratapsasan has earned recognition for its production of bell metal products, a tradition that has been prevalent since the fourteenth century, making it a place of historical significance (Census, 1981). The history of bell metal production in Pratapsasan has contributed to its cultural and economic significance in the region. Bell metal is an alloy of copper and tin, and its traditional production has likely been passed down through generations, making it a prominent craft in the area. Bell metal products have been traditionally manufactured by household industries, where skilled and unskilled workers engage in strenuous physical labor, involving heating and beating the material as can be seen in Fig. 8. The process utilizes indigenous tools and equipment. In the earlier days, this industry thrived due to the affordable raw materials accessible to a limited number of artisans, along with the abundant supply of fuel. Moreover, the high demand for these products among the people contributed significantly to its growth and development. But presently, peri-urbanization has brought both positive and negative effects on livelihood of bell metal artisans and their traditional craft. Employment opportunities have laid way for the craftsmen to other work sectors. The youth of the area is not ready to take it forward as an occupation because it is labor intensive, raw material procurement is difficult and lacks marketing facilities. In order to protect the loss of this indigenous knowledge and practice, it is necessary to revitalize the crafts sector. Few master craftsmen have received government assistance because of their advantageous circumstances like income data, entrepreneurial skill, artisan card, etc. and have a higher status and better quality of life than the others.

Both forward and backward linkages of the brass and bell metal industry need to be streamlined. There is very high dependence (72%) on Mahajan for raw material. Hence, there is a need for raw material store managed by the government to prevent exploitation at the hands of the Mahajan. Also the sales distribution channel

Fig. 8 Brass and bell metal workshop. (Source: Authors, 2021)

dependence on middlemen is very high, so the artisan association should be strengthened to handle the sales and distribution. There is a need to revive the cooperatives, most of which have become defunct. There are no training facilities and promotion centers (display, marketing, and exhibition) available in Pratapsasan for which the artisans have to go to Bhubaneswar and Cuttack. There needs to be product diversification and proper training centers for design development in products other than household utensils. They can diversify into home decoration items that are in great demand using professional services of designers and architects and market them through their own websites for the export market. Some processes can be mechanized to enable bulk orders. Casting and moldings can be done using mass production techniques to reduce the delivery time.

The present working sheds and working conditions not suitable as per standards. Not only is the light and ventilation poor but also the space is constrained. The Kansari (brass workers) have their workshop in a permanent or semi-permanent constructed house. The workshop is situated at the front or rear part of the site but never in the courtyard for privacy. A suitable workshop is always provided with a verandah for better accommodation. It accommodates the artisans, furnaces, tools, and equipment.

Food Production and Livestock

The primary challenge for agriculture production in Pratapsasan is the rapid conversion of valuable agricultural land into urban developments, particularly for residential purposes, as it serves as a cheaper alternative to housing in the city. This conversion leads to the unavailability of crucial agricultural land, resulting in reduced agricultural production and lower living standards for the local population (Rondhi et al., 2018). Unfortunately, this trend is likely to worsen, leading to the displacement of more farm households leading to consequences such as food insecurity and social instability in the surrounding communities. Though the sale of agricultural land can be seen as an opportunity to improve the standard of living for the farmers by switching to other means of livelihood, it is very important for sustainable development of the city that food production be an integral part of the city.

At present, 91% of the agricultural production (rice) is Kharif crop (520 hectares). Only 9% of the total sown area are having double crop of moong dal (rabi crop). Livestock rearing and vegetable growing are done for self-consumption only.

Informal Workers

The informal workers of Pratapsasan are engaged in multiple occupations as per the skill, requirement, and availability of jobs. The livability for informal workers in Pratapsasan is significantly higher compared to farmers and bell metal artisans. This is attributed to the plethora of diverse job opportunities available to them, ensuring more stable income and improved economic prospects. Notably, during the farming season, they collaborate with farmers, while the rest of the time, they are engaged in various other service-oriented activities. The proximity to Bhubaneswar further boosts their quality of life, granting easy access to urban amenities and thriving markets, which greatly enhances their overall well-being and convenience.

5 Results and Discussion

The strength, weakness, opportunity, and threat analysis of Pratapsasan (as can be seen in Fig. 9) shows that to capitalize on its strength and opportunities and create a sustainable planned development, the peri-urban area of Pratapsasan should be developed as a tourist node and manufacturing hub for brass and bell metal wares by leveraging on its heritage significance and proximity to capital city and tourist circuit.

5.1 *Development of the Area as a Tourist Hub*

Owing to its strategic position in the tourist circuit and handicraft circuit (see Fig. 10), it is proposed to develop the ring roads (Bainchua Road) and Banamalipur Road. The encroachment on the Banamalipur Road should be removed to allow the road widening. Commercial use is proposed on either side of Banamalipur Road. A transport and logistic hub with bus and auto stand is proposed to take care of tourist changeover to Konark. It will be well equipped with motels, eateries, and refueling stations. On the peripheral road, it is proposed to have a tourist information huvb, exhibition grounds, and display of brass and bell metal wares.

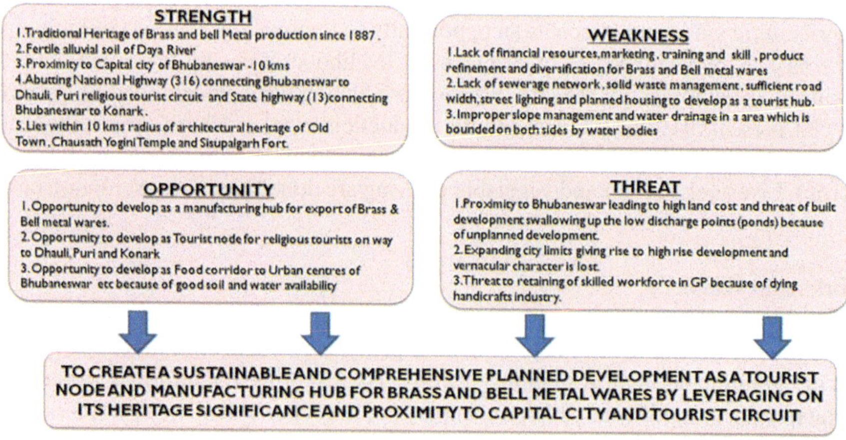

Fig. 9 SWOT analysis of Pratapsasan. (Source: Authors, 2021)

Link the Area with the Potential Tourist Circuit

At present, it is linked to NH 316 via the bridge. The entrance to the Pratapsasan can be widened at the start of Banamalipur Road with proper signage and welcome boards.

Creating a Handicraft Corridor and Connecting It with the Existing Handicraft Circuit

It is proposed to create a handicraft corridor linking Pratapsasan to Pipili and Raghurajpur. There will be stopover for tourists to explore the traditional brass and bell metal handicrafts.

Integration of the Area with the Bhubaneswar Expansion Program

Since Pratapsasan falls in the peripheral villages of the Bhubaneswar Development Plan, it will be part of the Bhubaneswar expansion program in another 20 years, so it has to be planned to retain the vernacular character and prevent the rapid urbanization by very high-rise structures and become the breathing lungs of the city expansion program taking care to retain its essence of environmental sustainability. The projected housing requirement of the peri-urban area as worked earlier in Sect. 4.4 (as per RADPFI guidelines) together with the projected spillover housing requirement of city can be accommodated in the portions shown in yellow.

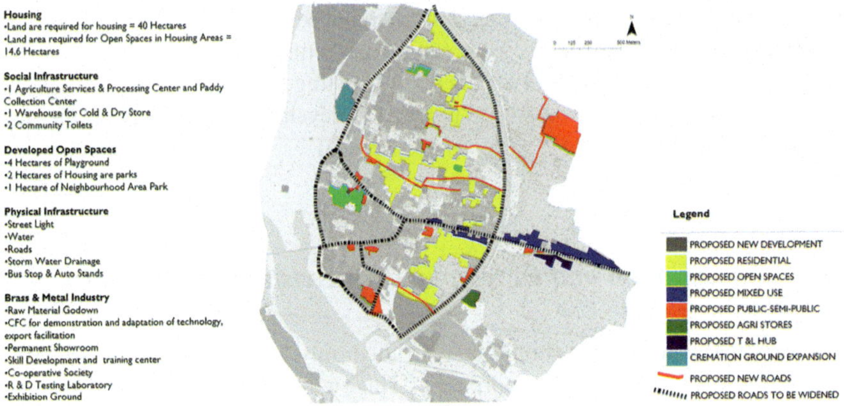

Fig. 10 Proposed land use plan by 2051

5.2 Quantitative and Qualitative Improvement of the Existing Social and Physical Infrastructure

There is a need to increase the qualitative aspect of existing social and physical infrastructure in terms of maintenance of educational and health infrastructure and following of norms in terms of qualified staff. There is a need for quantitative and qualitative improvement in terms of physical infrastructure like electrical, water, drainage, and solid waste disposal for the projected population as discussed earlier.

5.3 Enhance the Socioeconomic Development and Improve the Quality of Life

To boost the brass and bell metal industry, it is proposed to have a:

(a) Brass and bell metal development center having:

- Raw material store
- Facilitation center for demonstration and adaptation of technology and export facilitation
- Permanent showroom
- Skill development and training center
- Cooperative society
- Research and development laboratory
- Exhibition ground

To boost agriculture and related industry, it is proposed to have a:

- Agriculture center having dry and cold warehouses for agricultural, fishery, and poultry produce to encourage commercial marketing
- Market complex for giving permanence to existing haat
- Fishery at the several existing ponds by developing them under the fish pond scheme

5.4 Conservation and Development of Existing Water Bodies

The existing water bodies should be restricted only for eco parks and recreational purposes or for fishery purposes. This will prevent the environmental degradation.

5.5 *Incorporation of Riverfront Development Program*

The riverfront of Daya River should be developed with promenades and tourist cottages.

5.6 *Conservation of Agricultural and Forest Land for Sustainability*

The agricultural lands and grazing lands have been maintained at its present status to retain agriculture, dairy, and poultry as a significant form of livelihood for economic sustainability and also environmental sustainability.

5.7 *Environmental Sustainability by Treatment of Contaminants from Brass and Bell Metal Industry*

The pollution from brass industry impacts both air and water quality. Levels of water quality parameters are not within the prescribed safe limits. Besides this, the effluence has been found to contain lead and arsenic too. Besides having socioeconomic impact, brass industry has serious adverse health impacts on workers, and they suffer from tuberculosis, bronchitis, burns, skin diseases, etc. Majority of workers are illiterate and belong to poor section; thereby, they are unaware of the hazardous nature of brass industry, which makes them more vulnerable to health problem.

6 Conclusions

The objective of this research was to understand the issues faced by peri-urban areas taking the case of Pratapsasan and frame the planning principles of peri-urban areas by adhering to the Sustainable Development Goal 11, which talks about the environmental, social, and economic linkages between urban settlement areas and their rural hinterland. The spatial flow of people, products and services, knowledge, and culture to these future cities should be seamless.

As can be seen in the proposed development of Pratapsasan, an effort is made to integrate the area with the Bhubaneswar expansion program. To retain the environmental balance and vernacular character of the rural hinterland, the skyline should be regulated to have a low-rise, low-density character so that it acts as the breathing lung of the city with eco parks and recreational spaces. The agricultural lands and grazing lands have been maintained in its present status to retain agriculture, dairy, and poultry as a significant form of livelihood for economic and also environmental sustainability.

To make the planning of peri-urban areas community centric, the aspirations of people residing in these areas are the prime consideration. In 2011, around 89% of the residents were involved in brass and bell metal and other household industries, but now hardly 500 households are involved in these traditional activities. Pratapsasan is a place of historical, cultural, and economic significance in that region because of its bell metal industry, which started in the fourteenth century. The centuries-old legacy is phasing out, and the present generation is not finding it sustainable. But a multipronged effort has to be made now for a complete turnaround. To get back the cultural identity, it is important to connect it with the existing handicraft corridor of Pipili and Raghurajpur so that it can leverage its traditional craft to the religious tourists and holiday tourists moving from Bhubaneswar to Puri. The metal craft also needs to be revamped and reinvented to include it as decorative and utilitarian use in the building industry like door handles, lamps, finials, interior decoration, and furniture detailing. The market linkages to the brass and bell metal industry needs to be streamlined with certain facilities located near the tourist cottages built on the riverfront of Daya River.

Under the "Atma Nirbhar Bharat Scheme," these MSME/cottage industries will get a lot of impetus and also give a sense of identity and pride among the residents who are abandoning their heritage to migrate as labor to other parts of the country.

The other major occupation is agriculture, which is also practiced by the bell metal workers. The primary challenge for agriculture production in Pratapsasan is the rapid conversion of valuable agricultural land into housing schemes as it serves as a cheaper alternative to residing in the city. This conversion leads to the unavailability of crucial agricultural land leading to consequences such as food insecurity and social instability in the surrounding settlements. The area has good irrigation prospects and fertile soil, which should be capitalized for commercial vegetables and fruit production in addition to the existing rice (kharif) production. All these activities will make food production an integral part of the city.

Another cause of concern in peri-urban areas is integrating geomorphology, ecosystem, riverbeds, and watersheds while planning. There are several ponds in Pratapsasan, which are draining points to the existing natural drainage channels, which should be preserved by fishery development or creating recreation spaces around them. This will go a long way in preventing the flooding, which happens when there is a sudden downpour. A comprehensive regulatory framework for peri-urban areas being vigilantly implemented by the city planning authority will go a long way in preventing encroachments and bring about a planned development. Provision of urban amenities like street lighting, storm water drainage, and solid waste collection will be easier if they are regulated by city planning authorities as funding can be sourced.

A lot has to be done to prevent environmental degradation because of the city's cremation activities and solid waste dumping in Pratapsasan. Less intrusive alternatives have to be worked out. Environment-friendly solutions that don't harm the air and water quality by effluents from the brass and bell metal industry need to be factored in. By adhering to the sustainable development goals and addressing the concerns of the local community, taking a case of Pratapsasan, this paper tries to address the general issues faced by peri-urban areas and come up with some solutions that can give guidelines for framing the planning principles of these hitherto unregulated areas. Peri-urban development is a multidimensional concept, which caters to improving the life of peri-urban population through growth of agricultural and related activities, traditional crafts, cottage industries, community services and amenities, socioeconomic infrastructure, and the ecosystem. To derive the planning principles of spatial development in these transition areas, all settlement typologies in the periphery of urban areas need to be considered. This research has taken only one kind of typology where the main economic activities are handicraft and agriculture.

References

Aijaz, R. & Observer Research Foundation. (2019). India's peri-urban regions: the need for policy and the challenges of governance. In *ORF ISSUE BRIEF* (Issue 285) [Journal-article]. https://www.orfonline.org/wp-content/uploads/2019/03/ORF_Issue_Brief_285_Peri-urban.pdf

Camagni, R., & Brunetta, G. (1994). *Processi di utilizzazione e difesa dei suoli nelle fasce periurbane: dal conflitto alla cooperazione fra città e campagna*. Fondazione Cariplo per la ricerca scientifica.

Carey, R., Larsen, K., & Sheridan, J. (2015). *The role of cities in climate-resilient food systems: A Foodprint Melbourne briefing*. Foodprint Melbourne.

Cattivelli, V. (2021). Planning peri-urban areas at regional level: The experience of Lombardy and Emilia-Romagna (Italy). *Land Use Policy, 103,* 105282. https://doi.org/10.1016/j.landusepol.2021.105282

Census of India. (1981). *Handicraft survey report, bell-metal products*. Census of India.

2011 Census of India. Registrar General and Census Commissioner of India. Retrieved on 2021-05-04.

Dekkers, J. E. C. (2010). *Externalities, land use planning and urban expansion*. [PhD-thesis - research and graduation internal, Vrije Universiteit Amsterdam]. VU.

Hoggart, K. (Ed.). (2016). *The city's hinterland: Dynamism and divergence in Europe's peri-urban territories*. Routledge.

Laquinta, D., & Drescher, A. W. (2000). *Defining peri-urban: Understanding rural-urban linkages and their connection to institutional contexts*. The Tenth World Congress of the International Rural Sociology Association, Rio-de-Janerio, August 2000.

Pradhan, K. C. (2017). Unacknowledged urbanization: The new census towns in India. In E. Denis & M. H. Zérah (Eds.), *Subaltern urbanisation in India. Exploring urban change in South Asia*. Springer. https://doi.org/10.1007/978-81-322-3616-0_2

Pentland, A. (2015). *Social physics: How social networks can make us smarter*. Penguin.

Qviström, M. (2013). Searching for an open future: Planning history as a means of peri-urban landscape analysis. *Journal of Environmental Planning and Management, 56*(10), 1549–1569. https://doi.org/10.1080/09640568.2012.734251

Rauws, W. S., & de Roo, G. (2011). Exploring transitions in the peri-urban area. *Planning Theory & Practice, 12*(2), 269–284. https://doi.org/10.1080/14649357.2011.581025

Rondhi, M., Pratiwi, P. A., Handini, V. T., Sunartomo, A. F., & Budiman, S. A. (2018). Agricultural land conversion, land economic value, and sustainable agriculture: A case study in East Java, Indonesia. *Land, 7*(4), 148. https://doi.org/10.3390/land7040148

Wandl, A., & Magoni, M. (2017). Sustainable planning of peri-urban areas: Introduction to the special issue. *Planning Practice & Research, 32*(1), 1–3.

Sudha Panda, Professor at School of Architecture and Planning, KIIT University, has more than 30 years of experience in industry and academics in the field of architecture, planning, and urban management in both an industrial setup and civic city. She has worked on several research like urban flooding in Nayapalli Area for GIZ (Germany), spatial development for Gram panchayats of Odisha by the Ministry of Panchayati Raj (Govt. of India), peri-urban development by the Council of Architecture (Govt. of India), weaver settlement documentation by INTACH, and heritage bye-laws for the National Monument Authority, Ministry of Culture.

Soumyendu Shankar Ray is a leading professional Indian architect and educator. As the former Head of the Faculty of Architecture, Biju Patnaik University of Technology, Odisha, and now the Founding Director of School of Architecture and Planning, KIIT University, he has consistently championed context-sensitive architecture. His firm "Vastukar" is now one of the largest and most respected firms in the region for architecture, planning, and urban design. His book *Odisha: An Architectural Odyssey* is a meticulously researched timeline perspective of the diverse and rich architectural heritage of the state.

Design of GIS-Based Optimization Approach for Transportation Route Selection in Municipal Solid Waste Management

Dhiren Barman and Gouri Sankar Bhunia

1 Introduction

Human activity will inevitably result in the creation of MSW (Khan et al., 2022). The gathering and disposing of large amounts of MSW has become a critical issue as a result of growing urbanization, which has significantly increased the rate at which MSW is generated in developing countries (Rada et al., 2013). According to estimates, there will be a 1% growth in human population, growing urbanization, and the building of infrastructure, businesses, and residences. Solid waste production rises by 1.04% for every 1% increase in population and by 0.34% for every 1% boost in per capita income, according to estimates (Fu et al., 2015). According to estimates from the World Bank, by 2050, there might be 3400 million tonnes of MSW produced globally (Kaza et al., 2018). Urban residents are adversely affected by MSW in terms of the environment, society, and economy (Shafik et al., 2022). When it comes to automated waste management and planning, comprehensive geographic information systems (GIS) are among the most promising methods (Karadimas & Loumos, 2008). Geographical data is the cornerstone of collecting system scheduling, and GIS offers an effective means to import, organize, and analyse it. Combining mathematical methods with GIS can also be used to create vehicle routes (Viktorin et al., 2023).

The collection/transport element is the showpiece for any sustainable framework, and its repercussions are simple to reassess the system's progress and costs. The operation entails the expulsion and exchange of waste from points of production or assembly to transfer stations or from transfer stations to processing or final landfill sites (Ritesh et al., 2008). As a result, it is the most persuasive and costly

D. Barman · G. S. Bhunia (✉)
Department of Geography, Secom Skills University, Kendradangal, Bolpur, Birbhum, West Bengal, India

A. Rai et al. (eds.), *Contemporary Social Physics*, Springer Geography,
https://doi.org/10.1007/978-3-031-77596-3_19

element, consuming the majority of the budget allocated by municipalities for SWM at the expense of other operational processes in the waste management practices (Kallel et al., 2016). Several investigations have already studied the transportation patterns of municipal solid waste management to suggest garbage collection from user ends to dumpsites that can lower operational costs. According to Apaydin and Gonullu (2007), the distance required for waste collection was reduced by 4–59%. Karadimas et al. (2007) used an ant colony system method to simulate the transportation path in Attica Prefecture (a suburb of Athens) and reported a 25.6% reduction in collected tour length. Singh et al. (2014) tested Dijkstra's algorithm-based optimization model for eight waste points interlinked in a mesh network in Kanpur, and the findings showed a reduction in travelled distances of 12–15%.

Solid waste management (SWM) is a serious hurdle for any urban space, but particularly for a city in the developing world that is rapidly urbanizing with little or no planning. Researchers have increasingly used optimization algorithms including network analysis tools, optimization software, and self-modified programs to handle complicated route network challenges occurring in the collection and disposal of municipal solid waste (MSW) in urban contexts in recent years. In developed cities, transportation management of MSW accounts for 50% of overall expense, while in developing cities, it accounts for 85–90% of total budget (Singh & Behera, 2018). According to a Sri Lankan study, network design can cut weekly travel distance by 36 and weekly trip frequency by 19% (Rambandara et al., 2022). The shortest possible path technique was employed to evaluate route optimization in an experiment conducted in the United Arab Emirates (UAE) that produced Smart Waste Collection (SWC) techniques based on GIS, leading to a 19% reduction in expenses (Abdallah et al., 2019). Using ArcGIS Network Analyst, Sanjeevi and Shahabudeen (2016) found that mileage was lowered by 9.93% in India.

Zsigraiova et al. (2013) found that applying a GIS pattern also resulted in a 43% reduction in fuel use. Patna, the state capital of Bihar, is one of India's fastest-growing cities in terms of population density and urbanization (Alakshendra, 2019). Patna is the populous town in Bihar, which remains one of the country's least urbanized states. Quickly growing cities result in fast expanding waste mountains. According to Pandey (2014), the city of Patna generates approximately 1200 tonnes of waste per day, including both organic and inorganic wastes, and this amount is projected to double by 2036. Household waste accounts for 40% of waste produced by the city. This paper aims at investigating the ideal routes for MSW transportation in the platform of GIS to bridge a knowledge gap in the literature and present a modelling framework that meets the optimization model for collection paths of solid waste collection in Patna City.

2 Selection of Study Area and Existing Scenario of Solid Waste Management Practice

Patna is a historic city with a rich history dating back to the ancient, mediaeval, and colonial periods (Fig. 1). It saw substantial population and geographic expansion after independence, notably in recent decades. The city serves a variety of purposes for its citizens, as well as the state's rural and urban inhabitants, and as a result, it continues to receive in-migration. According to the Bihar Urban Development Ministry, Patna Municipal Corporation (PMC) encompasses 99.45 square kilometres (Patna Master Plan, 2013). According to Census 2011, the city is divided into 72 wards, which are organized into 6 circles: New Capital Circle (Western Zone), Kankarbagh Circle (Southern Zone), Bankipur Circle (Central Zone), Azimabad Circle, Patliputra Circle, and Patna City Circle (Eastern Zone). The city has a population of 1.7 million (1,683,200). The rapid rise of urbanization leads to an increase in population and an increase in the number of urban dwelling units in Patna Municipal Corporation (PMC), both of which are generally unplanned, exacerbating different solid waste problems. Improper planning causes traffic issues in the streets, causing delays in waste collection since waste vehicles are unable to access the location, leaving rubbish to remain uncollected. Patna residents have a habit of

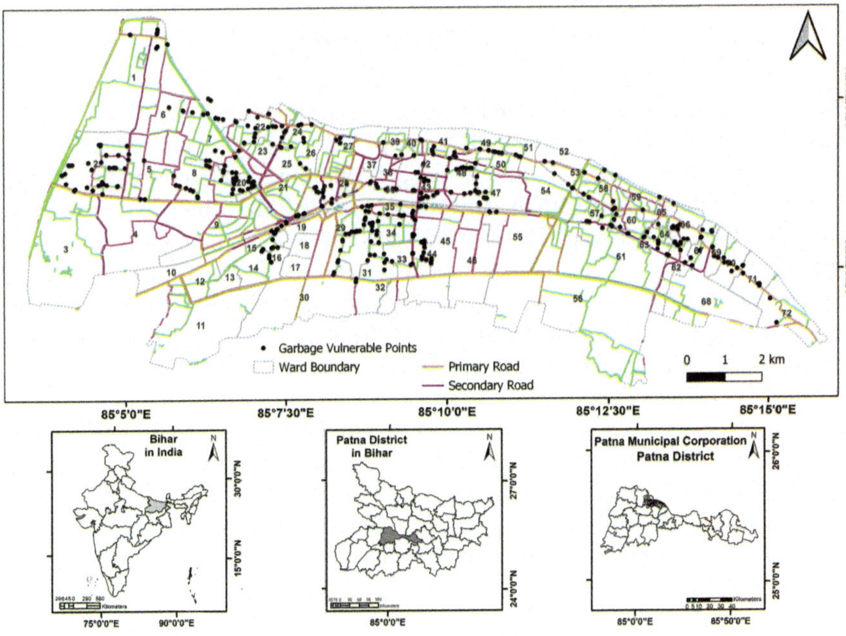

Fig. 1 Location map of the study area

throwing trash indiscriminately along the sides of roadways, roads, and open spaces/plots, as well as near vats/bins. PMC *Safai karmachari* (sanitation workers) gather this waste or garbage and transport it to roadside bins or pickup locations (Singh, 2012). Following that, these wastes are transferred to disposal locations in open vehicles or by tractor and trailer with the aid of the next group of sanitation workers.

Apart from the dumpsites, the PMC body dumps all of the garbage collected from the city's 72 wards without any treatment or scientific procedures in low-lying areas and the city's periphery. The garbage collected through door-to-door collection services in Kankarbagh Circle is disposed of in the AgamKuan Secondary Landfill Site. Garbage obtained through door-to-door collection services from New Capital Circle neighbourhoods is disposed of in the Gardanibagh Secondary Landfill Site. Currently, the state government's principal trash dumpsite is Raman Chak-Bairiya (Swarup et al., 2020). On average, roughly 1000 MT of waste per day is dumped here from the city's secondary disposal sites. The city's waste collection and transportation machinery, trucks, and staff are insufficient (Raj & Raj, 2015). Due to a lack of compactors at PMC, mechanized loading is done with the help of huge compactors, although only in some regions (Anand et al., 2018). The majority of PMC's trash delivery vehicles or machines are either rented or over a decade old, worn out, and in desperate need of replacement.

2.1 Data Collection and Database Creation

PMC was chosen as the study domain since it has 6 zones and 72 wards with 483 MSW collection points. A mobile GPS device was used to record the geographic locations of all the communal disposal and collection sites within the entire municipality through the leadership of the city manager of the PMC. This was done through the use of Topo Survey app. The data was downloaded in comma-separated value formats and imported into QGIS. It was also used as a test bed for the modelling experiment. The GIS-based approach used in this research included numerous steps to achieve the best results. In the first step, thematic layers depicting the current scenario and patterns in the city were produced in the GIS interface for the locations of collecting points, administrative boundaries, and road network comprising both primary and small roads. The road network was digitized onscreen from the Google Earth and OpenStreetMap. The geographic co-ordinate of waste collection (garbage vulnerable) points was collected through an online mobile app within the study site and denoted by a circular feature and in red colour. The road network in the research area is made up of both principal and secondary roads, which are represented by brown and red line features.

2.2 K-means Clustering Analysis

The K-means algorithm in data mining starts with a first group of randomly selected secondary collection points, which are used as the beginning points for each centroid, and then performs iterative (repetitive) calculations to optimize the secondary collection points and transfer station placements (Khalid & Herbert-Hansen, 2018). The K-means clustering algorithm requires the initialization of k cluster centroids $\mu_1, \mu_2, \ldots, \mu_k$. The total number of clusters is represented by k. Initial centroids were chosen at random from the dataset to avoid ending up in local minima. The randidx = randperm(size(Z,1)) function in QGIS was used to randomize the indices of the Z, i.e. reduced feature matrix. The first k indices were chosen as SCPs and TS using the function centroids = z(randix(1:K):). Individual values $z^{(i)}$ of matrix Z, i.e. selected feature matrix, and distances between centroids 1, 2,... k are determined using the equation below. The index $c^{(i)}$ of the centroid k is then assigned to the example $z^{(i)}$ for which the distance is the shortest, whereas $c^{(i)} = 1, 2, \ldots k$.

$$\left\| z^{(i)} - \mu_k \right\|^2 \quad \forall i = 1,2,3\ldots.m \;\&\; k = 1,2\ldots\ldots k \tag{1}$$

The average of all the data points $z^{(i)}$ with the same index value $c^{(i)}$ is used to compute new centroids in the next phase. And centroids for secondary collection points (SCPs) and transfer stations (TSs) are allocated to these mean values. The preceding strategy is continued until the centroids do not change after several iterations. Data that has been clustered is evaluated and compared. Each centroid for SCPs and TSs is demarcated as a collection of prospective facility locations.

2.3 Estimation of Road Optimization

To find the ideal route lengths in the research network, the computations were carried out using a GIS-based network analysis tool in QGIS software. Dijkstra's method (Dijkstra, 1959), a greedy algorithm that addresses the single-source shortest path problem when all edges have nonnegative weights, is one of the intrinsic optimization strategies in this tool (Singh & Behera, 2018). Based on Dijkstra's algorithm method, the generated findings from the integrated QGIS network analysis tool were evaluated. This tool was used in this study for optimization purposes, with '*s*' for the starting node and '*d*' for the destination node. The shortest path connecting the nodes was determined after selecting the 'solve feature' in QGIS 'network analyser tool' and then shown on the interface.

3 Results

Figure 1 depicts the various places of existing solid waste disposal plants. As shown in Fig. 1, most disposal sites are not correctly positioned in terms of people's environmental and social well-being, as evidenced by a field survey. The statistic also revealed that more than 80% of existing disposal sites are within 30–50 m of road networks (both SCPs and TSs), indicating that they are not adequately sited. Nonetheless, the condition of waste disposal facilities within the research area indicates that more than half of the disposal sites are unlicensed. Consequently, as one moves away from the city centre, the density of garbage disposal sites diminishes. This was due to the city centre's significant concentration of economic operations and population contrasted to the metropolis' fringes. Unauthorized garbage dumping sites pose a serious concern for many local governments, notably in the municipality's suburbs and residential areas. It is also our stance that unlawfully disposing of on public or private open plots, or even dumping in drains or water bodies, is an environmental felony that can lead to obstruction of drains, degradation of water resources, and worsening of sanitation facilities in municipal areas.

We conduct a series of comparative experiments in this part, using both simulated data to mimic known spatial processes and real-world geographical data. The values of the local K-function are computed using the formulae from the preceding section. To account for test synchronicity, we use a significance level of 0.1% for local outcomes. The size of the analysis area is shrunk by a distance equal to the maximum distance band to be used in the research to eliminate edge effects. The K-function is calculated using only the points within the smaller area. A total of 22 SCPs have been demarcated in the study area (Fig. 2).

The GIS offered through data on road network connectivity between various waste-sensitive areas. The algorithm defined the shortest path and presented it on a GIS map, along with route distance (Fig. 3a–c), for each of the intended links after

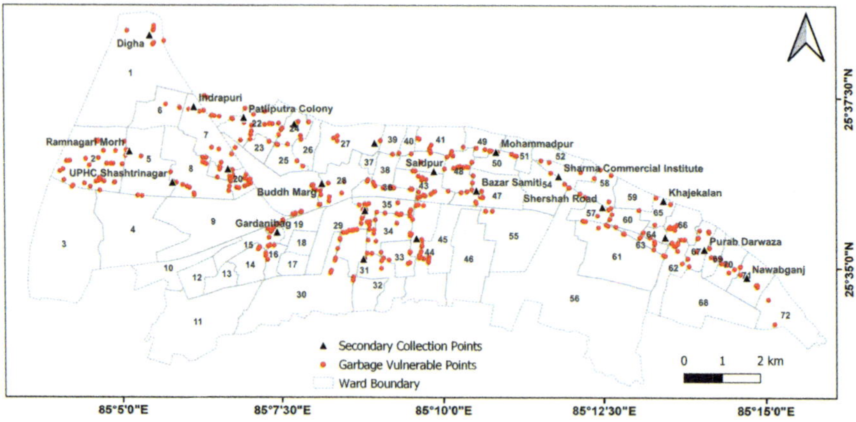

Fig. 2 Spatial distribution of secondary collection points

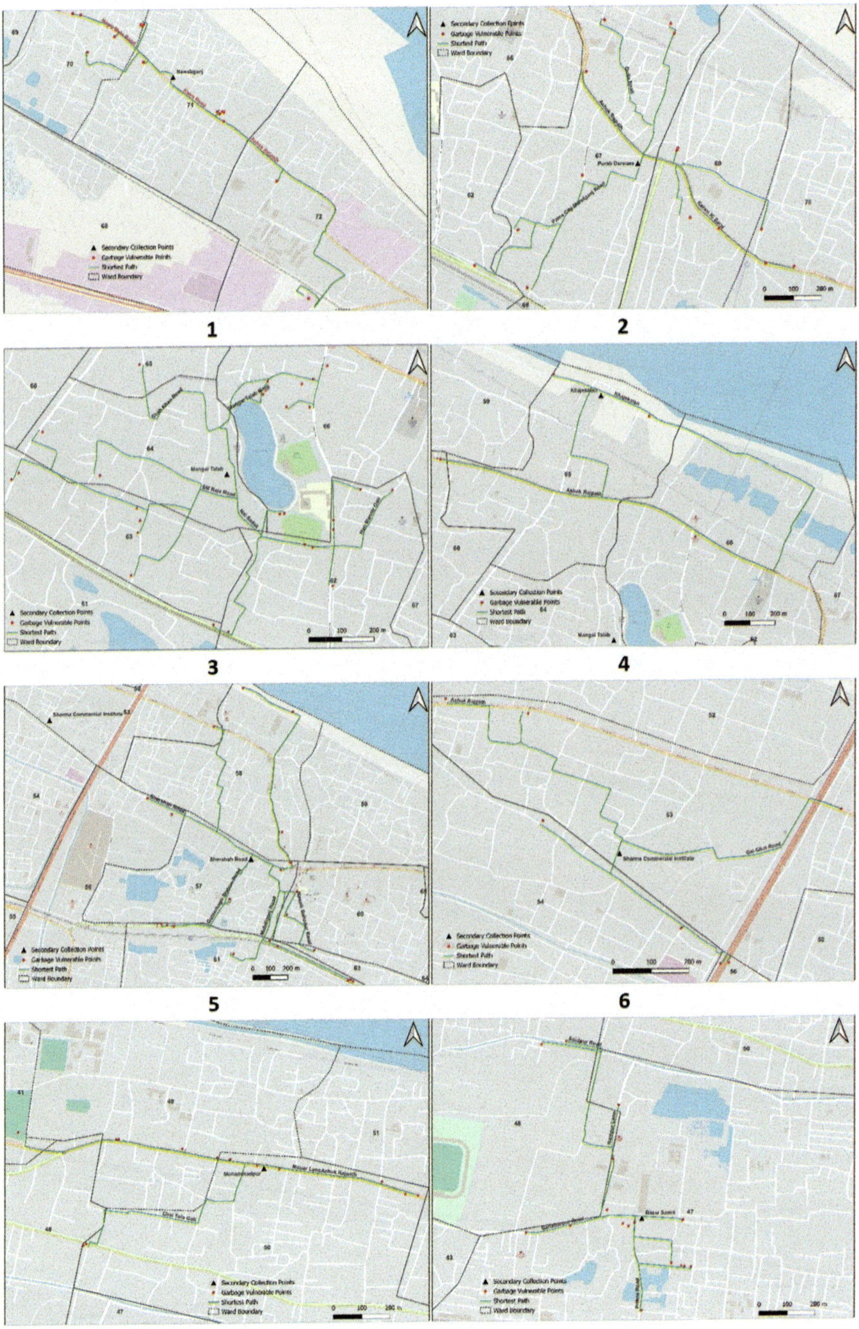

Fig. 3 (**a**) Shortest path analysis for secondary collection points (from 1 to 8). (**b**) Shortest path analysis for secondary collection points (from 9 to 16). (**c**) Shortest path analysis for secondary collection points (from 17 to 22)

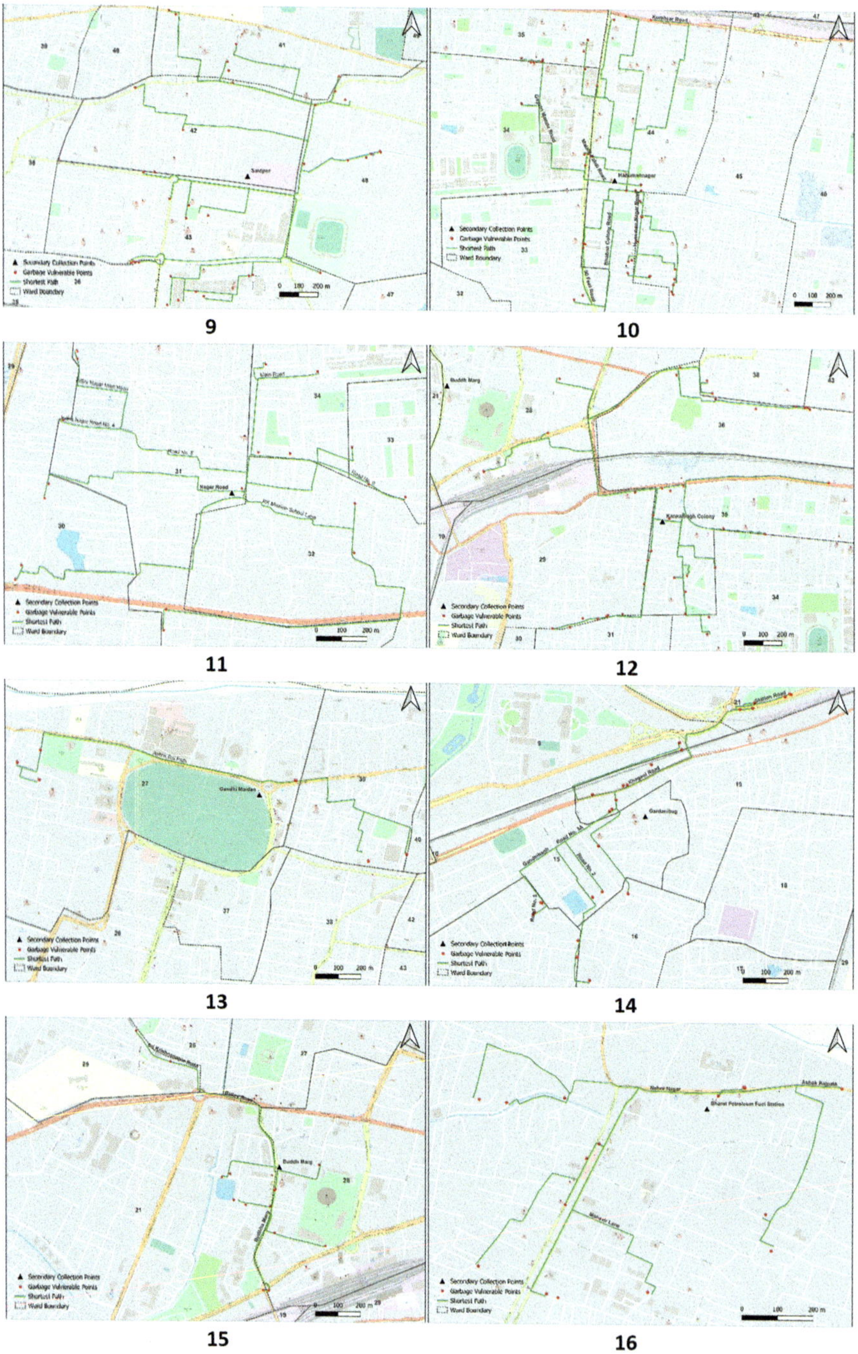

Fig. 19.3 (continued)

Fig. 19.3 (continued)

using QGIS 'shortest path analysis tool' and providing the starting nodes, i.e. numerous collection points one by one, and the destination address, i.e. secondary collection point 1 (SCP1) and secondary collection point 2 (SCP2).

The optimal routes developed by this method were compared with the existing routes at SC1 and SC22 (Table 1), which calculated the distance between individual collecting points and disposal sites. The current route length was evaluated in the QGIS-optimized route length for each SC point. Table 1 provides detailed information for each of the interconnections created using the current method. As illustrated in Fig. 4, the QGIS shortest path analysis tool was used to determine the shortest

Table 1 Results from the approach by comparing current distance (km) versus optimal distance

Transfer station	SCP ID	Practising distance (km)	Optimized distance (km)	Saving indistance(km)	Saving in km (%)
Golwara Plaza (Ashok Rajpath)	1	4.66	3.01	1.64	35.28
	2	2.62	1.57	1.05	39.91
	3	1.05	0.53	0.52	49.07
	4	1.44	0.83	0.60	41.95
	5	2.72	1.80	0.92	33.87
	6	4.69	3.33	1.36	28.99
Vaishali Circle (Road No. 12)	7	3.72	3.08	0.64	17.25
	8	2.5	1.71	0.79	31.41
	9	1.17	0.91	0.27	22.93
	10	2.15	1.53	0.62	28.92
	11	3.4	2.86	0.54	15.93
	12	3.63	2.41	1.22	33.74
	13	3.57	2.37	1.20	33.57
	14	5.57	4.74	0.83	14.96
	15	3.69	3.24	0.45	12.13
AN College, Bailey Road	16	3.65	3.03	0.62	17.04
	17	2.87	1.97	0.90	31.29
	18	2.71	1.18	1.53	56.47
	19	3.87	2.82	1.05	27.16
	20	5.22	3.51	1.71	32.78
	21	1.71	1.30	0.41	24.01
	22	5.29	3.98	1.31	24.84
Total		**71.9**	**51.71**	**20.18**	
Average		**3.27**	**2.35**	**0.92**	**29.70**
Standard deviation		**1.32**	**1.11**	**0.42**	**11.09**

SCP secondary collection point

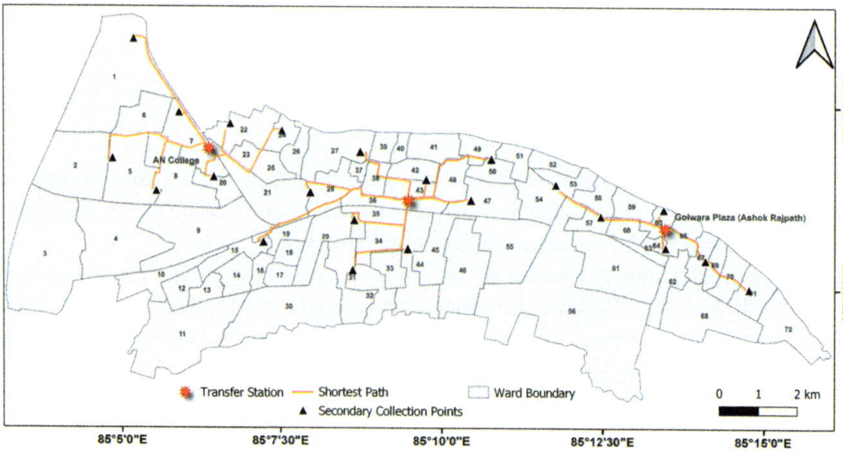

Fig. 4 Shortest path analysis between transfer stations

path between the transfer station and the SCPs. The shortest path between TS1 and secondary collecting points was estimated to be 51.71 km (optimized distance), as opposed to 71.90 km (existing route distance) as used by municipal vehicles.

Concurrently, this approach was used to discover the shortest route distances within the designated network of research regions, which included 72 wards and 3 transfer stations (TS1 and TS3), as shown in Fig. 4. In a PMC network, connecting the garbage vulnerable points of 72 wards (as a beginning location) and 3 transfer stations, as well as among the centroid of SCPs in a mesh network, the QGIS-optimized route exhibited a weighted average saving of 29.70% of the total haul distance (in km).

4 Summary and Conclusion

Solid management and disposal services are social programmes, and their implementation should be the responsibility of local governments or municipal agencies. Even though private businesses and individuals are participating in this duty, municipal authorities must remain in command in order to maintain a metropolitan-wide uniformity in solid waste management accomplishment. Funding in solid waste management does not entail justification in terms of 'positive return on investment' or 'minimum profits' because it is a 'public utility' and an essential service. Nevertheless, such an investment must be rationalized as the 'lowest cost technologically feasible option' for achieving the desired level of efficiency. The use of a geographic information system (GIS) in evaluating a spatial analysis, particularly when representing real-world phenomena, aids in the integration and simplification of the understanding of a facility distribution, particularly visual analysis and interpretations.

The K-means is widely recognized as one of the most successful approaches for detecting non-random patterns in a geospatial distribution of points, especially in the case of significant concentrations of point occurrences. Despite the fact that it is labelled as a "multi-distance spatial cluster analysis", we contend that it is incapable of accurately reflecting point pattern cross-scale alterations. Consequently, we use one of its variants, the incremental K-means, to solve this problem. The results of a series of comparison experiments on both replicated and real-world datasets were given in this manuscript. The outcomes of simulated datasets show that the incremental K-means can identify the scales at which it exhibits the most important grouping patterns while avoiding the false-positive errors that plague the K-means. The true scales designed in our data modelling techniques correspond to these peak scales.

In addition to TS number (Yadav et al., 2016), MSW type (Asefi et al., 2015), routing (Asefi & Lim, 2017), overall cost-minimizing (Eiselt & Marianov, 2014), TS capacity (Kůdela et al., 2019), and time optimization (Monzambe et al., 2021), the optimization models consider TS location as one of the diverse mechanisms of MSW. The economic component of ISWM system efficiency is critical. The majority of researchers in this discipline think about this category in terms of cost. As a result, Kůdela et al. (2019) include the cost of transfer with and without TS, as well as the costs of building, processing, operation, and maintenance. In contrast, Höke and Yalcinkaya (2021) employ fuel, vehicle, and labour expenses to do a quantitative economic analysis because the siting and routing difficulties are linked.

Furthermore, the results from the PMC Secondary Collection Point dataset provide real-world evidence that K-means clustering is highly effective in modeling the underlying heterogeneous point process and demonstrates greater reliability compared to alternative methods. It is also possible to acquire practical utility. The local K-means can serve to pinpoint these spots on the city map for further research. To sum up, we expect that this work will draw emphasis to the K-means as a useful SCP and TS detection approach, particularly when coping with complicated spatial point processes. The outcomes were found for optimal pathways that considered both one-way and two-way communication, as well as the case where there is no direct connectivity between the nodes. Future research should concentrate on quantifying fuel and emission savings, sectorizing larger garbage collection areas based on geographic evaluation rather than empirical approaches, and modifying the collection system to accommodate the implementation of distinct collection strategies for various water schemes for various categories of waste.

Acknowledgement The authors would like to express their gratitude to the Patna Municipal Corporation Team for freely providing data.

References

Abdallah, M., Adghim, M., Maraqa, M., & Aldahab, E. (2019). Simulation and optimization of dynamic waste collection routes. *Waste Management & Research, 37*, 793–802. https://doi.org/10.1177/0734242X19833152

Alakshendra, A. (2019). City profile: Patna, India. *Environment and Urbanization ASIA, 10*(2), 374–392.

Anand, S., Raman, E., & Rani, U. (2018). Role of Swachh Bharat Mission in municipal solid waste management in Patna, Bihar. *Advance in Environmental Waste Management & Recycling, 1*(1), 1–5.

Apaydin, O., & Gonullu, M. T. (2007). Route optimization for solid waste collection: Trabzon (Turkey) case study. *Global NEST Journal, 9*, 6–11.

Asefi, H., & Lim, S. (2017). A novel multi-dimensional modeling approach to integrated municipal solid waste management. *Journal of Cleaner Production, 166*, 1131–1143. https://doi.org/10.1016/j.jclepro.2017.08.061

Asefi, H., Lim, S., & Maghrebi, M. (2015). A mathematical model for the municipal solid waste location-routing problem with intermediate transfer stations. *Australasian Journal of Information Systems, 19*, S21–S35. https://doi.org/10.3127/ajis.v19i0.1151

Dijkstra, E. W. (1959). A note on two problems in connection with graphs. *Numerische Mathematik, 1*, 269–271.

Eiselt, H. A., & Marianov, V. (2014). A bi-objective model for the location of landfills for municipal solid waste. *European Journal of Operational Research, 235*, 187–194. https://doi.org/10.1016/j.ejor.2013.10.005

Fu, H. Z., Li, Z. S., & Wang, R. H. (2015). Estimating municipal solid waste generation by different activities and various resident groups in five provinces in China. *Waste Management, 41*, 3–11.

Höke, M. C., & Yalcinkaya, S. (2021). Municipal solid waste transfer station planning through vehicle routing problem-based scenario analysis. *Waste Management and Research, 39*, 185–196. https://doi.org/10.1177/0734242X20966643

Kallel, A., Kallel, A., & Damak, M. (2016). Sustainable water resource management: Insights from the application of integrated models in arid regions. *Journal of Water Resources Planning and Management, 142*, 126–134.

Karadimas, N. V., Papatzelou, K., & Loumos, V. G. (2007). Optimal solid waste collection routes identified by the ant colony system algorithm. *Waste Management & Research, 25*, 139–147.

Karadimas, N. V., & Loumos, V. (2008). GIS-based modeling for urban planning and disaster management: Case studies and methodologies. *Urban Planning International Journal, 15*, 45–60.

Kaza, S., Yao, L., Bhada-Tata, P., & Van Woerden, F. (2018). *What a waste 2.0: A global snapshot of solid waste management to 2050*. World Bank Publications.

Khalid, W., & Herbert-Hansen, Z. N. L. (2018). Using k-means clustering in international location decision. *Journal of Global Operations and Strategic Sourcing, 11*(3), 274–300. https://doi.org/10.1108/JGOSS-11-2017-0056

Khan, S., Anjum, R., Raza, S. T., Ahmed, B. N., & Ihtisham, M. (2022). Technologies for municipal solid waste management: Current status, challenges, and future perspectives. *Chemosphere, 288*, 132403. https://doi.org/10.1016/j.chemosphere.2021.132403

Kůdela, J., Šomplák, R., Nevrlý, V., Lipovský, T., Smejkalová, V., & Dobrovský, L. (2019). Multi-objective strategic waste transfer station planning. *Journal of Cleaner Production, 230*, 1294–1304. https://doi.org/10.1016/j.jclepro.2019.05.167

Monzambe, G. M., Mpofu, K., & Daniyan, I. A. (2021). Optimal location of landfills and transfer stations for municipal solid waste in developing countries using non-linear programming. *Sustainable Futures, 3*, 100046. https://doi.org/10.1016/j.sftr.2021.100046

Pandey, M. K. (2014). *Solid waste management in Patna*. National Solid Waste Association of India.

Patna Master Plan (2013). Patna Master Plan 2031, Government of Bihar, Urban Development and Housing Department.

Rada, E. C., Ragazzi, M., & Fedrizzi, P. (2013). Web-GIS oriented systems viability for municipal solid waste selective collection optimization in developed and transient economies. *Waste Management, 33*, 785–792. https://doi.org/10.1016/j.wasman.2013.01.002

Raj, P., & Raj, A. (2015). An analysis of waste production in Patna. *isa.e-Forum* (pp 1–13). Available at: https://www.iitp.ac.in/images/faculty_profile/human/Aditya/No.2.pdf

Rambandara, R. D. S. S., Prabodanie, R. A. R., Karunarathne, E. A. C. P., & Rajapaksha, R. D. D. (2022). Improving the efficiency of urban waste collection using optimization: A case study. *Process Integration and Optimization for Sustainability, 6*, 809–818. https://doi.org/10.1007/s41660-022-00232-8

Ritesh, V., Gautam, A., Kalamdhad, A., Gupta, A., & Devotta, S. (2008). GIS-based locational analysis of collection bins in municipal solid waste management systems. *Journal of Environmental Engineering and Science, 7*(1), 39–43.

Sanjeevi, V., & Shahabudeen, P. (2016). Optimal routing for efficient municipal solid waste transportation by using ArcGIS application in Chennai, India. *Waste Management & Research: The Journal for a Sustainable Circular Economy, 34*, 11–21. https://doi.org/10.1177/0734242X15607430

Shafik, A., Elkhedr, M., Said, D., & Hassan, A. (2022). Environmental impacts of MSW collection route optimization using GIS: A case study of 10th of Ramadan City, Egypt. *arXiv preprint arXiv:2209.02652*.

Singh, V. R. (2012). Quantification of municipal solid waste for Patna city. *International Journal of Environmental Technology and Management, 15*, 516–523. https://doi.org/10.1504/IJETM.2012.049242

Singh, S., & Behera, S. N. (2018). Development of GIS-based optimization method for selection of transportation routes in municipal solid waste management. *Advances in Waste Management*, 319–331. https://doi.org/10.1007/978-981-13-0215-2_22

Singh, G., Singh, B., Rathi, S., & Haris, S. (2014). Solid waste management using shortest path algorithm. *International Journal of Engineering Science Invention Research & Development, 1*(2), 60–64.

Swarup, S., Verma, U., & Kumar, R. (2020). Implementation of new technologies in solid waste management of Patna: An appraisal of Patna municipal corporation. *Pollution Research, 39*(4), 1122–1130.

Viktorin, C., Mueller, L., Fischer, H., & Smith, T. (2023). Advances in hydrological modeling for river basin management: Lessons from interdisciplinary approaches. *Environmental Science and Policy Journal, 29*, 78–94.

Yadav, V., Karmakar, S., Dikshit, A. K., & Vanjari, S. (2016). A feasibility study for the locations of waste transfer stations in urban centers: A case study on the city of Nashik, India. *Journal of Cleaner Production, 126*, 191–205. https://doi.org/10.1016/j.jclepro.2016.03.017

Zsigraiova, Z., Semiao, V., & Beijoco, F. (2013). Operation costs and pollutant emissions reduction by definition of new collection scheduling and optimization of MSW collection routes using GIS. The case study of Barreiro, Portugal. *Waste Management, 33*, 793–806.

Dr. Dhiren Barman is completed his Ph.D. degree from Seacom Skills University, West Bengal, India. He is working as an Assistant Teacher of Geography in Govt. School of West Bengal. He has published more than 10 papers in an International Journals and Books.

Dr. Gouri Sankar Bhunia has more than sixteen years of experience in geospatial technology and involved in several Government and Private Organizations in India. He received his Ph.D. from the Department of Geography, University of Calcutta, India in 2015. He also worked as a visiting Faculty in Private University, West Bengal, India. His Ph.D. dissertation work focused on environmental control measures of infectious disease using Geospatial technology. Currently, he is involved in Smart city and urban planning project. He has received Senior Research Fellowship Award from Indian Council of Medical Research. His research interests include environmental modelling, Health GIS, risk assessment, natural resources mapping and modelling, data mining and information retrieval using Geospatial technology. He has published Ten books with International Publishing Group. He is currently the editor of the GIScience and Geo-environmental Modelling (GGM) Book Series, Springer-Nature.

Deconstructing the Impact of Government Interventions Among the Marginalized Populations Using GIS: A Study of Burgula Cluster, Telangana

Sonal Mobar Roy and Prasad N. S. R.

1 Introduction

In recent years, the swift progress of technology has brought about substantial changes in the manner in which we engage, communicate, and conduct ourselves within our communities. A discipline known as social dynamics, which integrates aspects of the social sciences and data analytics, has arisen as a potent instrument for comprehending and influencing human conduct. Through the utilization of advanced computational methods, social dynamics enables us to decode the complex structures and fluctuations inherent in human connections, offering valuable insights that can catalyse constructive transformations in society. Henri de Saint-Simon, a forward-thinking French scholar in the realms of politics and economics, was one of the trailblazers who introduced the concept that society could be comprehended and elucidated using principles analogous to those encountered within the field of physics (Ball, 2012).

It is well-understood that human behaviour is extremely complex and influenced by a multitude of factors, some consciously and some subconsciously. According to Bhattacharya and Kaski (2019), over the past few years, social physics has evolved into a multidisciplinary area of research interests that draws attention to human sociality embedded in social network and societal structures. Understanding and predicting human behaviour has historically been extremely difficult, in large part

S. M. Roy (✉)
Centre for Wage Employment and Livelihoods, National Institute of Rural Development and Panchayati Raj, Hyderabad, Telangana, India
e-mail: sonalmobarroy.nird@gov.in

P. N. S. R.
Centre for Geo-Informatics Applications in Rural Development, National Institute of Rural Development and Panchayati Raj, Hyderabad, Telangana, India
e-mail: nsrprasad.nird@gov.in

A. Rai et al. (eds.), *Contemporary Social Physics*, Springer Geography,
https://doi.org/10.1007/978-3-031-77596-3_20

due to the dearth of complete data and analytical tools. However, the advent of the digital age has sparked a data explosion, providing an unmatched chance to study and shape society dynamics. The study of social dynamics aims to close the gap between quantitative data analysis and qualitative social science. Social dynamic experts may examine enormous datasets that cover anything from social media interactions and mobile phone activity to transportation trends and financial transactions by using the potential of large data resources and machine learning methods. Researchers can uncover hidden trends, correlations, and causal relationships using this data-centric approach, which leads to a deeper understanding of human behaviour. Conte et al. (2012) opines that the increasing integration of technology into our lives has created unprecedented volumes of data on society's everyday behaviour.

Social physics is a collection of dynamic research topics that are designed to resolve problems that rig our society and to which scientists with formal training in physics have contributed and continue to contribute considerably (Marco et al., 2022). In the context of social physics, a real social system emerges as a maze of intricate details that is always changing and bursting with life. It includes a wide range of intricately intertwined relationships and distinct functions, creating a vibrant tapestry of human interactions. People can access resources and information that influence their behaviour and decisions owing to the collective social network, which acts as a map of these connections (Miguel et al., 2012). However, when attempting to represent the intricacy of this dense network, typical artificial society models encounter substantial difficulties, largely because changes in behaviour within an open and unpredictable environment are unexpected in nature. It becomes crucial to combine a dynamic social network model with an agent-based artificial society model in order to comprehend it more deeply. This merger makes it possible to explore social dynamics and the formation of numerous phenomena in more detail.

Each person has a specific place within both a localized information network and a larger social network. These roles have an impact on how they arrive at choices and force them to modify their plans in light of the information at hand. Individual adjustments and decisions continuously modify the social network, creating a dynamic interplay of forces that can either direct the system's overall evolution or lead to the emergence of unique collective phenomena. The study of these phenomena is at the heart of social physics.

2 Importance of Social Physics

Social physics allows us to comprehend the collective behaviour of communities, cities, and even entire societies (Pentland, 2014). Social physics provides invaluable insights into various societal phenomena, including the spread of ideas, adoption of innovations, and establishment of cultural norms. Policymakers, urban planners, and organizations aiming for positive transformations must grasp this concept thoroughly. Furthermore, predictive modelling techniques within social physics help anticipate future behaviours and outcomes, with applications ranging from predicting traffic congestion to forecasting infectious disease outbreaks and economic trends. By enabling proactive decision-making and resource allocation, social physics facilitates more efficient responses to societal challenges. Geospatial technology, including geographic information systems (GIS) and mobile sensing technologies, plays a crucial role in understanding and influencing social behaviour. These tools enable real-time data collection on human movement, social interactions, and environmental variables, allowing for the identification of behavioural patterns and assessment of intervention effectiveness. For instance, GIS aids in infrastructure design and congestion relief, optimizing transportation networks for improved efficiency.

Leveraging geospatial technologies, organizations and policymakers can implement tailored interventions suited to specific contexts. For example, mobile applications can promote energy-efficient travel routes and offer location-specific recommendations for healthy lifestyle choices, nudging individuals towards more sustainable behaviours. Additionally, these tools facilitate citizen engagement through participatory decision-making, empowering communities to voice their opinions, report issues, and collaborate on initiatives. By involving citizens in decision-making processes, social physics fosters a sense of ownership and encourages collective action. These interventions can nudge individuals towards more sustainable and socially desirable behaviours. Geospatial tools facilitate citizen engagement by enabling participatory decision-making and empowering communities (Barnes & Wilson, 2014). By involving citizens in decision-making processes, social physics can create a sense of ownership and foster collective action. India, with its burgeoning population and dynamic economy, has undergone a significant economic transformation in recent decades, transitioning from an agrarian economy to a hub of innovation and technology. This transformation underscores the nation's immense potential and unique challenges, reflecting its emergence as a prominent global player.

In this chapter, the authors draw focus towards the 65% of the population that inhabits the rural areas of the country, depending on agriculture and allied sectors (Census, 2011). While economic growth is crucial, India recognizes that true development encompasses social progress and human well-being. Significant strides have been made in critical areas such as healthcare, education, and poverty alleviation. The nation has achieved remarkable success in reducing poverty levels, expanding access to education, and improving healthcare outcomes. Initiatives like the National Rural Health Mission, Sarva Shiksha Abhiyan, and Pradhan Mantri Jan Dhan Yojana have propelled India's progress in improving the lives of its citizens.

3 Significance of the Study

India's growth journey is not without its challenges. Income inequality, regional disparities, and the need for sustainable development remain pressing issues all across the country. Furthermore, addressing infrastructure gaps, enhancing the ease of doing business, and ensuring social inclusivity require sustained efforts and effective policies. However, these challenges also present opportunities for innovative solutions, public-private partnerships, and inclusive growth models that can propel India's development trajectory to new heights. This chapter brings out results based on a mixed-method approach. A detailed comprehensive framework of mapping was used wherein ethnographically mapping of government assets was done along with use of GIS data to corroborate the same.

4 Methodology

The authors used both qualitative and quantitative approaches. Crowdsourcing approach was also made use of for community ownership on data and collaboration. Help was sought from youngsters who volunteered along with some assistance through an NGO. Open Data Kit (ODK) forms were used to collect data of existing locations of infrastructure facilities and settlements, including pictures, videos, etc., and geo-tagging the same. The data was also collected through tools such as interview schedules and detailed questionnaires, and focus group discussions (FGDs) were used to collect in-depth data from the beneficiaries, government officials working at village level, and households. Door-to-door surveys were conducted, and data was gathered by the authors in an attempt to map the whole village with ODK. Excel sheets and maps were generated using ODK, subsequently converting raw data into inferences. Moreover, census data was used as a reference point, while secondary data was drawn from Mission Antyodaya (Mission Antyodaya, 2022), which is a convergence and accountability framework aimed at bringing optimum use and management of resources allocated by 27 ministries/departments of the Government of India for a multitude of interventions focussing on rural development. All of the 141 indicators were used in the mapping and designing of the dashboard. GIS has been used to generate maps regarding development indicators of Mission Antyodaya and locations of cultural resources and land-use details (Fig. 1).

5 Study Area

This chapter discusses about Burgula Panchayat situated at Farooqnagar mandal in Mahbubnagar district of Telangana State, India. When Telangana came into formation, Burgula village in Farooqnagar mandal was reorganized from Mahbubnagar

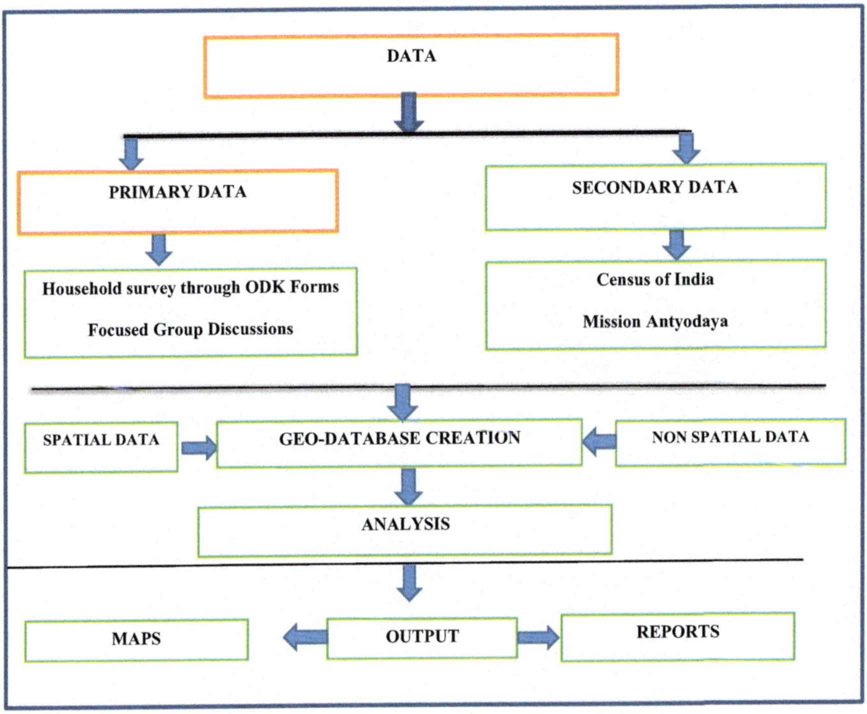

Fig. 1 Methodology flow chart

district to Rangareddy district. It is located 39 km towards the north of district headquarters Mahbubnagar and 8 km from Farooqnagar. Burgula's postal head office is located in Shadnagar, and the PIN code is 509216. Balanagar (3 km), Gouthapur (4 km), Raikal (4 km), Hemajipur (5 km), and Gunded (5 km) are the neighbouring villages and Thandas. Surrounded by Farooqnagar mandal towards the north, Keshampeta mandal towards the east, Kothur mandal towards the north, and Nawabpet mandal towards the west, Burgula is situated on the border of the Mahbubnagar and Rangareddy districts. Farooqnagar, Badepalle, Mahbubnagar, and Vikarabad are the nearby cities. People here speak Telugu. The study area map is shown in Fig. 2.

As per Census 2011, the entire populace is about 5993, and the number of houses is 1452 in about 738 ha. The female population is 48.6%, the village literacy rate is 52.7%, and the rate of literacy for women is 21.1%. The other details are presented in Table 1.

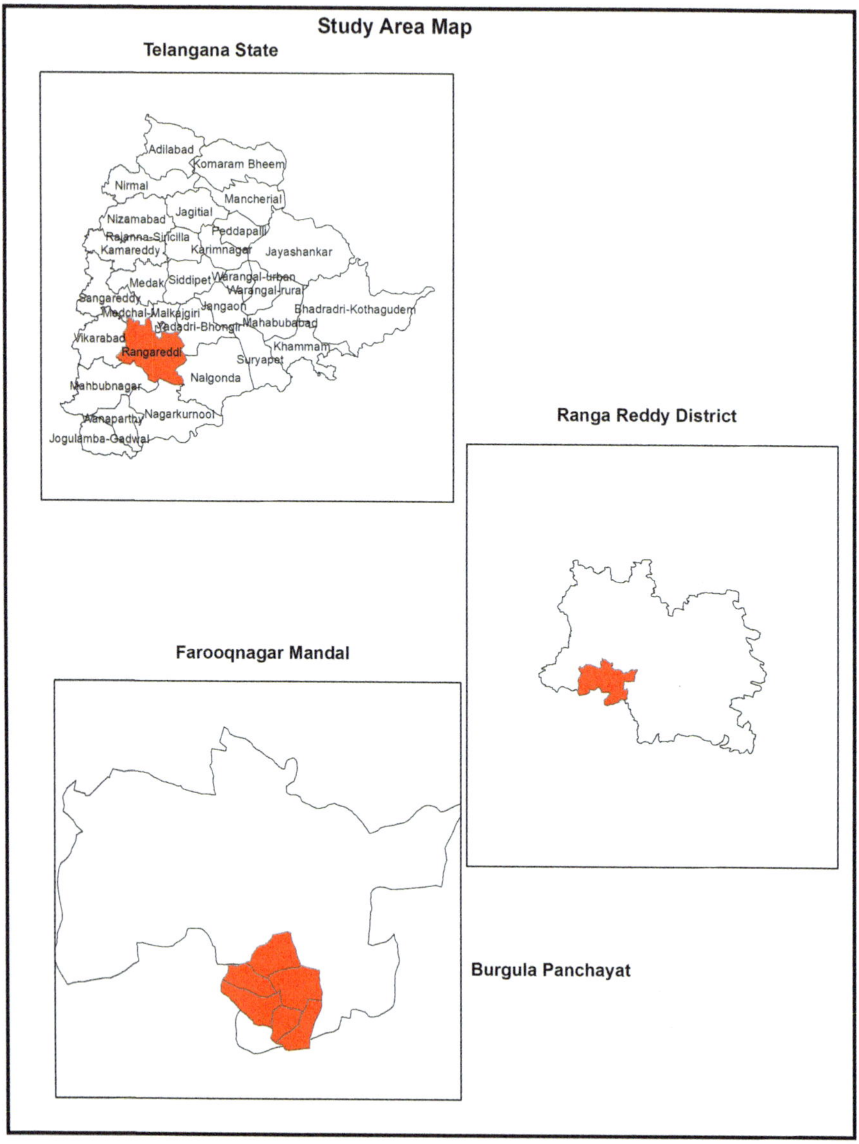

Fig. 2 Study area map

6 Role of GIS and Spatial Planning

A geographic information system (GIS) is carefully crafted to capture, preserve, manipulate, scrutinize, govern, and visually display spatial or geographic data. Empowered by GIS applications, users can explore interactive enquiries, delve into

Table 1 Socio-economic details of Burgula

S. no.	Parameter	No./units
1.	Total population	5993
2.	Total no. of houses	1452
3.	Female population (in %)	48.6% (2915)
4.	Total literacy rate (in %)	52.7% (3158)
5.	Female literacy rate (in %)	21.1% (1264)
6.	Scheduled Tribes population (in %)	14.3% (858)
7.	Scheduled Caste Population (in %)	23.7% (1418)
8.	Working population (in %)	50.3%
9.	Child (0–6) population by 2011	702
10.	Girl child (0–6) population % by 2011	49.6% (348)

Source: http://www.onefivenine.com/india/villages/Mahbubnagar/Farooqnagar/Burgula

spatial information, edit data on maps, and skilfully exhibit the outcomes of these procedures. The fusion and coordination of many activities with geographical (geo-referenced) and non-spatial properties become essential when it comes to planning.

Lately, geomatics-based methodologies have attracted considerable attention due to their impressive efficiency and viable solutions. Moreover, the rapid strides in hardware and software technologies, coupled with intense competition among industry players, have substantially slashed the cost of geomatics/GIS technology. Consequently, it has become highly affordable for large-scale deployment, particularly in the realm of decentralized planning. India operates under a decentralized governance pattern, employing a three-tier system. At its core lies a crucial emphasis on addressing the needs of the grassroots level, requiring careful contextualization and prioritization during the planning process. However, this approach is not widely implemented.

In a decentralized setup, development plans originate from the grassroots level as draft proposals, gradually progressing to the sub-district and district levels. Ideally, these ascending plans should reflect the local requirements while aligning harmoniously with the broader vision and strategic plans of the region or state. Visual representation of such plans plays a key role in reducing spatial disparities. To achieve this, the framework of spatial strategic planning should be interconnected across all three tiers of Panchayati Raj. Bhuvan Panchayat offers a platform for consolidating and linking spatial data. When spatial planning is integrated with sectoral planning, it is referred to as 'spatial strategy' possessing the ability to harmonize with individual sectoral plans. Spatial planning encompasses the strategies to influence how people and activities are distributed in spaces of varying scales, with the ultimate goal of enhancing the built, economic, and social environments of communities. This multifaceted discipline involves various professional fields, including land use, urban renewal, regional, transportation, and economic and community planning. Spatial planning operates at four tiers—the local level, regional level, national level, and international levels—often resulting in the creation of comprehensive plans.

Incorporating spatial planning into local self-governance can foster transparency and accountability in the functioning of local governing bodies (GPs). Leveraging GIS and satellite imagery enables the meticulous visual documentation of projects, easily accessible at any time. As a result, anyone, from any location, can conduct physical verifications of these projects. By utilizing GIS, the legitimacy and acceptance of Panchayati Raj Institutions (PRIs) among their stakeholders can be significantly enhanced. Planned spatial development is conspicuously absent in rural areas, resulting in significant repercussions for regional development, particularly in villages. Spatial data analysis is a diverse discipline encompassing land and water resources, topography, urban and rural planning, hydrology, and earth sciences. These datasets can be sourced from a variety of mediums, including texts, maps, charts, aerial photos, satellite images, and ground information. To manage and analyse the vast volumes of spatial data effectively, a computer-based system known as geographic information system (GIS) comes into play. GIS is specifically designed to collect, organize, analyse, and retrieve extensive spatial data and associated attributes derived from various sources. It proves invaluable in solving complex geographical, hydrological, and planning problems (Garg, 2008). GIS is defined as a system of computer hardware and software, designed to allow users to collect, manage, analyse, and retrieve large volumes of spatial referenced data and associated attributes collected from a variety of sources (Burrough & McDonnell, 1997).

The Rural Area Development Plan Formulation and Implementation (RADPFI, 2021) guidelines emphasize the need for the preparation of rural spatial plans, integrated with the overall development. The RADPFI guidelines aim to provide direction for the preparation of spatial plans for Gram Panchayat and also mention the required alterations and additions in the existing statutory provisions of planning. The need for the RADPFI guidelines arises from the fact that there are 6.49 lakh villages in the country and 68.84% of the total population lives in the rural areas (Gram Panchayat Spatial Development Plan, SPA, Bhopal, 2018). These villages have varied characteristics and show different degrees of transformation from rural to urban. The villages that show high degree of transformation due to their spatial location may attain urban characteristics in the future, and therefore, it becomes indispensable to plan these settlements.

During the process of regional planning, the integration of various spatial data and their attributes becomes crucial to explore different alternatives. In this context, GIS is a useful tool for integrating and analysing data from many domains for particular purposes. As a result, the policymakers and those who intend to chalk out plans gain the necessary tools to generate new visions from existing thematic layers of data tailored to their unique needs. The versatility of GIS lies in its ability to integrate both spatial (e.g. satellite images, maps) and non-spatial (e.g. census surveys) data, allowing for the independent or combined analysis of spatially registered layers. It ensures data integrity and security while minimizing redundancy and supporting multiple views of data. Additionally, GIS provides efficient processing of user transactions and concurrent access for numerous users.

7 Results and Discussion

This study was conducted at Burgula Panchayat. Apart from ethnographically collecting data from all the villages, GIS mapping was also done to map the assets digitally. It is to be noted that Burgula was divided into six Gram Panchayats, namely:

1. Burgula
2. Chintaguda
3. Kadiyala Kunta Tanda
4. Kasireddyguda
5. Kundel Kunta Tanda
6. Nerella Cheruvu

Burgula and Thimmajipalle are the revenue villages. The Burgula Cluster is dominated by tribals, the Lambadas. They live in tribal hamlets or Tandas. The seven Tandas that encircle the Burgula cluster are listed in Table 2.

These Tandas are mainly inhabited by the Lambadas also known as Lambadis, Sugalis, Banjaras, Banjari, and so on. They are originally said to have come from Rajasthan as salt traders and, over the period of decades, settled in parts of South India. They mainly speak Telugu along with their own tribal dialect of Rajasthani. The women wear extravagant jewellery, head gear, and heavy ornamented dresses with glasswork and beadwork. They adorn white ivory bangles in full hands reflecting their cultural roots from Rajasthan. The Tandas are marked in Fig. 3.

The authors conducted an extensive ethnographic study encompassing all development indicators, including health, water and sanitation, education, roads, and infrastructure, among others, in the villages of Burgula Cluster. The findings have been compiled in this chapter to showcase that while certain sectors in the cluster have seen development, there is a pressing need for targeted interventions in other areas.

During this study, a thorough analysis of various socio-economic indicators was undertaken for the entire cluster. The essence of human development revolves around the understanding that human well-being is contingent on multiple dimensions, with education and health emerging as primary welfare indicators. These dimensions play a pivotal role in shaping the overall welfare and progress of the

Table 2 No. of households in each Tanda in Burgula Cluster

S. no.	Tandas	No. of households
1.	Raikal Tanda	20
2.	Thimmajipalle Tanda	30
3.	Kundel Kunta Tanda	30
4.	Devinigunta Tanda	18
5.	Nallametha Tanda	30
6.	Munchenguda Tanda	20
7.	Pothurajuguda Tanda	18

Source: *Primary data from authors' fieldwork*

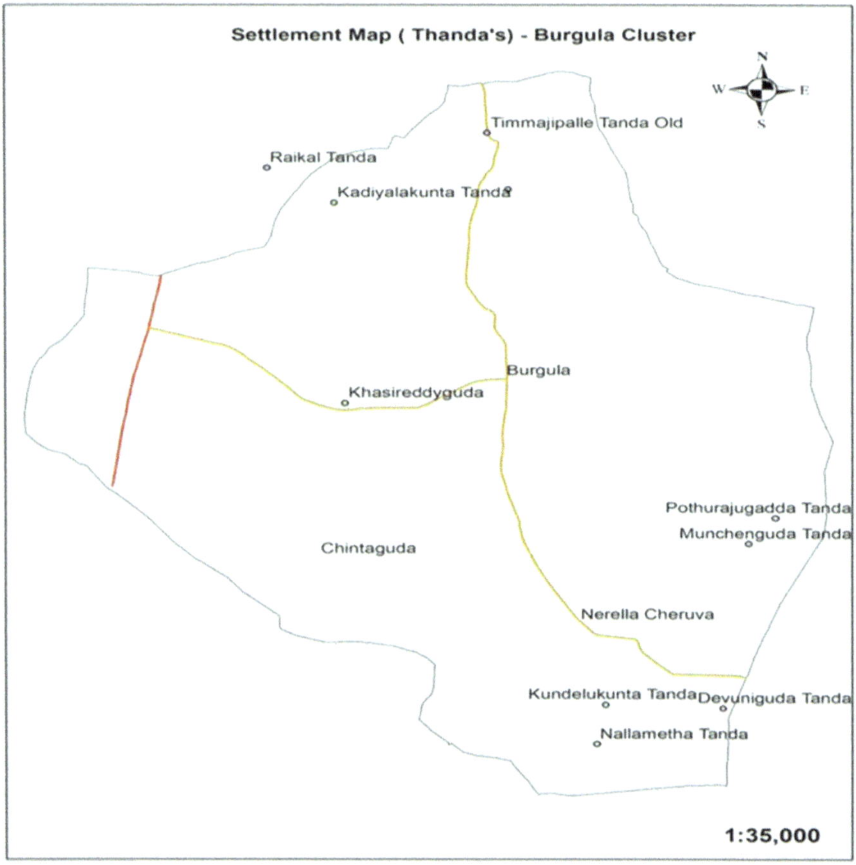

Fig. 3 Tandas in Burgula Panchayat

community. The United Nations' Sustainable Development Goals (SDGs) were taken as a lead for the study (SDGs, 2023).

The results hereby are discussed in accordance with the SDGs they are mapped to.

(i) For SDG 1 that focus on no poverty, the Mahatma Gandhi NREGA assets were mapped to see the dent on poverty by creating assets as envisaged by the government. The map below shows the various assets mapped. As can be seen, majority of the assets are located at Burgula Panchayat only, and the Tandas on the periphery are devoid of any such assets. The below maps show the assets created under Mahatma Gandhi NREGA in the study area. The major works implemented under this programme are with respect to drought proofing, individual land works, land development, and rural sanitation. Marking all the assets constructed under the scheme gave an idea for planning for the future. It would also lead to reduction of redundancy in proposed works and help in the proper allocation of resources. The various assets marked included Aanganwadi centres, banks, borewells, bus stops, substation, etc (Fig. 4).

Fig. 4 The MGNREGA asset map at Burgula Panchayat

(ii) SDG 2 deals with zero hunger. The authors mapped the Aanganwadi centres. These are centres run under the Integrated Child Development Services (ICDS) scheme of the government that aims to curb malnutrition among infants and lactating mothers by providing milk, eggs, and other nutritious food that include Balamrutham packets. The Integrated Child Development Services (ICDS) centres look after the nutrition of infants. One of India's government's major flagship schemes, it was first implemented in 33 blocks in 1975. It focuses on early childhood development and caters to nutritional aspects. The primary objectives of the programme are to enhance the nutritional and health status of young children (0–6); lay the foundation for the child's proper psychological, physical, and social development; lower the incidence of mortality, morbidity, malnutrition, and school dropout; attain efficient policy formulation and execution coordination among the various departments for promotion of child development; and improve the capacity of the various departments to properly carry out the programme in a smooth manner. The programme's main beneficiaries include children below 6 years of age, pregnant women, and lactating mothers. The authors visited two Aanganwadis at the Burgula GP and found that the Aanganwadi workers were given digital tablets to enter data and were imparted requisite training to operate the same. Through digital tablets and smartphones, the workers can enter the data more precisely and accurately and save precious time. The heavy load of maintaining 14 registers was done away with. As can be seen from the map below, the Aanganwadi centre is located at Burgula Panchayat, and beneficiaries from adjacent villages and Tandas could draw benefit from it, but there was no exclusive Aanganwadi for any Tanda (Fig. 5).

(iii) SDG 3 focuses on good health and well-being. The authors found one functional primary health centre (PHC) at Burgula. In rural India, primary health

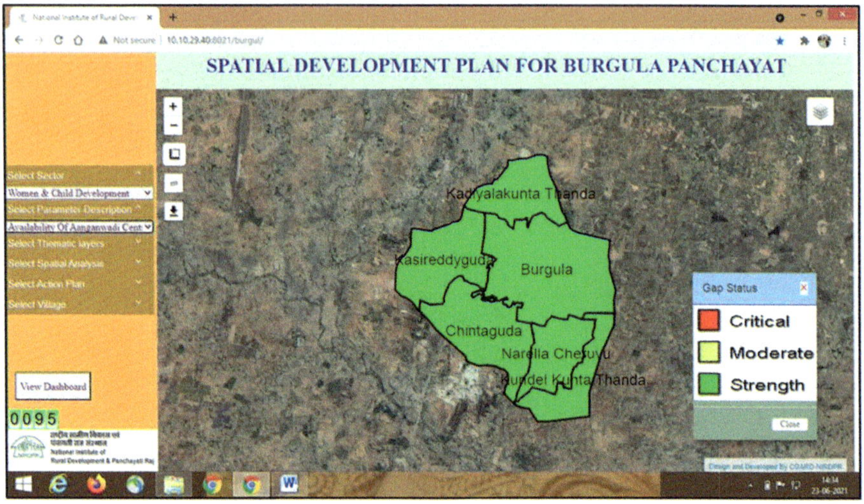

Fig. 5 Availability of Aanganwadi centre

centres (PHCs) play a crucial role in providing basic healthcare services to the local population. These healthcare facilities are the cornerstone of India's public healthcare system and are designed to bring accessible and affordable medical services to remote and underserved regions. It is observed that PHCs are strategically situated in rural areas to ensure maximum coverage and proximity to the population they serve. These facilities are commonly situated within rural areas, often within close proximity to villages or groups of villages. They provide a comprehensive array of vital medical services, encompassing general check-ups, maternal and paediatric healthcare, vaccination programmes, family planning services, and fundamental diagnostic assessments. These centres are staffed by proficient healthcare personnel, such as physicians, nurses, pharmacists, and auxiliary healthcare workers, who are equipped to address the various healthcare requirements of the local community. The primary responsibilities of ANMs (auxiliary nurse midwives) and ASHAs (accredited social health activists) involve raising awareness about preventive healthcare measures, such as providing health education, organizing awareness campaigns, and conducting vaccination initiatives to reduce the incidence of preventable diseases. PHCs (primary health centres) actively interact with local communities to gain insight into their specific healthcare needs and collaborate with them to devise effective solutions. These facilities receive support and financial backing from the government, reflecting the government's dedication to enhancing the healthcare infrastructure and services in rural areas nationwide. Despite the critical role they play, primary health centres often face challenges like inadequate resources, a shortage of medical staff, and limited infrastructure. The government continues to work towards improving these centres' capacity and expanding their reach to ensure better healthcare

Fig. 6 Map showing PHC and its coverage in Burgula Cluster

access for rural communities. The authors observed that those living in Burgula, Kasireddyguda, Chintaguda, and Nerella Cheruvu still have better access to the PHC, whereas the tribals residing in Kadiyala Kunta Tanda and Kundel Kunta Tanda do not have proper road connectivity to avail medical help; also, the PHC is quite far for them (Fig. 6).

(iv) SDG 4 deals with quality education and is meant for providing access to free education to all children without any discrimination. Under the flagship interventions like the Sarva Shiksha Abhiyaan (SSA) and Right to Education (RTE) Act, all children aged 6 to 14 are entitled to receive free and compulsory education. To ensure barrier-free access to schools, norms allow for a primary school to be within a radius of 3 km from a village, whereas a middle school should be within a 5-km radius, and a high school has to be within a 10-km radius. The children living in rural areas need to be provided with adequate educational facilities. This necessitates having a fully functional school building with a boundary wall, playground, classrooms with adequate lighting and ventilation, seating arrangement, blackboards, teaching and learning material (TLM), drinking water facility, library, spare rooms for teachers and headmaster/headmistress, toilets with water availability and sewage system, etc. Enough staff should be hired to run the day-to-day administration and teaching at schools. The authors observed a good number of schools in villages of Burgula Cluster. Since all the primary and upper primary schools cater to the needs of the GPs, no child was left out of school. For attending college, i.e. post high school, students attend intermediate college at Shadnagar (nearly 13 km away) or degree college at Jadcherla (30 km away). The authors realized that as they moved towards the Tandas, a gradual decline in the quality of infrastructure of schools was evident (Fig. 7).

Fig. 7 Primary school service area

(v) SDG 6 ensures access to water and sanitation for all, and this is achieved at Burgula Cluster too. The past and present Sarpanchs have ensured that all the GPs maintain hygiene and sanitation. There is an underground drainage system, but it is only confined to Burgula and Kasireddyguda. Elsewhere, the wastewater is simply discharged in the open, but it is being controlled by the Sarpanchs. As there are no designated dumping yards, a place was identified for the purpose. Moreover, the authors observed that under the Mahatma Gandhi NREGA scheme, many tribal households have constructed individual household latrines (IHHL). These can be seen at Kadiyala Kunta Tanda, Kundel Kunta Tanda, and Raikal Tanda. The green colour denoted availability of piped water in all the Panchayats. This is due to the Mission Bhagiratha programme of the Government of India, which aims to provide safe drinking water to every household in Telangana (Fig. 8).

(vi) SDG 11 focuses on sustainable cities and communities. Herein, it is envisaged that there would be access to safe, affordable, accessible, and sustainable transport systems as well as improved road safety for all (UN, SDG). The authors found that almost all Tandas had un-metalled roads, and the terrain was also rough. This poses a high risk to the commuters, especially the elderly, pregnant women, and children. Except for the main areas in Burgula and Kasireddyguda, all other villages and Tandas are connected through un-metalled roads. The map below shows the roads at Burgula. Whereas Kasireddyguda, Burgula, and Nerella Cheruvu are connected with all-weather roads and are shown in green on the corresponding map, Kadiyala Kunta Tanda, Chintaguda, and Kundel Kunta Tanda are not that well-connected and are shown in yellow in the map (Fig. 9).

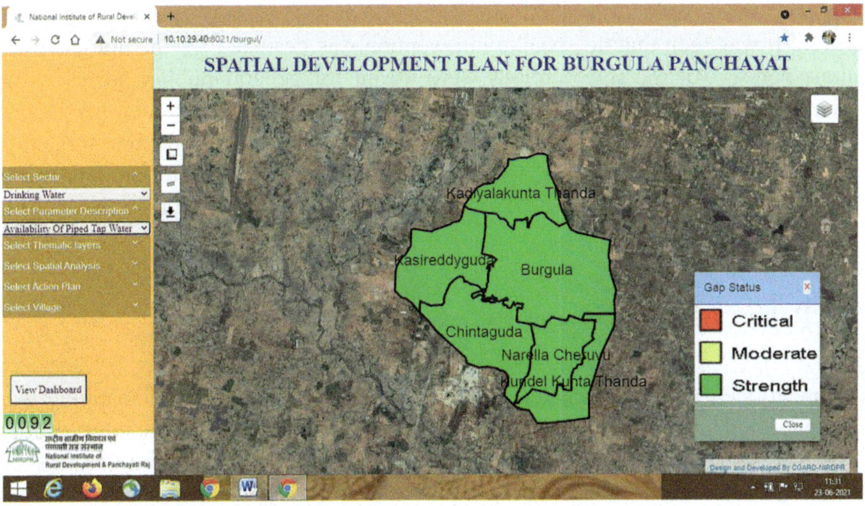

Fig. 8 Availability of piped tap water

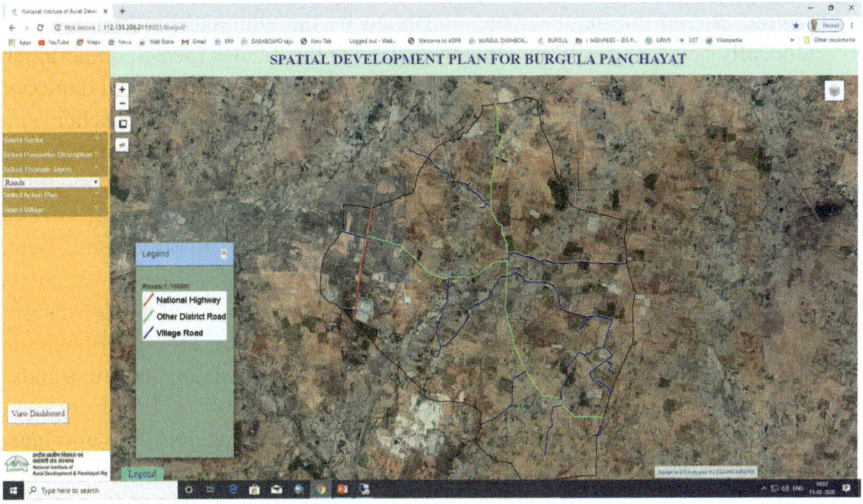

Fig. 9 Availability of connectivity through all-weather road

8 Limitations of the Study

Studying social behaviours is very subjective in nature. It is not easy to understand the social dynamics of a village just by mapping assets and infrastructure alone. The socio-economic and geographic context needs to be taken into account too. In this study, the authors attempted to deconstruct the impact of government interventions meant for development of the marginalized. A comprehensive framework for mapping the interventions, both ethnographically and through satellite, was followed. It was seen that no specific intervention was designed for the marginalized alone per se. All the interventions were meant for general public; however, the Panchayats could have prioritized their GPDPs in accordance with their needs. The limitation here would be that it is an ongoing process and very flexible in nature. The discourse keeps changing with every single day. Political interference, crowd mobilization, and lack of awareness have emerged as determining factors.

9 Policy Implications of the Study

As mentioned above, a detailed and comprehensive framework of mapping was designed and followed by the authors. Having identified the gap in the interventions designed for community development, this approach of mapping and geo-tagging not only helps in bringing about transparency but also helps in prioritizing needs while planning of budget. It is to be noted that policy advocacy for reducing the rising inequalities is a necessary approach. The study of Burgula has clearly highlighted centre-periphery dichotomy. As observed and discussed, all interventions are centred around the main villages inhabited by the elite and the like, who are economically better in position, while the tribals in the Tandas located on peripherals of main village are devoid of such government interventions. Spatial depiction of plans helps in reducing spatial disparities. If this is mandated in all schemes, a better picture of distribution of resources and allocation of funds can be determined. The authors strongly advocate this idea.

10 Conclusion

Primary health centres are indispensable pillars of healthcare delivery in rural India, contributing significantly to the nation's goal of achieving universal health coverage and reducing health disparities between urban and rural areas. On the same lines, Aanganwadi centres cater to childcare, malnourishment, and vaccination. No child should be out of school is the primary goal of the government. Hence, it is mandatory that schools, be it primary or secondary, are in peripheral areas wherein transportation, road connectivity, and habitation are not a hindrance for the child to attend school. Drinking water facility is now provided at every doorstep. It is ensured that the drainage pattern is feasible with terrain and climatic change. Proper connectivity through all-weather roads is facilitated. It is to be noted that India's growth and development story is a testament to its resilience, entrepreneurial spirit, and commitment to progress. The nation has made significant strides in unleashing its vast potential and improving the lives of its citizens. However, there are ongoing challenges that must be addressed to ensure sustainable and inclusive growth. By leveraging its demographic dividend, investing in human capital, and fostering an enabling environment for innovation, India can continue on its path towards becoming a global powerhouse, driving prosperity, and uplifting the lives of millions. A robust foundation for understanding, predicting, and influencing human behaviour is provided by social physics. It gives us the ability to reveal patterns and dynamics that were previously obscure of, by making use of the vast data resources readily available in today's digital world. With the help of geospatial technologies, we can improve our understanding of and influence over social behaviour, thus influencing changes around us and in our ecosystem. A path to a wealthier, sustainable, and inclusive future can be initiated by embracing the possibilities of social physics and using technology for social good and betterment of one and all.

Acknowledgement The authors would like to express their gratitude to the National Institute of Rural Development and Panchayati Raj, Hyderabad, for funding this study.

References

Ball, P. (2012). *Why society is a complex matter*. Springer.

Barnes, T. J., & Wilson, M. W. (2014). Big data, social physics, and spatial analysis: The early years. *Big Data & Society, 1*(1). https://doi.org/10.1177/2053951714535365

Bhattacharya, K., & Kaski, K. (2019). Social physics: Uncovering human behaviour from communication. *Advances in Physics, 4*(1), 1527723. Taylor and Francis Online.

Burrough, P. A., & McDonnell, R. A. (1997). *Principles of geographical information systems*. Oxford University Press.

Census of India. (2011). *Office of the Registrar General and Census Commissioner, New Delhi.*

Conte, R., Gilbert, N., Bonelli, G., Cioffi-Revilla, C., Deffuant, G., Kertesz, J., Loreto, V., Moat, S., Nadal, J.-P., Sanchez, A., et al. (2012). Manifesto of computational social science. *The European Physical Journal Special Topics, 214*, 325–346.

Garg, P. K. (2008). *Spatial planning of infrastructural facilities in rural areas around Roorkee, Uttarakhand, India*, FIG working week 2008, Stockholm.

Marco, J., et al. (2022). Social physics. *Physics Reports, 928*, 1–148. Elsevier.

Miguel, M. S., Johnson, J. H., Kertesz, J., Kaski, K., Daz-Guilera, A., MacKay, R. S., Loreto, V., Érdi, P., & Helbing, D. (2012). Challenges in complex systems science. *The European Physical Journal Special Topics, 214*, 245–271.

Mission Antyodaya. Retrieved June 26, 2022., from https://missionantyodaya.nic.in/

Pentland, A. (2014). *Social physics: How good ideas spread: The lessons from a new science*. Penguin. ISBN 9781594205651.

RADPFI. (2021). *Rural area development plan formulation and implementation (RADPFI) guidelines government of India*. Ministry of Panchayati Raj.

SPA Bhopal Report. (2018). *Gram Panchayat Spatial Development Plan.*

United Nations Sustainable Development Goals (SDGs). Retrieved August 2, 2023., from https://sdgs.un.org/goals

Dr. Sonal Mobar Roy is an Assistant Professor at the Center for Wage Employment and Livelihoods at the National Institute of Rural Development and Panchayati Raj, Hyderabad, India. Dr. Roy earned her master's in Social Anthropology from Lucknow University and her doctorate in Sociology from the Department of Humanities and Social Sciences, Indian Institute of Technology Kanpur, one of the premier institutes in India. Her doctoral study focused on stigma related to TB and HIV in the Ladakh region of India, and the fieldwork was sponsored by the Parkes Foundation, UK.

Her research interests include sociology of health and education, culture and tribal studies, rural development, and ethnography. She is a member of international editorial boards and has been involved in teaching, training, and research at her institute. She has national and international publications to her name and strives to bridge the gap between theory and practice through her writings and research. She has presented papers at universities of repute, such as the Oxford University, the Imperial College London, and the University of Berkeley, California. She can be reached at smobar2@gmail.com.

Dr. Prasad N. S. R. is working as an Assistant Professor at the Centre for Geo-informatic Application in Rural Development at NIRDPR, Hyderabad. He has a Ph.D. in Earth Sciences from Osmania University and Masters in Remote Sensing and GIS from JNTU, Hyderabad. He has more than a 20 years' experience in the field of Application of Geo-informatics in Rural Development. His area of interest is spatial planning for village development, natural resource management, rural infrastructure development, and Geo-ICT applications. He has multiple national and international publications to his name.

Selection of Potential Site for Solid Waste Disposal Using Geospatial Techniques and Analytic Hierarchy Process (AHP)

Virendra Kumar, Anoop Kumar Maurya, Vashi Ahmad, Mohd Jaid, Shivam Pandey, and Sarvajeet Kumar

1 Introduction

In almost developing and low-income countries like India, the challenge of garbage disposal and environmental degradation is significant. Factors such as haphazard or low levels of urbanization and rural-to-urban migration driven by the promise of employment and better facilities in large cities contribute to the problem. According to reports from the Central Pollution Control Board (2000) and the National Environmental Engineering Research Institute (NEERI) (1996), million-plus population cities in India generate between 400 and 500 grams of solid waste per person per day. The prevalent method of waste disposal in many developing nations involves landfilling in low-lying areas or vacant plots in urban centres. Hamer (2003) has suggested that organizing economic and social activities systematically in densely populated urban regions can help reduce municipal solid waste (MSW) generation and protect the environment from harm. Researchers such as Saaty (1980, 1990, 2003, 2008) and Cheng et al. (2003) have highlighted the utility of geospatial techniques coupled with the analytic hierarchy process (AHP) for identifying and selecting potential landfill sites for solid waste disposal. This approach involves assigning weights to various criteria on a scale of 1–9 and creating a priority ranking based on pairwise comparisons of these factors (Saaty, 2008; Cheng et al., 2003). Saaty and Niemira (2006) further emphasized the effectiveness of the AHP method in decision-making processes related to waste management and site selection. The state of Uttar Pradesh in India is the most populous state having around 22 crores of population as per estimated population in 2021, and this state has around 762 urban local bodies.

V. Kumar (✉) · A. K. Maurya · V. Ahmad · M. Jaid · S. Pandey
Remote Sensing Applications Centre, Lucknow, Uttar Pradesh, India

S. Kumar
National Atlas and Thematic Mapping Organization, Salt Lake City, Kolkata, India

A. Rai et al. (eds.), *Contemporary Social Physics*, Springer Geography,
https://doi.org/10.1007/978-3-031-77596-3_21

Table 1 Indian population (in crore)

	2001	2011	Difference
India	102.9	121.0	18.1
Rural	74.3	83.4	9.1
Urban	28.6	37.6	9.0

Source: Census of India (2011)

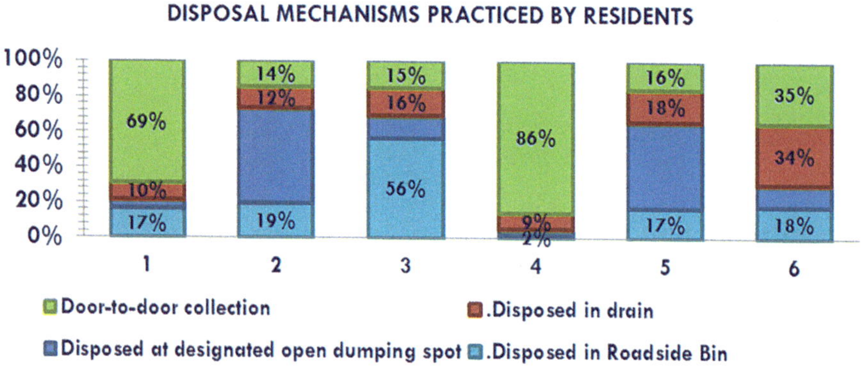

Fig. 1 Solid waste disposal mechanism. (*Source*: Kanpur Nagar Nigam)

Kanpur is an industrial town situated in the central part of the state having around 28 lakhs of population as per Census of India 2011 (Table 1). Hasana et al. (2009) have suggested the landfill demand and allocation of area for disposal of city's waste. Currently, the Kanpur Municipal Corporation is producing 919 tons of solid waste daily while lacking a suitable location for proper waste disposal (Figs. 1 and 2)

Moreover, the importance of community engagement and participation in landfill site selection processes is emphasized by scholars such as Susan Owens and David S. Jones. Their research highlights the significance of incorporating the perspectives and preferences of local residents and stakeholders to mitigate conflicts and foster collaborative decision-making (Owens & Jones, 2016). Social physics also allows for the consideration of socio-economic factors in landfill site selection. Research by scholars like Robert D. Putnam and Richard Florida emphasizes the impact of income levels, employment opportunities, and access to essential services on waste generation rates and disposal behaviour. Integrating these socio-economic considerations into site selection criteria ensures that facilities are located in areas that effectively serve the needs of the community while minimizing negative social environmental impacts (Putnam, 2000; Florida, 2002).

2 Objective

The primary goal of the ongoing investigation is to pinpoint and designate a viable landfill area for the purpose of solid waste management using Open Source (Google data) satellite 1 m ground resolution and IRS series satellite LISS-III sensor (23.5 m

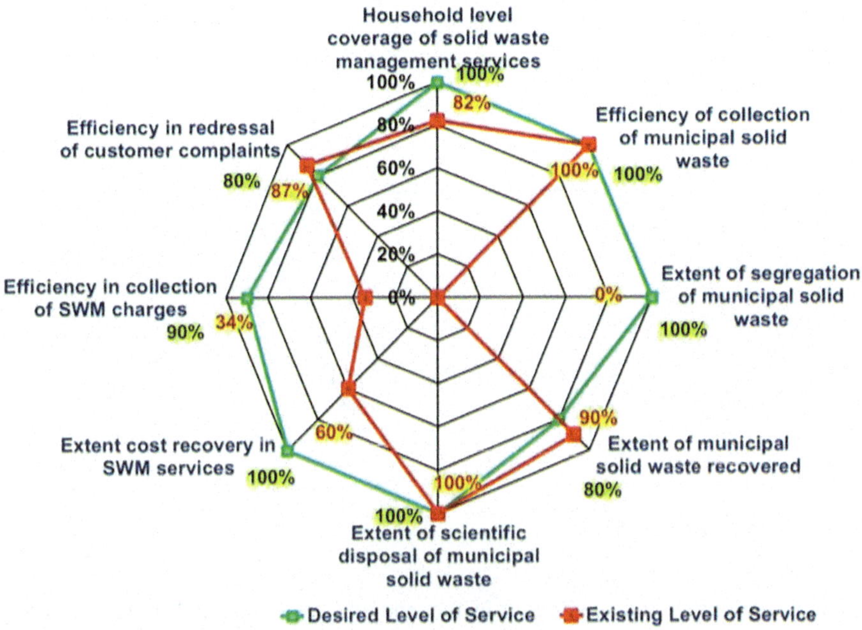

Fig. 2 Variation in desired and existing services of SWM in Kanpur

spatial resolution) data in GIS and AHP method for Kanpur City. The detailed objectives of the study are:

- To prepare different thematic layers using remote sensing data.
- To identify important criteria for waste disposal site selection.
- To identify possible appropriate locations for waste disposal site in environs of Kanpur City using AHP.
- To explore the efficacy of the analytic hierarchy process (AHP) in modelling site suitability processes. This empirical study aims to demonstrate the pairwise comparison procedure. This procedure facilitates the capture of relative and consistent judgements between two factors simultaneously.

3 Study Area

Kanpur is a district and commissionary and an industrial town in the state of Uttar Pradesh, India. It is located on the right bank of holy Ganga River (Fig. 3). Kanpur Municipal Corporation area is spread over an area of around 261 sq. km. (Kanpur Nagar Nigam), and including the industrial area, it is having around 630 sq. km. This city has a very good connectivity through railway, national/state highways, and air traffic networks (Gupta, 2015). Kanpur City is well-known for industries in northern part of India. Different types of industries, i.e. leather, textile, cotton, woolen, and some defence-related material and goods, are being manufactured at Kanpur. Being

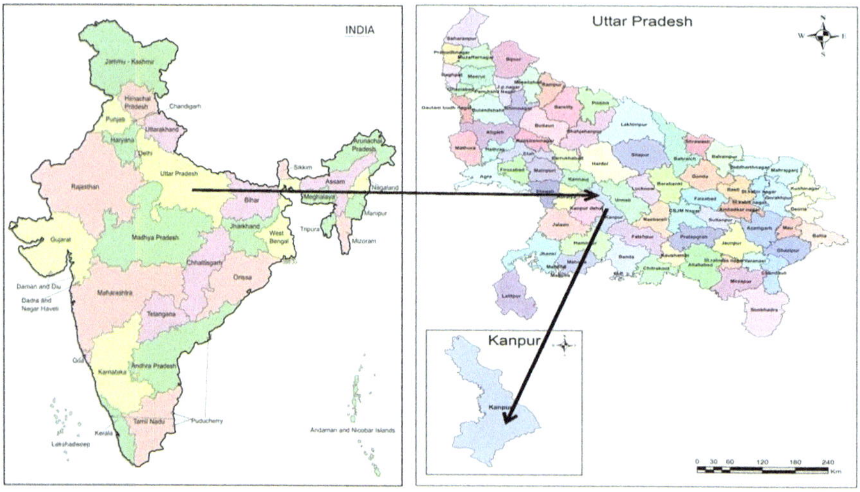

Fig. 3 Location map of study area

an industrial hub, so many goods of leather and other material are exported to foreign countries from this city. Kanpur City has a major contribution in economy of Uttar Pradesh. So many famous technical institutes are already established at this city (Jain & Sharma, 2017).

4 Material and Data Used

4.1 To Fulfil the Set Objectives of Study Area, the Material and Datasets Used Are Given Below

- Open Source (Google) satellite 1 m ground resolution data downloaded from Google Earth (Fig. 4)
- Indian Remote Sensing Satellite (IRS series) data having 23.5 m ground resolution for various map layers
- Guide map of Kanpur City on 1:20,000 scale
- Cartosat data for preparation of slope/digital elevation model
- Non-spatial data related to quantity of waste collected from Kanpur Nagar Nigam
- Groundwater (GW) table data from Nagar Nigam and from local people during field survey
- GPS/ground proofing data collected during field verification

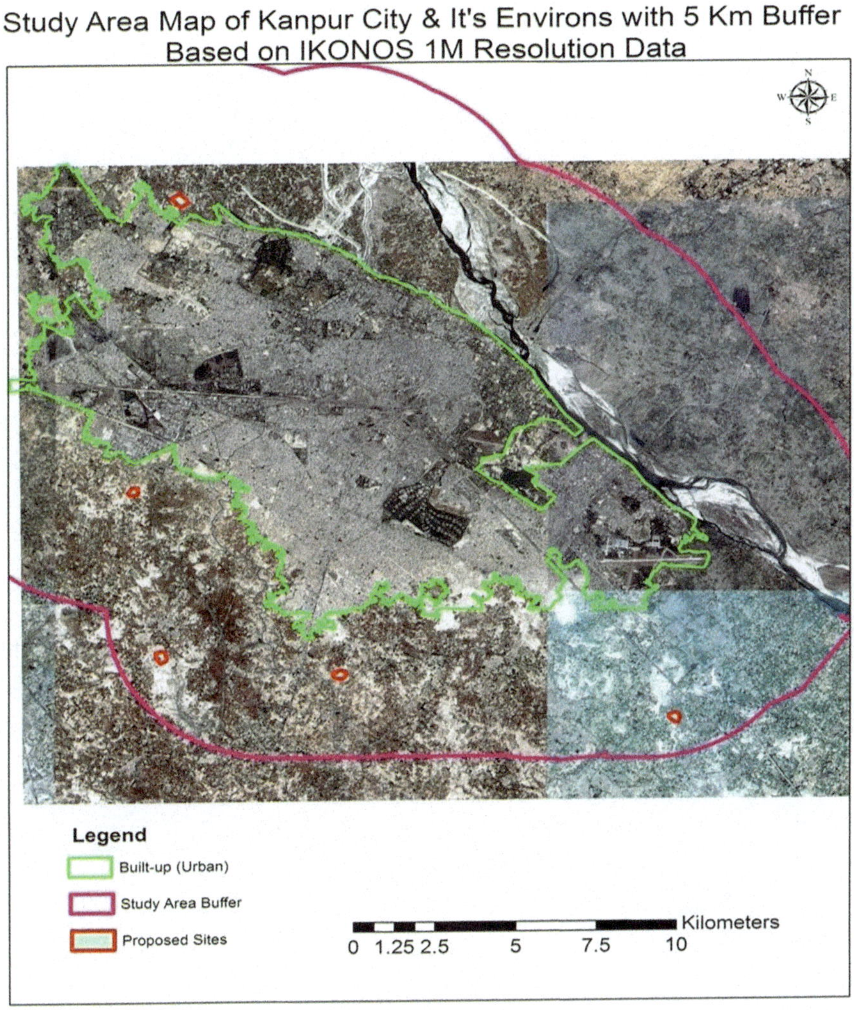

Fig. 4 Open Source satellite image with 5 km buffer from urban area

4.2 Siting Locations for Waste Disposal Site Selection

Siting locations for urban development are selected by field survey. There are five siting locations that are identified as most suitable area for waste disposal (Table 2). Accordingly, the field verification data is collected in each site location deliberated by using GPS coordinates and digital imagery (Table 3, Fig. 5).

Table 2 Proposed site physical area details

Siting location	Area (ha)	Village name
S-1	9.25	Mardanpur
S-2	8.21	Pratappur Sarsai
S-3	6.38	Patehuri
S-4	9.89	Kharagpur
S-5	5.36	Chakeri

Table 3 Geographical location of sites

Siting location	Latitude			Longitude			Elevation
	Deg	Min	S	Deg	Min	S	
S-1	26	22	22	80	15	50	421 ft.
S-2	26	21	39	80	25	4	400 ft.
S-3	26	25	33	80	15	3	410 ft.
S-4	26	30	59	80	16	8	410 ft.
S-5	26	21	0	80	21	14	407 ft.

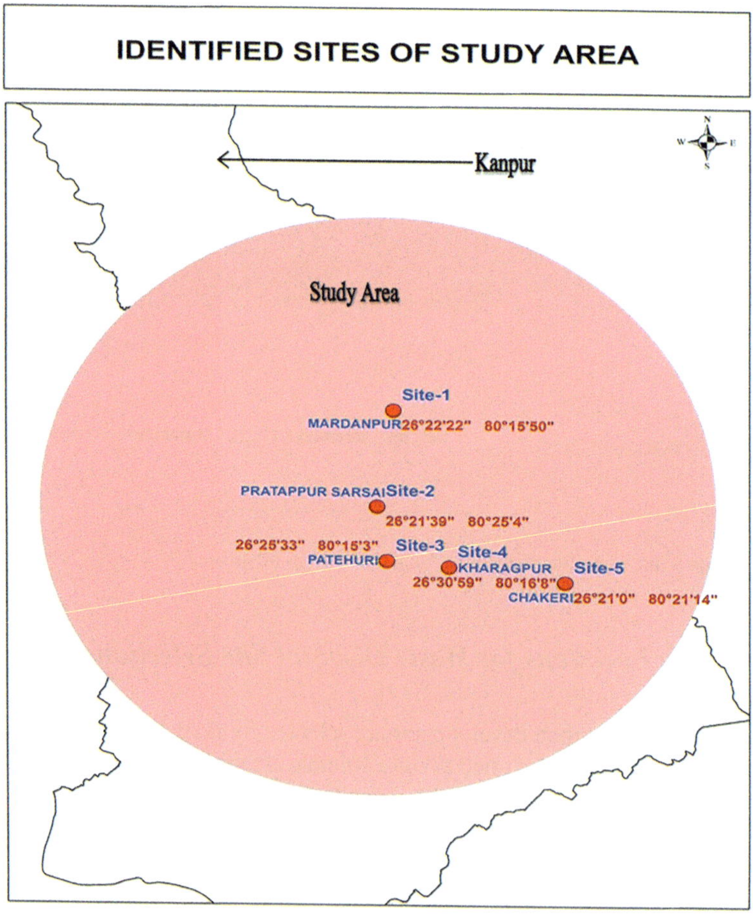

Fig. 5 Identified sites of study area

4.3 Software Used

- ArcGIS 10 for GIS applications for the preparation of thematic maps
- Super Decisions for decision-making applications
- Mobile GPS Mapper 10, for collecting the GPS coordinates of proposed sites
- Microsoft Office Word 2007 and Microsoft Office Excel 2007 for writing and compilation works

5 Methodology

The methodology for potential landfill site selection through various criteria decision-making analysis in GIS environment has been categorized into the following five parts (Al Khalil, 2002; Solnes, 2003):

1. Identification of suitable criteria for waste disposal site selection (Cirucci et al., 2015)
2. Preparation of thematic maps derived from satellite data on the basis of selected criteria to perform the objective of the study
3. Multi-criteria decision-making evaluation (Rezaei-Moghaddam & Karami, 2008)
4. Computational analysis for overall weight obtained by each site
5. Suitability analysis of final site for waste disposal site selection

5.1 Data Extraction

Overall eight factors were taken into consideration including geomorphology, soil texture, transport network, land use/land cover, surface water, groundwater, lithology, and slope. Thematic layers of these criteria were generated and collected from the centre (e.g. Asef & Lim, 2017b; Li et al., 2019; Shyamali & Kumar, 2021).

5.2 Registration/Geo-referencing

Registration is the process of transforming one set of image data to register with a second set of data. The basic data created by remote sensing is subjected to geometric distortion. The process of geo-referencing comprises of two major parts: spatial and spectral rectification (Table 4). Complete rectification and restoration was done using geometric correction method (Smith & Johnson, 2018).

In the present study, the following steps are considered for correction and accuracy of the image using ArcGIS 10.5 software (https://support.esri.com) (Fig. 6).

Table 4 Steps considered for correction and accuracy of the image using ArcGIS 10.5

S. no.	Steps
1	Projection: WGS_1984_UTM_ZONE_44
2	Importing the image from CD-ROM to ArcGIS
3	Selecting polynomial transformation of first order
4	Selecting projection as Universal Transverse Mercator
5	Specifying the spheroid as D_WGS_1984
6	Taking the ground control points (GCPs)
7	Taking the GCP tool, geo-referencing of the similar points (control points) of that entering their respective X and Y coordinates

Minimum four distributed ties points required taken from SOI topographical to ensure the RMS errors should be <1 pixel (Chandio et al., 2012).

5.3 *Multi-criteria Decision-Making Evaluation*

The analytic hierarchy process (AHP) method was taken into consideration for comparison and correlation of multiple criteria for selection of factors and option to recognize and rank the most potential sites through following formula:

Enumerate the CI as deviation

$$CI = \frac{(\lambda_{\max} - n)}{(n-1)}$$

where, CI = Consistency index.

RCI value proposed by Saaty are used to the size of the matrix and to find out the answer related to Consistency ratio by dividing the consistency index (CI) value over the random consistency index (RI), the consistency ratio (CR) are numerated by using the formula as given below:

$$CR = \frac{CI}{RI}$$

where, CR = Consistency ratio.

Overall composite weights are numerated through normalization of combination of linear and multiplication between selection criteria of related weights and alternatives by using formula as given below:

$$SI = \sum W_i X_i$$

where, SI = Suitability index.

If the value of CR is ≤ 0.1, the consistency is correct.

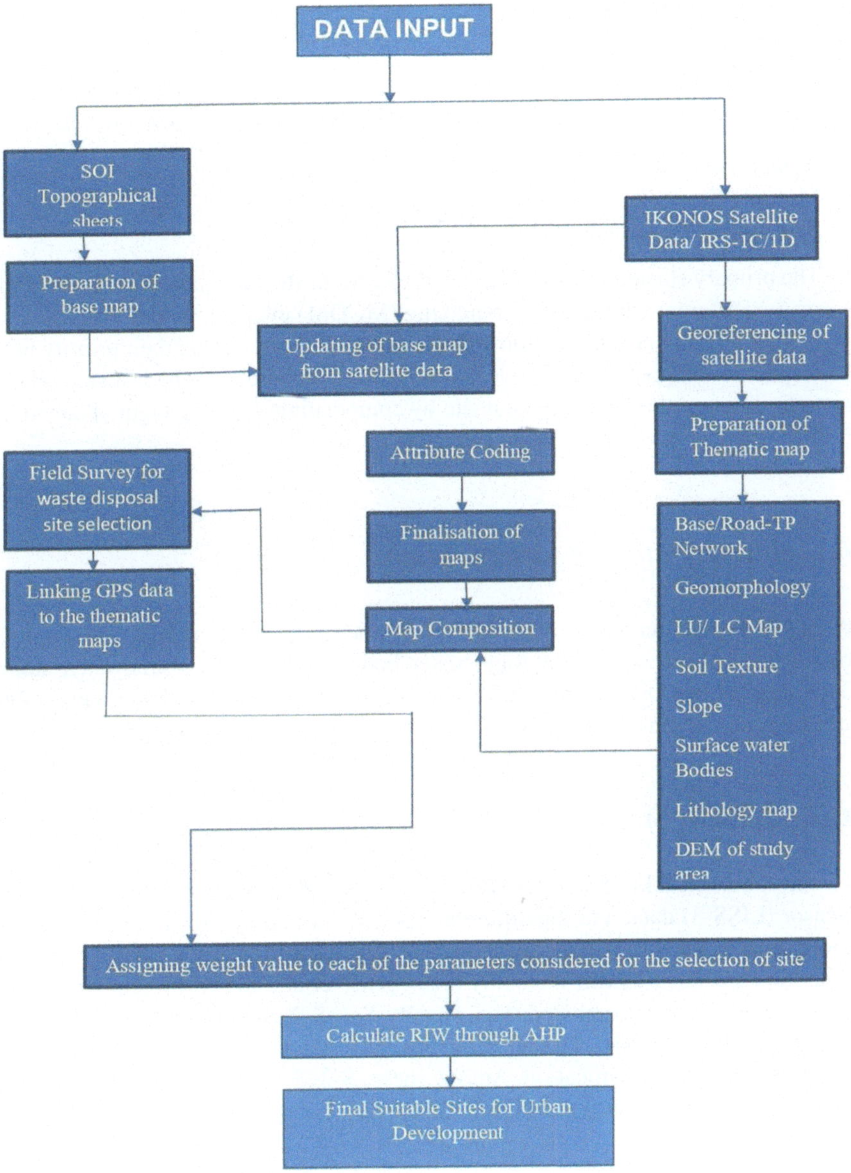

Fig. 6 Flow chart of methodology

6 Result and Discussions

The site suitability analysis model for waste disposal basically included three steps to find out the most potentiality of options for waste disposal as given below:

- Primary analysis
- Multiple theme decision-making
- Selection of best potential site

- The primary stage includes the creation of various thematic layers as input vector maps. Different theme decision-making (MCDM) evaluation comprises assigning the weight to each criterion and alternative and calculating their priority with respect to the goal of using selected modelling techniques. The last step relates to finding out the most potential site by compositing overall weight obtained by alternatives with respect to site selection.

6.1 Interpretation of Criteria Maps

A total of eight criteria have been selected for the study including soil, geomorphology, land use/land cover, lithology, water bodies, subsurface/groundwater, slope, and transport network as discussed earlier. Interpretation of these criteria maps discussed in succeeding section.

Geomorphology Map

Geomorphological layer was derived by IRS series satellite Linear Image Scanning Sensor (LISS-3) data, and the different types of classes, i.e. older alluvium that is more suitable for waste disposal due to its less permeability as compared to younger alluvium, are identified in the study area (Şener et al., 2011). In geomorphology layer, out of the five designated sites, namely, S-1, S-2, S-3, S-4, and S-5, which all fall under the category of older alluvium, S-1 stands as the exception with its younger alluvium classification. Among these, S-2, S-3, S-4, and S-5 are deemed appropriate for waste disposal (Table 5 and Fig. 7).

Table 5 Geomorphology suitability with respect to goal

S. no.	Site no.	Geomorphology	Suitability
1	S-1	Younger alluvium	Low
2	S-2	Older alluvium	High
3	S-3	Older alluvium	High
4	S-4	Older alluvium	High
5	S-5	Older alluvium	High

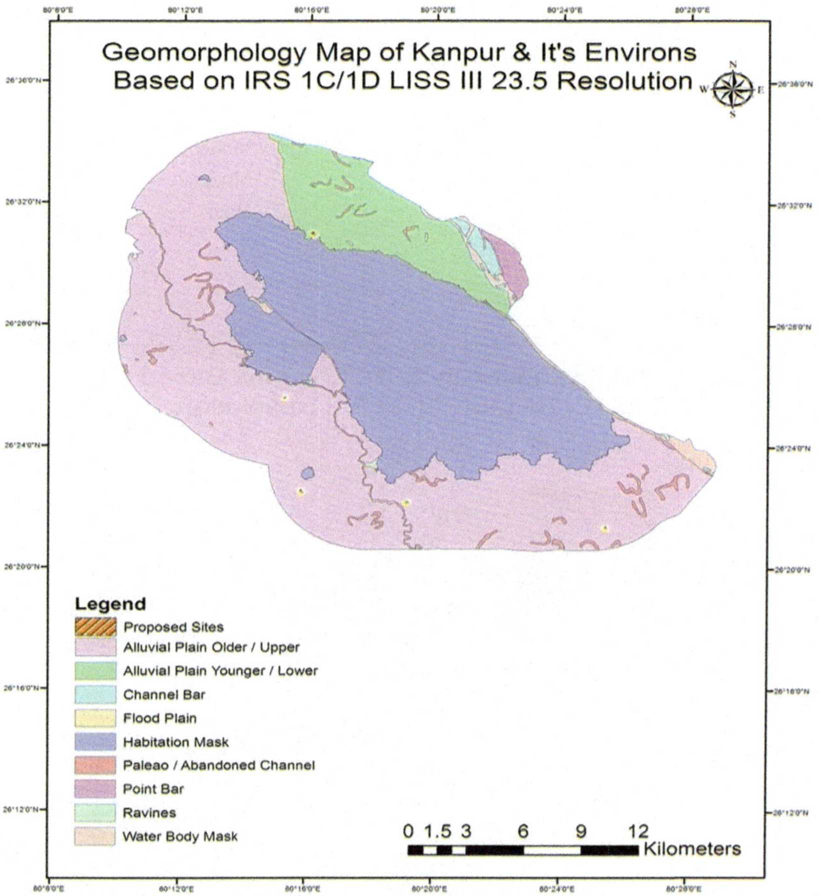

Fig. 7 Geomorphology map

Soil Texture Map

Soil texture information is crucial for landfill planning in the Kanpur region, and it has been derived from IRS series satellite's LISS-3 data. The major soil types identified in the area include coarse loamy, fine loamy, and fine silty (Table 6 and Fig. 8). Within these categories, soils with finer textures are regarded as more compatible with landfill usage in comparison to those with coarser textures. Thus, S-5 appears to be highly suitable for waste disposal as it is found on fine silty soil. As outlined by Brady (1996), the spectrum includes soils spanning from silt to highly fine-textured silty clay and clay. Sandy soil, mixed soil, clay, and clean, dense sand/gravel are ranked as having very high, high, moderate, low, and unsuitable suitability for waste disposal, respectively. This system of categorization aids in assessing the suitability of diverse soil compositions for landfill applications (Şener et al., 2011; Sumathi et al., 2008).

Table 6 Soil suitability with respect to goal

S. no.	Site no.	Suitability
1	S-1/fine loamy	Moderate
2	S-2/fine loamy	Moderate
3	S-3/fine loamy	Moderate
4	S-4/fine loamy	Moderate
5	S-5/fine silty	High

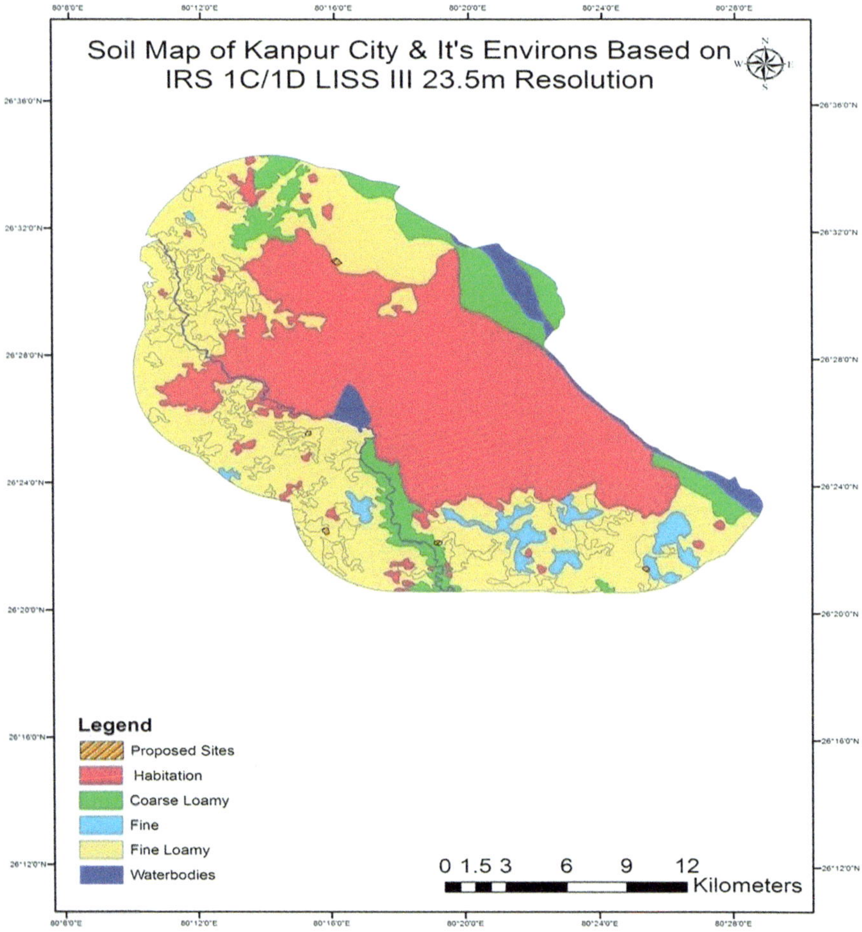

Fig. 8 Soil texture map

LU/LC Map

The LU/LC map was created using Open Source (Google) data of 2012. Within the study region, seven primary categories are depicted including cultivated land, constructed areas (urban/rural), unused land-waterlogged, open scrub wasteland, industrial zones, restricted area (airport/military area), and orchard, parks, and surface water features (Fig. 9). Thoso, in 2007, proposed that degraded and scrub wastelands are potentially more appropriate for waste disposal due to inadequate water availability (Table 7). Siddiqui (1996) has suggested that built-up, urban/rural, and crop land is important for environmental issues, and this category is not suitable for landfill (Table 8) (Jaybhaye et al., 2014; Guler and Yomralioglu 2017).

Surface Water Body Map

Surface water is also an important criterion for protecting the sources of potable water (Gorsevski et al., 2012). In the study area, four categories are derived from high-resolution satellite data. They are lake/pond, river, canal, and drains (Fig. 10). It has been observed that S-1, S-4, and S-5 are suitable sites (Paul, 2012). S-2 and S-3 are taken to be moderately suitable for disposal of waste (Table 9) (Chang et al., 2008; Gorsevski et al., 2012; Ebistu & Minale, 2013).

Fig. 9 LU/LC map

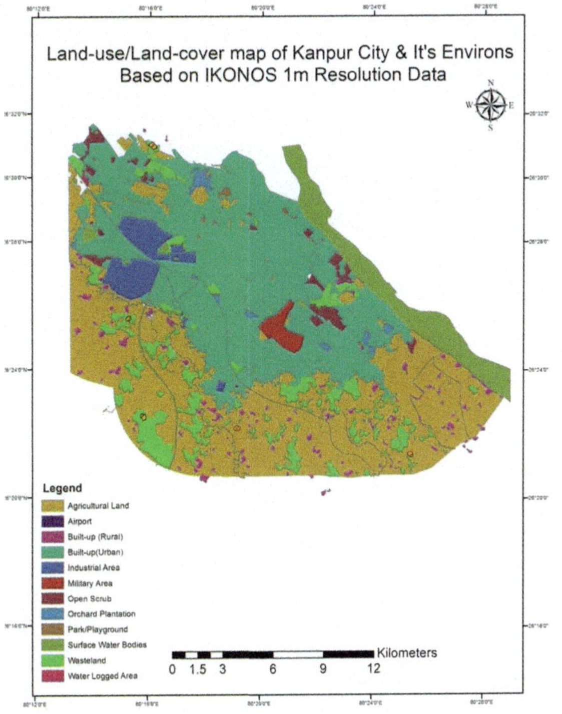

Table 7 Land use suitability with respect to goal

S. no	Site no.	Suitability
1	S-1	High
2	S-2	High
3	S-3	High
4	S-4	High
5	S-5	Moderate

Table 8 Proximity of habitation with respect to each proposed site

Habitation	S-1	S-2	S-3	S-4	S-5
Built-up (urban)	800 m (>1000 m)	1750 m (>1000 m)	1250 m (>1000 m)	4250 m (>1000 m)	3790 m (>1000 m)
Built-up (rural)	810 m (>500 m)	390 m (<500 m)	670 m (>500 m)	650 m (>500 m)	800 m (>500 m)

Fig. 10 Surface water map

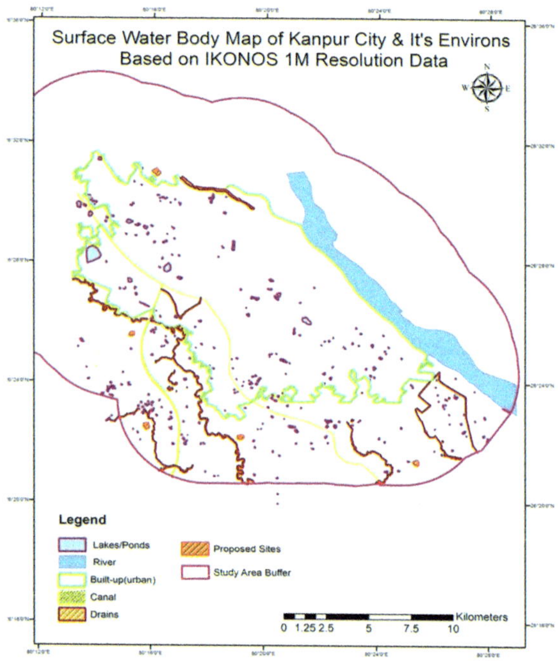

Table 9 Proximity of surface water bodies with respect to each proposed site

Surface water bodies	Site 1	Site 2	Site 3	Site 4	Site 5
River	7680 m (>1000 m)	16,384 m (>1000 m)	12,400 m (>1000 m)	11,450 m (>1000 m)	5890 m (>1000 m)
Canal	4600 m (>1000 m)	695 m (>1000 m)	2400 m (>1000 m)	2530 m (>1000 m)	2380 m (>1000 m)
Drain	1520 m (>500 m)	735 m (>500 m)	450 m (>500 m)	3300 m (>500 m)	1450 m (>500 m)
Water body	3450 m (>200 m)	885 m (>200 m)	740 m (>200 m)	1370 m (>200 m)	440 m (>200 m)

Road/Network Map

The connectivity of road/network layer was generated using Open Source (Google) high-resolution satellite data having the sub-meter ground resolution data (Fig. 11). S-1, S-2, and S-4 are found to be suitable because these sites are located near NH, CMjR, and CMnR, and S-3 and S-5 are unsuitable due to not having the adequate distance from NH and city major and minor roads. The proximity of the connectivity of transport network has been observed from Table 10. Lin and Cantwell (1999)

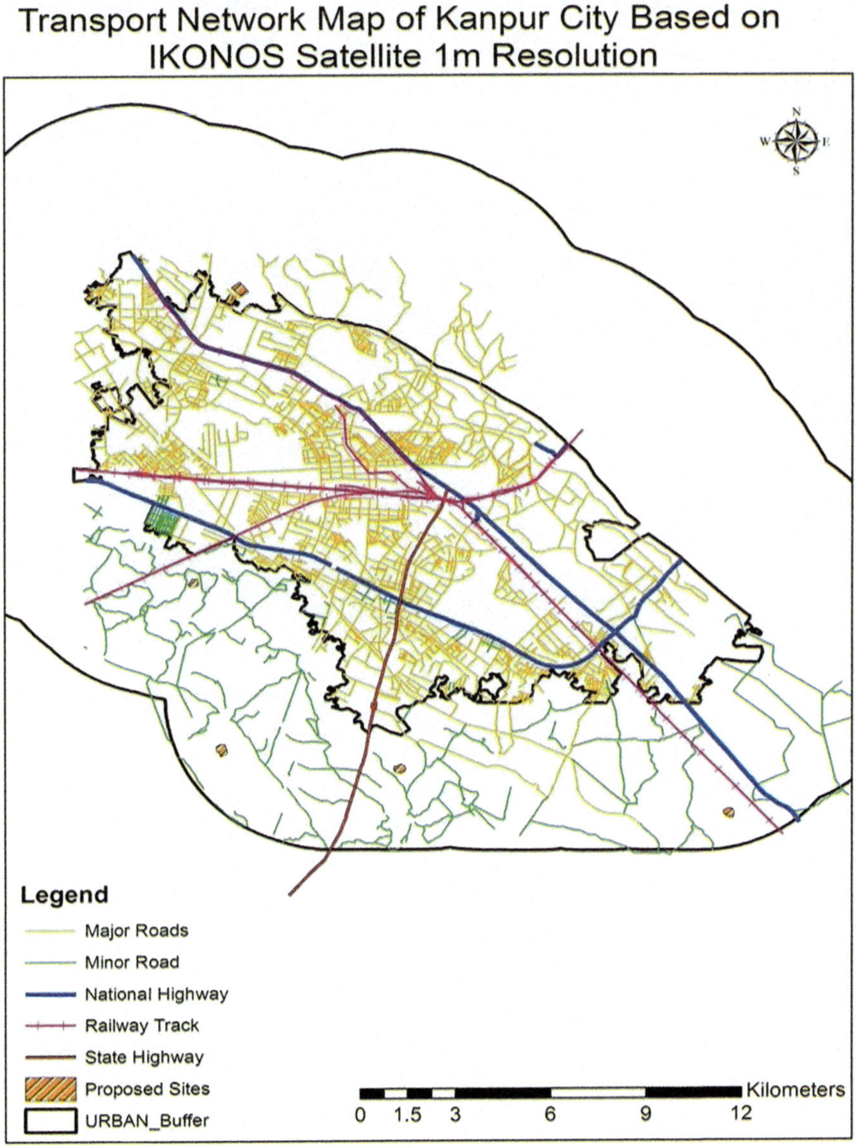

Fig. 11 Road/transport network map

Table 10 Proximity of road network as compared to each proposed site

Road network	S-1	S-2	S-3	S-4	S-5
Rail track	2250 m (>1000 m)	865 m (>1000 m)	7430 m (>1000 m)	6310 m (>1000 m)	760 m (>1000 m)
NH	2160 m (>1000 m)	2090 m (>1000 m)	5160 m (>1000 m)	7015 m (>1000 m)	1305 m (>1000 m)
Major road	980 m (>500 m)	730 m (>500 m)	1060 m (>500 m)	390 m (>500 m)	1280 m (>500 m)
Minor road	280 m (>200 m)	240 m (>200 m)	250 m (>200 m)	245 m (>200 m)	1010 m (>200 m)

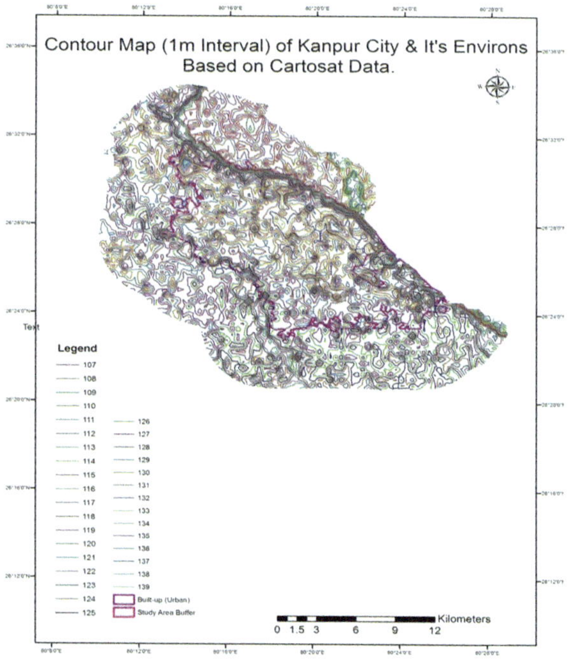

Fig. 12 Contour map

have advised that the road network is a more considerable factor in sanitary landfill to reduce the transportation cost of solid waste to carry it up to the waste disposal sites (Aziz & Khodakarami, 2013).

Slope Map

Slope gradient was calculated using contour and digital elevation data from Cartosat in the Kanpur City surroundings (Guiqin et al., 2009) (Fig. 12). The slope values in this area range from 0° to 20°. Slope is a critical factor for waste disposal, as steeper slopes increase the risk of waste runoff from the sites. Lin and Kao (1999) found that sites with slopes less than 12% are suitable for waste disposal. In this study, S-1, S-2, S-3, S-4, and S-5 are located in areas with slopes less than 5°, making them highly suitable for landfill stability (Fig. 13). Areas with slopes between 5° and 10° are considered moderately suitable for waste disposal. However, regions with slopes greater than 12° are considered low suitable for landfill stability. In conclusion, the study suggests that regions within 5° slope are highly suitable,

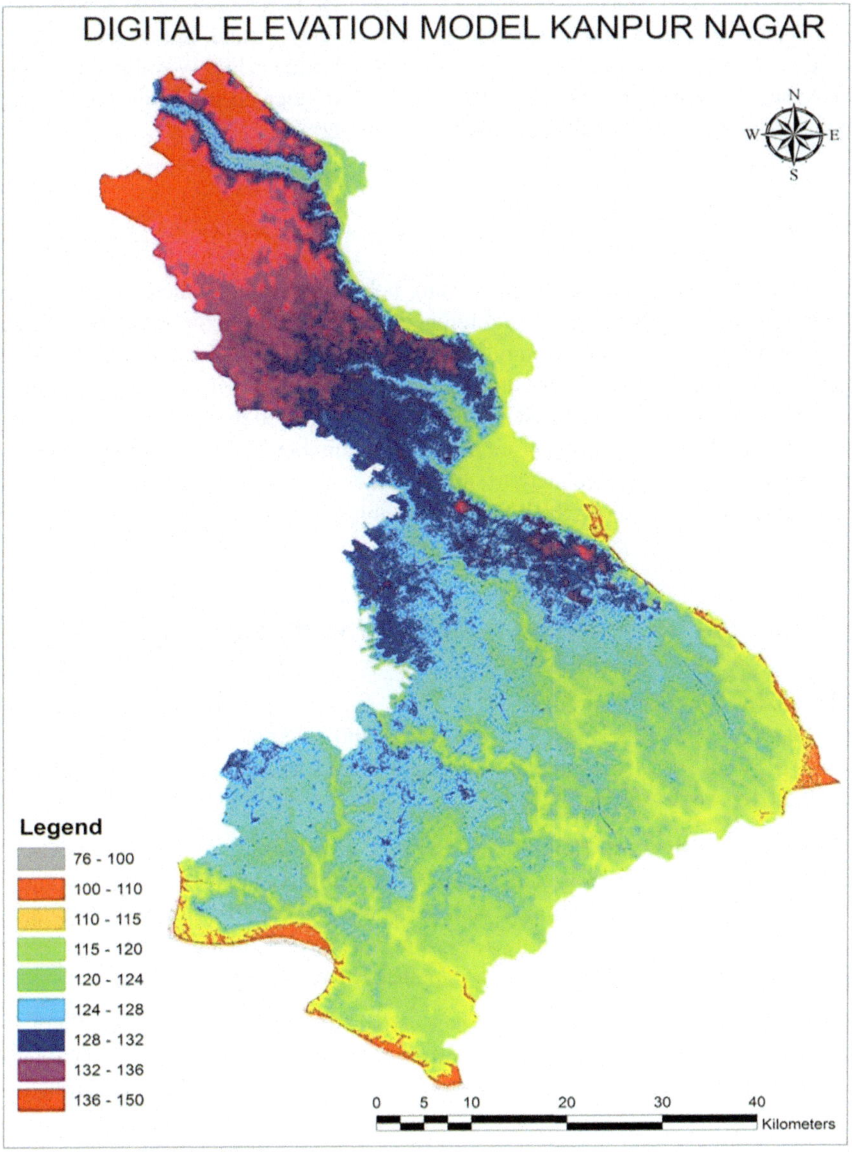

Fig. 13 Digital elevation model

regions within 5–10° slope are moderately suitable, and regions with slopes greater than 12° are less suitable for waste disposal sites in the Kanpur City surroundings. This information can be valuable for making informed decisions about waste management and landfill site selection (Ebistu & Minale, 2013; Ali & Ahmad, 2020).

Lithology Map

The lithology layer has been created through IRS-1C/D satellite data, and three main categories are found in the Kanpur region as clayey sand, gravel/sand/silt, and sandy clay (Fig. 14). S-1 is located in sandy clay, which is highly suitable for waste disposal; on other hand, S-2, S-3, S-4, and S-5 are located in clayey sand region, which is moderately suitable for waste disposal purpose.

Groundwater Prospect Map

A water table map has been created using field survey data gathered from several proposed sites. Additionally, data references were obtained from reputable sources such as the Central Water Commission (CWC) and Groundwater Boards. Water table should be as deep as possible at the site (>25 m) so that waste penetration should not occur at the proposed site. It has been observed that S-1, S-2, and S-5 are highly suitable and S-3 and S-4 are moderately suitable. Table 11 and Fig. 15 show the suitability of sites related to groundwater from the above discussed data (Domingo & Nadal, 2009; Sankoh, 2013) (Table 12).

6.2 Multi-criteria Decision-Making Evaluation

The analytic hierarchy process (AHP) with pairwise comparison is a structured methodology extensively used in evaluating and selecting landfill sites for solid waste disposal due to its systematic approach. Here's an analytical overview (Kharat et al., 2016; Ali & Ahmad, 2018):

Criteria Identification Relevant criteria such as proximity to population centres, environmental impact, geological stability, accessibility, and regulatory compliance are identified. These criteria may include factors such as proximity to population centres, geomorphology, hydrogeology, land use, environmental regulations, transportation infrastructure, and social acceptance (Saaty, 1980).

Pairwise Comparison Decision-makers conduct pairwise comparisons between criteria to establish their relative importance. This comparison is often facilitated using 1–9 scale to assign values based on perceived importance (Saaty, 1980).

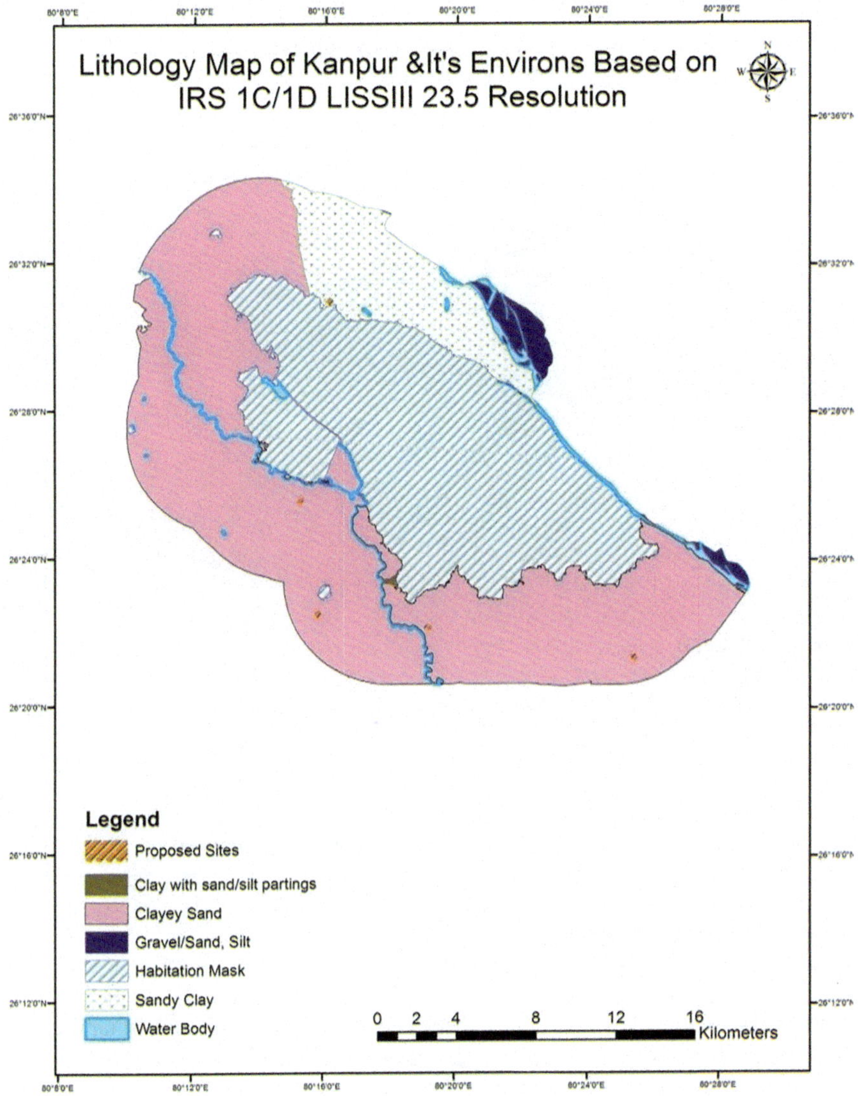

Fig. 14 Lithology map

Table 11 Groundwater table at proposed sites (field survey)

S. no.	Site	Water table (depth in ft.)	Suitability
1	S-1	150–160	High
2	S-2	120	High
3	S-3	100–120	Moderate
4	S-4	80–100	Moderate
5	S-5	130	High

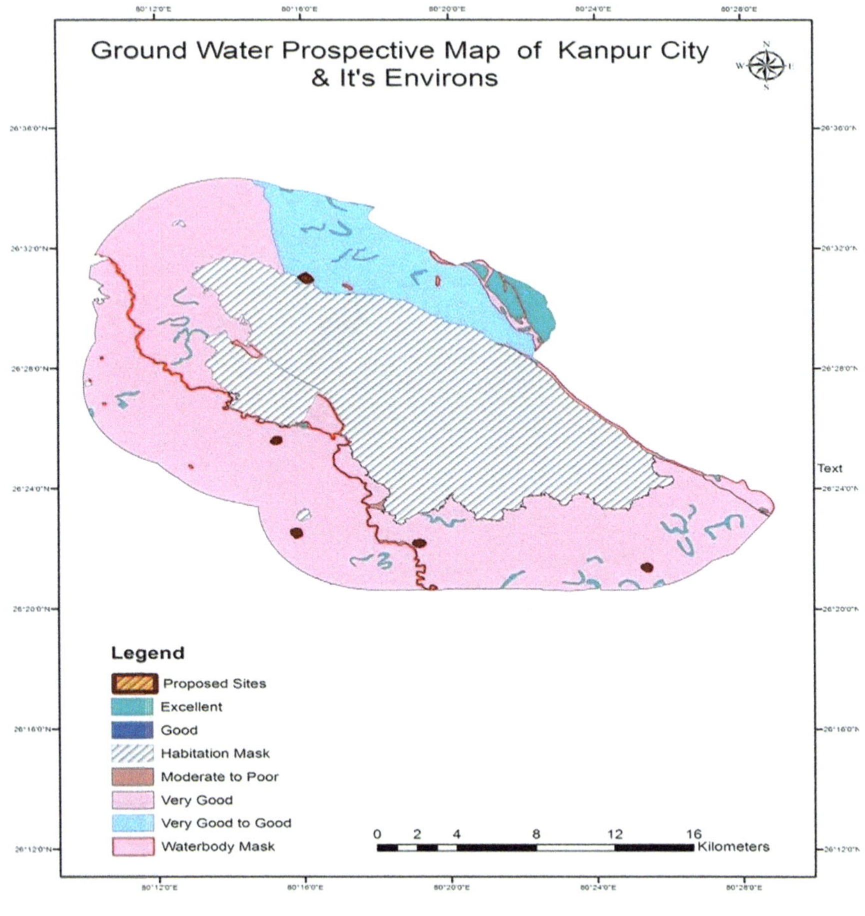

Fig. 15 Groundwater prospect map

Consistency Check A consistency check is performed to ensure the reliability and coherence of pairwise comparisons. Inconsistencies are resolved through adjustments to enhance the robustness of the process (Saaty, 1980).

Weight Aggregation Using established pairwise comparison scores, mathematical algorithms such as the eigenvector method are applied to aggregate weights for each criterion, reflecting their overall importance in the decision-making process (Saaty, 1980).

Site Evaluation Potential landfill sites are evaluated against the identified criteria, incorporating factors such as topography, land availability, environmental assessments, and community acceptance.

Table 12 Overall attributes of sites

Selection		Alternatives				
Criteria		S-1	S-2	S-3	S-4	S-5
Geomorphology		Alluvial plain Younger	Alluvial plain Older	Alluvial plain Older	Alluvial plain Older	Alluvial plain Older
Lithology		Sandy	Clayey	Clayey	Clayey	Clayey
Soil texture		Fine loamy	Fine loamy	Fine loamy	Fine loamy	Fine silty
Land use/land cover		Wasteland	Wasteland	Wasteland	Wasteland	Agriculture land
Transport networ k	RL	2250 m	865 m	7450 m	6310 m	760 m
	NH/SH	2160 m	2090 m	5160 m	7015 m	1305 m
	CMjR	980 m	730 m	1060 m	390 m	1280 m
	CMnR	280 m	240 m	250 m	245 m	1010 m
Groundwater table		150–160 ft.	120 ft.	100–120 ft.	80–100 ft.	130 ft.
Surface water	River	7680 m	16,384 m	12,400 m	11,450 m	5890 m
	Canal	4600 m	695 m	2400 m	2530 m	2380 m
	Drain	1520 m	735 m	450 m	3300 m	1450 m
	WB	3450 m	885 m	740 m	1370 m	440 m
	UBA	800 m	1750 m	1250 m	4250 m	3790 m
	RBA	810 m	390 m	670 m	650 m	800 m
Slope (in deg)		2–3°	0–1°	5°	1–2°	0–1°
Area		9.25 ha	8.21 ha	6.38 ha	9.89 ha	5.36 ha

Result Analysis Scores from the evaluation are aggregated, and a ranking of potential landfill sites is generated. The site with the highest score is considered the most suitable option for solid waste disposal based on the defined criteria.

Through this analytical process, AHP with pairwise comparison provides a transparent and structured approach to landfill site selection, aiding decision-makers in making informed and objective choices (Mu & Rojas, 2017) (Figs. 16 and 17).

The AHP can be implemented in two simple steps:

- Matrix of weighted value
- Ranking of the criteria (Table 13)

Relative Weights for Alternatives

In this comparative analysis of multi-thematic layers for selection, we evaluated various natural resource databases to determine their relative weights (RWs) for different criteria. The RW of S-2, S-3, S-4, and S-5 with respect to geomorphology is equal and valued at 0.23, indicating that they have similar importance in relation to this criterion. On the other hand, S-1 received a very low priority with an RW of 0.077 compared to S-2, S-3, S-4, and S-5.

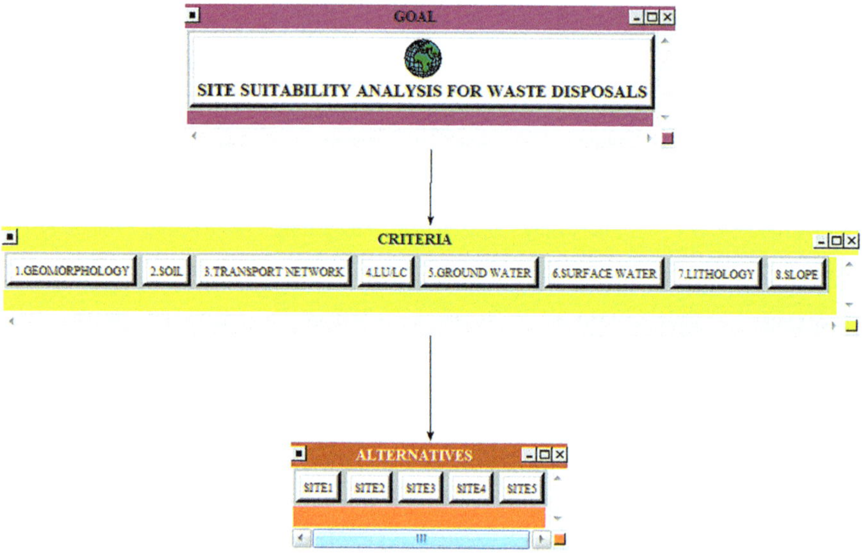

Fig. 16 Levels of decision hierarchy (Super Decisions software)

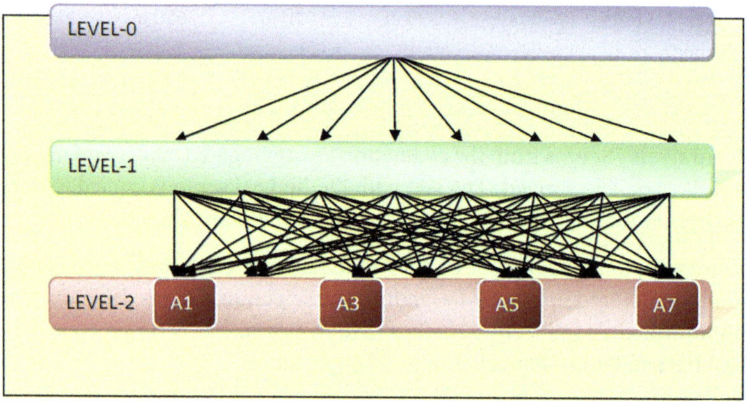

Fig. 17 Comparative analysis graph

Table 13 Levels of decision hierarchy for waste disposal site selection

Level 0 (goal)	Level 1 (selection criteria)	Level 2 (alternatives)
Site selection for waste disposal	Geomorphology Soil Land use/land cover Road network GW table Surface water Lithology Slope	S-1 S-2 S-3 S-4 S-5

For soil texture, both S-2 and S-3 were found to be equally important, each having an RW of 0.094. However, S-5 stands out as highly important with an RW of 0.472. Regarding land use/land cover (LU/LC), S-2 and S-3 also share identical RW of 0.187.

In terms of road network, S-4 emerged as the dominant option with the highest RW of 0.436. S-1 and S-2 were found to be more suitable than S-3, with RWs of 0.197 and 0.071, respectively.

Considering the groundwater table criterion, S-1 has the highest RW of 0.306, followed by S-2 and S-3, which have equal importance with an RW of 0.187. S-4 and S-5 are less significant, with RWs of 0.124 and 0.195, respectively.

For surface water, S-4 obtained the highest RW of 0.384, followed by S-1 with a slightly lower RW of 0.251. S-2, S-3, and S-5 received lower RWs of 0.06, 0.106, and 0.197, respectively, indicating their lower ranking in comparison.

In terms of lithology, S-1 has the highest RW of 0.428, while S-2, S-3, S-4, and S-5 have a relatively low RW of 0.143.

With respect to slope, all sites—S-1, S-2, S-3, S-4, and S-5—received an identical RW of 0.2, indicating that they are equally important when considering this criterion.

Based on this analysis, S-4 seems to be the most suitable option, as it received the highest RW for two criteria (road network and surface water). S-1 also performed well, obtaining high RWs for groundwater table and lithology. However, further considerations should be made based on the specific requirements and priorities of the project (Table 14).

This table provides a clear comparison of the relative weights assigned to each criterion for the different sites.

Enumeration of Maximum Priority Value, Index and Ratio of Consistency, and Alternatives

Maximum priority value (λ_{max}), index of consistency (CI), and ratio of consistency (CR) for option with respect to each and every criterion have been enumerated using the equations.

Relative Weights with Respect to Selection Criteria

Among the different ranking weights (RWs) for the selection criteria, slope holds the most substantial value (0.298), surpassing the rest and emerging as the predominant factor. Notably, surface water bodies (0.258) carry a higher level of significance compared to land use/land cover (0.119). Moreover, surface water bodies are moderately more influential than soil (0.044), lithology (0.046), groundwater table (GWT) (0.144), and geomorphology (0.046), signifying that these latter criteria are less suitable in the context of waste disposal site selection (Ali & Ahmad, 2018) (Fig. 18 and Table 15).

Table 14 Summary of relative weights (RWs) for each criterion of different sites

Criterion	S-1	S-2	S-3	S-4	S-5
Geomorphology	–	0.23	0.23	0.23	0.23
Soil texture	–	0.094	0.094	–	0.472
Land use/land cover	–	0.187	0.187	–	–
Road network	0.197	–	–	0.436	–
Groundwater table	0.306	0.187	0.187	–	–
Surface water	0.251	0.06	0.106	0.384	0.197
Lithology	0.428	–	–	–	–
Slope	0.2	0.2	0.2	0.2	0.2

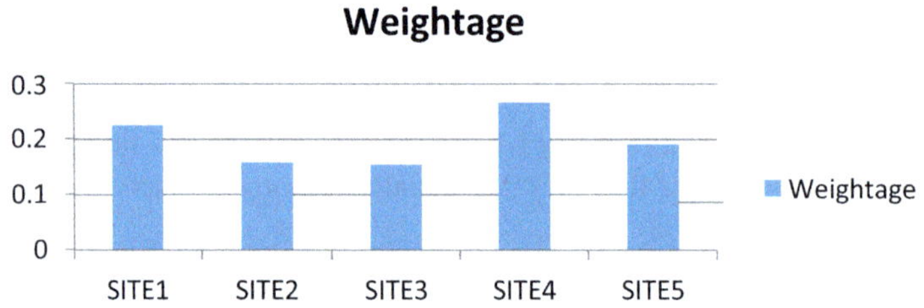

Fig. 18 Overall weightage chart

Table 15 Overall suitability index of alternatives

W_i	Geom	Soil	LU/LC	Road	GWT	SWB	Litho	Slope	X_i	SI = $\sum W_i X_i$
Site 1	0.077	0.133	0.118	0.197	0.306	0.251	0.428	0.2	0.046	0.226186
Site 2	0.23	0.094	0.269	0.197	0.187	0.06	0.143	0.2	0.044	0.159149
Site 3	0.23	0.094	0.269	0.071	0.187	0.106	0.143	0.2	0.119	0.155979
Site 4	0.23	0.206	0.269	0.436	0.124	0.385	0.143	0.2	0.046	0.267176
Site 5	0.23	0.472	0.074	0.099	0.195	0.197	0.143	0.2	0.143	0.19151
									0.257	
									0.046	
									0.298	

Enumeration of Overall Composite Weight

The overall suitability is enumerated by a summation of a result:

$$SI = \sum W_i X_i$$

where, SI = Suitability index.

7 Conclusion

The evaluation of landfill sites based on their suitability index (SI) provides valuable insights into the potential for waste disposal within the study area. Here's an analytical breakdown of the findings:

Suitability Index Analysis The suitability index values range from 0.155 to 0.267, indicating varying degrees of suitability among the selected sites. S-4 (Kharagpur) emerges as the most suitable location, with the highest

SI value of 0.267. This site fulfils all selection criteria comprehensively, including geomorphology, lithology, soil type, land use, and slope as shown in Table 12.

Moderately Suitable Sites S-1 (Mardanpur), S-2 (Pratappur Sarsai), and S-5 (Chakeri) are identified as moderately suitable for waste disposal, with SI values of 0.159 and 0.191, respectively. While they meet some criteria satisfactorily, they may have limitations in certain aspects such as lithology and land use (Table 12).

Least Suitable Site S-3 (Patehuri) has the lowest suitability index value of 0.155, indicating its limited potential for waste disposal. The main factor contributing to its low suitability is its relatively high slope compared to other sites (Table 12).

Factors Contributing to Suitability Factors such as geomorphology, lithology, soil texture, land use, and slope significantly influence the suitability of each site for waste disposal. Site-4 (Kharagpur) excels in meeting all these criteria, making it the optimal choice for waste management efforts.

Recommendations Based on the suitability index rankings, the study recommends focusing waste disposal efforts primarily on S-4 (Kharagpur) to ensure effective and environmentally responsible waste management. S-2 (Pratappur Sarsai) and S-5 (Chakeri) may also be considered for waste disposal, albeit with some limitations. S-3 (Patehuri) is deemed the least suitable and may not be ideal for waste disposal activities due to its higher slope.

Through this analytic analysis, decision-makers gain a comprehensive understanding of the suitability of each landfill site, enabling informed and strategic decision-making in waste management planning and implementation.

References

Al Khalil, M. I. (2002). Selecting the appropriate project delivery method using AHP. *International Journal of Project Management, 20*, 469–474.

Ali, S. A., & Ahmad, A. (2018). Using analytic hierarchy process with GIS for dengue risk mapping in Kolkata Municipal Corporation, West Bengal, India. *Spatial Information Research, 26*(4), 449–469. https://doi.org/10.1007/s41324-018-0187-x

Allen, A. R. (2002). Attenuation: A cost effective landfill strategy for developing countries. In *Proceedings of 9th congress of the International Association for Engineering Geology and the Environment, Durban, South Africa, 16–20 September 2002.*

Asef, H., & Lim, S. (2017a). A bi-objective optimization approach to a municipal solid waste management system. In *15th international conference on environmental science and technology*.
Asef, H., & Lim, S. (2017b). A novel multi-dimensional modeling approach to integrated municipal solid waste management. *Journal of Cleaner Production, 166*, 1131–1143.
Aziz, R. S., & Khodakarami, L. (2013). Application of GIS models in site selection of waste disposal in an urban area. *WIT Transactions on State-of-the-art in Science and Engineering, 77*(9), 27–35.
Brady, N. C. (1996). *The nature and properties of soils* (11th ed., p. 6). Macmillan.
Chandio, I. A., Matori, A. N. B., WanYusof, K. B., Talpur, M. A. H., Balogun, A. L., & Lawal, D. U. (2012). GIS-based analytic hierarchy process as a multicriteria decision analysis instrument: A review. *Arabian Journal of Geosciences, 6*(8), 3059–3066.
Chang, N., Parvathinathan, G., & Breeden, J. B. (2008). Combining GIS with fuzzy multi-criteria decision-making for landfill siting in a fast-growing urban region. *Journal of Environmental Management, 87*, 139–153. https://doi.org/10.1016/j.jenvman.2007.01.011
Cirucci, J. F., Miller, D. A., & Blanford, J. I. (2015). Retrospective GIS-based multi-criteria decision analysis: A case study of California waste transfer station siting decisions. In *Proceedings of the international symposium on sustainable systems and technologies*.
Cointreau, S. (1996). *Sanitary landfill design and siting criteria*. World Bank as an Urban Infrastructure Note.
CPCB. (2000). *Management of municipal solid waste*. Central Pollution Control Board, MOEF.
Domingo, J. L., & Nadal, M. (2009). Domestic waste composting facilities: A review of human health risks. *Environment International, 35*(2), 382–389. https://doi.org/10.1016/j.envint.2008.07.004
Ebistu, A. T., & Minale, A. S. (2013). Solid waste dumping site suitability analysis using geographic information system (GIS) and remote sensing for Bahir Dar Town, North Western Ethiopia. *African Journal of Environmental Science and Technology, 7*(11), 976–989. https://doi.org/10.5897/AJEST2013.1589
Erkut, E., & Moran, S. R. (1991). Locating obnoxious facilities in the public sector: An application of the hierarchy process to municipal landfill siting decision. *Socio-Economic Planning Science, 25*(2), 89–102.
Florida, R. (2002). *The rise of the creative class: And how it's transforming work, leisure, community and everyday life*. Routledge.
Gorsevski, P. V., Donevska, K. R., Mitrovski, C. D., & Frizado, J. P. (2012). Integrating multi-criteria evaluation techniques with geographic information systems for landfill site selection: A case study using ordered weighted average. *Waste Management, 32*, 287–296. https://doi.org/10.1016/j.wasman.2011.09.023
Guiqin, W., Li, Q., Guoxue, L., & Lijun, C. (2009). Landfill site selection using spatial information technologies and AHP: A case study in Beijing, China. *Journal of Environmental Management, 90*(8), 2414–2421. https://doi.org/10.1016/j.jenvman.2008.12.008
Gupta, S. (2015). Urbanization and socio-economic change in Kanpur City. *Journal of Social and Economic Development*.
Hamer, G. (2003). Solid waste treatment and disposal: Effects on public health and environmental safety. *Biotechnology Advances, 22*, 71–79.
Hasana, M. R., Tetsuob, K., & Ataharul, S. (2009). Landfill demand and allocation for municipal solid waste disposal in Dhaka city—An assessment in a GIS environment. *Islamic Journal of Civil Engineering (IEB), 37*(2), 133–149.
India Census Population Data. (2011).
Jain, N., & Sharma, A. (2017). Impact of industrialization on socio-economic development: A case study of Kanpur. *International Journal of Social Sciences and Humanities Research*.
Jaybhaye, R., Mundhe, N., & Dorik, B. (2014). Site suitability for urban solid waste disposal using geoinformatics: A case study of Pune Municipal Corporation, Maharashtra, India. *International Journal of Advanced Remote Sensing & GIS, 3*(1), 769–783. http://technical.cloud-journals.com/index.php/IJARSG/article/view/Tech-317

Kanpur Nagar Nigam. *Waste quantity data*.

Kao, J. J., & Lin, H. (1996). Multifactor spatial analysis for landfill siting. *Journal of Environmental Engineering, 122*(10), 902–908.

Kharat, M. G., Kamble, S. J., Raut, R. D., Kamble, S. S., & Dhume, S. M. (2016). Modeling landfill site selection using an integrated fuzzy MCDM approach. *Modeling Earth Systems and Environment, 2*(2), 53. https://doi.org/10.1007/s40808-016-0106-x

Li, Y. P., Huang, G. H., Cui, L., & Liu, J. (2019). Mathematical modeling for identifying cost-effective policy of municipal solid waste management under uncertainty. *Journal of Environmental Informatics, 34*, 55–67. https://doi.org/10.3808/jei.201900417

Mu, E., & Rojas, M. P. (2017). *Understanding the analytical hierarchy process* (SpringerBriefs in operations research). https://doi.org/10.1007/978-3-319-33861-3_2

NEERI. (1996). *Strategy paper on solid waste management in India*. MUD and PA.

Oweis, I. S., & Khera, R. P. (1990). *Geotechnology of waste management* (p. 273). Butterworths.

Owens, S., & Jones, D. S. (2016). *Policy change, public attitudes, and social control: A socio-psychological perspective*.

Paul, S. (2012). Location allocation for urban waste disposal site using multi-criteria analysis: A study on Nabadwip Municipality, West Bengal, India. *International Journal of Geomatics and Geosciences, 3*(1), 74–87. www.ipublishing.co.in/jggsvol1no12010/volthree/EIJGGS3107

Putnam, R. D. (2000). *Bowling alone: The collapse and revival of American community*. Simon & Schuster.

Rezaei-Moghaddam, K., & Karami, E. (2008). A multiple criteria evaluation of sustainable agricultural development models using AHP. *Environment, Development and Sustainability, 10*, 407–426.

Saaty, T. L. (1980). *The analytic hierarchy process*. McGraw-Hill.

Saaty, T. L. (1990). How to make a decision: The analytic hierarchy process. *European Journal of Operational Research, 48*(1990), 9–26.

Saaty, T. L. (2003). Decision aiding-decision-making with the AHP: Why is the principal eigenvector necessary. *European Journal of Operational Research, 145*, 85–91.

Saaty, T. L. (2008). Decision making with the analytic hierarchy process. *International Journal of Services Sciences, 1*(1), 83.

Saaty, T. L., & Niemira, M. P. (2006). A framework for making a better decision; how to make more effective site selection, store closing and other real estate decisions. *Research Reviews, 13*(1).

Şener, S., Sener, E., & Karaguzel, R. (2011). Solid waste disposal site selection with GIS and AHP methodology: A case study in Senirkent–Uluborlu (Isparta) Basin, Turkey. *Environmental Monitoring and Assessment, 173*, 533–554.

Smith, J., & Johnson, A. (2018). Geometric correction methods for remote sensing image registration. *International Journal of Remote Sensing*.

Sumathi, V. R., Natesan, U., & Sarkar, C. (2008). GIS-based approach for optimized siting of municipal solid waste landfill. *Waste Management, 28*, 2146–2160.

Thoso, M. (2007). *The construction of a geographic information systems (GIS) model for landfill site selection* (Thesis of M.Tech). Department of Geography, University of the Free State, Bloemfontein.

Dr. Virendra Kumar, PhD (Geography), Scientist-SE and Head, Landuse & Urban Survey Division, RSAC-UP. He has completed PhD (Geography) in 1996 at MJP Rohilkhand University, Bareilly, Uttar Pradesh. He joined Remote Sensing Applications Centre, Uttar Pradesh, Lucknow, as a Scientist-SC (Landuse & Urban Survey) in August 1997. Presently, he is a Scientist-SE and Head the Landuse & Urban Survey Division as well as serve as a Faculty-Associate Professor for the M. Tech (RS and GIS) programme, affiliated to Dr. A.P.J. Abdul Kalam Technical University, Lucknow. He has completed 60 R&D and consultancy projects based on geospatial technologies for state and central government departments and published 50 research papers in national and international journals, including nine at international conferences abroad, and authored nine books/chapters of international acclaim.

Anoop Kumar Maurya, M. Tech (RS and GIS), Lecturer (Civil Eng.) Govt. Polytechnic, Lucknow. He has completed B. Tech in Civil Engineering and M. Tech in Remote Sensing and GIS at Dr. A.P.J. Abdul Kalam Technical University, Lucknow, in 2016. Presently, he is working as Lecturer (Civil Eng.) at Govt. Polytechnic, Lucknow.

Vashi Ahmad, M. Tech (RS and GIS), Project Scientist, RSAC-UP. He has completed B. Tech in Civil Engineering and M. Tech in Remote Sensing and GIS at Dr. A.P.J. Abdul Kalam Technical University, Lucknow, in 2019. Presently, he is working as a Project Scientist at Remote Sensing Applications Centre, Uttar Pradesh, Lucknow.

Mohd Jaid, M. Sc (Ag) Agronomy, Project Scientist, RSAC-UP. He has completed M. Sc in Agronomy at Integral University, Lucknow, in 2021. Presently, he is working as a Project Scientist at Remote Sensing Applications Centre, Uttar Pradesh, Lucknow.

Shivam Pandey, M. Sc (RS and GIS), Project Scientist RSAC-UP. He has completed M. Sc in Remote Sensing and GIS at Mahatma Gandhi Chitrakoot Gramodaya University, Chitrakoot, Satna, Madhya Pradesh, in 2019. Presently, he is working as a Project Scientist at Remote Sensing Applications Centre, Uttar Pradesh, Lucknow.

Sarvajeet Kumar, M. A (Geography), Research Assistant. He has completed M. A (Geography) in 2005 at Purvanchal University, Jaunpur, Uttar Pradesh, and qualified UGC NET-JRF in 2013. He has joined NATMO in 2010 as a Research Assistant and have continued to date. He is also working as a Research Associate at National Atlas and Thematic Mapping Organisation (NATMO), Kolkata.

Glossary

Agent-Based Modeling Agent-based modeling simulates complex systems by modeling individual entities (agents) and their interactions. Each agent follows defined rules, and their collective behaviors produce emergent system dynamics. It is widely used in fields like economics, urban planning, and social sciences to explore scenarios and policy impacts.

AHP Method (Analytic Hierarchy Process) The AHP method is a decision-making tool that uses pairwise comparisons to prioritize options based on various criteria. It is commonly applied in urban planning, resource allocation, and environmental assessments to determine optimal solutions systematically.

Border Dispute A border dispute arises from conflicting claims over boundary demarcations between countries or regions. These disputes can stem from historical agreements, natural resource distribution, or cultural differences, often requiring negotiation or legal resolution.

Centroid A centroid is the geometric center of a shape, dataset, or distribution. In spatial analysis, it represents the average location of points, providing insights into spatial patterns and serving as a reference for distance measurements and clustering analysis.

Civilization A civilization is an advanced human society characterized by cultural achievements, technological advancements, governance structures, and social organization. Civilizations often emerge around geographic and resource-rich areas, shaping historical development and influencing modern societal structures.

Colonial Colonial refers to the period or system of control where one nation governs and settles another territory. It often involves economic exploitation, cultural exchange, and political dominance, with lasting impacts on the colonized region's development and identity.

Containment Zone A containment zone is a restricted area established during health crises like pandemics to prevent disease spread. These zones are monitored for infection control, ensuring public health safety and reducing transmission risks.

A. Rai et al. (eds.), *Contemporary Social Physics*, Springer Geography,
https://doi.org/10.1007/978-3-031-77596-3

COVID-19 COVID-19 is a global pandemic caused by the SARS-CoV-2 virus, first identified in 2019. It has significantly impacted public health, economies, and societies worldwide. The pandemic underscored the need for robust healthcare systems, international collaboration, and proactive crisis management strategies.

Cyber-attack A cyber-attack refers to deliberate attempts to damage, disrupt, or gain unauthorized access to computer systems, networks, or data. These attacks pose significant threats to businesses, governments, and individuals in the digital age.

Digital Crime Digital crime involves illegal activities conducted using digital technologies or the Internet, including hacking, identity theft, and cyber fraud. Addressing these crimes requires advanced cybersecurity measures and legal frameworks.

Drugging Drugging refers to administering a substance to someone without their knowledge or consent, often for malicious purposes. It poses serious risks to victims' health and safety, requiring legal and medical interventions.

Factor Analysis Factor analysis is a statistical method used to identify underlying relationships between variables by grouping them into factors. It simplifies complex datasets, aiding in research fields like social sciences, marketing, and geography.

Geographically Weighted Regression Geographically weighted regression (GWR) is a spatial regression technique that accounts for geographic variations in relationships between variables. It provides localized insights, enhancing the accuracy of spatial analyses in fields like health, urban planning, and environmental studies.

Geospatial Analysis Geospatial analysis involves examining spatial data to identify geographic patterns, trends, and relationships. By integrating tools like GIS and remote sensing, it supports decision-making in urban planning, environmental conservation, disaster management, and resource allocation.

GIS (Geographic Information System) GIS is a powerful tool for mapping, visualizing, and analyzing spatial data. It combines geographic information with databases to study patterns, relationships, and trends. GIS is widely used in fields like urban planning, environmental management, and disaster response, enabling data-driven decision-making and spatial insights.

Google Earth Engine Google Earth Engine is a cloud-based platform for analyzing and visualizing geospatial and remote sensing data. It facilitates large-scale environmental monitoring, urban studies, and land-use mapping through its powerful computational capabilities.

Heritage Heritage encompasses cultural, historical, and natural treasures inherited from past generations. It includes monuments, traditions, languages, and biodiversity. Preserving heritage fosters identity, continuity, and education, ensuring that future generations benefit from and appreciate these valuable legacies.

Hot Spot A hot spot refers to a location with a high concentration of specific activities or events, such as biodiversity, crime, or tourism. Identifying hot spots is crucial for targeted interventions, conservation efforts, and optimizing resource allocation.

Kriging Kriging is a geostatistical interpolation method used to predict spatially distributed variables. It considers both the distance and spatial correlation between data points, providing precise estimates and uncertainty assessments for applications like environmental monitoring and resource mapping.

K-Means Clustering K-means clustering is a machine learning method used to partition data into distinct groups or clusters based on similarities. It is widely applied in spatial analysis, market segmentation, and image processing to identify patterns and groupings.

Machine Learning Machine learning is a subset of artificial intelligence that enables systems to learn from data and improve their performance without explicit programming. It is widely used in applications like predictive modeling, image recognition, and natural language processing.

MNDWI (Modified Normalized Difference Water Index) MNDWI is an enhanced version of NDWI that improves the detection of water bodies in remote sensing imagery. By emphasizing water features and suppressing non-water features, it is useful for hydrological studies and water resource management.

Multicriteria Layers Multicriteria layers combine multiple spatial datasets to evaluate suitability or prioritize decisions. Common in GIS-based analysis, these layers facilitate comprehensive planning in fields like urban development, environmental conservation, and resource allocation.

NDVI (Normalized Difference Vegetation Index) NDVI is a remote sensing index used to measure vegetation health and density. By analyzing the difference between near-infrared and visible light reflectance, it provides insights into plant growth, agricultural productivity, and ecosystem monitoring.

NDWI (Normalized Difference Water Index) NDWI is an index that monitors water bodies using remote sensing data. It helps identify water features and their changes over time, supporting applications in hydrology, disaster management, and environmental monitoring.

Perception Mapping Perception mapping visualizes public opinions, attitudes, or preferences about specific issues or areas. It integrates survey data with geographic tools, enabling planners and policymakers to understand societal perspectives and make informed decisions.

Peri-urban Area Peri-urban areas are transitional zones on the outskirts of urban centers, blending urban and rural characteristics. These regions face challenges like rapid urbanization, infrastructure gaps, and environmental degradation, requiring balanced planning and management.

Phytoremediation Phytoremediation uses plants to clean contaminants from soil, air, or water. This ecofriendly method leverages natural processes to remove or neutralize pollutants, contributing to environmental restoration and sustainable practices.

Pickpocketing Pickpocketing is the act of stealing items, such as money or valuables, from someone's pocket or bag, typically unnoticed. It is a common petty crime requiring vigilance and security measures in crowded public spaces.

Random Forest Regressor Random forest regressor is a machine learning algorithm that uses an ensemble of decision trees for predictive analysis. It provides robust and accurate predictions in applications like regression modeling, spatial analysis, and environmental studies.

Riparian Buffer System A riparian buffer system consists of vegetated areas along waterways that protect water quality by filtering pollutants, stabilizing banks, and supporting biodiversity. These systems are vital for ecosystem health and sustainable land management.

Route Optimization Route optimization calculates the most efficient paths for travel or transport, minimizing time, cost, or distance. It is essential for logistics, urban mobility, and delivery services, improving efficiency and reducing environmental impacts.

SDG (Sustainable Development Goals) The Sustainable Development Goals (SDGs) are a set of 17 global objectives established by the UN to promote sustainable development. They address issues like poverty, education, climate action, and equality, aiming to create a better future by 2030.

Shifting Occupation Shifting occupation refers to changes in land or space use over time, driven by factors like urbanization, economic shifts, or environmental conditions. Understanding these changes is crucial for planning sustainable and adaptable landscapes.

Smart Border Management Smart border management employs advanced technologies like biometrics, AI, and surveillance systems to enhance security and streamline cross-border movement. It balances national security with efficient trade and travel, adapting to global challenges like migration and pandemics.

Social Behavior Social behavior encompasses the actions, interactions, and reactions of individuals within a society. It reflects cultural norms, social structures, and interpersonal relationships, influencing cooperation, conflict, and communication.

Social Inequalities Social inequalities refer to disparities in access to resources, opportunities, and rights within a society. Addressing these inequities is crucial for promoting social justice, economic stability, and inclusive development.

Social Learning Social learning involves acquiring knowledge or skills through observation, interaction, and imitation within social contexts. It is fundamental to human development, influencing behavior, cultural transmission, and innovation.

Social Physics Social physics applies mathematical models and tools to study patterns of human behavior, social interactions, and collective dynamics. By analyzing data from diverse sources, it reveals insights into how societies function, adapt, and evolve.

Social Systems Social systems are networks of interconnected human relationships, institutions, and norms that organize societal functions. They encompass structures like families, governments, and economies, influencing behaviors, roles, and interactions.

Spatial Environments Spatial environments are physical spaces and geographic contexts that shape human activities and natural processes. They include urban areas, rural landscapes, and ecosystems, influencing behaviors, cultural practices, and environmental interactions.

Spatial Modeling Spatial modeling uses mathematical and computational techniques to represent and analyze geographic phenomena. These models simulate real-world processes, such as urban growth or climate change, enabling planners and researchers to predict outcomes and develop strategies for sustainable development.

Spatial Networks Spatial networks represent connections influenced by geographic relationships, such as transportation systems, communication lines, or utility grids. These networks model how space impacts connectivity, accessibility, and flow.

Spatial Simulations Spatial simulations use computational methods to replicate and analyze dynamic spatial processes over time. Applications range from urban growth and traffic patterns to environmental changes. These simulations help decision-makers test scenarios, assess impacts, and plan for future developments effectively.

Suitability Index A suitability index is a metric used to assess how appropriate a location is for a specific purpose. It integrates multiple criteria, often through GIS analysis, to support decision-making in fields like urban planning, agriculture, and conservation.

Sustainable Agriculture Sustainable agriculture involves farming practices that maintain productivity while conserving natural resources and promoting environmental health. It emphasizes biodiversity, soil fertility, and water efficiency to ensure long-term food security.

Urban Mobility Urban mobility focuses on the movement of people and goods within cities using transportation systems like buses, trains, and bicycles. Efficient mobility is critical for economic growth, environmental sustainability, and quality of life.

Python Python is a versatile programming language widely used for data analysis, machine learning, and application development. Its simplicity and extensive libraries make it a popular choice for both beginners and experts in various domains.

Index

A. Rai et al. (eds.), *Contemporary Social Physics*, Springer Geography,
https://doi.org/10.1007/978-3-031-77596-3

The manufacturer's authorised representative in the EU is Springer Nature Customer Service Centre GmbH, Europaplatz 3, 69115 Heidelberg, Germany. If you have any concerns regarding our products, please contact ProductSafety@springernature.com

Printed and bound by CPI Group (UK) Ltd, Croydon, CR0 4YY
30/09/2026
02225586-0018